ENGINEERING MATERIALS

Research, Applications and Advances

ENGINEERING MATERIALS

Research, Applications and Advances

K.M. Gupta

CRC Press
Taylor & Francis Group
Boca Raton London New York

CRC Press is an imprint of the
Taylor & Francis Group, an **informa** business

CRC Press
Taylor & Francis Group
6000 Broken Sound Parkway NW, Suite 300
Boca Raton, FL 33487-2742

First issued in paperback 2020

ISBN-13: 978-1-4822-5797-7 (hbk)
ISBN-13: 978-0-367-73925-6 (pbk)

Library of Congress Cataloging-in-Publication Data

Gupta, K. M.
  Engineering materials : research, applications and advances / author, K.M. Gupta.
    pages cm
  "A CRC title."
  Includes bibliographical references and index.
  ISBN 978-1-4822-5797-7 (alk. paper)
  1. Materials. I. Title.

TA403.G85 2014
620.1'1--dc23   2014021519

Dedicated to the loving memory of my nephew

Jayant (Babul)

Contents

Preface

This book is intended to cover the vast and fast-growing field of materials and their science in accordance with modern trends. The book covers the syllabi being taught at the undergraduate and postgraduate levels in engineering institutes in India and abroad. It also covers the syllabi of various competitive and other national-level examinations. This book will be very helpful to postgraduate students and to those appearing for various professional examinations such as civil services, engineering services and GATE. The book is very suitable for students involved in research work at master's and PhD levels. It serves both as a textbook and reference material for undergraduate students studying materials science and engineering, postgraduate students and research scholars in materials and mechanical engineering disciplines. It is taught primarily to senior undergraduate and first-year postgraduate students in a variety of disciplines, but primarily to mechanical engineering and materials science students.

The field of materials science has advanced considerably. A key feature of this book is the inclusion of the latest developments in various fields of materials and their sciences and processes. Latest topics such as functionally graded materials, auxetic materials, whiskers, metallic glasses, biocomposite materials, nanomaterials, superalloys, superhard materials, shape-memory alloys and smart materials have been included. Recent advances in futuristic plastics, sandwich composites, biodegradable composites, special kinds of composites such as fire-resistant composites and marine composites and biomimetics have also been described. A review of various researches underway in the field of biocomposites is given. The book has been organized in a user-friendly manner. In addition, there are several enhancements to the book's pedagogy that make it more appealing to both instructors and students.

Illustrations, examples and sciences of different disciplines of engineering such as mechanical, production and industrial engineering, automobiles, electronics, chemical and interdisciplinary branches are presented. Topics including powder metallurgy, nanotechnology, intermetallic compounds, amorphous materials ferrous and nonferrous materials are explained with about 160 figures, 85 tables and a few equations and numericals. I gratefully acknowledge the authors and publishers of the books and of the research papers of the journals quoted in the references, who have provided guidelines in preparing this book.

I acknowledge the inspiration and blessings of my respected mother Bela Devi, brother-in-law Jawahar Lal, sister Savitri Lal, elder brother Gopal Das Gupta and other family members. I am full of gratitude to my daughter Nidhi, son-in-law Ritesh, son Nishu, wife Rita, grandson Akarsh (Ram) and granddaughter Anshika (Gauri), for their patience and encouragement to complete this venture. I also pay homage to my loving nephew Jayant (Babul) who left us for his heavenly abode at a premature age.

My heartfelt thanks are due to Anurag Sant of Umesh Publications, Delhi, the publisher of my other books; my friends Er. Ranjeet Singh Virmani, AGM (retd.) Punjab National Bank; and Er. K.R.D. Tewari, chartered civil engineer and consultant, Allahabad, for their support in vivid ways. I am very thankful to my postgraduate student Kishor Kalauni, MTech (materials science), for helping me with the compilation, typesetting and correcting of the manuscript, and for research information and computer support, without which it would have not been possible to prepare this manuscript. I also acknowledge my students Saurabh Kumar Singh, Ashwani Kumar and Sanjeev Kumar, all MTech (materials science), for helping me with some compilation and typing tasks. I acknowledge everybody else of MNNIT who helped me directly or indirectly to establish my calibre for the present task.

Every effort has been made to avoid errors and mistakes; however, their presence cannot be ruled out. Any suggestions to improve the standard of this book, indication towards errors, omissions or mistakes will be greatly appreciated.

Acknowledgments

The author acknowledges with heartfelt gratitude S.C. Sant and Anurag Sant, Umesh Publications (Daryaganj, Delhi), India, for being kind enough to provide literature support and valuable information that proved very useful in preparing this book. He is especially thankful to Anurag Sant for his courteous gesture in promoting quality book writing at the international level.

Author

Dr. K.M. Gupta is a professor in the Department of Applied Mechanics, Motilal Nehru National Institute of Technology, Allahabad, India. He has over 39 years of teaching, research and consultancy experience. He earned his diploma in mechanical engineering (Hons), bachelor of engineering (AMIE) in mechanical engineering and ME (Hons) in 1977. He earned his PhD from the University of Allahabad. Although a mechanical engineer, Professor Gupta has also specialised in automobile engineering. He has authored 28 books and edited 2 books on engineering, and 1 chapter in the Scrivener Wiley publication *Handbook of Bioplastics and Biocomposites Engineering Applications*. He has also authored 120 research papers in reputed national and international journals. Professor Gupta has presented his research papers at 16 international conferences in the United States, United Kingdom, Japan, China, France, Muscat, Bangkok, South Africa and Hong Kong. He has also chaired 8 international conferences in China, Singapore, Dubai and Bangkok. He is editor-in-chief of two journals, *The International Journal of Materials, Mechanics and Manufacturing (IJMMM)*, Singapore, and *International Journal of Materials Science and Engineering (IJMSE)*, San Jose, California, and has edited many international journals. He has worked as a reviewer for various national and international journals, and has been a member of several editorial boards.

In recognition of his academic contribution, Marquis Publication (USA) included him in the list of *World Who's Who in Science and Engineering 2007* and *Who's Who in the World 2008*. The International Biographical Centre, a leading research institute (Great Britain), selected him as one of the 2000 Outstanding Scientists—2009 from across the world; and Rifacimento International Publisher included his biographical note in *Reference Asia: Asia's Who's Who of Men and Women of Achievement*.

A recipient of many gold medals and prizes for his outstanding career from diploma to doctorate (a rare achievement), Professor Gupta has served as head of the automobile engineering department at the Institute of Engineering and Rural Technology, Allahabad. He masterminded the development of several laboratories, namely, automobile-related labs, materials science labs, strength of materials lab and hydraulics lab at various institutes/colleges. He was a trailblazer in establishing auto garages and repair workshops as well.

Dr. Gupta has undergone extensive industrial training at many reputed industries and workshops. He is endowed with vast experience in curriculum development activities and consultancy. He has served as dean of research and consultancy, head of the applied mechanics department at Motilal Nehru National Institute of Technology, Allahabad. He has acted as chairman of various research selection committees, research project monitoring committees and other administrative committees of his institute and other universities. He has also served as chairperson, Community Development Cell (CDC) of MNNIT for several years.

Presently, Dr. Gupta is teaching materials science, engineering mechanics, thermodynamics of materials, and electrical and electronic materials. His research interests are in the fields of materials science, composite materials, stress analysis and solid mechanics.

Basic Preliminary Information the Readers Need to Know

SI Prefixes of Multiples and Submultiples

Factor	Symbol	Prefix
10^{-1}	d	deci
10^{+1}	da	deca
10^{-2}	c	centi
10^{+2}	h	hecto
10^{-3}	m	milli
10^{+3}	k	kilo
10^{-6}	μ	micro
10^{+6}	M	mega
10^{-9}	n	nano
10^{+9}	G	giga
10^{-12}	p	pico
10^{+12}	T	tera
10^{-15}	f	femto
10^{+15}	P	peta
10^{-18}	a	atto
10^{+18}	E	exa

Greek Alphabets

Name	Symbol	Analogous English Sound
Alpha	α	a
Beta	β	b
Gamma	Γ	G
	γ	g
Delta	Δ	D
	δ	d
Epsilon	ε	e

(Continued)

Name	Symbol	Analogous English Sound
Zeta	ζ	z
Eta	η	z
Theta	θ	Th
Iota	ι	i
Kappa	κ	k
Lambda	Λ	L
	λ	l
Mu	μ	m
Nu	ν	n
Xi	Ξ	X
	ξ	x
Omicron	o	o
Pi	Π	P
	π	p
Rho	ρ	r
Sigma	Σ	S
	σ	s
Tau	τ	t
Upsilon	υ	u
Phi	Φ	Ph
	φ	ph
Chi	χ	kh
Psi	Ψ	Ps
	ψ	ps
Omega	Ω	O
	ω	O

Physical Quantities and Derived SI Units

Quantity	Unit		Other Units
	Read as	Write as	
Amount of substance	kilomole	kmol	mol
Capacitance	farad	F	CV^{-1}
Coefficient of viscosity	pascal second	Pa s	poise
Conductance	siemens	S	mho
Electric charge	coulomb	C	—
Electric current	ampere	A	—
Electric potential	volt	V	$W\,A^{-1}$

(Continued)

Quantity	Unit		Other Units
	Read as	Write as	
Force	newton	N	—
Frequency	hertz	Hz	*els*, s⁻¹
Inductance	henry	H	Wb A⁻¹
Length	metre	m	—
Luminous intensity	candela	cd	—
Magnetic flux	weber	Wb	V s
Magnetic flux density	tesla	T	Wb m⁻²
Mass	kilogram	kg	—
Pressure, stress	pascal	Pa	N m⁻²
Power	watt	W	J s⁻¹
Resistance	ohm	Ω	VA⁻¹
Temperature	kelvin	K	°C, °F
Time	second	s	min, h, d, y
Work, energy, heat	joule	J	N m

Conventions to Be Followed while Using SI Units

1. Full stop, dot, dash or plural is not used while writing unit symbols:
   - Write 10 mm and not 10 m.m.
   - Write 5 kg and not 5 kgs
   - Write 20 MN and not 20 M-N
2. Prefix symbol is used in continuation (without gap) with the unit symbol, e.g., for megawatt:
   - Write MW and not M W.
3. Two symbols should be separated by a single space. For example, for metre second:
   - Write m s and not ms (millisecond)
4. For temperature:
   - Write K and not °K
5. Write proper names in small letters when used as a word, and their symbols with a capital letter. For example:
   - Write coulomb and not Coulomb
   - Write T (tesla) and not t or Tesla
6. Use single prefix instead of double prefixes:
   - For terahertz write THz and not MMHz

7. Prefix should be attached to the numerator instead of the denominator:
   - Write MWb In^{-2} and not Wb min^{-2}

8. Group three digits together on either side of the decimal point. Do not group four digit numbers in this way. For example, write 20 465 and not 20465:
   - Write 1398 and not 139 8 or 1 398
   - Write 4.061 872 34 and not 4.0618 7234

9. Write 10 MV and not 10^7 V:
   - Write 1.602×10^{-19} and not 1.602×10^{-16}

Physical Constants

Quantity	Symbol Used in This Book	SI Unit
Acceleration due to gravity	g	9.807 m/s^2
Atomic mass unit	*(amu)*	1.660×10^{-27} kg
Avogadro number	N_A	6.023×10^{23}/gm-mol
		$= 6.023 \times 10^{26}$ particles/kg-mol
Bohr magneton	β	9273×10^{-24} Am^2
Boltzmann constant	k	1.381×10^{-23} J/K
		$= 8.614$
Charge of electron	e	1.602×10^{-19} C
Electron rest mass	m_o	$9.109 \times 1Q^{-31}$ kg
Faraday constant	—	9.649×0^7 C/kg-znol
Gas constant	R	8.314 Jimol K
Molar gas volume at STP	—	2.241×10^{-2} m^3/gin-mol
		$= 22.4$ m^{3}fkg-mol
Permeability of vacuum	μ_o	$4\,Tc \times 10^{-7}$ H/m (henry/meter)
Permittivity of vacuum	ε_o	8.854×10^{-12} F/m
		6.626×10^{-34} Js
Planck constant	h	1.672×10^{-27} kg
Proton mass	m_p	2.998×10^8 m/s
Speed of light in vacuum	c	3×10^8 rn/s

Conversion Factors

Quantity	Unit and Symbol	Converted Value
Area	1 m^2	$= 100$ dM$^2 = 10000$ CM$^2 = 10.764$ ft^2
	1 hectare (ha)	$= 10000$ m$^2 = 2.47$ acre
Density	1 ton	$= 2240$ lb $= 1016$ kg
	1 kg/m^3	$= 10^{-3}$ gin/cm$^3 = 10^{-3}$ kg/litre
		$= 1$ gm/litre
		$= 0.102$ k $= 10^5$ dyne $= 1$ kg-m/s^2
		$= 0.2248$ lbf
Dynamic viscosity	1 (poise) P	$= 0.1$ N s/m$^2 = 0.1$ Pa s $= 1$ dyne-s/cm^2
Electrical		
Conductivity	1 W/(m-K)	$= 1$ mho/m $= 1/$(ohm m) $= 10^2$ ohm cm
Resistivity	1 (siemen) S/m 1 ohm m	
Energy	1 eV	$= 1.602 \times 10^{-19}$ J
	1 J	$= 10^7$ erg $= 1$ W s
	1 kWh	$= 3.6$ MJ $= 3412$ Btu $= 860$ kcal
	1 kcal	$= 1000$ cal $= 4187$ J $= 3.967$ Btu
Force	1 N	$= 9.807$ N—10 N $= 2.2 \times 10^{-3}$ kip
	1 kgf	$= 133.3$ Pa $= 1$ torr $= 1.333$ mbar
Frequency	1 Hz	$= 1/s = 1$ *els*
Kinematic viscosity	1 (stoke) St	$= 1$ cm^2/s $= 10^{-4}$ m^2/s
Length	1 A	$= 10^{-10}$ m $= 10^{-8}$ cm $= 10^{-4}$ pm
		$= 0.1$ nm
	1 in	$= 3.281$ ft $= 39.37$ inch
	1 km	$= 0.621$ mile $= 1000$ m
	1 inch	$= 2.54$ cm $= 25.4$ mm $= 1000$ mils
	1 mile	1.609 km $= 1609$ m $= 1760$ yd
		$= 5280$ ft.
Magnetic		
Field strength	1 A/m	$= 0.01257$ oersted
Flux density	1 Wb	$= 10^8$ maxwell $= 10^8$ lines $= 1$ Wb/m$^2 = 10^4$ gauss
	1 nautical mile	$= 6080$ ft $= 1.852$ km
Mass	1 tonne	$= 1000$ kg
	1 kg	$= 1000$ gm $= 5000$ carat $= 2.196$ lb
Neutron cross- section	1 barn	$= 10^{-24}$ cm^2/nucleus

(Continued)

Quantity	Unit and Symbol	Converted Value
Nuclear	1 mil	= 0.001 inch = 0.0254 mm
Plane angle	1 (radian) rad	= 57.30°
Power	1 J	= 1000 W = 1000 J/s = 3.6 MJ/h
	1 kW	= 0.9478 Btu/s
	1 hp	= 746 W = 10.69 kcal/min = 550 ft-lb/s
Pressure intensity	1 mm of Hg at	= 101.325 kPa = 760 mm of Hg at 0°C
	0°C	= 1 N/m^2 = 10^{-6} Nym.^{2} = 10^{-5} bar
	1 atm	= 1.45 × 10^{-4} psi
	1 Pa	= 1000 psi = 6.895 MPa
Quantity of electricity	1 Ah	= 3600 C = 3.6 kC
Stress, strength	1 Pa	= 10 dynes/cm^2 = 1.0197 × 10^{-2} gm/cm^2
Surface tension	1 N/m	= 10^{-3} dyne/cm
Temperature		
	0°C	= 273.15 K– 273 **K**
	°K	= −273.15°C– 273°C = (1.8 × t + 32)°F
	t °C	
Thermal conductivity	1 (tesla) T	= 0.00237 cal/(cm-s-°C)
Volume	1 litre	= 1000 mL = 1 dm^3 = 1000 cm^3
	11/1^3	= 1.308 yd^3 = 35.31 ft^3 = 10^6 cilia
		= 1000 litre
Work	1 ksi	= 1 N m = 0.2389 cal

List of Abbreviations

Ah   ampere-hour
Btu   British thermal unit
ft   foot/feet
kip   kilo pound
ksi   kip per sq inch
lb   pound
ml   millilitre
psi   pound/sq. inch unit
s   second
W   watt
yd   yard

1

Introduction to Some Recent and Emerging Materials

1.1 Historical Perspective of Materials

Materials have been deeply involved with our culture since precivilization era. Historically, the advancement of societies was intimately tied with the development of materials to fulfil the needs of those eras. That is why the civilizations have been named by the level of their material development, for example, Stone Age, Bronze Age, steel age and plastic age. Stone, clay, wood, etc. were the common materials in historical days, but the scenario of present era is completely changed.

1.1.1 Modern Perspective

The advancement of any engineering discipline is not possible without the development of materials and their science, engineering and technology. Rapid advancement in electron-based computers or probable light-based computers in the future, changes in electronics engineering from vacuum valves to very-large-scale microchips (VLSCs), cement concrete (CC) to polymer-reinforced concrete (PRC) in civil engineering, pure metals to duplex stainless steel (DSS) in mechanical engineering, wood to ferroelectrics and ordinary steel to ferrites in electrical engineering are some illustrations which became possible due to developments in materials science.

1.2 Different Types of Engineering Materials

The materials are broadly classified as follows (Figure 1.1).

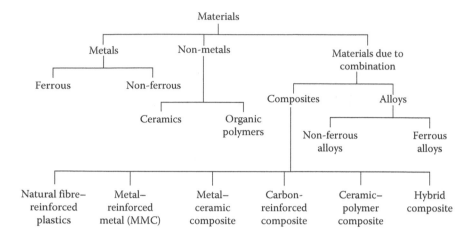

FIGURE 1.1
Broad classification of engineering materials.

1.2.1 Metals

Metals are elemental substances capable of changing their shape permanently. They are good conductors of heat and electricity. These may be of ferrous or non-ferrous type. The behaviour and properties of ferrous metals depend on the percentage and the form (phase and constituents) of carbon present in them. The difference between steel and iron and their specific names according to percentage of carbon are given in Table 1.1.

TABLE 1.1

Steel and Iron on the Basis of Percentage of Carbon[a]

Percentage of Carbon	Specific Names	Abbreviated Name	Steel or Iron
Purest to 0.05	Wrought iron	WI	↑
0.05–0.10	Dead mild steel	DMS	
0.10–0.30	Mild steel	MS	Plain carbon steel
0.30–0.70	Medium carbon steel	MCS	
0.70–1.50	High carbon steel	HCS	↓
1.50–2.00	Semisteel or semi-iron	—	—
2.00–4.50	Cast iron (grey, white, mottled, spongy, malleable, chilled, spheroidal, nodular)	CI	↑
			Iron
4.50–6.67	Pig iron	PI	
6.67 or more	Ore	—	↓

[a] Data are averaged and reasonably approximated.

1.2.2 Nonferrous Metals

Non-ferrous metals do not contain Fe and C as their constituents. Aluminium, copper, silver, nickel, zinc, tin, chromium, etc. are some examples. Al, Cu, Ag and Au are good conductors of electricity; Ag is the most malleable, Au is the most ductile, and chromium is corrosion resistant. Zinc is used in metal plating, tin is used to make bushes, and nickel imparts strength and creep resistance.

1.2.3 Ceramics

Ceramics are generally metallic or non-metallic oxides. Physically separable and chemically homogeneous constituents of materials consisting of phases are also ceramics. Rocks, glasses, fireclay and firebricks, cements and limes are ceramics. Ferrites, garnets, ferroelectrics and ceramic superconductors are the latest developments in this area.

1.2.4 Organic Polymers

Organic polymers are relatively inert and light and generally have a high degree of plasticity. These are derived mainly from hydrocarbons. These consist of covalent bonds formed by carbon, chemically combined with oxygen and hydrogen. The word *mer* in Greek means a unit, *mono* means one and *poly* means many. Thus, polymers are obtained from monomers bonded by a chemical reaction (a process called polymerization). In this process, long molecular chain having high molecular weight is generated. Bakelite, polyethylene, nylon and Teflon are some examples.

1.2.5 Alloys

An *alloy* is a combination of two or more metals. They possess properties quite different from those of their constituent metals. An alloy is prepared for a specific purpose to meet the particular requirement of an application. Alloys may be ferrous alloy or non-ferrous depending on the base metal used.

1.2.6 Composites

Composites may be inorganic or organic. They have two or more constituents of dissimilar properties. The two major constituents may be metals and ceramics, or metals and polymers, or ceramics and polymers or other combinations. Alloys may also be used instead of metals to make composites. One of the constituents (called reinforcing constituent) may be in particulate form, fibrous form or flake form. Fibrous composites are more common in present-day applications. Whisker-reinforced composites are likely to be the future material.

TABLE 1.2

Classified Groups of Materials with Examples and Uses

S. No.	Group of Materials	Examples	Uses
1.	Ferrous metals	Iron and steel	Structures, machines, alloy-making tools
2.	Non-ferrous metals	Cu, Al, Si, Sb, Co	Conductors, semiconductors
3.	Ferrous alloys	Invar, stainless steel, alnico, high-speed steel (HSS)	Precision measuring tapes, magnets, cutting tools
4.	Non-ferrous alloys	Phosphor bronze, brass, duralumin, babbits	Bushes of bearings, springs, utensils
5.	Ceramics	ZrO_2, SiO_2, B_2O_3, Na_2O, Al_2O_3, glasses	Refractories, furnace linings, lens
6.	Organic polymers	Polyethylene, nylon, epoxy, polystyrene	Domestic articles, biomedical uses, incomposite constructions
7.	Inorganic composites	Reinforced cement concrete (RCC)	Bridge structures, multistoreyed buildings
8.	Metal–ceramic particulate composites	Cermet	Space-going vehicles, rockets, missiles
9.	Fibrous, polymeric, composites	Polymer reinforced with the fibres of glass, carbon, boron, graphite, Kevlar	Automobiles, sports, structures, aircrafts, bodies of computer and radar
10.	Metal–metal composites	Aluminium conductor steel reinforced (ACSR)	Conductors in high-voltage transmission and distribution

1.2.7 Classified Groups of Materials and Their Examples and Uses

To illustrate the importance of materials, various groups of materials with examples and uses are summarized in Table 1.2.

1.3 Scale of Materials and Size of Devices

There are different levels of material structure that vary from the bulk size dimensions of metre to mm and down to 10^{-8} m or less. These are macro level (10^{-4} to 10^{-6} m), sub-micro level (10^{-6} to 10^{-8} m), crystal level (10^{-8} to 10^{-10} m), electronic level (10^{-10} to 10^{-12} m), nuclear level (10^{-12} to 10^{-15} m) materials, etc. Based on dimensions of these levels, the technology and devices involved are generally called as follows:

- *Macro technology* which is of the order of 10^0 to 10^{-3} m or more. Devices of these scales are known as *macro or bulk devices* such as brute machinery, genetics and heavy machineries.
- *Microtechnology* which is of order of 10^{-3} to 10^{-6} m. Devices of these scales are known as *micro devices* such as ICs and microbots.

- *Nanotechnology* which is of the order of 10^{-9} m or so. Devices of these scales are known as *nano devices* such as molecular computers.
- *Pico technology* which is of the order 10^{-12} m or so. Devices of these scales are/will be known as *pico devices* such as artificial elements, atomic precision manipulators and quantum effective bionanomachinery.
- *Femto technology* which is of the order of 10^{-15} m or so. Devices of these scale are/will be known as *femto devices* such as quark structures, super computational machines and subatomic devices.
- *Atto technology* which is of the order of 10^{-18} m or so. Devices of these scale are/will be known as *atto devices* such as enormous speed computational machines, quantum particle class devices and inconceivable intelligence (i.e. super brains).
- *Planck-scale technology* which is of order of less than 10^{-18} m or so. Devices of these scales will be such as space-time engineering machines, baby universe, closed time-like loops and exotic space-time topologies.

1.4 Requirements of Materials

Materials science does not mean just knowing the physics and chemistry of materials and their behaviour and properties. It is also essential to know how a material can be suitably and economically put to practical use under wide range of conditions. These conditions may relate to the operation or to the fabrication or to the stability of materials. An engineering material is used in one or all of the following areas:

1. *Machines* (such as IC engine, alternator, lathe)
2. *Structures* (such as building, bridge, tower)
3. *Devices* (such as strain gauge, integrated circuit, control switch)

1.4.1 Important Properties of Materials

Each material possesses a number of properties. Some most important properties are as follows:

1. *Mechanical*: creep, fatigue, toughness, hardness, impact, ductility, malleability, resilience and brittleness
2. *Physical*: density, melting point, colour, shape, size, finish and porosity

3. *Thermal*: expansion, conductivity, specific heat, thermal fatigue, thermal stress, thermal shock and latent heat of fusion

4. *Magnetic*: hysteresis, retentivity, permeability, susceptibility, coercive force and reluctivity

5. *Electrical*: resistivity, conductivity, dielectric constant, dielectric strength, relaxation time, loss angle and power factor

6. *Chemical*: corrosion resistance, passivity, atomic number, molecular weight, acidity, alkalinity and oxidation

7. *Nuclear*: half-life period, decay constant and radiation absorptivity

8. *Optical*: reflection, refraction, transmission, fluorescence, lustre and luminescence

9. *Acoustical*: sound reflection, absorption, damping and transmission

10. *Metallurgical*: phase rule, solid solution, crystallization rate and diffusion

11. *Cryogenic*: ductile–brittle behaviour, low-temperature impact behaviour and very-low-temperature phase changes

12. *Structural*: strength, stiffness, elasticity and plasticity

13. *Technological*: weldability, machinability, formability, castability, fabrication ability and hardenability

14. *Surface*: friction, abrasion, wear and erosion

15. *Aesthetic*: feel, texture, appearance and lustre

1.5 Present Scenario of Advanced Materials

To meet the challenges of advancing technology, material scientists are trying hard to develop more advanced materials. Some of them are given as follows:

- Rolled armour steel for military tanks
- Maraging steel for motor casing of booster rockets
- Zircaloy tube for nuclear reactors
- Cubic zirconia ($ZrO_2 + Y_2O_3$), a ceramic alloy for gem industry, artificial diamond and wear-resistant cutting tools
- Carbon–carbon composite for nose cone of missiles and fuselage of space shuttles
- Hybrid composites for aircraft components
- Epoxy–Kevlar composite for space-going vehicles and satellites
- Nickel-based cryogenic steel for cryogenic industries and cryogenic engines

- Diamond film (phosphorus doped) *n*-type semiconductor for diamond transistors and other microelectronic devices (fibre optics for telecommunication and in fibre-optic endoscopy to scan the intestine, lever, gall bladder, etc.)
- Low-temperature superconductors (LTS) for magnets below 25 K
- High-temperature superconductors (HTS) above 25 K, also known as perovskite oxides (La_x Ba_y Cu_z O_4) for hybrid magnets, switches, transmission, etc.
- Ceramic and metallic whiskers for ultrahigh-strength applications
- New-generation *Al, Ga and As* chips in hyper-high-speed computers
- Ferrites of Cu–Mn, Mg–Mn and Zn–Mn for computer memory cores
- Gadolinium–gallium–garnet (Gd_3 Ga O_{12}) for magnetic refrigerators operating below 10 K
- Iron garnet material grown on Sm_3 Ga_5 O_{12} for magnetic bubbles (memory devices)
- Oxygen-free high-conductivity (OFHC) copper for use as low-temperature conductors
- Aluminium diboride for planar reinforcement in flaked composites
- Polytetrafluoroethylene (PTFE), commercially known as *Teflon*, for seats in ball and butterfly valves and non-sticking coating on utensils
- Magnesium alloy plates for missile wings
- Hg–Cd–Te crystals having sensor elements for guided missiles
- Rochelle salt (Na K $C_4H_4O_6.4H_2O$), a ferroelectric for insulation
- Selenium in solar battery for electricity generation
- High-density concrete for preventing radiation (nuclear and electromagnetic) hazards
- Ferro-cement and fibre-reinforced concrete (FRC) for canal lining, road pavements and overlay of airstrip
- Cellular concrete (foamed and lightweight) for lightweight structures and collapsible bridges

1.5.1 Futuristic Materials

Industrial progress in the future will depend on the ability to produce newer materials. Fast-breeder nuclear reactors, lightweight road–airspace transportations, biomedical, robotics, supercomputers, telecommunications and levitated trains are some areas which will necessitate newer generation of materials. These are likely to emerge from hybridization of metals, polymers, ceramics and whisker composites. Some of the probable materials and their targeted applications are enumerated in Table 1.3.

TABLE 1.3

Future Materials and Their Anticipated Applications

S. No.	Material	Application
1.	SMAs such as Ni–Ti, Pd–TI	Robotics, biomedicals, sensors
2.	Amorphous lanthanum–nickel base alloy (hydrogen storage alloy)	Hydrogen storage, Ni–hydrogen battery, Freon-free air-conditioning
3.	Aluminium–lithium alloys	Corrosion reduction and lightweight aircrafts
4.	Microalloyed steels	Automobile structure
5.	Composites such as	Nuclear fusion reactor
	a. TiC and Ni, ZrO_2 and Ni	Rockets and missiles, x-ray equipment, military weapons
	b. Ti–alloy and carbon fibres	
	c. Al–alloy and ceramic fibres	Space vehicles
	d. Lithium aluminosilicate and silicon carbide fibres	Gas turbines
6.	High-performance plastics	Gears, bearings, body of autovehicles
7.	Liquid crystal polymers	Optical fibres
8.	Separation membranes	Medical, biotechnology, sewage and water treatment, chemical and petrochemical industries
9.	Synthetic diamond	Semiconductor lenses and mirrors for high-power lasers
10.	Single-crystal diamond Heterojunction microelectronic multilayered device (each layer a couple of atoms thick)	Light absorption and optics, negative resistivity
11.	Hybrid chips of GaAs and Si	Integrated circuits
12.	New-generation semiconductor chips for 16 MB dynamic random access memory (DRAM)	Computers
13.	HTS	Medical imaging machines, levitated train (which ride on a magnetic field), magnetic resonance imaging (MRI), magnets
14.	U–Pu–Zr, UAl_4 and PuAl, Hafnium carbide (HfC)	Nuclear fuels, super refractory

In addition to these, fibre-reinforced metals (FRMs), high-nitrogen steels, conducting plastics, high-temperature plastics, whiskers and whisker-reinforced composites are other probable materials.

1.6 Recent Advances in Materials Technology

Advances in materials and their technology are occurring very fast. On one hand, they are acquiring nanodimensions, while on the other, they are showing peculiar behaviour. Consequently, there are too many newer fields of study. To name a few, these are as follows:

- Nanomaterials
- Smart materials
- Functionally graded materials (FGMs)
- Porous materials
- Biomaterials
- Whiskers
- Metamaterials
- Photonic materials
- Photorefractive polymers
- Left-handed materials
- MAX phase materials
- Foamed materials
- Biomimetic materials
- Green materials

Industrial progress in the future will depend on the ability to produce newer materials. The following are some areas which will necessitate newer generation of materials:

- Fast-breeder (nuclear) reactors
- Micro-electromechanical systems (MEMs) and nano-electrome-chanical systems (NEMs) devices
- Lightweight road–airspace transportations
- Fire resistance structure
- Biomedical
- Robotics
- Supercomputers
- Telecommunications
- Levitated trains
- Marine applications
- Alternative energy sources
- Thermo-acoustic insulation purposes

These are likely to emerge from hybridization of metals, polymers, ceramics and whisker composites.

Although it is not possible to cover all aspects of these materials here, yet introductory details are given in the following to acquaint the readers with the latest trends.

1.7 Smart Materials (or Intelligent Materials)

Smart material has one or more properties that can be dramatically altered. Most everyday materials have physical properties which cannot be significantly altered; for example, if oil is heated, it will become a little thinner, whereas a smart material with variable viscosity may turn from a fluid which flows easily to a solid. A variety of materials already exist and are being researched extensively. These include

1. Piezoelectric materials
2. Magneto-rheostatic (MR) materials
3. Electro-rheostatic (ER) materials
4. Shape memory alloy (SMA)

Some everyday items are already incorporating smart materials (coffeepots, cars, eyeglasses), and the number of applications for them is growing steadily. A smart fluid developed in the labs at the Michigan Institute of Technology has made amazing developments in the design of electronics and machinery using standard materials (steel, aluminium, gold). Imagine the range of possibilities which exist for special materials can manipulate the properties. Some such materials have the ability to change shape or size simply by adding a little bit of heat or to change from a liquid to a solid almost instantly when near a magnet. Each individual type of smart material has a different property which can be significantly altered such as viscosity, volume and conductivity. The property that can be altered influences the type of applications the smart material can be used for.

Response to stimulus is a basic process of living system. Based on the lesson from nature, scientists have been designing useful materials that respond to external stimuli such as temperature, pH, light, electrical field and chemical's ionic strength. These responses are manifested as changes in one or more of the following: shape, surface characteristics, solubility, formation of an intricate molecular self-assembly, a sol-to-gel transition and so on. Such structures are called smart structures and such materials are referred to as smart materials. Active materials and adaptive materials of such kinds have evolved over the past decades, with increasing pace during the 1990s.

1.7.1 Classification of Smart Materials

Smart materials can be divided into two groups. One group comprises the *classical* active materials and is characterized by the materials response with a change in shape and/or in length of material. Thus, input is always transformed into strain, which can be used to introduce motion or dynamics into system. Smart polymers responsive to solvent composition, pH and temperature are being utilized for potential applications like chemical valves, shape memory and biomedical applications including artificial organs and drug delivery systems. The second group consists of materials that respond to stimuli with a change in a key material property, for example, electrical conductivity or viscosity. A classification is shown in Table 1.4.

1.7.2 Piezoelectric Materials

Piezoelectric materials have two unique properties which are interrelated. When a piezoelectric material is deformed, it gives off a small but measurable electrical discharge. Alternately, when an electrical current is passed through a piezoelectric material, it experiences a significant increase in size (which can be up to a 4% change in volume). Piezoelectric materials are most widely used as sensors in different environments; they are often used to measure

- Fluid compositions
- Fluid density
- Fluid viscosity
- The force of an impact

TABLE 1.4

Classification of Smart and Other Materials

Category	Fundamental Material Characteristics	Fundamental System Behaviours
Traditional materials: 1. Natural materials (stone, wood) 2. Fabricated materials (steel, aluminium, concrete)	Materials have given properties and are *acted upon*.	Materials have no or limited intrinsic active response capability but have good performance properties.
High-performance materials: 1. Polymers 2. Composites	Material properties are designed for specific purposes.	—
Smart materials: 1. Property-changing materials 2. Energy-exchanging materials	Properties are designed to respond intelligently to varying external conditions or stimuli.	Smart materials have active responses to external stimuli and can serve as sensors and actuators.

An example of a piezoelectrical material in everyday life is the airbag sensor used in car. The material senses the force of an impact on the car and sends and electric charge deploying the airbag.

1.7.3 Electro-Rheostatic and Magneto-Rheostatic

ER and MR materials are fluids, which can experience a dramatic change in their viscosity. These fluids can change from a thick fluid (similar to motor oil) to nearly a solid substance within the span of a millisecond when exposed to a magnet or electric field. The effect can be completely reserved just as quickly when the field is removed. The MR fluid is liquid when no magnetic field is present, but turns solid immediately after being placed in magnetic field on the right. MR fluids experience a viscosity change when exposed to magnetic field, while ER fluids experience similar changes in an electric field. The composition of each type of smart fluid varies widely. The most common form of MR fluid consists of tiny iron particles suspended in oil, while ER fluid can be as simple as milk chocolate or cornstarch and oil. MR fluids are being developed for use in

- Car shocks
- Damping washing machine vibration
- Prosthetic limbs
- Exercise equipment
- Surface polishing of machine parts

ER fluids have mainly been developed for use in clutches and valves as well as engine mounts designed to reduce noise and vibration in vehicles. Nature is full of magical materials which are to be discovered in forms suitable to our needs. Such magical materials are also known as intelligent or smart materials. These materials can sense, process, stimulate and actuate a response.

Functions: Their functioning is analogous to human brain, slow and fast muscle action. These materials are the wonder materials that can feel an action and suitably respond to it just like any living organism. Analogous to human immune system, the intelligent material comprises three basic components which are given as follows:

1. Sensors such as piezoelectric polymers (polyvinylidene) and optical fibres
2. Processors such as conductive electroactive polymers and microchips
3. Actuators such as SMAs (NiTi, i.e. nitinol) and chemically responding polymers (polypyrrole)

Characteristics: These components in the form of optical fibres or electrorheological (ER) fluids are embedded or distributed in materials. They possess

TABLE 1.5

Intelligent (or Smart) Materials

S. No	Material	Characteristics	Example	Application
1	Piezoelectric ceramics	*Linear* and shear deformations occur along longitudinal, transverse and thickness directions.	Quartz, Pb Zr, titanate	Aircraft airfoils, identifying Braille alphabet (an aid for blinds)
2	Viscoelastic (VE)	They relax any stress produced in it by external strain.	—	Damping in spacecrafts, earthquake-prone structures, aircrafts
3	ER fluids	These are like suspended fine polarizable particles cohesive and tend to coalesce. They form new chains even when old chains are broken.	Zeolite in silicone oil, starch in corn oil	In filling of graphite–epoxy beams to variable stiffness in them
4	SMAs	Below a critical transition temperature, they can deform plastically to their memorized valves, robotics shape.	NiTi (nitinol)	Fire alarm due to change of shape at transition temperature

ability to change with environmental radiation, stress, temperature, pressure, voltage, etc.

Examples, salient characteristics and applications of intelligent materials are given in Table 1.5.

1.8 Shape Memory Alloys

SMAs are metals that exhibit two very unique properties, namely, pseudo-elasticity and the *shape memory effect*. The most effective and widely used SMAs include NiTi (nickel–titanium), CuZnAl and CuAlNi. The two unique properties described earlier are made possible through a solid-state phase change, that is, a molecular rearrangement, which occurs in SMA. Typically, when one thinks of a phase change, it is similar in that a molecular rearrangement is occurring, but the molecules remain closed packed so that the substance remains a solid. In most SMAs, a temperature change of only about 10°C is necessary to initiate this phase change. The two phases, which occur in SMAs, are martensite and austenite. Martensite is the relatively soft and easily deformed phase of SMAs which exists at lower temperatures. The load molecular structure in this phase is twinned while the configuration of this phase takes on the second form

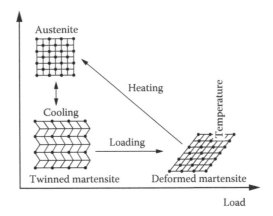

FIGURE 1.2
Microscopic diagram of the shape memory effect.

as shown in Figure 1.2. Austenite, the stronger phase of SMAs, occurs at higher temperature.

1.8.1 Shape Memory Effect (SME)

The shape memory effect is observed when the temperature of a piece of SMA is cooled below the temperature of martensite formation. At this stage, the alloy is completely composed of martensite which can be easily deformed. After distorting the SMA, the original shape can be recovered simply by heating the wire above the austenite formation temperature. The heat transferred to the wire is the power driving the molecular rearrangement of the alloy, similar to the heat that melts ice into water, but the alloy remains solid. The deformed martensite is now transformed to the cubic austenite phase, which is configured in the original shape of the wire. The shape memory effect is currently being implemented in following applications:

- Coffeepots
- Thermostat
- Hydraulic fittings for airplanes
- Space shuttle
- Vascular stents

1.8.2 Material Systems of Different Shape Memory Alloys [1]

Since the discovery of NiTi, at least 12 different binary, ternary and quaternary alloy types have been discovered that exhibit shape changes and unusual elastic properties consequent to deformation. These are

- Nickel–titanium
- Nickel–zirconium–titanium
- Nickel–titanium–copper
- Nickel–iron–zinc–aluminium
- Titanium–niobium
- Titanium–niobium–aluminium
- Titanium–palladium–nickel
- Iron–manganese–silicon
- Iron–zinc–copper–aluminium
- Copper–aluminium–iron
- Zirconium–copper–zinc
- Hafnium–titanium–nickel

1.8.3 Preparation of SMA

The use of the one-way shape memory or superelastic property of NiTi for a specific application requires a piece of SMA to be moulded into the desired shape. The characteristic heat treatment is then done to set the specimen to its final shape. The heat treatment methods used to set shapes in both the shape memory and the superelastic forms of NiTi are similar. Adequate heat treatment parameters are needed to set the shape and properties of the item. The two-way memory training procedure can be made by shape memory effect (SME) training or stress-induced martensite (SIM) training. In SME training, the specimen is cooled below M_f and bent into the desired shape. It is then heated to a temperature above A_f and allowed freely to take its austenite shape. The procedure is repeated 20–30 times which completes the training. The sample now assumes its programmed shape upon cooling under M_f and to another shape when heated above A_f. In SIM training, the specimen is bent just above M_s to produce the preferred variants of SIM and then cooled below M_f temperature. Upon subsequent heating above the A_f temperature, the specimen takes its original austenitic shape.

1.8.4 Applications of SMA in Different Fields

Application in aircraft manoeuvrability [2]: Aircraft manoeuvrability depends heavily on the movement of flaps found at the rear or trailing edge of the wings. The efficiency and reliability of operating these flaps are of critical importance. Most aircraft in the air today operate these flaps using extensive hydraulic systems. These hydraulic systems utilize large centralized pumps to maintain pressure and hydraulic lines to distribute the pressure to the flap actuators. In order to maintain reliability of operation, multiple hydraulic lines must be run to each set of flaps. This complex system of pumps and lines is often relatively difficult and costly to maintain. Many alternatives to the hydraulic systems are being explored by the aerospace industry. Among the most promising alternatives are piezoelectric fibres, electrostrictive ceramics and SMAs. *Smart* wings which incorporate SMAs, are typically much more compact and efficient in that the shape memory wires require an electric current only for movement.

Application in medical applications [1]: The variety of forms and the properties of SMAs make them extremely useful for a range of medical applications. For example, a wire that in its deformed shape has a small cross section can be introduced into a body cavity or an artery with reduced chance of causing trauma. Once in place and after it is released from a constraining catheter, the device is triggered by heat from the body and will return to its original memorized shape. Increasing a device, volume by direct contact or remote heat input has allowed the development of new techniques for keyhole or minimally invasive surgery. This includes instruments that have dynamic properties, such as miniature forceps, clamps and manipulators. SMA-based devices that can dilate, constrict, pull together, push apart and so on have enabled difficult or problematic tasks in surgery to become quite feasible.

Stents [1]: The property of thermally induced elastic recovery can be used to change a small volume to a larger one. An example of a device using this is a stent. A stent, either in conjunction with a dilation balloon or simply by self-expansion, can dilate or support a blocked conduit in the human body. Coronary artery disease, which is a major cause of death around the world, is caused by a plaque in-growth developing on and within an artery, inner wall. This reduces the cross section of the artery and consequently reduces blood flow to the heart muscle. A stent can be introduced in a deformed shape, in other words with a smaller diameter. This is achieved by travelling through the arteries with the stent contained in a catheter. When deployed, the stent expands to the appropriate diameter with sufficient force to open the vessel lumen and reinstate blood flow.

Vena-cava filters [1]: Vena-cava filters have a relatively long record of successful in vivo application. The filters are constructed from NiTi wires and are used in one of the outer heart chambers to trap blood clots, which might be the cause of a fatality if allowed to travel freely around the blood circulation system. The specially designed filters trap these small clots, preventing them from entering the pulmonary system and causing a pulmonary embolism. The vena-cava filter is introduced in a compact cylindrical form about 2.0–2.5 mm in diameter. When released, it forms an umbrella shape. The construction is designed with a wire mesh spacing sufficiently small to trap clots. This is an example of the use of superelastic properties.

Dental and orthodontic applications: Another commercially important application is the use of superelastic and thermal shape recovery alloys for orthodontic applications. Archwires made of stainless steel have been employed as a corrective measure for misaligned teeth for many years. Owing to the limited stretch and tensile properties of these wires, considerable forces are applied to teeth, which can cause a great deal of discomfort. When the teeth succumb to the corrective forces applied, the stainless steel wire has to be re-tensioned. Visits may be needed to the orthodontist for re-tensioning every 3–4 weeks in the initial stages of treatment. Superelastic wires are now used for these corrective measures. Owing to their elastic properties and extend-ibility, the level of discomfort can be reduced significantly as the SMA applies a continuous, gentle pressure over a longer period. Visits to the orthodontist are reduced to perhaps three or four per year.

Bone plates [2]: Bone plates are surgical tools, which are used to assist in the healing of broken and fractured bones. The breaks are first set and then held in place using bone plates in situations where casts cannot be applied to the injured area. Bone plates are often applied to fractures occurring to facial areas such the nose, jaw or eye sockets. Repairs like this fall into an area of medicine known as osteosynthesis. Currently, osteotomy equipment is made primarily of titanium and stainless steel. The broken bones are first surgically reset into their proper position. Then a plate is screwed onto the broken bones to hold them in place, while the bone heals back together. This method has been proven both successful and useful in treating all manner of breaks; however, there are still some drawbacks. After initially placing the plate on the break or fracture, the bones are compressed together and held under some slight pressure, which helps to speed up the healing process of the bone. Unfortunately, after only a couple of days, the tension provided by the steel plate is lost, and the break or fracture is no longer under compres-sion, slowing the healing process. Bone plates can also be fabricated using SMAs, in particular nickel–titanium. The NiTi plates are first cooled to well below their transformation temperature, and then they are placed on the set break just like titanium plates.

However, when the body heats the plate up to body temperature, the NiTi attempts to contract applying sustained pressure on the break or fracture for far longer than stainless steel or titanium. This steady pressure assists the healing process and reduces recovery time. There are still some problems to consider before NiTi bone plates will become common place. Designing plates to apply the appropriate amount of pressure to breaks and fractures is the most important difficulty, which must be overcome. However, a badly fractured face can be reconstructed using bone plates.

Robotics [2]: There have been many attempts made to re-create human anat-omy through mechanical means. The human body, however, is so complex that it is very difficult to duplicate even simple functions. Robotics and elec-tronics are making great strides in this field, of particular interest are limbs such the hands, arms and legs. SMAs mimic human muscles and tendons very well. SMAs are strong and compact so that large groups of them can be used for robotic applications and the motion with which they contract and expand is very smooth creating a lifelike movement unavailable in other systems.

1.8.5 Current Examples of Applications of Shape Memory Alloys [3]

- Aircraft flap/slat adjusters
- Anti-scald devices
- Arterial clips
- Automotive thermostats
- Braille print punch
- Catheter guide wires
- Cold start vehicle actuators
- Contraceptive devices
- Electrical circuit breakers
- Fibre-optic coupling
- Filter struts
- Fire dampers
- Fire sprinklers
- Gas discharge
- Graft stents
- Intraocular lens mount
- Kettle switches
- Keyhole instruments
- Keyhole surgery instruments
- Mobile phone antennas
- Orthodontic archwires
- Penile implant
- Pipe couplings
- Robot actuators
- Rock splitting
- Root canal drills
- Satellite antenna deployment
- Scoliosis correction
- Solar actuators
- Spectacle frames
- Steam valves
- Stents
- Switch vibration damper
- Thermostats
- Vibration dampers

1.8.6 Future Applications of SMA

There are still some difficulties with SMAs that must be overcome before they can live up to their full potential. These alloys are still relatively expensive to manufacture and machine compared to other materials such as steel and aluminium. Most SMAs have poor fatigue properties; this means that while under the same loading conditions (i.e. twisting, bending, compressing), a steel component may survive for more than 100 times more cycles than an SMA element.

There are many possible applications for SMAs. Future applications are envisioned to include engines in cars and airplanes and electrical generators utilizing the mechanical energy resulting from the shape transformations. Nitinol with its shape memory property is also envisioned for use as car frames. Other possible automotive applications using SMA springs include engine cooling, carburettor and engine lubrication controls.

1.9 Advances in Smart Materials

Until relatively recent times, most periods of technological development have been linked to changes in the use of materials (e.g. the Stone, Bronze and Iron ages). In more recent years, the driving force for technological change in many respects has shifted towards information technology. This is amply illustrated by the way the microprocessor has built intelligence into everyday domestic appliances. However, it is important to note that the IT age has not left engineered materials untouched and that the fusion between designer materials and the power of information storage and processing has led to a new family of engineered materials and structures. A few of them are briefly described in the following.

Dumb materials: Most familiar engineering materials and structures until recently have been *dumb*. They have been preprocessed and/or designed to offer only a limited set of responses to external stimuli. Such responses are usually non-optimal for any single set of conditions, but *optimized* to best fulfil the range of scenarios to which a material or structure may be exposed. For example, the wings of an aircraft should be optimized for takeoff and landing, fast and slow cruise, etc.

Biomimetics: *Dumb* materials and structures contrast sharply with the natural world where animals and plants have the clear ability to adapt to their environment in real time. The field of biomimetics, which looks at the extraction of engineering design concepts from biological materials and structures, has much to teach us on the design of future man-made materials. The process of balance is a truly *smart* or intelligent response, allowing in engineering terms a flexible structure to adapt its form in real time to minimize the effects of an external force, thus avoiding catastrophic collapse. The natural world is full of similar properties including the ability of plants to adapt their shape in real time (e.g. to allow leaf surfaces to follow the direction of sunlight), limping (essentially a real-time change in the load path through the structure to avoid overload of a damaged region) and reflex to heat and pain.

ER and MR: ER and MR materials are fluids, which can experience a dramatic change in their viscosity. These fluids can change from a thick fluid (similar to motor oil) to nearly a solid substance within the span of a millisecond when exposed to a magnetic or electric field; the effect can be completely reversed just as quickly when the field is removed. MR fluids experience a viscosity change when exposed to a magnetic field, while ER fluids experience similar changes in an electric field. The composition of each type of smart fluid varies widely. The most common form of MR fluid consists of tiny iron particles suspended in oil, while ER fluids can be as simple as milk chocolate or cornstarch and oil.

Photochromic materials: Photochromic materials change reversibly colour with changes in light intensity. Usually, they are colourless in a dark place, and when sunlight or ultraviolet (UV) radiation is applied, the molecular structure of the material changes and it exhibits colour. When the relevant light source is removed, the colour disappears. Changes from one colour to another colour are possible by mixing photochromic colours with base colours. They are used in paints and inks and mixed to mould or casting materials for different applications.

Thermochromic materials: Thermochromic materials change reversibly colour with changes in temperature. They can be made as semiconductor compounds, from liquid crystals or using metal compounds. The change in colour happens at a determined temperature, which can be varied by doping the material. They are used to make paints and inks or are mixed to moulding or casting materials for different applications.

Electroluminescent materials: Electroluminescent materials produce a brilliant light of different colours when stimulated electronically (e.g. by AC current). While emitting light, no heat is produced. Like a capacitor, the material is made from an insulating substance with electrodes on each side. One of the electrodes is transparent and allows the light to pass. The insulating substance that emits the light can be made of zinc sulphide. They can be used for making light stripes for decorating buildings or for industrial and public vehicles' safety precautions.

Fluorescent materials: Fluorescent materials produce visible or invisible light as a result of incident light of a shorter wavelength (i.e. x-rays, UV rays). The effect ceases as soon as the source of excitement is removed. Fluorescent pigments in daylight have a white or light colour, whereas under excitation by UV radiation, they irradiate an intensive fluorescent colour. They can be used for paints and inks or mixed to moulding or casting materials for different applications.

Other smart materials: Some other smart materials are listed as follows:

1. Phosphorescent materials
2. Conducting polymers
3. Dielectric elastomers
4. Polymer gels
5. Magnetostrictive alloys
6. Electrostrictive materials
7. Piezoelectric materials
8. Sensual materials

Thermoelectric materials: Thermoelectric materials are special types of semiconductors that, when coupled, function as a *heat pump*. By applying a low-voltage DC power source, heat is moved in the direction of the current (+ to −), (see Peltier effect). Usually, they are used for thermoelectric modules where a single couple or many couples to obtain larger cooling capacity are combined. One face of the module cools down while the other heats up, and the effect is reversible. Thermoelectric cooling allows for small size and light devices, high reliability and precise temperature control and quiet operation. Disadvantages include high prices and high operating costs, due to low energy efficiency.

By making our buildings smarter, we can improve our comfort and safety. The difference between a regular brick and a smart brick is a compartment on one side of the smart brick. Its inside has stuffed advanced wireless electronic sensors, signal processors, a wireless communication link and a battery, all packaged in one compact unit. The sensors also are embedded in construction materials such as plastic, wood and steel-reinforced concrete. The sensors are designed to monitor a building's temperature, vibration and movement. They send the information wirelessly to a remote computer. Several of bricks embedded at different locations in a building act as a kind of network, working together to provide a picture of overall stability of the structure. This information would be vital to fire fighters battling a blaze or to rescue workers determining the soundness of an earthquake-damaged structure. Even in the absence of a disaster, the building managers and homeowners could use the data to manage and repair the buildings. For example, the data could pinpoint for where the foundations need to be reinforced or walls replaced. Homeowners could tailor heating and air-conditioning throughout a house for maximum energy efficiency.

1.9.1 Biomedical Applications as Smart Material Application

Recent advantages in the design of stimuli responsive have created opportunities for novel biomedical applications. Stimuli-responsive changes in shape, surface characteristics, solubility, formation of an intricate molecular self-assembly and a sol–gel transition enable several novel applications in the delivery of therapeutics, tissue engineering, cell culture, bioseparations, biomimetic actuators, immobilized biocatalysts, drug delivery and thermo-responsive surfaces.

Smart shirt: *Smart shirt* is a T-shirt that functions like a computer, with optical and conductive fibres integrated into the garment. The shirt monitors the wearer's heart rate, ECG, respiration, temperature and a host of vital functions, alerting the wearer or physician if there is a problem. The smart shirt also can be used to monitor the vital signs of law enforcement officers, firemen, astronauts, military personnel, chronically ill patients, elderly persons living alone, athletes and infants.

Smart suture: A smart suture that ties into the perfect knot is a potential medical application for new biodegradable plastics with *shape memory*. The materials are also biocompatible, that is, safe for use in a living animal.

Smart pressure bandages: Polyethylene glycols bonded to various fibrous materials such as cotton and polyester posses the intelligent properties of thermal adaptability and reversible shrinkage. Reversible shrinkage involves imparting a *dimensional memory* to the material such that when material is exposed to a liquid (e.g. water), it shrinks in area. Such materials could be used for pressure bandages that contact when exposed to blood therapy putting pressure on a wound.

Hydrogel: Hydrogels exhibit plastic contraction with changes in temperature, pH and magnetic or electrical field and have a vast number of applications, for example,

1. Soft actuators in the biomedical field
2. For controlled drug release

1.9.2 Textile Applications as Smart Material Application

Highly visible polyester yarns for improved visibility: The vira (D) spun-dyed filaments in radiant orange and radiant yellow have been used for hazard warning clothing. With spin dyeing, the colour fastness is excellent and the yarns are also convincing in terms of textile and ecological properties.

Electrically conductive cellulose filaments: A variant of Lycra has been developed which incorporated electrically conductive cellulosic fibres and can be used to reduce electrostatic charge in textile material.

1.9.3 Biotechnological Applications as Smart Material Application

Protein partitioning: Smart polymers are attractive for biotechnological applications. Smart polymers can be used in downstream processing because they facilitate preferential partitioning of protein between two phases by undergoing phase transition with little change in the environment's properties.

The advantage of these techniques is that the principal chemical composition of solution remains unchanged, thus eliminating a step to remove salts and specific effluents.

Protein purification techniques: Smart polymers undergo fast and reversible changes in microstructure triggered by small changes of medium property (pH, temperature, ionic strength, etc.). These properties of smart polymers are exploited for the development of new protein purification techniques like affinity partitioning and temperature-induced elution in dye-affinity chromatography.

1.9.4 Other Smart Material Applications

Biomimetics: Ideas abstracted from pinecone scales and stomata have been used to develop a smart textile, which can change its thermal and moisture vapour-permeable properties according to the weather.

Improving recyclability of polymers: Some polymer matrices include smart fibres which have chemicals that ultimately assist in denaturing, degrading or destroying the polymeric structures by depolymerization or chemical reaction to improve recyclability of polymer materials.

Road repairing: Smart matrix materials may be used to repair roads and potholes. Smart-release fibre-containing uncured material is added to a pothole and an agitation or pressure-release curative agents from the interior of fibres provided in the matrix material to facilitate adhesion and curing of the pothole repair mass to the subtract road surface.

Countering radioactive rays: Composite containment structure can be used to counter radioactive or chemical waste materials. Fibres with chemically sensitive coating or radiation-sensitive coating may be provided which are adapted to release scavenger compounds when radiation or chemical waste is detected.

1.10 Nanotechnology

Meaning: Nano is the prefix of a unit. As one nano is equal to 10^9, so 1 nm means 10^{-9} m (i.e. 1 nm $= 10^{-9}$ m). Any technology involving processes, operations, applications and development of devices of such small dimension is referred to as nanotechnology. Nanotechnology is undergoing a revolutionary change. It has initiated beginning of a new era in industrial revolution. Its advent in day-to-day life will soon be felt and realized.

History: The word nanotechnology is originated from the Greek word *nano*, which means extremely small. This name was evolved by a Japanese scientist Norio Taniguchi in 1976. The interest of scientists in nanotechnology started increasing in 1986 when a *scanning tunnelling microscope* was developed by Nobel Prize winner West German scientist Gerd Binnig and

Switzerland scientist Heinrich Rohrer. This is the same microscope with the help of which the scientists could easily visualize the structure of an atom and a molecule for the first time.

Working philosophy: The atoms and molecules of materials are almost of the nano size. Therefore, they influence the nano region of materials, consequent upon which the newer changes appear in them. Hence, under nanotechnology, the size of material is reduced to the size of nano-domain (one nano-domain generally refers to 1–100 nm). By doing so, the changes in different properties of material, namely, mechanical, thermal, electrical and optical, occur at each level. For so-small-sized materials, the surface-to-volume ratio plays important role, and most of the molecules make the material hyperactive by coming on the surface. The consequence of this effect is to change the *chemical activeness* of material due to which the probability exists for origination of newer alloys, composites, sensors and catalysts.

1.10.1 Processes to Prepare Nanomaterials

Nanomaterials are prepared by following two methods:

1. Larger to smaller doing method
2. Smaller to larger doing technique

Since each and every object on earth is constructed of molecules, the prepared object obeys the characteristics of its chemical nature. This nature can be changed again and again. The nanomaterials thus prepared are much lighter, strong, transparent and completely different from their parent materials.

Construction of different structures of carbon nanotubes: Construction of some possible structures of carbon nanotubes is shown in Figure 1.3a through c. These configurations are the following:

1. Armchair structure (Figure 1.3a)
2. Zigzag structure (Figure 1.3b)
3. Chiral structure (Figure 1.3c)

The particular type of structure depends on the style of rolling of graphite sheets.

1.10.2 Uses of Nanotechnology

The scope of unique and peculiar products made from nanotechnology is expanding very fast. It ranges from clothes to cosmetics, hard disc of computer to industries, medical sciences to water purification, etc. In this regard, different major fields of application of nanotechnology are enumerated as follows:

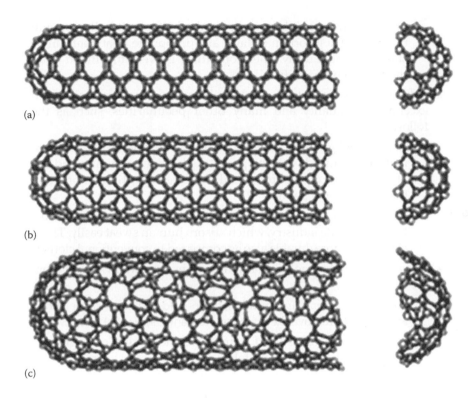

FIGURE 1.3
Illustration of some possible structures of carbon nanotubes: (a) armchair, (b) zigzag and (c) chiral structures.

1. Computer stream
   a. *In supercomputer*: The development of supercomputer is the outcome of nanotechnology. For that, a *nanowire* is used instead of a computer *connector*. Consequently, the memory and speed of computer enhances many times. Increase in memory may be as high as 1 million.
   b. *In chips and ICs*: Computer chips, circuits, transistors, registers, etc. can be made more useful and of much improved quality as compared to existing chips.
2. *Medical sciences*: Presently, the maximum use of nanotechnology is in the field of medical sciences. Main among them are the following:
   a. *Cancer treatment*: Bacteria tumour cells of gold particle are being developed for changing the structure of cancer. It is likely to destroy the dangerous element of tumour.
   b. *Disease-detecting devices*: Nanotechnology-based devices such as *wristwatch* are being developed, which are likely to predict the

probable diseases/illnesses of the body. Since the disease will be detected at its original stage, an appropriate treatment can be started without delay. Therefore, there is every likelihood of an increase in human life.

3. *Industries*: The industrial products/items can be produced with excellence of quality and many other peculiarities, such as the following:

   a. *Nanotechnology-based paints*: If titanium dioxide (TiO_2) as nanomaterial is mixed with the paint, the hardness, shining and life of the paint thus prepared are likely to be many times more than the conventional paint.

   b. *Nanotechnology-based clothes*: Such clothes are being manufactured by textile industry, which absorb human sweat easily. These clothes are more durable also as compared to conventional clothes.

4. *Other uses*: Nanotechnology has many other uses. Main among them are as follows:

   a. *Water purification*: Nano-mineral, nano-gold and nano-silver nitrate are useful to remove the harmful element *arsenic* (As) from water.

   b. *Nanofoster material* is used to improve the quality of brightness and contrast of pictures on television.

5. In future

   a. It will become possible to produce lighter and stronger materials.

   b. The nanostructured pharmaceuticals can be delivered into the body's circulatory system in a very short duration.

   c. The storage capacity of magnetic tapes can be increased.

   d. It will be possible to manufacture scratchproof glasses.

1.10.3 Future Prospects

The scope of nanotechnology is expanding very fast. It is getting increasing applications in physics, chemistry, robotics, biotechnology, etc. Some major fields in which the future prospects of nanotechnology lies are as follows.

Nanomaterials in different forms such as the following:

- Quantum dots
- Nanoparticles
- Nanocluster
- Nano-wells and -wires
- Bulk nanostructured materials
- Magnetic clusters
- Carbon nanostructures

- Superfluid clusters
- Nanostructured ferromagnetism

Nanodevices such as the following:

- Atom switches
- Drug delivery systems
- NEMs
- Molecular machine
- Nanorobots
- Single-electron transistors
- Nanobiometrics

Atomic-size devices: In future, the development in computer technology will be based on devices that will use atom switches. As the size of an atom (of a material) is of the order of nanodimension, the technology of manufacturing atom switches, atomic-size devices and other related researches/developments will fall in the category of nanotechnology. Actions of computers based on nanotechnology will be very fast and responses will be utmost quick.

1.10.4 Nano-Electromechanical Systems

Progress in nanotechnology is poised for a great jump. The development of ultra-miniature systems, namely, NEMs, will make the maintenance, safety and updating of computers and other hardware an easy task.

- Systems will provide almost trouble-free services. The large organizations will be able to squeeze their installations/equipments into a smaller chamber.
- Many electronic systems may be mounted—on or housed—in the watches, shirt pockets, pens, other body parts and their belongings.

The future expectations of nanotechnology can be viewed in the light of the following development.

Japanese company *Nippon Denso* has made an ultra-micro car of unimaginable 4.78 mm size. Made of aluminium and nickel, it contains a motor of 0.67 mm size. The 3 V motor consists of a coil of 1000 conductors. This car will run inside the human veins for medical treatment, such as of the heart.

Nanotechnology is undergoing a revolutionary change. Its advent in day-to-day life is yet to be felt and realized.

1.11 Functionally Graded Materials

Definition: An FGM may be defined as a homogeneous material in which the physical, chemical and mechanical properties change continuously from point to point. Such materials have no discontinuities inside. FGMs may be composites or not, but a vast majority of them are composite materials with a macroscopic microstructural gradient.

Characteristics: An FGM is one in which there is a gradual change of material properties with position. These are therefore also known as *gradient materials*. The property gradient is caused due to position-dependent parameters as the following:

- Micro-structure
- Atomic order
- Chemical composition

The property variation can extend over a large part of material and on to the surface of the material or may be limited to a smaller interfacial region only.

Usefulness of FGMs: FGMs are useful in the sense that the microstructural gradients produce optimum functional performance with minimum material use. Examples of natural and synthetic FGMs are as follows:

1. A natural example of FGM is the culm of bamboo in which the high-strength natural fibres are embedded in a matrix of ordinary cells. In that, the fibre content is not homogeneous over the entire cross section of the culm, rather decreases from outside to inside.
2. A synthetic example of FGM is case-hardened steel in which the material has a gradation, that is, the surface is hard but the interior is tough.

1.11.1 Types of FGMs

Depending upon the nature of gradient, FGMs (composites) may be grouped into the following types:

1. Fraction gradient type (Figure 1.4a)
2. Shape gradient type (Figure 1.4b)
3. Orientation gradient type (Figure 1.4c)
4. Size (of material) gradient type (Figure 1.4d)

Materials such as graded glasses or graded single crystals having one phase possess spatial property variation caused due to gradient in chemical

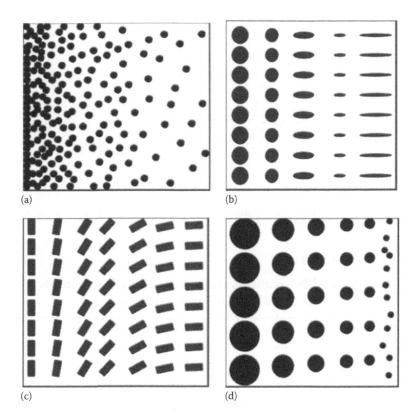

FIGURE 1.4
Based on vividities of gradients, different types of FGMs may be (a) fraction gradient type, (b) shape gradient type, (c) orientation gradient type and (d) size gradient type.

composition. A typical functionally graded bar is shown in Figure 1.5a. It may be Cu/Al_2O_3 bar, TiC/Ti bar, etc. The variation in stress as a function of position is depicted in Figure 1.5b.

1.11.2 Functional Properties

Gradient materials display functional properties that cannot be achieved by a directly joined material or by a homogeneous material. Some earlier uses of FGMs are found in the following applications:

- Graded refractive index lenses (in 1970s)
- Graded refractive index glass fibres for data transmission
- Glass/metal joints with a gradient in coefficient of thermal expansion (in 1960s)
- Polymer skin foams with a dense surface layer and increasing porosity towards the interior

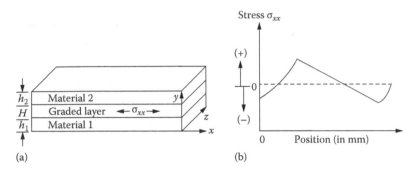

FIGURE 1.5
(a) Geometry of an FGM bar and (b) stress variation with position, that is, linear gradation across the entire thickness of the bar.

1.11.3 Processing of FGMs

FGMs are produced by creating variation in materials by already existing methods. Mainly, these include the following methods:

1.	Powder processing	The example is ZrO_2/stainless steel bulk FGM.
2.	Thermal spraying	The example is YSZ*/NiCrAlY thermal barrier coating (TBC).
3.	Melt processing	The example is W/Cu heat sink material.
4.	Coating processes	The example is TiC/TiN wear-resistant coating.

The suitability of these methods for producing a particular product depends upon the material combination, geometry of desired component and type of required transition function.

- *In powder processing,* the gradient is obtained by mixing different powders in variable ratios and stacking the powder mixtures in separate layers.
- In *thermal spraying,* the gradation profile is obtained by combining highly refractory phases with low-melting metals.
- In *melting processing,* the gradient is produced by centrifuging or sedimentation of refractory phase particles with low-melting metal.
- In *coating process,* a hard surface layer in a compressive stress state is applied on the tools with a gradual variation in binder content from the surface to the interior.

1.11.4 Applications of FGMs

The concept of FGM came into existence in 1990s. However, there were examples of its application in as early as 1960s. Currently, the major fields of applications of FGMs are found in the following areas:

1. *Thermal protection systems*
    i. *Space reentry vehicles*: Normally, a nose cone with a graded SiC protection interlayer on a C/C* composite exposed to high-temperature supersonic gas flow exhibits improved thermal protection.
    ii. *Gas turbine blades* made of superalloy and coated with a bond coat of NiCrA/Y and heat-insulating layer of ZrO considerably improve the resistance of the coating.
    iii. *TBC*: By using a 2 mm thick graded TBC on piston crowns and cylinder heads of diesel engines, about 5% reduction in fuel consumption can be achieved.

2. Wear protection

    Wear resistance of tools increases by hard surface layer in a compressive stress state, for example, a WC/Co FGM with a gradual variation in Co develops a compressive stress state at the surface.

3. Medical implants

    Cementless artificial joints: A coating of glass on a Ti–6Al–4V (titanium–aluminium–vanadium) substrate possessing hydroxyapatite and pores close to its outer surface is an excellent material for making cementless artificial joints and dental implants. The previous material is a graded composite material.

4. Other applications

    The following are some proposed applications, proposed FGMs and functions of gradient in other fields:
    i. *Energy conservation. Thermoelectric converter*: Bi_2Te_3/PbTe/SiGe for improved conversion efficiency
    ii. *Energy conservation. Radiation emitter*: A/N/W for improved emissivity
    iii. *Electronics. Graded capacitor*: $BaTiO_3$/$SrTiO_3$ for zero temperature coefficient of capacitance
    iv. *Acoustics. Acoustic coupler*: CuMn/WC for acoustic impedance matching
    v. *Optics. X-ray mirror*: Si/Ge for graded lattice constant

The use of FGMs avoids the problems associated with the presence of an interface in a material, poor adhesion, unwanted reflections, etc. They can provide an optimum response to an external field in functional applications. However, for processing techniques, the reliability of produced materials is not guaranteed. Additional cost for gradation also adds to the cost of component fabricated.

1.12 Introduction to Biomedical Materials

The biomaterials and biomedical devices are used in the following main applications in the human body:

Purpose	Specific Use
1. For load transmission and stress distribution	Bone replacement
2. Control of blood and fluid flow	Artificial heart
3. Electrical stimuli	Pacemaker
4. Articulation to allow movement	Artificial knee joint
5. Light transmission	Implanted lenses
6. Space filling	Cosmetic surgery
7. Sound transmission	Cochlear implant

1.12.1 Desired Functional Properties of Biomedical Materials

The biomedical materials are subjected to static and dynamic stresses during activities of human body. They degrade by various mechanisms such as the following:

   i. Corrosion

  ii. Dissolution

 iii. Wear

 iv. Swelling

  v. Chemical modification

 vi. Leaching

Consequently, the following biomaterial properties are adversely affected:

- Elastic modulus (or stiffness)
- Wear resistance
- Chemical stability
- Strength
- Surface roughness
- Fracture toughness

Therefore, the selection of biomaterials should be made such that the degradation in them is minimum and the desired properties are least deteriorated. In this regard, various favourable behaviour and harmful effects of different biomaterials are listed as follows:

Biomaterial	Favourable Behaviour	Harmful Effects
1. Biometals	High strength, good toughness and biocompatible	Generation of fine wear particles leading to implant loosening and inflammation
a. CoCrMo alloy	Excellent wear resistance and excellent corrosion resistance	Releases harmful Co, Ni and Cr ions into the body
b. Ti alloys	Allows bone growth at the interface	Unsatisfactory wear resistance, may produce wear debris
2. Bioceramics	Stiff, hard, chemically stable and very good wear resistant	Brittle and relatively difficult to process
a. Alumina	Excellent biocompatibility and inertness in the body	Brittle

1.12.2 Some Key Applications of Biomaterials in Various Biodevices and Allied Components

S. No.	Biodevices/Allied Components	Biomaterials
1.	Pacemakers	
	a. Sheathed wires acting as sensing and stimulating electrodes	Polyurethane as insulation on these wires, silicone
	b. Metal casing to house a battery (i.e. pacemaker housing)	Ti, Pt and Pt–Ir electrodes for providing resistance to galvanic corrosion
2.	Dental materials	
	a. Tooth replacement consisting of an implant that is fixed into the bone and a crown (or denture superstructure)	Alumina, pure Ti, single-crystal sapphire
	b. To promote bone growth into implant interface	Glass ceramic coatings, calcium phosphate ceramics
	c. Tooth crown	Dental porcelains: borosilicate, feldspathic glass
	d. Crown fusing metals	Au, Au-Pd, Ag-Pd, Cu-Pd and Ni-Cr
3.	Ophthalmology materials	Poly hydroxyethylmethacrylate (HEMA)
	a. Hydrogel soft contact lenses	Si-rubbers, fluoropolymers
	b. Rubbery soft lenses	Polymethylmethacrylate (PMMA)
	c. Rigid contact lenses	
4.	Orthopaedics	Ti–6Al–4V, stainless steel Ti–6Al–4V,
	a. Hip and knee implants	stainless steels
	b. Screws, fittings and wires	Alumina (polycrystalline)
	c. Hip replacement	Tricalcium phosphate $Ca_3(PO_4)_2$ ceramic
	d. Bone repair such as maxillofacial and periodontal defects	Polyorthoesters
	e. Bone plates	Austenitic 316L stainless steels Co-based alloys, pure Ti, Ti–6Al–4V, alumina,
	f. Bone cement	zirconia
		PMMA, UHMWPE (ultra-high-molecular-weight polyethylene)

(Continued)

S. No.	Biodevices/Allied Components	Biomaterials
5.	Cardiovascular devices	
	a. Kidney dialysis machines	
	i. Blood feeding catheters	Polyurethane, silicone
	ii. Membrane inside the dialyzer	Cellulosic
	b. Vascular grafts	Woven fabrics of expanded PTFE, polyesters and polyester terephthalate (PET)
	c. Membrane in oxygenator	Perfluorobutyl ethyl cellulose
	d. Cardiac valves	
	i. Metal valves	Ti–alloy, CoCrMo
	ii. Non-metal valves	Silicone elastomer
	e. Artificial heart	
	i. Heart valves	Polydimethyl siloxane, polyurethane
	ii. Pump bladders	Polyurethanes with layers of butyl rubbers inside
	iii. Blood contacting inlet and outlet connecters	PET
	iv. Housing for devices	Epoxy, Kevlar, polyurethane, polycarbonates, Ti
6.	Miscellaneous	
	a. Blood bags, tubings	PVC, polyurethanes, silicones
	b. Soft tissue reconstruction	Silicones
	c. Wound closure	Cyanoacrylates
	d. Tendon repair	Polylactic acid
	e. Drug delivery system	Ferrofluids

References

1. T. Anson, Materials world, shape memory alloys – medical applications, Vol. 7, No. 12, pp. 745–747, December 1999, http://www.azom.com/article.aspx?ArticleID=134.
2. 2001 SMA/MEMS Research Group, Shape memory alloys, August 17, 2001, http://webdocs.cs.ualberta.ca/~database/MEMS/sma_mems/flap.html.
3. R. Lin, Shape memory alloys and their applications, February 22, 2008, http://stanford.edu.

2

Peculiar Materials with Fascinating Properties

2.1 Introduction to Auxetic Materials

Auxetic polymers are very exciting materials of multidisciplinary field. These are advanced materials having extreme properties such as very high strength and very high modulus. In these fascinating materials, the main property is their volume expansion when subjected to tensile load. They possess negative Poisson's ratio. By *auxa*, we mean *enlargement*. So auxetic materials mean *enlarging materials*. This is just unbelievable as all the known materials (metals, alloys, ceramics, polymers, etc.) elongate in the direction of applied tensile load and contract in other two lateral directions (dimensions). That is why we obtain the 3D stress–strain relations for isotropic materials (non-auxetic) as

$$\varepsilon_x = \frac{1}{E}\left(\sigma_x - v\sigma_y - v\sigma_z\right) \tag{2.1a}$$

$$\varepsilon_y = \frac{1}{E}\left(\sigma_y - v\sigma_z - v\sigma_x\right) \tag{2.1b}$$

$$\varepsilon_z = \frac{1}{E}\left(\sigma_z - v\sigma_x - v\sigma_y\right) \tag{2.1c}$$

where
  ε and σ are linear strains and stresses, respectively
  E and v are modulus of elasticity and Poisson's ratio
  x, y, z correspond to three orthogonal axes

These equations indicate that the Poisson's ratio $v = \varepsilon_y/\varepsilon_x$ and $v = \varepsilon_z/\varepsilon_x$ is a positive quantity.

The stress–strain behaviours expressed by Equations 2.1a through c are not true for auxetic materials. In them, the Poisson's ratio is a negative quantity as the material enlarges in all three directions on application of uniaxial tensile load in any one direction. Hence, Equations 2.1a through c modify as

$$\varepsilon_x = \frac{1}{E}\left[\sigma_x - (-v)\sigma_y - (-v)\sigma_z\right] = \frac{1}{E}\left(\sigma_x + v\sigma_y + v\sigma_z\right) \qquad (2.2a)$$

$$\varepsilon_y = \frac{1}{E}\left[\sigma_y - (-v)\sigma_z - (-v)\sigma_x\right] = \frac{1}{E}\left(\sigma_y + v\sigma_z + v\sigma_x\right) \qquad (2.2b)$$

$$\varepsilon_z = \frac{1}{E}\left[\sigma_z - (-v)\sigma_x - (-v)\sigma_y\right] = \frac{1}{E}\left(\sigma_z + v\sigma_x + v\sigma_y\right) \qquad (2.2c)$$

These are fibrous forms of auxetic materials. They can be produced by a continuous process developed recently. Auxetic fibres possess unique characteristics. They can be used in a wide variety of structures as a single or multiple filaments or as a woven structure. Auxetic fibres are used to make biological drug-release systems. This is accomplished by extending the fibres that open the micropores. Consequently, certain dose of drug is released. These fibres are also used to prepare composites for crash helmet, sports clothing, fire retardant components, etc.

Re-entrant polymer foam materials having negative Poisson's ratio are a new class of materials with unique properties. Such materials are called auxetic materials. The thermodynamic restriction on compressibility of an elastic material limits the Poisson's ratio v to lie between −1 and +0.5. However, polymer foams are not governed by this restriction and develop negative Poisson's ratio. Negative values of v in them are a result of the size scale, highly anisotropic behaviour and many hidden variables. In this chapter, a review of such materials is presented to describe their present status and likely future.

2.1.1 Types of Auxetic Materials

Several kinds of auxetic materials are known to exist. Some of them are discovered from nature, and others have been developed by research efforts. Different known auxetic materials may be enumerated as follows:

1. Auxetic polymer, for example, re-entrant foam
2. Auxetic ceramics
3. Auxetic metals, for example, Cu foam
4. Auxetic piezoelectrics
5. Auxetic compound, for example, Ni_3Al

6. Auxetic microporous materials
7. Auxetic fibres
8. Auxetic laminates
9. Auxetic composite
10. Mechanical lungs

2.1.2 Positive, Zero and Negative Poisson's Ratio (Auxetic) Materials

Mostly conventional materials have positive Poisson's ratio ν. Isotropic materials can have Poisson's ratio up to a maximum of +0.5. Some materials have $\nu = 0$, and many others have $\nu < 0$, that is, a negative value. Poisson's ratios for some materials are shown in Table 2.1 for a ready reference. Materials having different Poisson's ratio are suitable for different applications. For illustration, the cork of a wine bottle is a good example of $\nu \approx 0$. Since it can be easily inserted and removed and can also withstand the pressure from within the bottle, it is an ideal application for bottle closing. In contrast to it, the rubber having $\nu \approx 0.5$ expands on compressing and may jam the neck of the bottle. Hence, it cannot be used for this purpose. Based on such explanation, it is anticipated that the re-entrant foams may be used for applications such as the following:

- Air filter
- Fasteners

TABLE 2.1

Elastic Moduli of Some Materials

Material	Tensile Modulus E (GPa)	Shear Modulus G (GPa)	Poisson's Ratio
Foamed plastics and soft rubbers	10^{-3}	—	—
Unbranched polyethylene	0.2	—	0.48
Lead	16.0	6.2	0.40
Glass–epoxy composite ($V_f = 60\%$), cross-plied	22	4.4	0.26
Cement concrete	45	—	0.11–0.21
Silicon glass	70.0	22	0.23
Aluminium	71.0	25	0.33
Cast iron	110	51	0.17
Kevlar fibre	130.0	2.2	0.34
Mild steel	210.0	82	0.30
Metals and alloys	200–600	70–210	0.25–0.33
Tungsten	415.0	145	0.43
Diamond	1140.0	—	—

- Sponge
- Biomaterials
- Shock-absorbing materials

2.1.3 Effect of Anisotropy on Poisson's Ratio

The analysis of tensorial elastic constants of anisotropic single-crystal cadmium suggests that the Poisson's ratio may attain negative value in some directions. Anisotropic, macroscopic 2D flexible models of certain honeycomb structures also exhibit negative Poisson's ratio in some directions. Negative Poisson's ratio depends on the presence of a high degree of anisotropy. This effect occurs in some directions and may be dominated by coupling between the stretching force and shear deformation.

The ratio of shear stress τ and the shear strain γ is defined as *shear modulus* or *modulus of rigidity G*. It is related to the Young's modulus E and Poisson's ratio ν by

$$G = \frac{E}{2(1+\nu)} \tag{2.3a}$$

K is related to E and ν by

$$K = \frac{E}{3(1-2\nu)} \tag{2.3b}$$

The elastic moduli are also related by

$$\frac{1}{E} = \frac{1}{3G} + \frac{1}{9K} \tag{2.3c}$$

The tensile modulus E and the shear modulus G for various materials are also shown in Table 2.1.

Example 2.1

Justify that the limits of the values of Poisson's ratio for isotropic materials are +0.5 and –1.0. Why are these not true for orthotropic or anisotropic materials?

Solution: Equations 2.3a and b relating the four elastic constants decide the minimum and maximum possible values of Poisson's ratio.

    i. In Equation 2.3a, the value of shear modulus G approaches infinity, when the value of ν approaches –1. This becomes the extreme case. Hence, the minimum value of Poisson's ratio $\nu = -1$.

ii. Similarly from Equation 2.3b, the value of K becomes infinity when $\nu = +0.5$. Bulk modulus will become negative if $\nu > +0.5$. Thus, the maximum possible value of Poisson's ratio $\nu = +0.5$.

iii. It should be clearly noted that Equations 2.3a through c are true for isotropic materials only. For anisotropic and orthotropic materials, these equations do not hold good. Hence, the limits of Poisson's ratio for anisotropic and orthotropic materials are not the same, +0.5 or –1.0.

2.1.4 Causes of Negative Poisson's Ratio

The causes of negative Poisson's ratio effects may be attributed to the following phenomena:

1. Cosserat (or micropolar) elasticity
2. Non-affine deformation
3. Certain chiral microstructures
4. Structural hierarchy
5. Atomic scale

This effect is not due to *Cosserat elasticity*. It is because the negative Poisson's ratio is classically attainable and does not require length scale which is present in Cosserat elasticity. It is well established that the Poisson's ratio in materials is governed by the following aspects of microstructure: (1) presence of rotational degrees of freedom, (2) non-affine deformation kinematics and (3) anisotropic structure. It is seen that the non-affine kinematics are essential for production of negative Poisson's ratio. Non-central forces combined with preload can also give rise to negative Poisson's ratio. A chiral microstructure with non-central force interaction or non-affine deformation can also exhibit a negative Poisson's ratio.

Consideration of conservation of volume: The materials having negative Poisson's ratio do not conserve volume. Rubbery materials ($\nu \approx 0.5$) are nearly incompressible, but the metals, hard plastics and conventional foams ($\nu < 0.5$) do not conserve volume. Although these materials do not obey law of conservation of volume (there is no such law), but they obey law of conservation of energy.

2.1.5 Applications of Auxetic Materials

Presently, the uses of auxetic materials are limited, and probably, they are not used to utilize the auxetic effect. Important among these applications are the following:

1. Extended form of polytetrafluoroethylene (PTFE) is used to make Gore-Tex.

2. Large single crystals of Ni_3Al are used to make vanes of gas turbine engines for aircrafts.

3. Pyrolytic graphite is used for thermal protection in aerospace applications.

4. Medical bandages prevent the swelling of wound on applying a wound healing agent.

5. The biomedical fibres made of auxetic materials are used as drug release agent for a specified quantity of drug to the patient.

2.1.6 Auxetic Polymers

A foam structure exhibits a negative Poisson's ratio. Foams are produced from conventional low-density open-cell polymer by causing the ribs of each cell to permanently protrude inward, resulting in a *re-entrant* structure. Normal polymer foams have a positive Poisson's ratio, but the re-entrant polymer foams possess a negative Poisson's ratio. These are called auxetic polymers (or anti-rubber or dilational polymers). On stretching, the auxetic polymer-like *chiral honeycomb* unrolls while the *re-entrant honeycomb* unfolds.
Creation of re-entrant structure: The reentrant structure is created as follows:

1. First, the conventional foam is compressed triaxially and placed in a mould.

2. The mould is then heated slightly above the softening temperature of foam material.

3. The mould is now cooled to room temperature and foam is extracted.

A typical polyester foam has the following properties:

- Specific gravity = 0.03
- Young's modulus = 71 kPa
- Cell size = 1.2 mm
- Poisson's ratio = 0.4

2.1.7 Characteristics of Foamed Materials

Foams with re-entrant structures exhibit greater resilience than conventional foams. They do not obey the classical theory of elasticity or viscoelasticity. The polymeric cellular solids (re-entrant foams) exhibit non-linear stress–strain relationship. In viscoelastic materials, the Poisson's ratio is not a material constant, rather it depends upon time. For polymeric solids, the shear modulus G relaxes much more than the bulk modulus K. Such time dependence is not necessarily the consequence of the theory of viscoelasticity. Auxetic foams exhibit enhanced indentation resistance. The resistance of ultrahigh-molecular-weight polyethylene (UHMWPE) may be enhanced by three times as compared with conventional UHMWPE. Auxetic foams also posses high shear resistance, sound and vibration damping, fracture toughness and ultrasonic energy absorption.

Auxetic materials possess novel behaviour under deformation. Various material properties can also be enhanced as a result of having negative Poisson's ratio. Typically, such properties are inversely proportional to $(1 + \nu)$ or $(1 - \nu^2)$. According to Saint–Venant's principle, they exhibit slow decay of stress. Their significant properties are expected to increase the sensitivity of piezoelectric devices, many times.

2.1.8 Auxetic Fibres' Future Opportunities and Challenges

Future scope for auxetic materials is tremendous and very bright. By utilizing the positive properties of negative Poisson's ratio materials, their applications have to expand in future. However, the key opportunities for future lie in developing the molecular and multifunctional auxetics. The potential of their success will depend upon the cheaper and easier synthesis.

2.2 Metallic Glasses

Most metals and alloys are crystalline, that is, their atoms are arranged in a regular, ordered pattern that extends over long distances (hundreds or thousands of atoms). These regions of ordered atomic arrangement are crystals. The regular arrangement of atoms in a crystalline material can be directly viewed using a transmission electron microscope (TEM). Many of the important properties of engineering alloys can be explained in terms of this sort of crystalline order or, in many cases, in terms of defects in the crystal structure. Metallic glasses, in contrast, are alloys that are non-crystalline or amorphous. There is no long-range atomic order. In this case, the atoms are more or less randomly arranged. This tells us that there are none of the long rows of atoms and the material is indeed amorphous.

2.2.1 Interesting Amorphous Metal

Making amorphous solids is nothing new. Many common materials including oxide glasses such as ordinary window glass and most polymers are amorphous. It is quite unusual, however, for a metallic material to be amorphous. The technique to making a metallic glass is to cool down a metallic liquid (which has a disordered structure as well) so rapidly that there is not enough for the ordered, crystalline structure to develop. In the original metallic glasses, the required cooling rate was quite fast as much as a million degrees Celsius per second! More recently, new alloys have been developed that form glasses at much lower cooling rate, around 1°–100° per second. While still fairly rapid, it is slow enough that we can cast bulk ingots of these metallic alloys and they will solidify to form glasses.

From an engineering point of view, our interest in metallic glasses stems from their unique structure. Since the structure of a material determines its properties, one might expect that a material with an unusual structure might have interesting properties. This is certainly true of metallic glasses. For instance, metallic glasses can be quite strong yet highly elastic, and they can also be quite tough (resistant to fracture). Even more interesting are the thermal properties. For instance, just like an oxide glass, there is a temperature (called the *glass transition temperature*) above which a metallic glass becomes quite soft and flows easily. This means that there are lots of opportunities for easily forming metallic glasses into complex shapes.

2.2.2 Unusual Properties of Metallic Glasses

These bulk metallic glasses (BMGs) have unusual properties. They are typically much stronger than crystalline metal counterparts by a factor of 2–3. They are quite tough (much more so than ceramics) and have very high strain limits for Hooke's law–obeying materials. Metallic glasses possess several other superior properties as compared to crystalline (conventional) metals. These are given as follows:

- High strength, about three times stronger than steel
- High springing nature, about 10 times to that of a best-quality industrial steel, high strength-to-weight ratio
- High hardness
- Becomes soft on heating and so are easily malleable

2.2.3 Materials Systems of Metallic Glasses

The rapid advance in science and technology has reached to the stage of gigahertz (GHz). Various types of electronic equipments such as mobile communication devices integrated circuit boards and local area network devices are already used in GHz frequency range. Now, the rapid development of high-frequency devices in GHz range calls for a class of functional materials that can solve the electromagnetic interference problems. Although traditional thicker-microwave ferrite absorber has better electromagnetic wave (EM) absorbing capability, but this absorber is too bulky to be useful in small device. In order to satisfy the development of novel EM absorber, polymer stealth materials (PSMs) have got more recognition due to low density, structural versatility, good processability and magnetism. Ferrocenyl organic magnetic material is one such kind of PSM. Besides many advantages, the absorbing intensity of ferrocenyl organic magnetic material is comparatively

TABLE 2.2

Various Material Systems of Metallic Glasses and Their Specialities

Material System	Specialities
• Cu-Zr	Binary alloy
• Cu-Zr-Al	Ternary alloy
• Cu-Zr-Al-Y	Quarter nary alloy
• Ta-Cu-Ni-Al	Plastic strain $\varepsilon_p = 4.5\%$
• Pt-Cu-Ni-P	Plastic strain $\varepsilon_p = 20\%$
• $Pd_{40}Ni_{40}P_{20}$	Palladium-nickel based alloy
• $Fe_{40}Ni_{40}P_{14}B_6$	Iron-nickel based alloy
• $W_{60}Ir_{20}B_{20}$	Tungsten-iridium-based alloy
• $Zr_{41.2}T_{13.8}Cu_{12.5}Ni_{10.0}Be_{22.5}$	Vitreloy (or liquidmetal) used for golf club heads
• $Al_{90}Fe_5Ce_5$	Aluminium-iron-cesium based alloy
• $La_{55}Al_{25}Ni_{20}$	Lanthanide based alloy

low. Various material systems of metallic glasses and their specialities are given in Table 2.2.

2.3 Whiskers

Definition. Whisker is the fibre form of a material which is obtained on elongation of a single crystal. Materials are generally available in bulk form. A piece of mild steel or copper, a tiny part of semiconductor, or a block of reinforced cement concrete (RCC) is available in bulk form. If bulk is changed to the form of wire, fibre or whisker, the strength of material improves considerably. However, the imperfections and dislocations are the major hindrances due to which theoretical strength is not achieved in the materials. To achieve the goal of $\sigma_{ult} \approx E$, material in the form of whisker is a must. Whiskers are available in metallic and non-metallic and organic and inorganic materials. History of whiskers is as old as 1970 when synthetic whiskers were produced. Whiskers, in the natural form, were known to the mankind since historical days. Spider's web, bamboo flakes, human bone flakes, some kinds of grass, etc. are the known natural whiskers.

2.3.1 Difference between Bulk, Fibre and Whisker Forms of Materials

Whiskers are obtained by elongating single crystal into fibrous form. By this action, shown in Figure 2.1a, the defects such as grain boundaries are fully eliminated. Figure 2.1a shows a bulk polycrystalline material. It may be

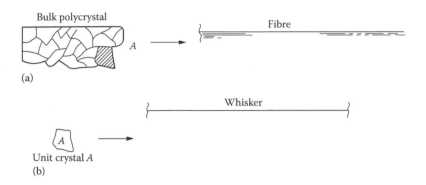

(a)

Bulk polycrystal

Fibre

A

Whisker

Unit crystal A

(b)

FIGURE 2.1
(a) Bulk polycrystal elongated to fibre form and (b) unit crystal elongated to whisker form.

elongated to the form of a fibre. A single crystal A is taken out of it (Figure 2.1b) and is elongated to produce a whisker. Thus, the differences among bulk, fibre and whisker forms of a material lie in their manufacturing that will further affect their strengths.

2.3.2 Effects of Size of Whiskers on Mechanical Properties of Materials

Diameter of whiskers varies between 2 and 20 µm. Greatest strengths are usually observed in whiskers of smallest diameters as shown in Figure 2.2. Strongest whiskers are almost free from dislocations. Strengths nearing 2.7 and 13.2 GPa have been achieved in copper and iron whiskers, respectively. A comparison between whisker and bulk forms for strength σ, Young's modulus E and elastic deformation is shown in Table 2.3.

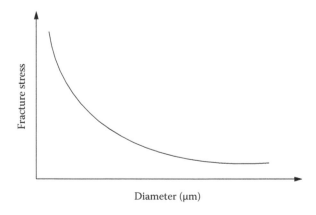

Diameter (µm)

FIGURE 2.2
Effect of whisker diameter on fracture strength.

TABLE 2.3

Strength and Young's Modulus of Some Whiskers and Their Bulk Form

Material	Strength (GPa)		Young's Modulus (GPa)		Elastic Deformation	
	Bulk	Whisker	Bulk	Whisker	Bulk (%)	Whisker (%)
Tungsten	1.5	14.0	400	2400	0.01	0.3
Mild steel	0.48	12.8	200	1000	0.2	5
Alumina	5.0	14.0	390	2250	0	3.5
Graphite	2.6	19.7	27	703	—	—
Silicon carbide	10.0	20.7	450	840	0	—

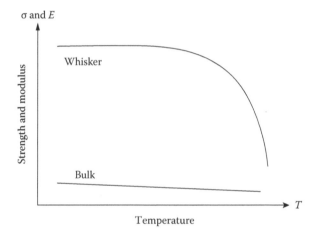

FIGURE 2.3
Temperature dependence of whisker and bulk forms for their strength and Young's modulus.

2.3.3 Effects of Temperature on Properties of Whiskers

The temperature dependence of the strength and modulus of the whisker and bulk forms are shown in Figure 2.3. It shows a much higher value for whisker as compared to the bulk form materials.

2.4 Intermetallic Compounds and Intermediate Compounds

Two or more metallic constituents under specific conditions form an *intermetallic compound*. They are ionically or covalently bonded. Copper aluminide ($CuAl_2$) and magnesium silicide (Mg_2Si) are the examples. Based on the specific conditions of their formation, the intermetallic compounds are sub-classified as

1. Valence intermetallic compounds
2. Electron intermetallic compounds
3. Definite radii ratio intermetallic compounds
4. Intermediate compounds

2.4.1 Valency Intermetallic Compounds

A metal with strong metallic nature combines with another metal of weak metallic properties to form a *valency intermetallic compound*. In doing so, they obey normal rules of valency. The carbides, fluorides, oxides, hydrides and combination of metals and metalloids (e.g. S, Te, Sb, Bi, Se, As) are the examples.

2.4.2 Electron Intermetallic Compounds

The *electron intermetallic compounds* are formed between two metals having atoms of comparable sizes but different valencies. The CuZn compound has a valency of 3 that includes valencies of copper and zinc as 1 and 2, respectively. Only two atoms are associated in the formation of compound, hence the ratio of

$$\frac{\text{Valence electrons}}{\text{No of atoms}} = \frac{3}{2}$$

Electron compounds, generally, exist with this ratio of
3:2 (i.e. 21:14), 21:13 and 21:12 (i.e. 7:4)

2.4.3 Definite Radii Ratio Intermetallic Compound

Compounds such as Cu_2Mg, LiZn and CaMg bear a definite atomic radii ratio of 1:25, 1:15 and 1:24 in their formation. Such compounds are known as *definite radii ratio intermediate compounds*. A favourable atomic packing is possible with them.

2.4.4 Intermediate Compounds (or Phases)

These compounds are also known as *intermediate phases*. Similar sounding terms, namely, intermediate phases and intermetallic compounds, are different from each other. Intermediate phases are formed in binary alloy systems when their mutual solubility is limited and the chemical

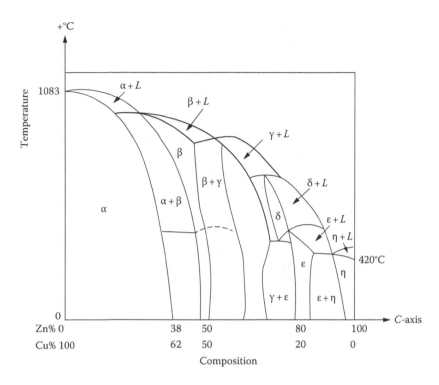

FIGURE 2.4
Binary copper–zinc phase diagram showing different phases including intermediate phases.

affinity is high. The intermediate phase formed in Cu–Zn system is shown in Figure 2.4.

α-, β-, γ-phase of brass, that is, α-brass, β-brass and γ-brass: Various solid phases are α, β, γ, δ, ε and η. A zinc-rich β-phase appears when solubility of copper exceeds in zinc. This copper–zinc alloy is called brass. This diagram shows a series of alloys (brass), namely, α-brass, β-brass and γ-brass. Transformation of α-brass takes place into β-brass when proportion of zinc exceeds 38%. β-brass is harder than α-brass. The intermediate phases have narrow to wider ranges of homogeneity. They are brittle and poor in electrical conduction.

2.5 Emerging High-Pressure Materials and Technologies for the Future

Researches in high-pressure technology are leading to the development of new families of materials, which possess the unique physical behaviour and characteristics. These high-pressure properties are different from the properties of the same materials at ambient pressures. Such newer kinds of materials being discovered are equipped with previously unknown types of physical and chemical properties and basic structures. The existing high-pressure techniques generally yield powdered form of materials that are useful for applications such as superhard abrasives, single-crystal semi-conductors, gems, electronic and optoelectronic devices and newer class of superconductors. In this section, a state-of-the-art review has been presented that discusses the changes in behaviour of materials under high pressures and the newer class of materials and their applications in mechanical, opto-electronical and superconducting fields. It is suggested here to enhance the investigations in high-pressure techniques to enable the commercial produc-tion of new materials.

2.5.1 High-Pressure Synthesis and Development of Fascinating Materials

High-pressure technology plays an important role in synthesizing of materi-als that develop unique properties, which otherwise are not obtained when produced under ambient pressure. The ambient pressure on earth is atmo-spheric, that is, 1.00132 bar = 101.325 kPa (or 0.000101 GPa), whereas the high pressure is many more times of it. Higher pressure generally means a pres-sure higher than 1 GPa (10 atmospheric pressure on earth). It may be as high as 500 GPa. The high-pressure technique not only gives the wonder materi-als as quoted later but also hints at some new substances that can exist in principle. The high-pressure synthesis (technology) is applied to obtain the following fascinating materials/material systems, useful for the undermen-tioned applications:

1. Synthetic diamonds to be used as superhard abrasive for cutting and shaping of hard metals and ceramics
2. Cubic boron nitride to be used as superhard abrasive for cutting and shaping of hard metals and ceramics
3. Alternative superhard materials such as polymer C_{60}–C_{70}, nanocrys-talline $TiNSi_3N_4$ composite and cubic BC_2N phase pure solid
4. Superconducting behaviour in condensed rare gases and ionic com-pounds such as in CsI
5. Many hypothetical new polymers from low-Z elements

However, the main limitation of using pressures more than 1 GPa is to pro-duce a small volume and small mass of materials. Moreover, they are avail-able in powder form. The present chapter reviews various new materials and the aspects of the production for newer materials under high pressures.

TABLE 2.4

High-Pressure Range in Different Applications

Application in	Areas	Useful Pressure Range (GPa)
• Autoclave presses	Chemical industry, biotechnology, crystal growth techniques	0.0001–0.2
• Chemical processes	Industrial production of materials	0.001–1.0
• Ocean depth	Oceanographic research Mineralogy	0.01–0.1
• Centre of the earth	Molecular research	—
• Molecular systems at planetary conditions in	H and He	340.0
(i) Jupiter	—	0.1–450
(ii) Saturn	—	100
(iii) Uranus and Neptune	—	2–70

Source: Brazhkin, V.V., *High-Pressure Synthesized Materials: A Chest of Treasure and Hints*, Institute for High Pressure Physics, Moscow, Russia.

2.5.2 Meaning of High Pressure

Not any formulated standard exists to demarcate the high and low pressures, in respect of synthesizing of materials. Yet a pressure higher than 1 GPa (1×10^9 N/m^2) is generally taken to be high pressure. Other meaningful pressures employed for various scientific and industrial purposes are compared in Table 2.4 along with other related details [1].

2.5.3 Advances in High-Pressure Methodology [2]

The efforts being carried out for high-pressure technology researches employ the following main equipment and systems:

1. Diamond-anvil cells (DACs) that allow to access a pressure of the order of 0.1 GPa to more than 500 GPa (1 Mbar = 100 GPa) within samples of nanograms to micrograms (10^{-9} to 10^{-6}) gram and sample volume of 10^{-6} to 10^{-9} cm^3.

2. *Large-volume synthesis presses* uses a pressure of 5–10 GPa.

3. Laser and resistive heating from a few −100°C to many −1000°C for the sample or its cooling to milli-Kelvin temperature at megabar pressures.

4. Dynamic shock waves generated by explosives, lasers, or compressed gas generate the pressure in the range of terapascal, that is, 10–50 TPa (100–500 Mbar) with simultaneous heating to few –100°C, few –1000°C, or few 10^{-6}°C. Materials produced by this process are Si_3N_4, SiAlON compounds, etc. [2].

2.5.4 Magical Effects of High-Pressure Techniques on Properties of Materials

Most elements exist in metallic state under normal condition of temperature and pressure. But their structures get changed under high-pressure states accompanied with temperature. Recent studies by high-pressure diffraction techniques have revealed many astonishing changes in metals. These are as follows:

1. Simple metals having 8–16 atoms in their unit cells recrystallize into complex structures having up to many hundreds of atoms in their unit cells, when compressed to 15–30 GPa range.
2. Newer intermetallic compounds and *elemental alloys* are produced which possess unusual electronic and magnetic properties.
3. New high-density amorphous solids are produced due to polymorphic transformations at high pressures. High-density amorphous silicon may be a useful material for the future.
4. Semimetallic transition-metal nitrides (TiN, MoNx) and nitrides of B, Al, Ga, In, Si and Ge elements are newer class of materials which are the outcome of high-pressure technology.
5. Newly synthesized materials such as the following are now available for advanced applications:
   a. Cubic spinel-structured Y-Si_3N_4, a high-hardness refractory ceramic, having Vickers hardness $H_v = 30$–43 GPa.
6. Spinel-structured ceramic SiAlON compounds are available due to high-pressure synthesis. This is a very useful material for use as gas and steam turbine blades.
7. P_3N_5 is a new form of nitride, which can be recovered to ambient conditions.

2.5.5 High-Pressure Mechanical (Superhard) Materials

High-pressure technology presents a very useful research field in searching of superhard materials. According to generally acceptable norms, the superhard materials are those whose hardness values are more than 45 GPa, and they are comparable with or even superior to diamond and c-BN. Superhard materials also possess improved thermal and chemical stability. Some of

TABLE 2.5

Some Recent Superhard Materials Which Are the Outcome of
High-Pressure Research

Name of New Superhard Material	Formula	Vickers Hardness H_v in (GPa)	Type of Structure
Cubic boron nitride	c-BN	—	Pure phase
	C–BC$_2$N	76	Compound form
Fullerite	C60, C70	—	Partly polymerized solids
Carbon nitride	C$_3$N$_4$	—	Tetrahedrally coordinated phases
Graphene	C$_x$N$_y$	—	Layered structure
Boron nitride	BN	—	Hexagonally layered phase
Boron suboxide	B$_6$O$_{1-x}$	35	—
Boron subnitride	B$_6$N	—	May be metallic

these materials, which are outcome of high-pressure research, are listed in Table 2.5 along with other related details.

2.5.6 Low-Compressibility and High Bulk Modulus Solid

The low-compressibility solids (i.e. highly incompressible solids) are usually hard. These possess high bulk modulus K and shear modulus G. Osmium (Os) possesses lowest compressibility ($K = 462$ GPa) among all known elements. It is even higher than that of the diamond ($K = 443$ GPa) which is the hardest known material. The bulk modulus of some highly incompressible superhard materials is compared in Figure 2.5. It depicts metals and compounds both. MoN has the highest bulk modulus ($K = 487$ GPa) of all known compounds (oxides, carbides, nitrides, transition metals and diamonds). Bulk modulus of other hard metals and compounds are listed in Table 2.6.

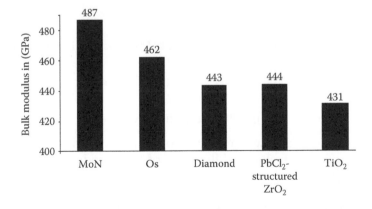

FIGURE 2.5
Comparison of bulk modulus of some highly incompressible superhard materials.

TABLE 2.6

Bulk Modulus of Some Metals and Compounds

CoN, WC, HfO$_2$	Between 470 and 400 GPa in decreasing order
TiN, Re, NiN	Between 400 and 300 GPa in decreasing order
Pt, ZrN	Between 300 and 200 GPa in decreasing order
Pd, Hf	Between 200 and 100 GPa in decreasing order
Zr	Below 100 GPa

Source: McMillan, P.F., *Nat. Mater.*, 1, 19, September 2002.

2.5.7 High-Pressure Electronic and Optoelectronic Materials

The high-pressure synthesis has resulted in the development of various electronic and optoelectronic materials also. Main among them are the following:

1. Coloured diamonds, which are the special class of optical materials. These are produced to prepare the gemstones.

2. Single-crystal *p*-and *n*-doped c-BN, which in future is likely to be suitable material for high-temperature semiconductor applications. CoN, WC, HfO$_2$ between 470 and 400 GPa in decreasing order; TiN, Re, NiN between 400 and 300 GPa in decreasing order; Pt, ZrN between 300 and 200 GPa in decreasing order; Pd, Hf between 200 and 100 GPa in decreasing order and Zr below 100 GPa.

3. GaN, (Ga–In–Al) N alloys as optoelectronic materials, which constitute the new families of semiconductor diodes and lasers. These devices operate efficiently in blue-green ultraviolet wavelengths.

4. Germanium nitride (Ge$_3$N$_4$), a wide-bandgap optoelectronic material ($Eg = 3.2$ eV), is a tetrahedrally coordinated compound at ambient pressure, which converts into a new spinel-structured form under high-pressure treatment.

5. AlN, a high-thermal conductivity substrate material, is also the outcome of high-pressure crystal growth.

2.5.8 Development of High-Pressure Superconductors [3]

The strategies in high-pressure technology are credited to have searched many new superconductors and the types of superconducting behaviour, as the following:

- Elemental superconductors have been observed in the pressure range of 10–100 GPa.

- The highest critical temperature (Tc) recorded in $HgBa_2Ca_2Cu_3O_8$ was 133 K at ambient temperature [3] and has increased to 164 K when subjected to a high pressure of 30 GPa.
- Magnesium diboride (MgB_2), many metal-silicides and aluminides and *Zintl-phase* compound $BaSi_2$ are the new members of superconductor family which have been yielded by high-pressure synthesis.
- New superconductors such as $(Ba, Na)_8Si_{46}$ are prepared under high-pressure conditions.

References

1. V.V. Brazhkin, *High-Pressure Synthesized Materials: A Chest of Treasure and Hints*, Institute for High Pressure Physics, Moscow region, Russia.
2. P.F. McMillan, New materials from high-pressure experiments, *Nature Materials*, 1, September 2002, 19–25.
3. A. Schilling et al., Superconductivity above 130 K in the Hg-Ba-Ca-Cu-O system, *Nature*, 362, 1993, 56–58.

3

Amorphous Materials and Futuristic Scope of Plastics

3.1 Introduction to Organic Materials

Materials derived from carbon are termed as *organic materials.* Range of organic materials *is* in thousands owing to vast variety of hydrocarbons and their derivatives. Organic materials may be natural or synthetic. Some of them are grouped as follows:

Natural Organic Materials	Synthetic Organic Materials
Petroleum	Oil
Rubber	Solvent
Wood	Lubricant
Coal	Adhesive
Biological fibres	Dye
Food	Synthetic rubber
Cotton	Plastics
Flax	Explosive

Organic materials are covalent (polar) bonded. Polymers such as plastics, synthetic rubber and wood are common organic materials of important engineering applications.

3.2 Difference between Monomers and Polymers

Monomers: A monomer is a small molecule, which on polymerization converts into a large molecule (macromolecule) called polymer. Polymerization is the process of linking of monomers under suitable conditions of temperature

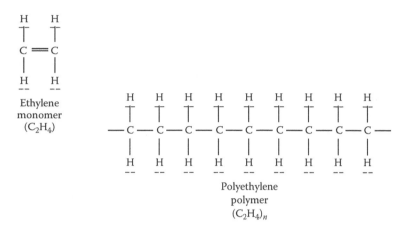

FIGURE 3.1
Ethylene monomer and polyethylene polymer.

and pressure. Polymers arc composed of large number *of monomers. Poly* means *many* and *mer* means *unit.* Thus, polymers are large number of repeating monomers. Their molecular configuration may be that of a long chain or network.

Polymers: Polymers are generally non-crystalline solids. Their mechanical properties are highly sensitive to molecular configuration, degree of polymerization (DOP) and cross-linking. Linking of an ethylene monomer (C_2H_4) to form polyethylene (C_2H_4)$_n$ is shown in Figure 3.1. Two single bonds (C–C) in polymer are formed from each double bond (C=C) in the monomer.

Polymers are mostly made from carbon compounds but can also be made from inorganic *silicates* and *silicons.* They are also found in natural form. Some of the natural and synthetic polymers are grouped as follows:

Natural Polymers	Synthetic Polymers
Protein	Polyethylene (polythene)
Cellulose	Polystyrene
Resin	Polypropylene
Starch	Nylon
Shellac	Polyvinyl chloride (PVC)
Lignin	Polymethylmethacrylate (PMMA) (Plexiglass)
Hemp	Polytetrafluoroethylene (PTFE) (Teflon)
Silk	Polyacrylonitrile (Orlon)

3.3 Degree of Polymerization

The size of a macromolecule depends on the number of repeating monomers constituting it. It is designated by *DOP* and is defined as

$$DOP = \frac{\text{Molecular weight of polymer}}{\text{Molecular weight of monomer Mrn}}$$

A lower value of DOP forms light oil, and a higher value results into solid plastics. Generally, a molecule contains about 75–750 mers in rubber and plastics; hence, their DOP = 75–750. The average molecular weight is up to 10,000 in rubber and up to 100,000 in plastics.

3.3.1 Geometry of Polymeric Chain

The bond angle varies with the type of atoms connected to the backbone atoms. Figure 3.2a through c show the bond angle between the carbon atoms of backbone chain connected to the substituent atoms. In a fully extended chain (Figure 3.2d), the carbon atoms acquire zigzag planer configuration. The length *l* of an extended chain may be found from the geometry as

$$l = nr_0 \sin \frac{\theta}{2} \tag{3.1}$$

where
  n is the number of bonds or links (or mers) in the chain
  r_0 is the bond length
  θ is the bond angle

The average end-to-end distance in a freely rotating polymeric chain (Figure 3.2e) is determined from

$$l^{\cdot 2} = \lambda n r_0^2 \tag{3.2}$$

where λ is a constant whose value lies between 1 and 2. The terminologies are explained in Figure 3.2d.

3.4 Additives in Polymers

Additives are the substances mixed with monomers to obtain desired properties in polymers.

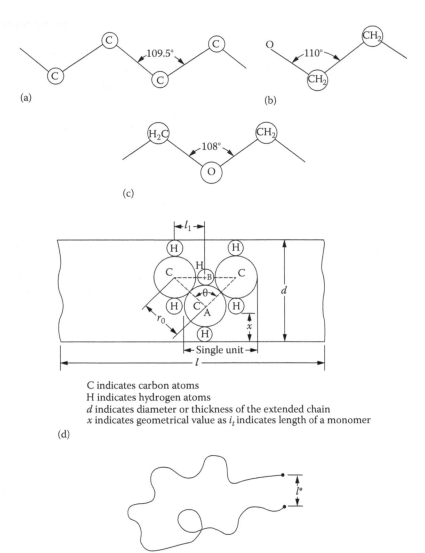

(a)

(b)

(c)

C indicates carbon atoms
H indicates hydrogen atoms
d indicates diameter or thickness of the extended chain
x indicates geometrical value as i_t indicates length of a monomer

(d)

(e)

FIGURE 3.2
Geometry of chain showing (a–c) bond angle between carbon atoms of backbone chain and substituent atoms, (d) fully extended chain with zigzag planer configuration and (e) end-to-end distance of a freely rotating chain.

This addition is done prior to or during the polymerization process. Additives usually added to the monomers are given in the following:

Additive	Example
Plasticizers	Tricresyl phosphate, camphor, resins, organic solvents
Fillers	Mica, asbestos, wood, slate powder, glass powder
Catalysts (i.e. accelerators or hardeners)	*Solid form*: benzoyl peroxide with calcium phosphate and camphor
	Paste form: benzoyl peroxide with tricresyl phosphate
	Liquid form: acetyl peroxide, dimethyl phthalate
Initiators	Hydrogen peroxide (H_2O_2)
Dyes and pigments	Metallic oxides

One or more of these are added in varying proportions. Their characteristics, properties and important applications are described as follows:

Plasticizers: Their properties and functions are given in the following:

i. These are organic compounds of low molecular weight.

ii. They are oily in nature.

iii. Plasticizers act as lubricants and improve the flow of material during processing.

iv. These impart flexibility and toughness to the materials.

v. They prevent crystallization (i.e. maintain non-crystalline structure) by keeping the chains separated from one another.

vi. They impart flexible nature to the thermoplasts.

vii. PVC, hard at room temperature, is plasticized with tricresyl phosphate to make it flexible.

viii. Synthetic *celluloid* made from *nitrocellulose* is plasticized with camphor.

Fillers: Their addition improves the following properties:

i. Heat resistance

ii. Strength

iii. Dimensional stability

Catalysts: The rate of polymerization reaction is increased due to their addition.

Initiators: Their addition helps to initiate the polymerization process. They also stabilize the reaction sites in molecular chains.

Dyes and pigments: Their addition is optional and is needed to impart desired colour to the materials. Red, blue, green and other colour shades may be given to the polymers by addition of dyes and pigments.

3.5 Various Types of Plastics and Their Applications

Long-chain polymers are subgrouped into three types:

1. Plastics (isotropic in nature)
2. Fibres (directional in nature)
3. Elastomers (rubberlike elastic nature)

In this section, we will discuss about plastics, and the other two subgroups are discussed in subsequent sections. Plastic is the name given to a material which is a mixture of resin, catalyst, accelerator, inhibitor and pigment. Resin forms a major part of this mixture, and hence, the properties of plastics are mostly governed by the properties of the resin used. Plastics are broadly classified into two categories:

1. Thermosetting (or thermoset)
2. Thermoplastic (or thermoplast)

They differ from each other in the way in which they are affected by heat.

3.5.1 Thermosetting Plastics

They have a 3D network of primary bonds in all the directions. These types of plastics, on application of heat, first become soft and then hard, and after that, they cannot be softened again by application of heat. This permanent hardening called *curing* is a chemical change. Common examples of thermosets are phenolics, epoxies, melamine, Bakelite and DAP. They generally require both heat and pressure to be moulded into any shape. If intensely heated, they break down by *degradation*. Important types of thermosetting plastics showing their properties and applications are illustrated in Table 3.1.

3.5.2 Thermoplastics

Their long-chain molecules are secondary bonded which break more readily due to increase in thermal energy. They become soft when heated (generally with pressure) but require cooling to set into a definite shape. Since no chemical hardening action takes place, the shaped articles resoften on reheating. Only a physical change is involved with them. They have excellent plasticity but low melting point. Common examples are cellulose, polystyrene, polyvinyl chloride, nylon, Teflon, polysulphone, acrylic, polyurethane and acetal. Important thermoplastics are shown in Table 3.2 depicting their properties and applications.

TABLE 3.1

Properties and Applications of Thermosets

Properties	Polyester	Epoxy	Phenolics	Melamine	Silicone
Specific gravity	1.2	1.2	1.2	1.4	1.3
Normal usable[a] temperature (°C)	80–180	90–120	250–300		250–300
Moulding pressure	Low to medium	Low to medium	Low to high	Medium to high	Low to medium
Mechanical properties	Very good	Very good	Very good	Very good	Fair
Electrical properties	Excellent	Excellent	Good	Excellent	Excellent
Water resistance	Very good	Excellent	Very good	Fair	Good
Heat resistance	Good	Fair	Good	Excellent	Excellent
Flammability	Burns slowly	Burns slowly	Self-extinguishing	Self-extinguishing	—
Price	Low to medium	Low	Medium to high	High cost	High cost
Limitations	Large	Poor mould release	Colour		
Advantages	Cure shrinkage fluid before cure	Low shrinkage properties	Good general	Arc resistance	Heat resistance
Shrinkage (%)	4	2	2	—	
Application	Synthetic fibres	Machine and structural components of composites	Telephone receivers, foams	Crockeries	High-temperature resisting components

[a] They do not have melting points as they do not melt.

TABLE 3.2

Properties and Applications of Thermoplasts

Polymer	Specific Gravity	Melting Point (°C)	Tensile Modulus (GPa)	Tensile Strength (MPa)	Application
Polyethylene	0.91–0.971	15–137	0.4–1.3	21–38	Bags, tubes, containers
Polypropylene	0.90	176	1.1–1.6	29–38	Ropes, vacuum flasks
Polystyrene	1.04–1.09	239	2.8–4.1	35–82	Soundproofing of refrigerators and buildings
PVC	1.35–1.45	212–273	2.4–4.1	35–62	Electrical insulation, piping, gramophone records
PTFE, i.e. Teflon	2.13–2.18	327	0.4–0.7	14–35	Biomedical implants, frying pan and ropes, other coatings
Polyhexamethylene adipamide (Nylon 66)	1.13–1.15	265	2.8–3.3	77–90	
Polyhexamethylene sebacamide (Nylon 610)	1.07–1.09	228	1.1–1.9	48–59	Flexible tubes
Polycaprolactam (Nylon 6)	1.13–1.14	225	1.4–3.1	65–85	Synthetic fibres
Cellulose acetate	1.28–1.32	—	2.1–4.1	31–55	Fibres

3.5.3 Comparison between Thermosets and Thermoplasts

Thermosets are non-recyclable plastics but thermoplasts can be recycled again and again. Their characteristics and properties are compared as follows:

Description	Thermoset	Thermoplast
Bonding network	3D, primary in all directions.	Chain molecules are secondary bonded.
Effect of heat	Becomes hard and cannot be soften again.	Bond breaks and becomes soft.
Consequence of curing	A chemical change occurs during permanent hardening.	No chemical change occurs.
Moldability to any shape	Both heat and pressure are required.	Requires cooling for setting.
Effect of intense heating	They break down by degradation.	They resoften 115°C–330°C.
Melting point	They do not melt.	
Normal usable temperature	80°C–300°C.	60°C–200°C.
Applications	Telephone receivers.	Gramophone records.

3.6 Polymeric Fibres

Polymeric fibres are long chains of molecules aligned in longitudinal direction. This imparts directional properties to the fibre. The elastic modulus and strength are much higher in the longitudinal direction than in the other transverse directions. These fibres may be further subgrouped as follows:

1. *Natural* fibres such as wool, cellulose, cotton and silk
2. *Synthetic* fibres such as nylon, terylene (Dacron), rayon, Orlon and Kevlar

Cellulose fibres are flexible and strong in tension. *Cotton* clothes shrink due to the presence of wrinkled molecules. *Sanforized* cotton clothes are non-shrinking as long chains are made to align in such clothes.

Polyester fibre: Nowadays, synthetic fibres find too much use due to their durability, dimensional stability and uniformity, etc. Synthetic polymeric fibres belong to the family of both, the thermosets and the thermoplastics. *Polyester* fibres (a thermoset) such as terylene are characterized by linkage. Oxygen provides flexibility to polyester. It, therefore, becomes soft more easily with increasing temperature.

Nylon fibres: Contrary to this, nylon 66 (a thermoplast) which is a polyamide fibre does not soften easily with increasing temperatures. The polyamide fibres are characterized by linkage.

3.6.1 Properties of Various Synthetic and Natural Fibres

Related details of common synthetic and natural fibres are given in Table 3.3.

3.7 Mechanical Behaviour of Plastics

Plastics are characterized by the following mechanical properties:

Strength: They have low tensile strength. Their strength in compression is better than in tension. Tensile strength of thermoplasts generally ranges between 15 and 90 MPa.

High-temperature suitability: They have low melting point and therefore are unsuitable for high-temperature usage, generally above 100°C.

Dimensional stability is poor due to creep effect at room temperature. At higher temperatures, this is poorer due to softening in them.

TABLE 3.3

Common Synthetic and Natural Fibres and Their Related Details

Fibre	Group to Which It Belongs	Specific Gravity	Tensile Strength (MPa)	Tensile Modulus (GPa)	Typical Use
Rayon	Regenerated cellulose	—	—	—	Tyre cords
Terylene (Dacron)	Linear aromatic Polyester	1.38	600	1.2	Clothing
Nylon 6	Polyamide	1.14	800	2.9	Tyre cords
Orlon (Acrilan)	Polyacrylonitrile	—	—	—	Synthetic fibres
Kevlar 29	Aromatic polyamide	1.45	3400	60	Ropes, radial tyres, cordage
Kevlar 49	Aromatic polyamide	1.45	3600	130.0	Polymer reinforcement
Cellulose	—	1.52	1100	2.4	Synthetic fibres
Cotton	—	1.50	350	1.1	Clothing
Silk	—	1.25	500	1.3	Clothing
Wool	—	1.30	350	6.0	Clothing
Wood	—	1.0	900	72.0	Building door

Weather effect: They are susceptible to weather conditions. They suffer from distortions in moist conditions and are highly affected by chemical (acidic and alkaline) environments.

Impact strength is low which further decreases at higher temperatures due to softening.

Hardness is lower than those of metals and ceramics. It is generally between 5 and 30 BHN which is comparable with hardness of talc and lead alloys.

Tear strength is moderate. It is defined as the ability to resist tearing, especially of thin films used in packaging. The magnitude of tear strength is related with the magnitude of tensile strength.

Fatigue failure: They have poor resistance to fatigue failure and generally fail at much lower stresses as compared to their yield strength. They do not exhibit endurance stress, and therefore, they are designed for a certain estimated number of cycles of failure.

Stress–strain profile: Some plastics are linearly elastic but most of them are non-linearly elastic. Therefore, their stress–strain profile is of mixed variety. Those can be seen in Figure 3.3.

Crack formation behaviour: They fail due to formation of cracks in the regions of localized stress concentration (e.g. notches, sharp flaws, scratches). It happens due to breaking of covalent bond in the network.

Fracture mode: The thermosetting plastics fail in *brittle mode* but the thermoplasts can fail in the *ductile, brittle,* or *ductile–brittle transition mode*.

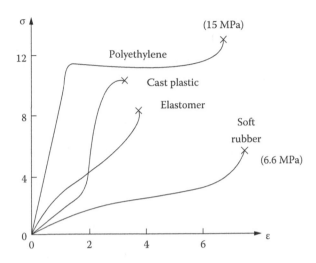

FIGURE 3.3
Stress–strain curves of different materials under tensile loading.

Thermal and electrical conductivity: They are poor conductors of heat and electricity. Therefore, they are used as insulators also. Bakelite, PVC, resins, plastic films, etc. are used for these purposes.

3.8 Rubber

Rubber is an organic polymer found in *latex* (i.e. sap) of certain plants. Latex is treated to obtain *raw* or *natural rubber*. Additive compounds are blended in raw rubber to get desired properties. These additives are

- Accelerators (i.e. catalysts) such as lime, magnesia and litharge
- Antioxidants (negative catalysts) such as complex organic compounds
- Reinforcing agents such as carbon black and zinc oxide
- Plasticizers such as stearic acid and vegetable oils
- Sulphur

3.8.1 Different Types of Processed Natural Rubber

Processed natural rubber is of the following three types:

1. Chlorinated rubber
2. Rubber hydrochloride
3. Cyclized rubber

TABLE 3.4

Synthetic Rubbers and Their Characteristics and Applications

S. No.	Name of Rubber	Characteristics	Application
1.	Polychloroprenes (Neoprenes)	Highly resistant to temperature, oil, grease and ageing	Oil seals, gaskets, adhesives, tank lining, low-voltage insulation
2.	Butyl	Highly resistant to oxidizing agents and impermeable to gases	Tyres, tubes, insulations for cables and wires
3.	Butadiene	Highly abrasion and weather resistant	Belts, soles of shoes, flooring
4.	Nitrile (Buna-N)	Excellent resistance to solvent, grease and oily situations	Conveyor belts, tank lining
5.	Polyurethane	High strength and highly abrasion resistant, poor to acidic and alkaline attacks	Expanded foams, tyres, conveyor belts
6.	Silicone	Mechanically weak, costlier, temperature stability up to about 600 K	Coatings, packaging, tubes, gaskets, cable insulation

They are used in the following applications:

  i. *Chlorinated rubber* is used in the production of adhesives and protective coatings.
 ii. Packing of delicate equipment is done by *rubber hydrochloride*.
iii. *Cyclized* rubber is used in manufacturing of papers with the help of paraffin wax (i.e. solid alkanes).

Various properties of natural rubber are given in Table 3.4.

3.8.2 Synthetic Rubber

Synthetic rubbers are superior to natural rubbers in many respects. These are manufactured from raw materials like coal tar, petroleum, coke, limestone, natural gas, alcohol, ammonia, salt and sulphur. Some synthetic rubbers and their characteristics and applications are given in Table 3.4.

3.9 Elastomer

The elastomers exhibit rubbery behaviour and undergo elastomeric (recoverable) deformation of 300%–1000%. Elastomer is basically a natural rubber whose monomer is *isoprene* (Figure 3.4a). It has two double bonds out of which one is used in chain formation while the other in producing an elastomer (Figure 3.4b).

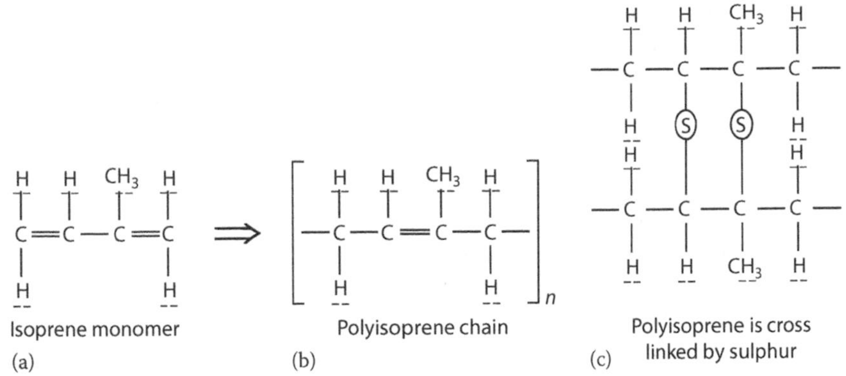

Isoprene monomer　　　　Polyisoprene chain　　　　Polyisoprene is cross
(a)　　　　　　　　　　　　(b)　　　　　　　　　　(c)　linked by sulphur

FIGURE 3.4
(a–b) Double bond in isoprene monomer breaks into two single bonds in polyisoprene chain and (c) sulphur cross-links the chain during vulcanization.

3.9.1 Method of Producing Elastomer from Raw Rubber

Production of elastomer requires heating of raw rubber in the presence *of sulphur* under a specified pressure. Sulphur forms covalent bonds with carbon and cross-links the chains. This process is called *vulcanization* and is shown in Figure 3.4c. The more the number of cross-links produced, the more will be the degree of vulcanization, which in turn will result in a stiffer and rigid rubber. The following examples illustrate the influence of varying degrees of vulcanization on rubber:

S. No.	Example	Degree of Vulcanization
1.	Puncture repairing of a scooter tube	Low
2.	Manufacturing of a pocket comb	More
3.	Production of hard and brittle ebonite	Maximum

3.9.2 Vulcanizing Agents

Various vulcanizing agents are used in different types of rubbers. These agents in

1. Neoprene rubber are S, MgO or ZnO
2. Hypalon are PbO
3. Silicone and Viton are peroxides and diisocyanates are in urethane rubber

Various properties of vulcanized natural rubber are compared with non-vulcanized natural rubber in Table 3.5.

TABLE 3.5

Comparison of Vulcanized and Non-vulcanized
Natural Rubber

Property	Vulcanized	Nonvulcanized
Tensile modulus (GPa)	1	0.01
Tensile strength (MPa)	30	3
Elongation when breaks (%)	600–800	1000–1200
Useful temperature range (°C)	−30 to +100	10–60
Hysteresis loss	Small	Large
Permanent deformation	Small	Large
Water absorption	Less	More

Example 3.1

Calculate the weight of sulphur required to fully cross-link 68 kg of polyisoprene starting from natural rubber. Which product will form due to full cross-linking?

Solution: The molecular weight of isoprene monomer = 68 g.

After vulcanization with sulphur, it is observed from Figure 3.4c that the two molecules of isoprene monomer require two molecules of sulphur. Hence, for full cross-linking, (68×2) g of isoprene requires (32×2) g of sulphur. Therefore, 68 kg of isoprene requires

$$\frac{32 \times 2}{68 \times 2} \times 68 = 32 \text{ kg of sulphur}$$

The fully cross-linked product will be *ebonite*.

Example 3.2

Constitutional composition of a rubber is isoprene 51%, butadiene 27%, sulphur 16% and carbon black 6%. If all the sulphur is consumed in cross-linking, determine percentage of possible cross-links during vulcanizations.

Solution: The monomer of butadiene is as shown in Figure 3.4 whose molecular weight is $(4 \times 12) + (1 \times 6) = 54$. The weights of isoprene, sulphur and carbon black are 68, 32 and 12 units, respectively.

Let the amount of rubber in question be 100 g. Then the amounts of isoprene, butadiene, sulphur and carbon black will be 51, 27, 16, and 6 g, respectively. Relative amounts of mers (atoms) will be

$$\frac{27}{54} \text{ in butadiene, } \frac{6}{12} \text{ in carbon black}$$

$$\frac{51}{68} \text{ in isoprene, } \frac{16}{32} \text{ in sulphur}$$

One atom of sulphur cross-links with each mer of isoprene or butadiene. Therefore, 16/32 atoms of sulphur will cross-link with 16/32 mers of isoprene and butadiene.

Hence, percentage of cross-linking $= \dfrac{(16/32)}{(51/68 + 27/54)} \times 100 = 40\%$

3.10 Behaviour of Polymers under Different Situations

Shelf (or pot) life: Thermosets have low shelf (or pot) life which necessitates that they should be consumed within this time. Therefore, they are unsuitable for high-volume (i.e. huge quantity) production. Thermoplasts are suitable for high-volume production that helps in lowering their costs.

Mixture and alloy form of plastics: Polymers in mixture and alloy forms are suitable for high-performance applications because of improved toughness in them. Important among these are polyether ether ketone (PEEK), polyphenylene sulphide (PPS), polyethersulphone (PES), acrylonitrile butadiene styrene (ABS), polyethylene terephthalate (PET), poly phenyl oxide (PPO), poly carbonate, polyetherimide (PEI), polytetrafluoroethylene (PTFE) and polyphthalamide.

3.11 Recent Advances and Futuristic Scope of Plastics

Plastics are assuming new dimensions in their development and applications. Newer kinds of plastics having properties and characteristics comparable to metals have emerged. Polyurethanes and polyacrylates are used as *shape memory polymers*, and *polypyrrole as chemically responding polymers* are the most recent developments. Some special purpose plastics are described as follows.

3.11.1 Expanding Plastics

Common plastics shrink when cured. *Expanding monomers (EMs) experience zero shrinkage during polymerization*. The polycyclic ring-opening monomers are combined with conventional monomers and oligomers to produce expanding plastics. These plastics have enhanced resistance to corrosion and abrasion. Expanding monomers (EMs) find use in coatings, adhesives, electro-optics, dentistry, medical prostheses, electronics and other industries. EMs may be used to eliminate internal stresses by expanding into micro-cracks in *advanced composites*; copolymerize with epoxies, acrylics and ure-thanes to generate unprecedented properties in *aviation* and *aerospace* plastics application; deliver biodegradable polymerized vesicles in *pharmaceuticals*; produce advanced *microchip sealants* and *conductive links* in electronics and combine with GMA for superior fillings, dentures and crowns *in dentistry*.

3.11.2 Conducting Polymers

The advent of *polyacetylene* (PAC) in the early seventies raised hopes and inter-est in conducting polymers. This system, however, is non-conducting and may be regarded as semi-conducting. The conducting fillers such as metal wires and powders, graphite powders and transition metal compounds are incorporated in non-conducting polymers to make them conducting.

Advancement: In a later development, the conductivity of PAC is enhanced by more than 10 times when *iodine* is doped into it. The doped PAC possesses conductivity much higher and almost equal to metallic range. It, however, loses conductivity rapidly when exposed to atmospheric conditions. Another limitation is the migration and loss of dopant from the surface.

Uses: Conducting polymers can generate electrical signals by ion-exchange mechanism. *Conducting electroactive polymers* belong to the class of smart materials and are used as *chemical sensors*.

3.11.3 Polymers in Electronics

Several polymers show excellent *optoelectronic* properties. Many polyimides are used for integrated insulation and crossovers for *microwave integrated cir-cuits*. *Aviation radome, anti-radar paints* and *heat shields* are some novel applica-tions of special polymers.

3.11.4 Thermoplast-Thermoset Plastics

The PPS developed by Shri Ram Fibres Ltd., Madras, and National Chemical Laboratory, Pune, is a unique polymer possessing both thermoplastic and thermosetting properties. The PPS is characterized by its following properties:

1. High-temperature resistance
2. Exceptional corrosion and chemical resistance
3. Inherent flame resistance
4. Low moisture absorption
5. Easy processability
6. Excellent electrical and mechanical properties

3.11.5 Liquid Crystal Polymers

These are made of rigid, rod-type, ordered molecules, which maintain their crystalline order even on melting. Based on the molecular orientation displayed, they are classified as follows:

1. Nematic liquid crystal polymers (LCPs)
2. Smectic LCPs

The main difference between these polymers, and the conventional liquid crystals used in electrical display devices, is their molecular weight. *LCPs have much higher molecular weight.*

Different types of LCPs and their examples and advantages. Depending on packing arrangement of rigid units, there are two types of LCPs:

1. Main chain LCPs (e.g. polyesters, polyamides)
2. Side chain LCPs (e.g. polyacrylate, polymethacrylate, poly-siloxane)

The advantages of these polymers over conventional plastics are their

- High impact strength
- Excellent dimensional stability
- Low thermal expansion
- Better chemical resistance
- Outstanding processability

They are used in highly specialized areas, namely, telecommunication, fibre optics and chemical and hostile environments. *Xydar* and *Vectra* are commercial LCPs developed in 1984.

3.11.6 Photocurable Polymers

These polymers find applications in *microlithography* for making printing plates, photoresist for microcircuits, video discs, optical fibre coatings, etc. In this highly specialized radiation curable system, *the photoresist* such as *polyvinyl cinnamate* and related materials are synthesized by using low-temperature interfacial transfer catalyzed technique.

3.11.7 Biomedical Polymers

Medical and pharmaceutical applications of synthetic polymers range from

- Catheters to vascular grafts
- Semiocclusive dressings to mammary implants
- Transdermal drug delivery systems to medicated patches

TABLE 3.6

Biomaterials for Medical Applications

Polymer	Application
Segmented polyurethane	Artificial heart, heart valve, vascular tubing
Polydimethyl siloxane	Artificial heart, heart valve, vascular tubing
Segmented copolydimethyl siloxane–methane	Artificial heart, heart valve, vascular tubing
Perfluorobutyryl ethyl cellulose	As membrane in oxygenator
Polyallyl sulphone	As membrane in oxygenator
Hydrogels	As grafted surface for polymers

Current activities of device designers, manufacturers and research physicians indicate that devices manufactured from synthetic polymers are being increasingly accepted as the biomaterial of choice in most applications requiring compliance with soft tissue and cardiovascular tissue or being non-irritating to the skin for transdermal applications.

Desired characteristics: The biomedical polymers should be such that the blood, enzymes and proteins should not cause adverse immune response, cancer and toxic and allergic reaction. Highly crystalline polymers such as nylon, polyethylene, polyurethane and protein are used after testing and treating for creep, stiffening, stress and strain, etc. Silastic rubber, Teflon, siloxane–urethane polymers, surface-modified elastomers by hydrogels, glass-reinforced plastic (GRP) composites, etc. seem promising materials.

Applications: A few selected biomaterials in blood contacting applications are given in Table 3.6.

3.11.8 Polymer Foams

The development of technology, especially in the aviation and building industries, created stringent requirements that could not be met by existing natural and man-made materials. Research began in the late 1930s and early 1940s in many countries with the aim of creating new gas-filled materials based on synthetic organic compounds.

Applications: At present hundreds of various elastic and rigid gas-filled materials such as *nomex honeycomb*, polyurethane are used in literally all branches of industry and produced on the basis of reactive oligomers and high polymers. Production of these materials is rapidly expanding at an exceptional rate.

Advances: The current technology for processing and production of foamed polymers, including thermosetting and thermoplastic polymers such as polyolefins, PVC, polyurethanes, structural foams and filled and reinforced foams (syntactic) are much advanced.

3.12 Photorefractive Polymers

Photorefractive effect is a light-induced change in optical properties of a material caused by absorption of light and transport of charged photocarriers. This effect is obtained from different photorefractive materials such as high-resistivity semiconductors, multilayer thin films, ferroelectric oxides and photorefractive polymers. The photorefractive materials are useful or likely to be useful for the following main applications:

- Optical computers
- Optical control systems
- Holographic optical memory
- Optical image processing
- Phase conjugation

Photorefractive polymers are high-performance photorefractive materials of new era. Compositional flexibility of these materials is similar to electrophotographic polymers and non-linear optical polymers. Their use is likely to be made in the aforementioned applications in the future. The photoconductivity and optical non-linearity necessary for photorefractivity are generally provided by combining the sensitizers, that is, photoinduced charge generators, charge transporters and non-linearly optical (NLO) chromophores. Their examples are the following:

1. *Sensitizers*: Charge transfer complexes, for example, PVK–TNF, dyes and dye aggregates and fullerenes (C_{60})
2. *Charge transporters*: Electron donors, for example, hydrazones, aryl amines, carbazoles, conjugated polymers and electron acceptors, for example, trinitrofluorenone
3. *NLO chromophores*: Conjugated molecules, for example, tolanes, stilbenes, styrenes, polyenes, azobenzenes and liquid-crystalline mesogens

3.13 Wood

Wood or timber is a common engineering material since pre-technological days. It is found from trees. Cellulose chain is its main constituent whose monomer is symbolized as in Figure 3.5. Wood possesses directional property due to arbitrary chain alignment. The strength and modulus in longitudinal

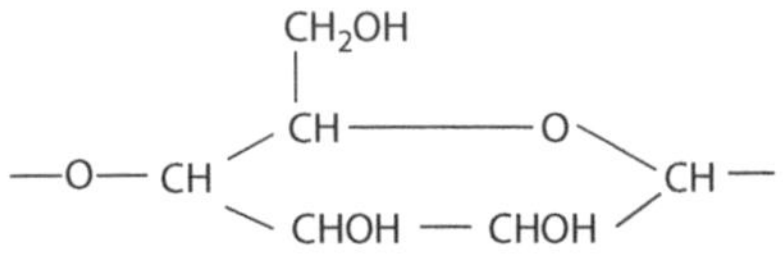

$$CH_2OH$$

FIGURE 3.5
Formula of wood monomer, a cellulose.

TABLE 3.7

Various Properties of Some Wood

S. No.	Properties	Firewood	Pinewood	Common Wood
1.	Young's modulus (GPa)			
	(Parallel to the grain)	14	7.6	6–16
	(Across the grain)	—	—	0.6–1.0
2.	Tensile strength[a] (MPa)	34	16	15–55
3.	Hardness (BHN)			
	(Across the grain)	—	—	6
	(Parallel to the grain)	—	—	3
4.	Work of fracture (kJ/m^2)			
	(Across the grain)			20
	(Parallel to the grain)			0.015
5.	Specific gravity	—	—	0.4–0.9
6.	Coefficient of thermal expansion ($\times 10^{-6}$)/°C	—	—	Up to 9

[a] Parallel to the grain. Long-term strength reduces to about half.

direction are more than 10 times its values in transverse directions. Various properties of wood are shown in Table 3.7.

Stress–strain diagram: The σ–ε diagram and other mechanical properties of wood. The wood is generally a bi-modulus material directions. The stress–strain behaviour in tension and compression are shown in Figure 3.6. Various properties of wood are shown in Table 3.7.

Types of timbers: Different kinds of woods/timbers used are as follows:

- Teak/sagaon
- Sakhu/sal
- Shisham
- Neem
- Mahogany
- Babool
- Chir

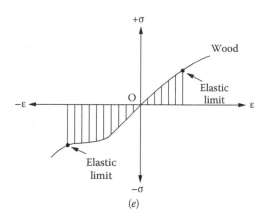

FIGURE 3.6
Stress–strain curve (*e*) in tension and compression both.

- Deodar
- Mango
- Semal

Standard forms of timber: Timbers are sold in market in various standard shapes and sizes such as the following:

1. Log
2. Balk
3. Post
4. Plank
5. Board
6. Batten
7. Scantlings
8. Deal

End products: Various end products manufactured are the following:

- Domestic and office furnitures
- Tables and chairs
- Beds
- Decorative items, designs and fittings
- Beams of houses
- Door and window

4

Structures and Applications of Ceramics, Refractories and Glasses, etc.

4.1 Ceramic Materials

Ceramics are non-metallic, inorganic, amorphous solids and are mostly metallic oxides. They have poor tensile strength and are brittle. They can be either crystalline or non-crystalline. Many ceramics are workable in extremely low (cryogenic) temperature range, while many others are able to sustain high temperatures. Silica, silicates, NaCl, rock salt, MgO, CaF_2, glasses and CsCl are some examples of ceramics.

4.1.1 Basic Ceramic Structure

Ceramic structures are more complex than those of metals. It is because they are composed of at least two elements, and the atomic bonding in them ranges from purely ionic to totally covalent. Many ceramics exhibit a combination of ionic–covalent bonding with varying degrees of ionic character. For example, the ionic character in CaF_2 is 89%, in NaCl 67%, in SiO_2 51% and in ZnS 18%.

For those ceramic materials which dominantly have ionic character, the basic structure is tetrahedral. It has ions of different sizes, namely, smaller circle and larger circles. They are of opposite nature. The smaller ion is called *cation*, and larger ion is known as *anion*. This is a tetrahedral configuration of cation–anion packing in which the cation radius r_c and anion radius r_a bear a ratio $r_c/r_a = 0.225$. Since a cation surrounds itself with four anions, its coordination number (or ligancy) is 4. Silica, silicate and ZnS have such ceramic structure.

4.2 Types of Ceramics

Ceramics may broadly be classified under various groups. Main among them are the following:

1. Refractories
2. Silicates
3. Glasses
4. Limes
5. Cements
6. Plain concretes
7. Prestressed concretes
8. Rocks and stones
9. Reinforced cement concrete (RCC)
10. Abrasives
11. Clay and clay products: bricks, tiles, etc.

Carbides of tungsten, titanium, zirconium, silicon, alkali halides and silicon nitride (Si_3N_4) are some important ceramic materials of present time. An exhaustive list of ceramics is given in Table 4.1 with their related details and applications.

4.3 Refractories

Refractories are ceramic materials of specific nature which are capable of withstanding high temperatures. Commercial refractories are made of complex solid oxides of elements such as silicon, aluminium, magnesium, calcium and zirconium. Refractories confine the heat in ovens and furnaces by preventing heat loss to the atmosphere. Refractories are unaffected by high temperatures. They are expected to resist mechanical abrasion, infusion of molten metals, slag or metallic vapours and also the action of superheated steam, sulphurous oxide, chlorine and other gases.

Examples: One of the most widely used refractories is based on alumina–silica composition, varying from nearly pure silica to nearly pure alumina. Other common refractories are silica, magnesite, forsterite, dolomite, silicon carbide and zircon.

TABLE 4.1

Ceramics, Their Related Details and Applications

S. No.	Material	Related Details	Application
1	Cement	Chemically active, non-organic, non-metallic	Plastering, preparing different grades of concrete
2	Lime	A kind of ceramic, non-organic, non-metallic	Foundation, mortar, plastering
3	Surkhi	Crushed to fine powder form from bricks	As mortar
4	Mud	Clay mixed with water	As mortar
5	Stone Marble Slate Granite Dolerite	Non-metal, obtained from rocks	Building masonry
6	Gravel	Obtained by crushing the stone	In making of concrete
7	Ceramics Refractories Superrefractories	Metallic oxides High-temperature resisting lightweight High melting point	Firebricks and fireclays Armour plate, cutting tools
8	Firebrick and fireclay	Magnesia, alumina, silica, zirconia, etc.	Chimney lining
9	Terra cotta	Baked earthenware	Tiles, sewage and drain pipes
10	Asbestos and glass wool	Non-metals	Heat insulation
11	Chemicals: sulphates of magnesium, sodium, calcium, etc.	Chemically active to limited extent	Maintenance of brickwork
12	Porcelain	Acid proof	Electric insulation
13	POP	Non-metal, ceramic nonorganic	Plastering
14	Bitumen	Organic non-metal	Damp proofing
15	Mastic asphalt	Organic non-metal	Damp proofing
16	Bituminous felt	Organic non-metal	Damp proofing
17	RCC	M.S. bar used for reinforcement	Beam, slab, column, etc.
18	Plain concrete	M10,... M40 mix used	In foundations
19	Prestressed concrete	M20,... M70 mix used	Beam and slab
20	Terrazzo	Kind of stone	Flooring
21	Mosaic	Kind of stone	Flooring
22	Resin-treated concrete	Non-conventional material	For very quick repairs

(Continued)

TABLE 4.1 (*Continued*)

Ceramics, Their Related Details and Applications

S. No.	Material	Related Details	Application
23	Glass Sodium Borosilicate High silica Lead	Non-metal, belongs to the family of ceramic	Panes of windows for lighting, bulbs
24	Rock Igneous Metamorphic Sedimentary	Hard, dense, strong foliated structure budding layered structure	Building stones are obtained for stone masonry works
25	Sand Coarse Medium Fine	Particle's size influences the quality	With cement and concrete Roofing Plastering

4.3.1 Refractoriness

It is the ability of a material to withstand the action of heat without appreciable deformation or softening. Refractoriness of a material is measured by its melting point. Some refractories, like fireclay and high-alumina brick, soften gradually over certain range of temperatures.

4.3.2 Types of Refractories

According to chemical behaviour, the refractories are classified as follows:

1. *Acidic refractories*: They readily combine with the bases and are therefore termed acidic. Silica is their chief constituent. Important acid refractories are quartz, sand and silica brick.
2. *Basic refractories*: They consist mainly of basic oxides without free silica and resist the action of bases. The most common basic refractories are magnesite and dolomite.
3. *Neutral refractories*: They consist of substances which do not combine with either acidic or basic oxides. With increasing alumina content, silica–alumina refractories may gradually change from an acidic to a neutral type. Examples are silicon carbide, chromite and carbon.

4.3.3 Properties of Refractories

Refractories in general possess the following properties:

1. They are chemically inactive at elevated temperatures.
2. They are impermeable to gases and liquids.

3. They have long life without cracking or spalling.

4. They are capable of withstanding high temperatures, thermal shocks, abrasion and rough usage.

5. Temperature variations cause minimum possible contraction and expansion in them.

6. They are able to resist fluxing action of slags and corrosive action of gases.

7. They are good heat insulators.

8. They have low electrical conductivity.

4.4 Silica and Silicates

The structures of silica and silicates consist of repeating units of silicate tetrahedron. Each unit has a silicon cation at the centre of tetrahedron accompanied with four oxygen anions at the four corners. Three-dimensional network of tetrahedral is formed in the structure of silica. Electrical neutrality is maintained in silica due to the arrangement of silicon cation and oxygen anions. The effective number of silicon is $(1 \times 1) = 1$ per tetrahedral unit and $(4 \times 1/2) = 2$ for oxygen.

4.4.1 Crystalline and Non-Crystalline Forms of Silica

Silica exists in both the crystalline and non-crystalline forms having the same coordination between silicon and oxygen. Quartz is an example of crystalline form, and silica glass of non-crystalline form. Quartz exhibits piezoelectric effect. Silica yields silicates when oxides dissolve in them. Glass of different types are examples of silicates and are described in the section that follows.

4.4.2 Configuration of Minerals

Instead of a single silicon cation at the centre of tetrahedron, other cations may be introduced to obtain different mineral structures. Based on the sharing of corners, these minerals may be classified as (Figure 4.1a through e)

1. Island such as olivine and hemimorphite (Figure 4.1a)

2. Single chain such as enstatite (Figure 4.1b)

3. Double chain such as asbestos (Figure 4.1c)

4. *Ring chain* such as *beryl* (Figure 4.1d)

5. *Sheet* such as *mica*, talc and clay (Figure 4.1e)

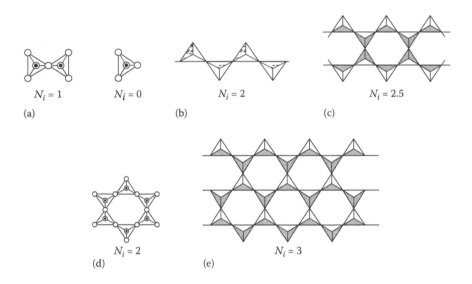

FIGURE 4.1
Configuration of tetrahedral structural unit: (a) island, (b) single chain, (c) double chain, (d) ring chain and (e) sheet. N_i = number of oxygen ions shared.

4.5 Applications of Ceramics

Ceramics are widely used materials. They find use in almost all fields of engineering, ranging from mechanical, civil and electrical to computer, biomedical and nuclear. Important applications of ceramics are listed in the following.

In conventional applications main among them are the following:

1. As firebricks and fireclay for lining of ovens and furnaces
2. As artificial limbs, teeth, etc., in biomedical/medical field
3. As insulators (dielectrics) in electrical transmission and distribution
4. As crockeries in domestic uses, sanitary wares, etc.
5. Components of chemical processing vessels
6. As radiation shield for nuclear reactor
7. As ferroelectric crystals and piezoelectric crystals to generate ultrasonic waves for non-destructive testing and other purposes
8. As nuclear fuel elements, moderators, control rods, etc.
9. As magnetic materials in electrical machines, devices, etc.
10. As cutting tools such as Borazon (CBN), in explosive forming, etc.

In recent applications, main among them are the following:

1. As powerful superconducting magnets to make magnetic rail for wheelless train (levitated train).
2. As ferrites in memory cores of computers.
3. As miniature capacitors for electronic circuits.
4. As garnet in microwave isolators and gyrators.
5. As components of sonar device that helps to locate a small object in large volume (e.g. searching of a black box from within the ocean).
6. As zirconia (ZrO_2) thermal barrier coating to protect superalloy made aircraft turbine engine.
7. Ceramic armour to protect military vehicles from ballistic projectiles.
8. As electronic packaging to protect ICs. Aluminium nitride (AlN), SiC and boron nitride (BN) are used as substrate materials for this purpose.

4.6 Mechanical Behaviour of Ceramics

Ceramic materials possess the following mechanical properties:

Strength: They are much stronger in compression than in tension. Therefore, they are suitable for compressive load applications. Brickwork in a building is one such example.

High temperature resistance: They can sustain high temperatures. Their resistance to abrasion and chemical attack and rigidity are also high at higher temperatures.

Brittleness, ductility and malleability: They are brittle in nature. They have negligible ductility and poor malleability. Therefore, they are not suitable for tensile load applications.

Melting point: They generally possess a high melting point. For example, the melting point of alumina (Al_2O_3) is ≈1700°C, of silica (SiO_2) is ≈2000°C and of zirconia (ZrO_2) is ≈3200°C.

Thermal conductivity and thermal expansion are poor in them. Therefore, they are suitable for thermal and electrical insulation purposes.

Stress–strain profile: Generally, they are linearly elastic materials. Their stress–strain curve is a straight line up to fracture point. Stress–strain curves of silicate, cement concrete and plaster of Paris (POP) are shown in Figure 4.2.

Hardness: They are very hard materials. Their hardness normally lies in the range of 500–2000 BHN.

Creep behaviour: Due to high melting point, ceramics do not normally creep up to workable high temperatures. Therefore, they are good creep-resistant

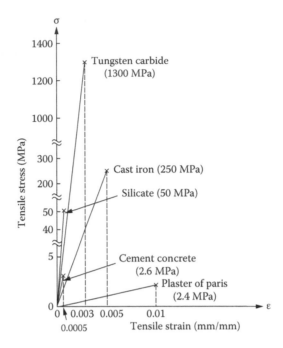

FIGURE 4.2
Stress–strain curves of different materials under tensile loading.

materials. Sialon (alloy of Si_3N_4 and Al_2O_3) is a recent material used for making gas turbine blades, which are workable up to 1300°C.

Propagation of crack: Under the applied loads when stress concentration is created, the cracks are formed in ceramics. These cracks propagate under tension but are not affected by compressive loads. The compressive load is transmitted across the cracks.

Fracture mode: Ceramics fail in brittle mode, also holds good for other ceramics.

4.6.1 Other Behaviour of Ceramics

Imperfections in ceramics are mostly of Frenkel and Schottky defect types. Their non-destructive testing may be performed by well-known techniques. Details of RCC and glasses are described in this chapter.

Characteristics of a good ceramic material: A good ceramic material should also possess the following electrical properties:

- High thermal stability
- High abrasion resistance
- High hardness

- High mechanical strength
- High volume resistivity
- Least water absorptivity
- Least effect of oil, acid and alkalies

Volume resistivity of porcelain is of the order of 10^9–10^{12} Ω m, and water absorptivity is up to 0.5% only. This makes the porcelain a very favourable insulating material.

4.7 Electrical Behaviour of Ceramics

Ceramics are widely used for insulation purposes in electric power transmission, electrical machines/equipments/devices, etc. Mica, porcelain, glass, micanite, glass-bonded mica, asbestos, glass tape, etc., are commonly used (ceramic) materials. They show the following characteristics in respect of their electrical behaviour:

Dielectric constant: Ceramics have high values of dielectric constant ε_r. It is 8 for mica, 7 for soda-lime glass and 6 for porcelain. That is why they are used for insulation purposes in electric motors, alternators, transformers, etc.

Dielectric strength of ceramics is high. It is defined as the voltage required per unit thickness to cause a breakdown in material. Dielectric strength of mica is 100 MV/m. This is pretty good strength that makes the mica a very good insulating material.

Dielectric loss: A good electrical insulator should have a low value of dielectric loss. It is characterized by power factor tan δ, where δ is the loss angle. Value of dielectric loss in mica is only 0.0005, which further supplements its use as a good insulating material.

4.8 Processing of Ceramics

Due to high melting temperatures, hardness and brittleness, the ceramics cannot be fabricated using conventional forming techniques as adopted for metals. Therefore, various other techniques shown in Figure 4.3 are adopted for their processing/fabrication. A brief account of these processes is given as follows.

4.8.1 Glass-Forming Processes

Glass products are produced by heating the raw materials above their melting points. As these products have to be homogeneous and pore free, a

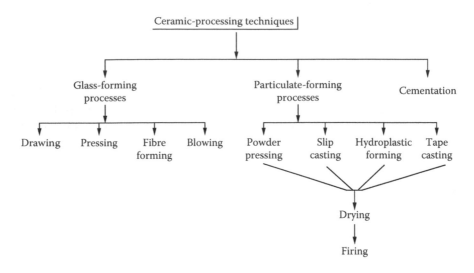

FIGURE 4.3
Various techniques of ceramic processing.

thorough mixing of raw ingredients and their complete melting is essential. To provide porosity in glass products, small gas bubbles are deliberately produced and blown into them in molten state. If porous spots are to be made in colours, the powders of desired colours are also blown along with the gas.

Different glass-forming processes are used to produce glass products of vivid varieties. These are

- Drawing process – to form long piece of rods, sheets, tubes, etc.
- Pressing process – to form thick-walled plates, dishes, etc.
- Blowing process – to form bulbs, jars, bottles, etc.
- Fibre forming process – to form glass fibres

Drawing process is similar to a continuous hot rolling method in which the glass sheet is drawn from molten glass over the rolls. To provide surface finish and flatness to the sheet, the glass sheet is floated on a bath of molten tin at elevated temperature. The sheet is then slowly cooled and heat treated by annealing.

Pressing process is accomplished by pressing the glass piece in a graphite-coated cast iron mould. The shape of mould has to be in conformity with the desired shape of product. The mould is heated also to obtain an even surface of the sheet.

Fibre forming is a kind of drawing operation. It is a sophisticated process in which the molten glass contained in a platinum heating chamber is drawn through several small orifices provided at the chamber base.

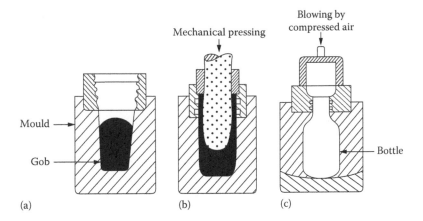

FIGURE 4.4
The *press and blow* technique to manufacture a glass bottle: (a) raw glass kept in a mould, (b) mechanical pressing and (c) blowing by compressed air.

Blowing of a glass piece kept in a blow mail is performed by the pressure created from air blast. It can be done manually or by an automated system. The *press and blow* technique used for producing a glass bottle is shown in Figure 4.4. In it, a raw glass (gob) is placed in the mould (Figure 4.4a), which is pressed mechanically (Figure 4.4b) and then blown by compressed air pressure (Figure 4.4c). The contour of the mould has to be prepared in conformity with contour of the bottle desired.

4.9 Particulate Forming Processes

Particulate forming processes are generally adopted for manufacturing of clay-based products, fireclay and refractories. Different particulate forming processes are used to produce vivid varieties of products as given in the following:

1. Powder pressing: to form firebricks, refractory abrasives, electronic ceramics, magnetic ceramics, etc.
2. Slip casting: to form sanitary ware, scientific laboratory ware, decorative objects, intricate articles, etc.
3. Hydroplastic forming: to form tiles, pipes, bricks, solid and hollow blocks, etc.
   a. High hardness
   b. High mechanical strength
   c. High volume resistivity
   d. Least effect of oil, acid and alkali

Volume resistivity of porcelain is of the order of 10^9–10^{12} Ω m, and water absorptivity is up to 0.5% only. This makes the porcelain a very favourable insulating material.

4.10 Glasses

Glass is an inorganic fusion product, cooled to a condition in which crystallization does not occur. Silica is a perfect glass-forming material. It has a very high melting point. To lower its fusion point and viscosity, some basic metal oxides are added to it. If sodium oxide is added to silica, the mixture obtained is called *sodium disilicate* ($Na_2O\ 2SiO_2$). If calcium oxide is added to it, *soda-lime glass* will be obtained. Glasses are highly resistant (chemically) to most corrosive agents. Borosilicate and high-silica glasses have much higher chemical resistance. Fused silica used in construction of chemical plants is even more chemically resistant.

4.10.1 Glass-Forming Constituents

Besides silica, oxides of boron, vanadium, germanium and phosphorous are the other constituents of glass. Elements and compounds such as tellurium, selenium and BeF_2 can also form the glasses. The oxide components added to a glass may be subdivided as follows:

1. Glass formers
2. Intermediates
3. Modifiers

They are grouped on the basis of functions performed by them within the glass.

Examples: *Glass formers* and network formers include oxides such as SiO_2, B_2O_3, P_2O_5, V_2O_6 and GeO_2. They form the basis of random 3D network of glass. *Intermediates* include Al_2O_3, ZrO_2, PbO, BeO, TiO_3 and ZnO. These oxides are added in high proportions for linking up with the basic glass network to retain structural continuity. *Modifiers* include MgO, LiO_2, BaO, CaO, SrO, Na_2O and K_2O. These oxides are added to modify the properties of glasses.

4.10.2 Devitrified Glass

The other additions in glass are the fluxes which lower down the fusion temperature of glass and render the molten glass workable at reasonable temperatures. But fluxes may reduce the resistance of glass against chemical attack. *Devitrified glass* is undesirable since the crystalline areas are extremely weak and brittle. Stabilizers are therefore added to this type of glass to overcome these problems.

4.11 Types of Glasses

Commercial glasses may be of the following four types:

1. Soda-lime glass
2. Lead glass
3. Borosilicate glass
4. High-silica glass

Chemical composition of these glasses is given in Table 4.2.

4.11.1 Soda-Lime Glasses

Soda-lime glasses mainly contain oxides of sodium and calcium and silica. Its compositional formula is $Na_2O CaO 6SiO_2$. Small amount of alumina and magnesium oxides are added to it to improve the chemical resistance and durability of the glass. Soda-lime glasses are cheap and resistant to water and devitrification. They are widely used as window glass, in electric bulbs, bottles and tablewares where high temperature resistance and chemical stability are not essentially required.

TABLE 4.2

Chemical Composition of Commercial Glasses

Component	Soda-Lime Glass (%)	Lead Glass (%)	Borosilicate Glass (%)	High-Silica Glass (%)
SiO_2	70–75	35–58	73–82	96
Na_2O	12–18	5–10	3–10	—
K_2O	0–1	9–10	0.4–1	—
CaO	5–14	0–6	0–1	—
PbO	—	15–40	0–1	—
B_2O_3	—	—	5–20	3
Al_2O_3	0.5–2.5	0–2	2–3	—
MgO	0–4	—	—	—

4.11.2 Lead Glasses

Lead glasses contain 15%–40% lead oxide. They are also known as *flint* glasses. They are used to make high-quality tableware, optical devices, neon sign tubing, etc. Glasses having a high lead content, up to 80%, have relatively low melting points, high electrical resistivity and high refractive index. They are used for extra-dense optical glasses, windows and protective shields to protect against x-ray radiations.

4.11.3 Borosilicate Glasses

Borosilicate glasses contain mainly silica and boron oxide. Small amounts of alumina and alkaline oxide are also added to them. They have low coefficient of thermal expansion and high chemical resistance. They are used in scientific piping, gauge glasses, laboratory ware, electrical insulators and domestic items. This glass is known as *Pyrex* glass by trade name.

4.11.4 High-Silica Glasses

High-silica glasses containing up to 96% silica are made by removing alkalies from a borosilicate glass. They are much more expensive than other types of glasses. They have very low thermal expansion and high resistance to thermal shock. High-silica glasses are mainly used where high temperature resistance is required.

4.11.5 Photochromic and Zena Glasses

Photochromic and zena glasses are also in common use. *Photochromic glasses* are used in making lens for goggles. Silver chloride is mixed in ordinary glass to make them. *Zena glass* is used to make chemical containers. One of the notable applications of glass is in the world's largest telescope situated on Mount Semivodrike in USSR, which utilizes 6 m diameter mirror of 1 m thickness and 70 tonnes weight.

4.12 Perovskite Structures (or Mixed Oxides)

When the oxides TiO_2, ZrO_2 or HfO_2 are fused with the oxides of other metals, we get titanates, zirconates and hafnates. These are called *mixed oxides*. Some examples of titanates are calcium titanate $CaTiO_3$, iron titanate (*ilmenite*) $FeTiO_3$, barium titanate $BaTiO_3$ and magnesium titanate Mg_2TiO_4. Titanates of Mn, Co, Zn, etc., do also exist.

When the two metals differ widely in size, the result is *a perovskite structure*. The name Perovskite comes from the name of its originator, von Perovski.

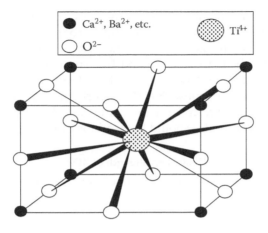

FIGURE 4.5
Perovskite structure.

Such structures may be natural or artificial. $CaTiO_3$ is a natural kind of perovskite, and $BaTiO_3$ an artificial type. A perovskite structure is shown in Figure 4.5. It has a face-centred cubic array of Ca and O with Ti occupying one-quarter of the octahedral holes. Coordination number of Ca is 12. In the structure of $BaTiO_3$, the barium atoms will occupy corner positions in Figure 4.5.

4.13 RCC

Concrete has high compressive strength but poor tensile strength. It is liable to crack when subjected to tension. Hence, to provide sufficient tensile strength, reinforcements in the form of steel bars are provided. The combination of concrete and steel reinforcement results in a material known as RCC. It contains favourable properties of both concrete and steel. Steel is used as reinforcing material due to its high tensile strength and elasticity and because its thermal coefficient is nearly equal to that of concrete.

4.13.1 Ingredients of RCC

Various ingredients of RCC are

1. Cement
2. Coarse aggregate
3. Fine aggregate (or coarse sand)
4. Water
5. Reinforcement

Cement used should be according to the specifications of BIS. Generally sand is used as fine aggregate. Sand should be coarse with fineness modulus 2.90–3.20. Coarse aggregate may be stone grit, gravel, or some other inert material. It should be hard, durable and free from other injurious materials such as saltpetre and salts. Generally crushed stone grit, granite tips, or stone ballast is used for this purpose. For RCC work, the size of coarse aggregate varies from 2 to 3 cm. Fresh drinkable water which is free from impurities like sulphates and chlorides should be used for RCC.

Grades and Strength of RCC

Concrete mix in RCC works is classified into the following categories:

M10...	1:3:6 mix
M15...	1:2:4 mix
M20...	1:1.5:3 mix
M25...	1:1:2 mix
M30...	1:4:8 mix, 1:5:10 mix, 1:6:12 mix
M35...	Post-tensioned prestressed concrete
M40...	Pre-tensioned prestressed concrete

The mix 1:3:6 indicates proportions of cement, aggregate and sand, respectively.

M10 and M20 grades of concrete are used in mass concrete work. M15 is generally used for all purposes of general nature such as RCC slabs and flooring. M20 and M25 are used for thin RCC members such as precast members, shed roofs and thin shells. M35 and M40 are used in prestressed concrete structures, that is, for post-tensioned and pre-tensioned concrete members, respectively. Strength of cement depends on the water/cement ratio.

4.13.2 Reinforcing Materials

The following reinforcing materials are commonly used in RCC works:

1. Mild steel bars (plain bars)
2. High yield strength deformed (HYSD) bars
3. Twisted bars
4. Steel wires (or tendons)

Mild steel bars (plain bars): The main drawback of these bars is their poor bond stress.

Deformed bars (HYSD bars): These bars are more suitable in comparison to plain bars.

Twisted bars: Cold worked twisted bars are used to improve the bond between concrete and reinforcing materials. These bars increase the yield stress by about 50% and thus save on reinforcing material by 33%. These bars have greater bond strength and hence are very useful for water-retaining structures such as retaining walls and dams. The size of bars varies from 5 to 50 mm.

Steel wires: Steel wires with high elasticity are used in prestressed concrete members. They have a diameter of about 6 mm and are known as *tendons*. They are also used in the form of wire mats.

4.13.3 Advantages of RCC

RCC has many advantages. It can be moulded into any desired shape due to its plastic properties. It is fire resistant, damp proof, rigid, durable and impermeable to moisture. Its structures can bear the shocks of earthquake effectively. Maintenance of RCC work is less, while its appearance is good.

4.14 Clays and Clay-Based Ceramics

Clays are natural inorganic amorphous materials obtained from the remains of rocks. Clay minerals are hydrated aluminosilicates derived from feldspar minerals present in the rocks. Clay minerals are the following:

1. Kaolinite ($Al2O_3\cdots 2SiO_2 \cdot 2H_2O$)
2. Montmorillonite ($Al2O_3\cdots 4SiO_2 \cdot nH_2O$)

Kaolinite, white in colour, is used in making of sanitary wares, firebricks paper, etc. Montmorillonite family of clays includes talc and is used in making of steatite for high-frequency electrical applications.

Use: Plasticity of clays varies with the water content. For example, the quantity of and the type of clays and technique employed in mixing them with graphite decide the characteristics of pencils such as 2H, H, HB, B or 2B. The 2H quality is harder than HB, and 2B is softer than HB. Drying, shrinking and vitrification properties of clay-based ceramics are influenced by adding non-plastic materials such as crushed and powdered quartz, talc and feldspar. Different clay-based ceramics and their characteristics are illustrated in Table 4.3.

4.15 Chemically Bonded Ceramics

Dramatic new materials of today and tomorrow, the chemically bonded ceramics (CBCs), are high-performance, low-cost materials. CBC materials

TABLE 4.3

Types and Characteristics of Clay-Based Ceramics

Clay-Based Ceramic	Temperature Range (K)	Apparent Porosity (%)	Application
Stoneware	Above 1500	1–2.5	Roof tiles, glazed pipes
Earthenware	1100–1300	5–15	Drainage pipes, water filters, bricks and wall tiles
Porcelain	1500–1700	0–1	Scientific and electrical items
China clay	1400–1600	<1	Tableware

are mostly silicates, aluminates and phosphates. Their low cost results from processing temperatures far below the 1100°C needed for producing conventional ceramics. CBCs represent a new technology that produces high-performance ceramic components at a lower cost. CBCs have 10–20 times the tensile strength of concrete and are about 32 times cheaper than aluminium, 20 times cheaper than steel, and 4–6 times less expensive than plastics. CBCs have opened a new materials arena where formed shapes are used in the aerospace, automotive, appliance, electrical, electronics and construction industries.

Applications: Great growth in CBCs is assured, not only due to their low cost and high performance, but also due to a worldwide movement away from materials that are petroleum based, toxic or flammable. Currently, the following CBCs are in use:

- CBC armour brake linings
- Ceramic coatings
- CBC roofing
- CBC wall panels
- CBC flooring
- CBC electrical fixtures

CBC materials may soon be used in following applications:

- Fireproof sound dampeners for auto engines and electric motor
- Engine blocks and parts
- Aircraft and rocket parts
- Medical prostheses
- Roadways
- Cryogenic containers
- Containers for superconducting materials

4.16 Applications of Ferroelectrics

The ferroelectric materials are employed in the following main applications:

- Matrix-addressed memories
- Shifts registers and switches (i.e. transpolarizers)
- Multiplate capacitors
- Thermistors
- Light deflectors, modulators and displays
- Holographic storage

5

Polymeric Composite Materials: Types and Mechanics

5.1 Introduction

Definition: Composite material is a material system composed of two or more dissimilar constituents, differing in forms, insoluble in each other and physically distinct and chemically inhomogeneous. The resulting product possesses properties much different from the properties of constituent materials. Composite materials, also referred to as *composites*, are broadly classified as

1. Agglomerated composite materials
2. Laminated composite materials
3. Reinforced composite materials

Grinding wheel is an example of agglomerated composite, sheet moulding compound is an example of laminated composite, and fibre-polymeric composite is an example of reinforced composite.

5.2 Laminated Composites

Lamina and laminate: Laminated materials also referred as laminates are layered composites made up of many laminae. A lamina also known as a ply or a layer is very thin, about 0.1–1 mm thick. A single lamina is unsuitable for any purposeful application. They are, therefore, joined or glued together to form a laminate of desired thickness. Thus a laminate is made up of an arbitrary number of lamina. The number of lamina, in a laminate, can be few to many tens. A few examples of common laminates are the following:

1. Plywood
2. Metal to metal laminate, namely, cladded metals
3. Sheet moulding compounds (SMCs)
4. Bulk moulding compounds (BMCs)
5. Linoleum
6. Tufnol

Now, we shall discuss main among them, one by one.

5.2.1 Laminate

A laminate is a stack of lamina with various orientations of the directions of the principal materials in the lamina. Laminates can be built up with plates or plies of different materials or of the same material, such as glass fibres. Shear stresses are always present between the layers of a laminate because each layer tends to deform independently of its neighbouring layers due to each layer having different properties. These shear stresses, including the transverse normal stresses, are a cause of delamination.

5.2.2 Bulk Moulding Compounds

BMCs are a premixed material of short fibres (chopped-glass strands) pre-impregnated with resin and various additives. A *dough moulding compound* (*DMC*) is an alternative term for a BMC. Some thermoset resins are quite thick and are called *moulding doughs*. Parts made by BMCs are limited to about 400 mm in their longest dimension due to problems with separation of the components of moulding compound during moulding.

5.2.3 Sheet Moulding Compounds

SMCs are non-metallic plastic-reinforced-composite laminates made up by pressing together many unidirectional (U/D) laminae one over the other. Laminate of desired properties may be prepared by placing U/D laminae in different orientations. Generally, a prepreg is used for that.

5.2.4 Prepreg Sheet Moulding Compound

SMC is an intermediate compound between raw material and the final product. Its *pot life* is 3–4 days and cannot be used after that. SMC is a blend of resin, hardener, fibres, catalyst and accelerator. Its curing time can be further decreased by adding more than the specified quantity of hardener.

proper orientation exhibit a more strengthening effect. Elastic modulus of a particulate composite may be obtained by the simple rule of mixture described in art. 5.14.

Applications: Particulate composites are made by sintering, which is a powder metallurgy technique. The W–Ni–Fe and W–Ni–Cu systems are the examples of particulate composites. Cermets and dispersion-strengthened composites described as follows are similar to particulate composites.

5.4.1 Dispersion-Strengthened Composites

They differ from particulate composites in the size of dispersed particles and their volume concentrations. In dispersion-strengthened composites, finely dispersed hard particles are less than 0.1 µm in size which are stronger than the pure metal matrix and increase the elastic limit of the composite thus formed. The particles may be metallic, intermetallic or non-metallic. Oxides are more frequently used due to their inertness, hardness and high thermal stability. SAP, an aluminium–aluminium oxide system, is an example of this kind.

Applications: The thorium-dispersed (*TD*) *nickel* is a system of nickel having 3% thoria (ThO_2) dispersed in nickel matrix. This is a very efficient creep-resisting material. Dispersion of lead particles in steel and copper alloys improves their machinability. Copper and silver matrices mixed with molybdenum, tungsten and their carbides are used to make composites for applications in electrical contacts.

5.4.2 Cermets

Cermet is made of *cer* and *met*. Here, *cer* stands for *ceramic* and *met* for *metal*. Cermet is a combination of ceramic and metal. A cermet consists of ceramic matrix and the reinforcing metal particles. They are generally made by powder metallurgy technique. We know from Chapter 4 that the refractories (a type of ceramics) are high-temperature-resisting, brittle materials. Contrary to this, metals are ductile and have melting points varying from low to very high. They also have higher mechanical strengths. A composition of ceramics and metals in different proportions gives desired thermal and mechanical properties in cermets. They are very good wear- and abrasion-resistant materials and possess high hardness.

Applications: Some notable applications of cermets are the following:

1. Rotary drills in mining industries
2. Cutting tools in metal-cutting industries
3. Shaping tools for refractories
4. Cemented carbide in high-speed cutting
5. Components in satellites and space-going vehicles

5.4.3 Rubber-Toughened Polymers

Polymers have low toughness, which can be improved by reinforcing them by glass or carbon fibres. These fibres act as crack stoppers and help in resisting the propagation of cracks. Toughness of polymers can be increased by other means too. One such way is the use of fillers in the form of small particles. These particles clamp the cracks and intersect and stretch them. Rubber-toughened polymers such as *ABS* are the examples. In these, the polymers derive toughness in them due to the presence of small rubber particles.

5.5 Flake Composites

These are composites of 2D nature and are preferred when planer isotropy is also desired in components of structures and machines. It should be noted that the composites generally possess orthotropy or anisotropy and not isotropy. Moreover, flakes of 2D geometry can be more closely packed than the fibres. These qualities make them suitable for various applications. Mica flakes–glass matrix composites are easily machinable and are used in heat and electrical insulating applications. Silver flakes are used where good conductivity is desired.

5.6 Whisker-Reinforced Composites

Whiskers are a form of materials possessing extremely high strength (HS) and high moduli (HM). We have described in art. 2.3.1 that the strength and moduli of whiskers are much superior than the bulk and fibre form of the same material. Hence, their reinforcement in composites will impart too higher strengths and moduli. Whisker-reinforced composites are in the initial stage of development. It is a likely material to be used in the near future.

5.7 Hybrid Composites

Meaning of hybrid: Fibre composites are made up of only a single type of fibre such as glass, carbon or Kevlar. Each type of fibre has its own limitations in terms of strength, cost and other materials properties. These limitations can be overcome by using a combination of two or more types of fibres in the same matrix. Mixing of two or more different types of continuous fibres in

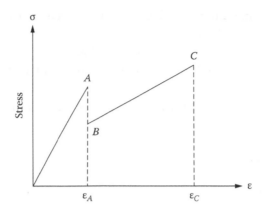

FIGURE 5.1
Idealized stress–strain curve of a hybrid composite showing hybrid effect.

the same matrix is called hybridization, and the resulting product is called hybrid composite. Improvement in properties of such composites is due to the hybrid effect.

Characteristics: Figure 5.1 shows an idealized stress–strain characteristic curve of two fibre systems. The initial slope OA gives initial modulus and slope BC gives the final modulus. The fall AB in the curve is due to failure of the first fibre system at strain ε_A. The second fibre system continues to take the load until final fracture occurs at strain ε_C. In a typical case, the first and second fibre systems may be those of glass and carbon.

5.7.1 Types of Hybrid Composites

Hybrid composites are subdivided into the following four types:

1. Interply hybrid composites
2. Intraply hybrid composites
3. Inter–intraply composites
4. Super hybrid

The *interply hybrid* composite consists of alternate lamina (layer or ply) of the same matrix but different fibres. *Intraply* hybrid composite contains each lamina having two or more kinds of fibre system. The *inter–intraply* composite is a combination of these two types.

Applications: Practical utilization of hybrid composites has been widely made. Out of these some important applications are given as follows:

1. Antenna dishes of carbon fibre–reinforced plastics (CFRPs) and aluminium honeycomb

2. CFRP and glass-reinforced plastic (GRP) leaf springs and drive shafts for automobiles

3. Busbars of aluminium, reinforced by CFRP

4. Squash racquets and golf clubs with shafts of CFRP/GRP/wood hybrid

5. Helicopter rotors and thin-walled tubes of CFRP and GRP

5.8 Sandwich Composites

Construction: A sandwich composite is constructed by sandwiching foam core between two skins of fibre-reinforced polymer (FRP) laminates as shown in Figure 5.2. The thickness of the skin is kept up to 3 mm, and the thickness of core t_c is kept deeper. The core is either foamed or made of honeycomb material so that its density is very less. As t_c is deep ($t_c \gg t_s$), the area moment of inertia of the cross section is enhanced too much, due to which the flexural rigidity of sandwich beam becomes higher. This higher flexural rigidity construction along with very lightweight makes the sandwich composite most suitable as a beam.

5.8.1 Honeycomb Materials

Sandwich constructions are widely employed as beam component of structures in airplanes, spacecrafts, satellites, etc. (Figure 5.3). A non-sandwiched, FRP composite beam will be much heavier than a sandwiched

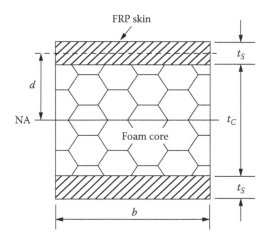

FIGURE 5.2
A sandwich composite construction showing sandwiched core between FRP skins.

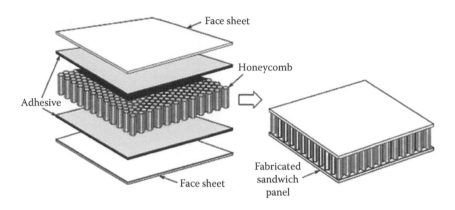

FIGURE 5.3
Constructional details of a sandwich panel.

TABLE 5.1

Mechanical Properties of Some Honeycombed Core Materials

Honeycomb Core Material	Density ρ (kg/m³)	Temperature (K)	Longitudinal Modulus (MPa)	Shear Modulus (MPa)	Compression Modulus (MPa)
Graphite–polyimide (Gr–PI)	30	293	223	86.2	135.1
		505	191.7	92.4	135.1
Aluminium (Al)	32	293	186.2	89.6	310.0
		505	117.2	56.5	195.1
Glass–polyimide (Gl–PI)	40	293	97	41.0	138.0
		505	79.3	33.8	113.0
Polyurethane foam	80	293	20.0	—	—

beam of the same dimensions, which can be understood more clearly from Example 5.1. Mechanical properties of some honeycomb materials are given in Table 5.1.

5.8.2 Flexural Rigidity of Sandwich Beam

The flexural rigidity D of the sandwich beam is the sum of flexural rigidities of the skin D_s and the core D_c, respectively. The flexural rigidity of a beam is expressed by the product of Young's modulus E of the material used and its area moment of inertia I:

$$D = EI = D_s + D_c$$

$$= 2E_sI_s + E_cI_c \quad \text{for two skins} \tag{5.1}$$

where E_s and E_c are the Young's moduli of skin and core, respectively. For a rectangular cross section,

$$I_s = \frac{bt_s^3}{12} + bt_s d^2 \tag{5.2a}$$

and

$$I_c = \frac{bt_s^3}{12} \tag{5.2b}$$

about neutral axis (NA). Here, b is the width of the beam, and

$$d = \frac{t_s + t_c}{2} \tag{5.2c}$$

There is an optimum proportion of the skin and core materials giving maximum stiffness at a given total weight per unit area. According to H.G. Alien, this optimum occurs when combined weight per unit area of skins is 1/3 of the total weight of the sandwich beam.

5.8.3 Historical Application

One of the important applications of sandwich composite was in the aircraft *voyager* which in 1986 completed non-stop flight around the world without refuelling. The core of self-supporting skin, sandwiched between layers of carbon fibre tape impregnated with an epoxy resin, had honeycombed paper-like material threaded with fibres. Its wings were supported by carbon–epoxy tubes. In fact, it was all composite aircraft except its engines and few fasteners.

Example 5.1

A sandwich construction employs two GRP skins each 3 mm thick made from polyester resin in combination with polyurethane foam core 24 mm deep. The construction is in the form of a 100 mm wide beam. Moduli of polyester skin and foam core are 7000 and 20 N/mm², respectively. Calculate (a) the flexural rigidity of the sandwich construction and (b) compare it with the flexural rigidity of 6 mm thick GRP beam of the same width.

Solution: (a) The values given are

$$t_s = 3 \text{ mm}, t_c = 24 \text{ mm and } b = 100 \text{ mm}$$

Hence,

$$d = \frac{t_s + t_c}{2} = \frac{3 + 24}{2} = 13.5 \text{ mm}$$

Values of I_s and I_c are found from Equations 14.2a and b.

$$I_s = 109,800 \text{ mm}^4, I_c = 115,200 \text{ mm}^4$$

Using Equation 14.1, we obtain

$$D = D_s + D_c$$

$$= \left(2 \times 7,000 \frac{N}{mm^2} 109,800 \text{ mm}^4\right) + \left(\frac{20 \text{ N}}{mm^2} \times 115,200 \text{ mm}^4\right)$$

$$= \left(768.6 \times 10^6\right) + \left(2.3 \times 10^6\right)$$

$$= 771.0 \times 10^6 \text{ N mm}^2$$

(b) Now, $t_s = 6$ mm

$$\therefore \quad D_s = E_s I_s = \frac{E_s b t_s^2}{12}$$

$$= \frac{7000 \times 100 \times 6^2}{12} = 12.6 \times 10^6 \text{ N mm}^2$$

On comparing Equations (i) and (ii), we find

$$\frac{D \text{ of sandwich construction}}{D_s \text{ of GRP beam}} = \frac{771 \times 10^6}{12.6 \times 10^6} \approx 61$$

Hence, the flexural rigidity of sandwich beam is 61 times higher.

5.9 Advantages and Limitations of Composites

Besides having favourable effects, composites also suffer from some short-comings and limitations which are discussed in the section that follows:

Advantages

1. They possess a combination of excellent mechanical, chemical, structural, electrical, optical and other desired properties.
2. They are lightweight materials possessing higher specific strength and specific modulus than the conventional materials.
3. Power-by-weight ratio in airplanes is approximately 5 with the conventional materials while it is about 16 with composites. This will

require prime mover of reduced power resulting in fuel economy or more payload-carrying capacity. Alternately, it helps in weight reduction.

4. Composites can be moulded to any shape and size and according to any desired specification.
5. They possess excellent anti-chemical and anti-corrosion properties.
6. Making, repairing and fabricating of composites are easier than in the metals and RCC.
7. Assembling and de-assembling of components is easy and quick.
8. Efficient utilization of material may be done. The fibres may be oriented in such a way so as to provide greatest strength and stiffness in the desired direction.
9. Seepage and weathering problems are negligible.
10. Composites may be designed to obtain aesthetic appearance.

Limitations

1. They have low flash and fire points.
2. They may develop undesired biological effects seen in polymers.
3. Polymeric composites are not suitable for high-temperature applications.
4. The cost of composites is still higher than many conventional materials.
5. On prolonged exposure to sunlight, the colours of composites generally fade out.

5.9.1 Comparison between Conventional and Composite Materials

Table 5.2 compares the properties of conventional and composite materials.

5.10 Various Types of Fibres and Their Aspect Ratio

Fibres are classified in different ways such as the following:

1. Natural and synthetic fibres
2. Organic and inorganic fibres
3. Continuous and short fibres

TABLE 5.2

Properties Showing Comparison between Conventional and Composite Materials

S. No.	Material	Tensile Modulus E (GPa)	Tensile Strength σ_{tu} (MPa)	Specific Gravity ρ	Specific Modulus E/ρ	Specific Strength σ_u/ρ	Nature
1.	Aluminium	70	60	2.7	26.3	22.22	I
2.	HS Al alloy	72	650	2.7	26.3	240.74	I
3.	Cast iron	55	360	7.26	20.9	49.58	I
4.	Glass sheet	70	70	—	—	—	I
5.	Concrete	50	3.5	2.4	20.3	1.45	I
6.	Epoxy	4.5	60	1.9	3.7	50.42	I
7.	Rubber (nature)	0.02	32	—	—	—	I
8.	E-glass–epoxy (V_f=57%) cross-ply	21.5	570	1.97	10.9	0.26	ANI
9.	Kevlar 49–epoxy (V_f=60%) cross-ply	40	650	1.40	29	0.46	ANI
10.	Carbon–epoxy (V_f=58%) cross-ply	83	380	1.54	53.5	0.24	ANI
11.	Mild steel	210	450–830	7.80	26.9	0.058–0.106	I
12.	Pure boron	440	—	—	—	—	I
13.	Boron–Al (40% B) composite	220	—	—	—	—	Particulate
14.	Boron–epoxy cross-ply	106	380	2.0	53	0.19	ANI

Notes: I, isotropic; E, modulus of elasticity; σ_u, ultimate strength; ρ, density; ANI, anisotropic.

Natural fibres are obtained from natural sources such as plants, animals and minerals. Their examples are

- Jute
- Hemp
- Silk
- Sugar cane
- Pineapple
- Cotton
- Flax
- Banana
- Ramie
- Sisal
- Bamboo
- Wood
- Palmyra
- Kenaf
- Coconut
- Oil palm

Synthetic fibres are produced in industries. They are cheaper and more uniform in cross section than the natural fibres. Their diameters vary between 10 and 100 μm. Their examples are the fibres of

- Glass
- Boron
- Carbon
- Graphite
- Kevlar

Organic fibres such as carbon and graphite fibres are light in weight, flexible, elastic and heat-sensitive. Commercial carbon fibres are available by the trade names such as

- Hyfil
- Grafil
- Fortafil
- Thornel
- Polyacrylonitrile (PAN)

Inorganic fibres have HS, low fatigue resistance and good heat resistance. Their examples are

- Glass
- Tungsten
- Ceramic

Continuous and short fibres: The strength of composite increases when it is made of long continuous fibres. A smaller diameter of fibres also enhances the overall strength of composite. Fibres of various cross sections such as square, rectangular, circular, hollow circular, hexagonal and irregular cross section are employed in composite constructions.

Advanced fibres and composites: Composites using Kevlar, graphite and boron fibres are termed as *advanced composites*.

5.10.1 Aspect Ratio of Fibres

Composites may be called as *particulate* composite, *continuous fibre composite* or *short-fibre composite* depending on the *aspect ratio (l/d)* given as follows:

$$\frac{l}{d} \approx 1 \quad \text{for particulate composite}$$

$$\frac{l}{d} \approx 10 - 1000 \quad \text{for short fibre composite}$$

$$\frac{l}{d} \approx \infty \quad \text{for continuous fibre composite}$$

where
  l is the length
  d the diameter or shorter dimension in a non-circular section fibre

Filament-wound continuous fibres are used for fabrication of bodies of revolution.

5.10.2 Glass Fibres

Glass fibres are made by molten *glass drop* through minute orifices and then lengthening them by air jet. The standard glass fibre used in glass-reinforced composite materials is *E-glass*, a borosilicate type of glass. The glass fibres produced, with diameters from 5 to 25 μm, are formed into strands having a tensile strength of 5 GPa. Chopped glass used as a filler material in polymeric resins for moulding consists of glass fibres chopped into very short lengths. *E*-glass is the first glass developed for use as continuous fibres. It is composed of 55% silica, 20% calcium oxide, 15% aluminium oxide and 10% boron oxide. It is the standard grade of glass used in fibreglass and has a tensile strength of about 3.5 GPa and high resistivity; fibre diameters range between 3 and 20 urn *S-glass* was developed for high-tensile-strength applications in the aerospace industry. It is about one-third stronger than *E*-glass and is composed of 65% silicon dioxide, 25% aluminium oxide and 10% magnesium oxide.

Types of glass fibres: Glass fibres classified as *A, E, S,* etc. have particular fields of applications. These are

1. *A*—glass fibre for acid resistance
2. *C*—glass fibre for improved acid resistance
3. *D*—glass fibre for electronic applications
4. *E*—glass fibre for electrical insulation
5. *S*—glass fibre for HS

5.10.3 Boron Fibre

Boron fibres are composites of the substrates tungsten, silica coated with graphite or carbon filaments upon which boron is deposited by a vapour-deposition process (CVD). The final boron fibre has a specific gravity of about 2.6, a diameter between 0.01 and 0.15 mm, a tensile strength of about 3.5 GPa and a tensile modulus of around 415 GPa. Boron is more expensive than graphite and requires expensive equipment to place the fibres in a resin matrix with a high degree of precision.

5.10.4 Carbon and Graphite Fibres

Carbon is a non-metallic element. Black crystalline carbon, known as graphite, has a specific gravity of 2.25. The terms *carbon* and *graphite* are often used interchangeably. However, a line of demarcation has been established between them in terms of modulus and carbon content. Carbon fibre usually has a modulus of less than 345 GPa and a carbon content between 80% and 95%. Graphite fibres have a modulus of over 345 GPa and a carbon content of 99% or greater. Another distinguishing feature is the pyrolyzing temperatures. For carbon, this temperature is around 1315°C, and for graphite, it is around 1900°C–2480°C. *Pyrolysis* is the thermal decomposition of a polymer.

5.10.5 Kevlar Fibre

Kevlar, an organic fibre is an aramid or aromatic polyamide fibre. This organic fibre is melt-spun from a liquid polymer solution. The aromatic ring structure results in high thermal stability. The rodlike nature of the molecules classifies Kevlar as a liquid-crystalline polymer characterized by its ability to form ordered domains in which the stiff, rodlike molecules line up in parallel arrays.

Grades of Kevlar fibres: There are three grades of Kevlar fibres. These are

1. Kevlar 29
2. Kevlar 49
3. Kevlar 149

Kevlar 29 provides high toughness with a tensile strength of about 3.4 GPa for use where resistance to stretch and penetration are important. Kevlar 49 has a high-tensile-strength modulus of 130 GPa and is used with structural composites. Kevlar 149 has an ultrahigh-tensile-strength modulus of 180 GPa.

5.10.6 Ceramic Fibres

The development of both continuous and discontinuous ceramic fibres based on oxide, carbide and nitride compositions was undertaken due to the need for high-temperature reinforcing fibres in composites for the aerospace industry. Most oxide fibres are compositions of Al_2O_3 and SiO_2, although a few are almost pure oxide of aluminium and silicon (alumina and silica). The average properties of continuous oxide fibres are as follows:

Density	$3 g/cm^3$
Diameter	$12 \mu m$
Tensile strength	2 GPa
Tensile modulus	20 GPa
Usable temperature	1300°C

5.10.7 High-Performance Fibres

Included in the high-performance category are fibres based on polyester, nylon, aramid and polyolefin. Fibres such as aramid (Kevlar and Nomex), polybenzimidazole (PBI), Sulfar and Spectra have increased the range of choices for materials engineers in designing materials with tailor-made properties. However, most do not possess the thermal properties found in ceramic- and metal-based fibres. Comparison of properties of different fibres is given in Table 5.3. Their stress–strain curves may be seen in Figure 5.10a through e.

5.10.8 Natural Fibres

Most of the natural leaf, stalk (bast) and seed fibres can be used in filling or reinforcing the thermoplastics. Bast fibres are typically the best for

TABLE 5.3

Comparison of Properties of Reinforcing Fibres

Material of Fibre	Specific Gravity	Tensile Modulus E (GPa)	Tensile Strength σ_u (GPa)	Specific Strength (σ_u/ρ)	Specific Modulus (E/ρ)	Max. Use Temperature (°C)
E-glass	2.6	72.0	3.5	1.18	27.6	350
S-glass	2.48	85.0	4.8	1.94	34.3	300
Boron	2.6	440.0	2.8	1.08	169.2	2000
Carbon (HM)	1.96	517.0	1.86	0.95	264	600
Carbon (HS)	1.8	295.0	5.6	3.11	164	500
Silicon carbide	3.3	427.0	4.0	1.1	129.3	1150
Alumina	3.25	210.0	1.8	0.55	64.6	1250
Kevlar 29	1.44	60.0	2.7	1.87	41.6	—
Kevlar 49	1.45	130.0	2.7	1.86	89.6	—
Ceramic	2.5	152.0	1.72	0.7	60.8	1200
Polyolefin	0.97	117.0	3.0	3.1	120	120
Steel (bulk)	8.0	210.0	0.35–2.1	0.25	26	900
Aluminium alloy (bulk)	2.7	70.0	0.14–0.62	0.23	25.5	250

Source: Bunsell, A.R., Fibre reinforcement-past, present and future, in *Proceeding ICCM 6,* London, U.K.

Notes: σ_u, ultimate strength; σ, density; HM, high modulus; HS, high strength.

improvements in tensile and bending strength and modulus. For toughness, the coarse fibres such as sisal and coir (coconut husk fibre) are best.

Ramie is one of the strongest natural fibres. It exhibits greater strength when wet. It is not as durable as other fibres and, therefore, is usually used as a blend with other fibres such as *cotton* or *wool*. It is stiff and brittle and breaks if folded repeatedly in the same place. It lacks resiliency and is low in elasticity and elongation potential.

Sisal fibre is fairly coarse and inflexible. It is valued for its strength, durability, ability to stretch affinity for certain dyestuffs and resistance to deterioration in saltwater.

Coir fibres are gained from coconut husks. Coir fibres are light in weight, strong and elastic.

Flax is obtained from flax fibre plant. It is commonly known as *patsan* which is a substitute of *jute*. The flax fibre is strong and wiry, longer and finer in nature. Flax fibre is valued for its strength, lustre, durability and moisture absorbency.

Jute is a bast fibre and one of the cheapest natural fibres. It possesses a poor resistance against moisture, brittles under the influence of light and easily absorbs the paint.

Hemp fibre is a bast fibre and is yellow brown in colour. It resembles flax in appearance, but is coarser and harsher. It is strong and lightweight and has very little elongation.

TABLE 5.4

Characteristic Values and Mechanical Properties of Various Natural Fibres

Fibre	Density (g/cm³)	Diameter (μm)	Tensile Strength (MPa)	Young's Modulus (GPa)	Elongation at Break (%)
Ramie	1.55	—	400–938	61.4–128	1.2–3.8
Sisal	1.45	5–200	468–700	9.4–22	3–7
Coir	1.15–1.46	100–460	131–220	4–6	15–40
Flax	1.5	40–600	345–1500	27.6	2.7–3.2
Jute	1.3–1.49	25–200	393–800	13–26.5	1.16–1.5
Hemp	1.47	25–500	690	70	1.6
Cotton	1.5–1.6	12–38	287–800	5.5–12.6	7–8

Source: Mohanty, A.K. et al., *Natural Fibers, Biopolymers and Biocomposites*, Taylor & Francis/CRC Press, Boca Raton, FL, 2005, p. 41, Table 2.1 modified.

Cotton fibre is a seed fibre. The length of its fibres is 8–50 mm, diameter 12–20 μm, and width 16–40 μm. For cotton fibre, the value of elongation at break is 3%–8%, and the water absorption capacity is 20%–100%.

Palmyra fibre is obtained from palm (toddy) plant. It is a bast fibre.

Characteristics and properties of various natural fibres are given in Table 5.4 for a ready reference.

5.11 Configurations of Reinforcing Fibres

The fibres alone are not suitable for any kind of reinforcement. They are likely to be subjected to twisting and abrasion, resulting in wearing out of their reinforcements. Moreover, because of their negligible weight and micro-dimensions, getting a desired reinforcement pattern may be a problem. Hence, fibres are transformed to other forms for reinforcement. Various forms obtained from continuous filaments are shown in Figure 5.4. Usually, the transformed forms are the following:

1. Strands
2. Yarns

5.11.1 Forms of Fibres

Strands: In order to minimize self-abrasion and mechanical damage to the fibres, they are collected, and size (starch) is applied. Then sized fibres are binded together to form a strand. Thus, a strand is a combination of large number of fibres which can be termed as unified multi-fibres.

Yarn: Yarn is produced from sized cake by first twisting the strand then followed by plying a number of twisted strands together to form a doubled

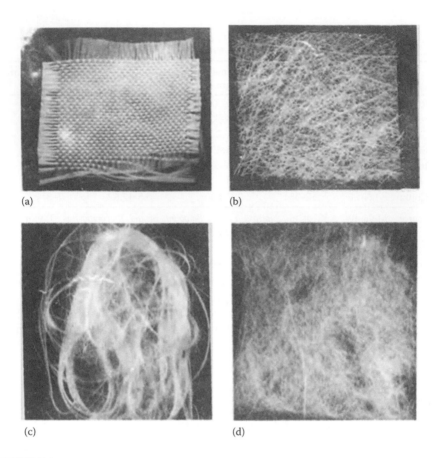

FIGURE 5.4
Forms of reinforcements (a) glass fabric, (b) chopped strand mat, (c) fibreglass roving and (d) fibre wool mat.

balanced yarn. It is designated by its *count* which is expressed in the unit of *Tex*. One Tex is a measure of linear density and is defined as weight in gram per kilometre. Yarns are used for manufacturing woven fabrics.

5.11.2 Forms of End Products

The strands and yarns are then converted into the following forms of end products:

1. Tapes and cloth.
2. Fabrics: Plain weave form, U/D form, square weave form, twill weave form and twill weave form. The satin weave may be of harness type and eight harness type.

3. Rovings: Filament winding, preform, weaving grade, translucent sheeting, chopping grade, woven roving fabrics and chopped strands.

4. Mats: Continuous strand type or swirl mat, weave mat, surfacing mat and needled chopped strand mat.

5. A complex form: These may have arrangements which are a combination of two or more of the aforementioned forms.

Rovings are available in densities of 280, 360 and 600 g/m^2 and widths of 1 and 1.4 m. Mats have densities ranging from 300 to 450 g/m^2.

5.12 Various Matrix Materials

We have already mentioned in art. 5.3 that the various matrix materials are metals, ceramics, polymers, elastomers, etc. A brief detail of them has already been given in Chapter 1. Some more information will now be given about them.

5.12.1 Mylar: A Form of Flake

Mylar, a polyester, is used as thin flake (sheet of 25–150 μm thickness) in magnets and as an insulating material. *Aluminium diboride* is a new fascinating material for planer reinforcement in flaked composites. Planer stiffness of its flake measures 265 GPa which is four times higher than the cross-plied planer reinforced graphite fibres. The flakes have excellent damage tolerance and are easy to process, and holes can be drilled into them for attachments. However, they suffer from limited stretchability under application of pressure.

5.12.2 MMC Composites

An MMC such as SiC in the matrix of aluminium alloy possesses HS and high specific modulus. It is useful as structural components of aircraft and aerospace applications. Another kind of composite composed of silicon (Si) matrix reinforced with SiC fibres has been used as liner (or sleeve) for combustion chamber of an IC engine.

5.12.3 Wood–Plastic Composite

Wood–plastics composite is made by impregnating natural untreated wood with liquid monomers such as methyl methacrylate, acrylonitrile or styrene.

5.13 Mechanics of Composite Laminates

The strengths and moduli of composite laminates are greatly influenced by their fibre orientation. Most laminates are made of several distinct layers of U/D laminae. The mechanics of a structural laminate, therefore, requires adequate knowledge of the properties of individual lamina. The study of properties of U/D lamina and lamina with various fibre orientations has been carried out in the articles that follow.

5.14 Rule of Mixture for Unidirectional Lamina [1]

A typical U/D lamina (Figure 5.5) has parallel fibres embedded in a matrix. The longitudinal, transverse and normal directions are shown by L, T and T' respectively. Consider volumes v_c, v_f and v_m for composite, fibre and matrix, respectively. Let w_c, w_f and w_m be corresponding weights of composite, fibre and matrix, respectively.

5.14.1 Volume Fraction and Weight Fraction [1]

The volume fractions V_f and V_m and weight fractions W_f and W_m of the fibres and matrix, respectively, are defined as

$$V_f = \frac{v_f}{v_c}, \quad V_m = \frac{v_m}{v_c} \tag{5.3}$$

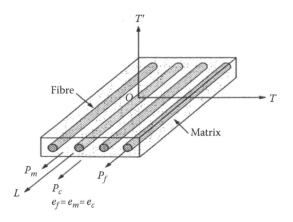

FIGURE 5.5
Representation of a U/D lamina.

where

$$V_f + V_m = 1 \tag{5.4}$$

and

$$W_f = \frac{w_f}{w_c}, \quad W_m = \frac{w_m}{w_c} \tag{5.5}$$

where

$$W_f + W_m = 1 \tag{5.6}$$

In these equations, tin volume of voids v_v is assumed to be zero.

5.14.2 Density of Composite [1]

The density of composite ρ may be obtained in terms of densities of its constituents by simple rule of mixture as

$$\rho_c = \rho_f V_f + \rho_m V_m \tag{5.7}$$

Here, ρ_f and ρ_m are densities of fibre and matrix, respectively.

The conversions between weight fraction and volume fraction in terms of density may be obtained from

$$W_f = \frac{\rho_f}{\rho_c} V_f, \quad W_m = \frac{\rho_m}{\rho_c} V_m \tag{5.8}$$

5.14.3 Load on Composite [1]

The resultant load P_c carried by the composite along longitudinal direction is the sum of loads P_t and P_m shared by the fibres and matrix, respectively. It is expressed by

$$P_c = P_f + P_m \tag{5.9a}$$

This load P_c induces an average stress σ_c in the composite of cross-sectional area A_c. Thus,

$$P_c = \sigma_c A_c = \sigma_f A_f + \sigma_m A_m \tag{5.9b}$$

where
A_f and A_m are the cross-sectional areas of fibres and matrix
σ_f and σ_m are the stresses produced in fibre and matrix, respectively

5.14.4 Longitudinal Strength and Modulus [1]

The longitudinal strength is obtained from Equation 5.9b as

$$\sigma_c = \sigma_f \frac{A_f}{A_c} + \sigma_m \frac{A_m}{A_c}$$

But for U/D composite, the volume fractions may be written as

$$V_f = \frac{A_f}{A_c} \quad \text{and} \quad V_m = \frac{A_m}{A_c}$$

Therefore,

$$\sigma_c = \sigma_f V_f + \sigma_m V_m \tag{5.10}$$

Now, differentiating Equation 5.10 with respect to strain and knowing that $\varepsilon_f = \varepsilon_m = \varepsilon_c = \varepsilon$ (say), we get

$$\frac{d\sigma_c}{d\varepsilon} = \frac{d\sigma_f}{d\varepsilon} V_f + \frac{d\sigma_m}{d\varepsilon} V_m$$

Assuming linear stress–strain curve of the materials, the slope $d\sigma/d\varepsilon$ is a constant and corresponds to elastic modulus. Hence,

$$\frac{d\sigma_c}{d_\varepsilon} = E_c \frac{d\sigma_c}{d_\varepsilon} = E_f \quad \text{and} \quad \frac{d\sigma_m}{d_\varepsilon} = E_m \tag{5.11}$$
$$\text{thus} \quad E_c = E_f V_f + E_m V_m$$

where E_f and E_m are the Young's modulus of fibre and matrix, respectively.

5.14.5 Transverse Strength and Modulus [1]

Tensile properties of composite are obtained using a mathematical model shown in Figure 5.6. Here, the composite is stressed in direction *T*. Let the elongations in composites, fibre and matrix be δ_c, δ_f, and δ_m and the thicknesses t_c, t_f, and t_m, respectively. In this case,

$$\delta_c = \delta_f + \delta_m \tag{5.12a}$$

Therefore,

$$\delta_c = \varepsilon_c t_c, \ \delta_f = \varepsilon_f t_f \text{ and } \delta_m = \varepsilon_m t_m \tag{5.12b}$$

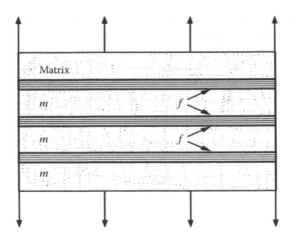

FIGURE 5.6
Mathematical model for predicting transverse properties in U/D composite. (From Gupta, K.M., Starch based composites for packaging applications, in *Handbook of Bioplastics & Biocomposites Engineering Applications*, Dr. S. Pilla (ed.), Scrivener Publishing LLC, Salem, MA, June 2011, Chapter 8, pp. 189–266.)

On putting Equation 5.12b in Equation 5.12a, we get

$$\varepsilon_c t_c = \varepsilon_f t_f + \varepsilon_m t_m$$

$$\varepsilon_c = \varepsilon_f \frac{t_f}{t_c} + e_m \frac{t_m}{t_c} \tag{5.13}$$

$$\text{or} \quad \varepsilon_c = \varepsilon_f V_f + \varepsilon_m V_m$$

where suffices c, f and m represent composite, fibre and matrix, respectively. Knowing that $\varepsilon_c = \sigma_c/E_c$, $\varepsilon_f = \sigma_f/E_f$ and $\varepsilon_m = \sigma_m/E_m$ in elastically deformed composite, Equation 5.13 can be written as

$$\frac{1}{E_c} = \frac{V_f}{V_f} + \frac{V_m}{E_m} \tag{5.14}$$

where E_c is the transverse modulus.

5.14.6 Poisson's Ratio [1]

Based on the rule of mixture of the constituents of U/D composite, the Poisson's ratio in longitudinal–transverse (*LT*) plane may be obtained from

$$\nu_{LT} = \nu_f V_f + \nu_m V_m \tag{5.15a}$$

where v_f and v_m are the Poisson's ratio of fibre and matrix, respectively.
Poisson's ratio in LT' plane can be determined by

$$V_{LT} = v_f V_f + \left[\frac{1 + v_{m-v_{LT'}} \left(E_m / E_c \right)}{1 - v_m^2 + v_m v_{LT'} \left(E_m / E_c \right)} \right] v_m V_m \tag{5.15b}$$

where E_c is the longitudinal modulus. Poisson's ratio in TT' plane may be
obtained from this equation by replacing TT' for LT'

5.14.7 Shear Moduli [1]

The fibres and matrix experience the same shear stress τ_{LT}; hence, the shear
strain γ_{LT} in composite is given by

$$\gamma_{LT} = \gamma_f V_f + \gamma_m V_m \tag{5.16}$$

where

$$\gamma_f = \frac{\tau_{LT}}{G_f}, \quad \gamma_m = \frac{\tau_{LT}}{G_m} \quad \text{and} \quad \gamma_{LT} = \frac{\tau_{LT}}{G_{LT}}$$

Now, the shear modulus GLT of composite may be obtained from

$$\frac{1}{G_{LT}} = \frac{V_f}{G_F} + \frac{V_m}{G_m} \tag{5.17}$$

where G_f and G_m are the shear moduli of fibres and matrix, respectively. The
same equations may be used to approximate G_{LT}.

5.14.8 Range of Poisson's Ratio in Composite Materials

Earlier in Example 2.1, we have proved that Poisson's ratio for isotropic materials has a bound of +0.5 to –1.0. The same is not true for composite materials
which are non-isotropic (i.e. orthotropic or anisotropic). It can be verified by
the following equations given by Wolf (1935):

$$G_{LT} = \frac{E_{CL} E_{CT}}{E_{CL} + E_{CT} (1 + 2v_{LT})} \tag{5.18a}$$

and Gola and Gugliotta (1982)

$$\frac{1}{G_{LT}} = \frac{1}{E_{CT}} + \frac{1}{E_{CT}} + \frac{2V_{LT}}{E_{CL}} \tag{5.18b}$$

Here, $E_{CL} = E_C$ is given by Equation 5.11 and $E_{CT} = E_C$ is given by Equation 5.14. By substituting two extreme values (i) zero and (ii) infinite, for G_{LT} in these equations, it can be established that v_{LT} can be greater than one and less than minus one. In fact, it depends on the ratio of E_{CL}/E_{CT} of the composite. Experimentally observed values, reported by various investigators, are in conformity with the these conclusions. Some of them are given as follows:

Investigator	Reported Value of Poisson's Ratio	Material System of Composite
K.M. Gupta, the author	0.66	In-plane arched glass fibre–epoxy laminate
Dickerson and DiMartino	1.97	Off-axis testing in filamentary materials
Herakovich	−0.21	Pyrolytic graphite–epoxy [±25] laminate
K.M. Gupta, the author	−0.41	Diagonally reinforced glass–epoxy, 12 ply laminate

Example 5.2

Determine the volume ratio of aluminium and boron in aluminium–boron composite which can have the same Young's modulus equal to that of iron. The Young's moduli of aluminium, iron and boron are 71, 210 and 440 GPa, respectively.

Solution: Let v be the volume and suffices a, b and c represent aluminium, boron and composite, respectively. Using Equation 5.11 in which

$$E_c = 210 \text{ GPa, i.e.} = E_{iron}$$

$$E_a = 71 \text{ GPa and } E_b = 440 \text{ GPa} \tag{i}$$

we get $210 = 71\ V_a + 40\ V_b$

Assuming void volume $v_v = 0$ and knowing that

$$V_a + V_b = 1 \tag{ii}$$

$$V_b = 1 - V_a$$

∴ substituting Equation (ii) in Equation (i), we obtain

$$210 = 761 \, V_a + 440 \, (1 - V_a)$$

$$\therefore V_a = 31.8$$

Equation (ii), therefore, yields $V_b = 68.2$:

$$\frac{V_a}{V_b} = \frac{31.8}{68.2} = 0.46$$

Example 5.3

A Kevlar 49–epoxy U/D composite has been constructed with 65% fibre volume fraction. Calculate its (a) longitudinal strength, (b) longitudinal modulus, (c) transverse modulus, (d) Poisson's ratio and (e) shear modulus. Take data from the given table.

Material	Tensile Strength (GPa)	Tensile Modulus (GPa)	Poisson's Ratio	Shear Modulus (GPa)
Kevlar	2.8	130	0.34	1.2
49–epoxy			2.2	
	0.0025	3.5	0.36	

Solution: As $V_f = 65\%$, therefore from Equation 5.4,

$$V_m = 1 - 0.65 = 0.35$$

(a) Longitudinal strength is determined from Equation 5.10 as

$$\sigma_C = (2.8 \times 0.65) + (0.0025 \times 0.35)$$

$$= 9.528 \times 108 \, \text{Pa} = 0.95 \, \text{GPa}$$

(b) Longitudinal modulus is obtained from Equation 5.11 as

$$E_C = (130 \times 0.65) + (3.5 \times 0.35)$$

$$= 58.572 \times 10^{10} \, \text{Pa} = 85.72 \, \text{GPa}$$

(c) Transverse modulus is estimated from Equation 5.14 as

$$\frac{1}{E_c} = \frac{0.65}{130} + \frac{0.35}{3.5}$$

$$= 1.05 \times 10^{-10} \, \text{Pa}$$

$$E_c = 9.52 \, \text{GPa}$$

(d) Equation 5.15a is used to get Poisson's ratio as

$$\nu_{LT} = (0.34 \times 0.65) + (0.36 \times 0.35)$$

$$= 0.347$$

(e) Shear modulus is calculated from Equation 5.17 as

$$\frac{1}{G_{LT}} = \frac{0.65}{2.2} + \frac{0.35}{1.2} = 5.87 \times 10^{-10} \ \text{Pa}$$

$$G_{LT} = 1.7 \ \text{GPa}$$

5.14.9 Modified Rule of Mixture for Non-Unidirectional Composites

Composite laminates are fabricated employing fibre orientations of various configurations. Some important fibre configurations are shown in Figure 5.7a through f. The rule of mixture for U/D composite does not predict them well. It is, therefore, modified to predict strengths and moduli of non-U/D composites by introducing an *efficiency factor* η. This factor η is called Krenchel's efficiency factor after the name of its investigator. Thus, Equations 5.10 and 5.11 may be rewritten as

$$\sigma_c = \eta \ \sigma_f \ V_f + \sigma_m \ V_m \tag{5.19}$$

$$E_c = \eta \ E_f \ V_f + E_m \ V_m \tag{5.20}$$

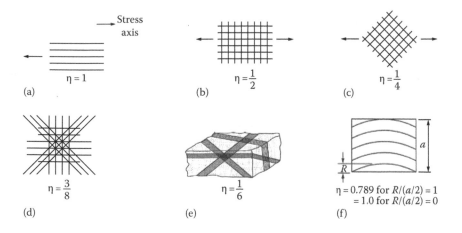

FIGURE 5.7
Composites with different fibre orientations: (a) U/D, (b) cross-plied, (c) angle plied, (d) uniform planer, (e) uniform 3D randomly distributed and (f) arched-in-plane fibres.

TABLE 5.5

Efficiency Factor for Fibre Orientation of Different Kinds

Value of η	Figure Number	Orientation of Fibres
1	Figure 5.7a	U/D
½	Figure 5.7b	Bidirectional (cross-ply)
¼	Figure 5.7c	Angular directional (angle ply)
3/8	Figure 5.7d	Uniform planer distribution
1/6	Figure 5.7e	Uniform 3D random distribution
0.789–1.0[a]	Figure 5.7f	Arching fibre in arched-in-plane fibres composite (AFC) conceived by K.M. Gupta, the author.

[a] Arching efficiency factor.

where values of η for different cases are given in Table 5.5. The value of η is known as *arching efficiency factor,* and may be found by

$$\eta = \frac{1}{\left[1 + \frac{2}{3}\{R/(a/2)\}^2 - \frac{2}{5}\{R/(a/2)\}^4 + \cdots\right]} \qquad (5.21)$$

where
   a is the span of the arch
   R is the rise of the arched fibres

Example 5.4

Construction of a composite employs HS graphite-polyester material system. Content of polyester is 45% in it. Estimate longitudinal strength and modulus for all possible fibre orientations. Consider data from the following table:

Material	Tensile Modulus, (GPa)	Tensile Strength (GPa)	Shear Modulus (GPa)	Poisson's Ratio
HS graphite	240	1.7	1.7	0.28
Polyester	4.5	0.0029	1.4	0.32

Solution: As $V_m = 0.45$, therefore, $V_f = 1 - 0.45 = 0.55$. Longitudinal strength and modulus are calculated using Equations 5.19 and 5.20, respectively. In these equations, Krenchel's efficiency factor η is taken from Table 5.5 for various fibre orientations. The calculated values are given as follows:

For η	Strength (GPa)	Modulus (GPa)
1	0.93	134
1/2	0.46	68
1/4	0.23	35
3/8	0.35	51.5
1/6	0.15	24

TABLE 5.6

Number of Elastic Constants in Composites

Types of Composite Material	Number of Non-Zero Elastic Constants	Number of Independent Elastic Constants
Three-dimensional case		
Anisotropic	36	21
Orthotropic	12	9
Transversely isotropic	12	5
Two-dimensional case		
Anisotropic	9	6
Orthotropic	5	4
Transversely isotropic	5	2

5.14.10 Number of Dependent and Independent Elastic Constants

The number of elastic constants in a highly anisotropic composite may be 81 which reduces to 54 due to symmetry of strains and reduces further to 36 on considering symmetry of stresses. Details of these reductions are omitted as those are beyond the scope of this text. The summary details of non-zero elastic constants and independent elastic constants for different class of materials are illustrated in Table 5.6. Many non-zero elastic constants bear relation among themselves, that is why the number of independent elastic constants are less than the non-zero constants.

5.15 Generalized Hooke's Law and Elastic Constants

Generalized Hooke's law is different from Hooke's law. Isotropic materials obey Hooke's law and non-isotropic materials such as composites follow generalized Hooke's law. Generalized Hooke's law can be expressed mathematically as

$$\sigma_L = A_{11}\varepsilon_L + A_{12}\varepsilon_T + A_{13}\varepsilon_{T'} + A_{14}\gamma_{LT'} + A_{15}\gamma_{TT'} + A_{16}\gamma_{LT}$$

$$\sigma_T = A_{21}\varepsilon_L + A_{22}\varepsilon_T + A_{23}\varepsilon_{T'} + A_{24}\gamma_{LT'} + A_{25}\gamma_{TT'} + A_{26}\gamma_{LT}$$

$$\sigma_{T'} = A_{31}\varepsilon_L + A_{32}\varepsilon_T + A_{33}\varepsilon_{T'} + A_{34}\gamma_{LT} + A_{35}\gamma_{TT'} + A_{36}\gamma_{LT}$$

$$\tau_{LT'} = A_{41}\varepsilon_L + A_{42}\varepsilon_T + A_{43}\varepsilon_{T'} + A_{44}\gamma_{LT'} + A_{45}\gamma_{TT'} + A_{46}\gamma_{LT}$$

$$\tau_{TT'} = A_{51}\varepsilon_L + A_{52}\varepsilon_T + A_{53}\varepsilon_{T'} + A_{54}\gamma_{LT'} + A_{55}\gamma_{TT'} + A_{56}\gamma_{LT}$$

$$\tau_{LT} = A_{61}\varepsilon_L + A_{62}\varepsilon_T + A_{63}\varepsilon_{T'} + A_{64}\gamma_{LT'} + A_{65}\gamma_{TT'} + A_{66}\gamma_{LT}$$

where coefficients $A_{11}, A_{12}, \ldots A_{23}, \ldots A_{44}, \ldots A_{55}, \ldots A_{66}$ are elastic constants. Some of these elastic constants are Young's moduli, some are shear moduli, many are Poisson's ratios, and others are coupling constants.

5.15.1 Different Moduli and Coupling Coefficients

Considering the first row and the first column, we can see that A_{11} is *Young's modulus in longitudinal direction* as $\sigma_L/\varepsilon_L = A_{11}$ is Young's modulus for uniaxial case. Considering the fourth row and the fourth column, we notice that A_{44} *is shear modulus for LT' plane* because $\tau_{LT}/\gamma_{LT} = A_{44}$ which is shear modulus in uniaxial case. Now, consider the third row and the sixth column which shows that a shear strain γ_{LT} is produced in *LT* plane on application of direct tensile stress $\sigma_{T'}$ in *T'*-direction. Here, the effect of direction *T'* has reached to plane *LT* due to coupling effect. Thus, A_{36} is a *coupling coefficient*.

5.15.2 Major and Minor Poisson's Ratio

From the first row and the second column, we observe that a longitudinal stress σ_L causes a transverse strain ε_T. Similarly, the first column of the second row reveals that stress in transverse direction σ_T produces a longitudinal strain ε_L. Thus, A_{12} and A_{21} are Poisson's ratios. Of these, A_{12} is called the major Poisson's ratio and A_{21} is known as the minor Poisson's ratio. The numerical value of the major Poisson's ratio may be less than that of the minor Poisson's ratio.

5.16 Applications of Composite Materials

Fields of applications: Nowadays, composites are used almost everywhere. Some important fields of applications are as follows:

1. Space vehicles and satellites
2. Aircrafts
3. Rockets
4. Automobiles
5. Pressure vessels and heat exchangers
6. Sports, music and amusements
7. Building constructions
8. Machine components
9. Electronic and computer components

Notable applications: Some notable applications where composites have been used are given as follows:

1. The roof of Montreal Olympic Stadium was built with Kevlar fabric-reinforced composite.
2. The air-conditioned, Nagar Mahapalika's underground market of London is fully made of composites.
3. Boeing 747 contains 929 m^2 surface area made of composites.
4. Fighter airplane F-18 has about 10% structure made of graphite–epoxy (Gr–Ep) composite that results in a weight saving of about 25%.
5. Moon-landing mission Apollo 15, 16 and 17 had tubular drill-bore stem made of boron–epoxy (B–Ep) composite.
6. Turbine and engine shafts, discs, and bearings are made of carbon fibre composites.
7. Satellites employ composites to work in the temperature range of −160°C to +95°C.
8. Voyager spacecraft used 3.7 m diameter Gr–Ep flight antenna.
9. Commercial acoustic guitars are made of graphite–epoxy composites.
10. Natural gas vehicle fuel cylinder and fibre-reinforced aluminium pistons are used in Toyota cars.
11. Printed circuit boards, switchgears, body of computers, etc. are made of composites.
12. G-10 is a laminated thermosetting glass fibre composite used in magnets and cryostats as a HS material where metals are unsuitable.
13. Carbon-fibre-reinforced composites are used in biomedical applications such as *total hip replacement* and *fracture fixation*.
14. Electrorheological fluids like zeolite in silicone oil and starch in corn oil are used to fill in the graphite–epoxy cantilever beam. They give damping and variable stiffness to beams.

The usage of composites in a typical aircraft and satellite are shown in Figures 5.8 and 5.9, respectively.

5.17 Stress–Strain Behaviour of Fibres, Matrix and Composites

It has already been described that composites possess the nature of orthotropy and anisotropy. They can be homogeneous or heterogeneous and obey generalized Hooke's law. The stress–strain behaviour of various composites and of the fibres impregnated in them are discussed here. These are shown

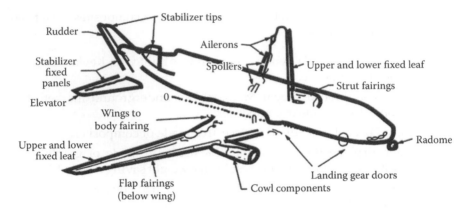

FIGURE 5.8
Composite components in a typical aircraft.

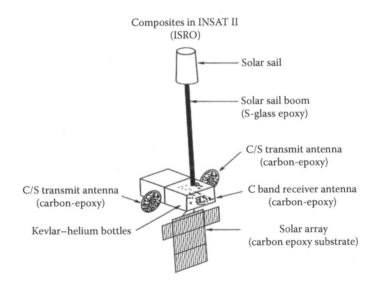

FIGURE 5.9
Composite components in a typical satellite.

in Figure 5.10a through e. The tensile stress–strain curve of various fibres in longitudinal direction, shown in Figure 5.10a, is linear. HM graphite fibre is stiffest among them. Figure 5.10b depicts tensile longitudinal behaviour of U/D boron–epoxy and graphite–epoxy composites. Based on the volume fraction of fibre in the matrix of linear or non-linear nature, the curves may assume the shapes as shown in Figure 5.10c through d. The stress–strain curve of a laminate is shown in Figure 5.10e in which the successive failures of different plies are marked by 1, 2, 3,..., etc.

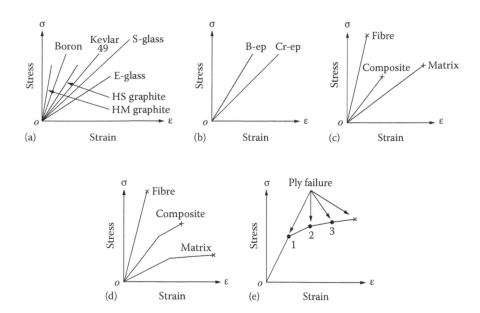

FIGURE 5.10
Stress–strain curves of (a) various fibres, (b) U/D composite, (c) linear matrix composite, (d) nonlinear matrix composite and (e) laminate.

5.18 Basic Composite Manufacturing Methods

Polymeric composites can be manufactured by various methods. These methods are different for thermoset composites and thermoplast composites. These are listed in the following:

Manufacturing processes for thermoset composites

1. Prepreg lay-up process or autoclave processing or vacuum bagging process
2. Hand lay-up (or wet lay-up) process
3. Spray-up process
4. Filament winding process
5. Pultrusion process
6. Resin transfer moulding process
7. Structural reaction injection moulding (SRIM) process
8. Compression moulding process
9. Roll wrapping process
10. Injection moulding process

Manufacturing processes for thermoplastic composites

1. Thermoplastic tape winding process
2. Thermoplastic pultrusion process
3. Compression moulding of glass mat thermoplastic (GMT)
4. Hot-press technique
5. Autoclave processing
6. Diaphragm-forming process
7. Injection-moulding process

A brief description of some of these processes is given in subsequent sections. Other processes can be referred to in any related text/reference.

5.18.1 Prepreg Lay-Up Process

This is a very common process in aerospace industry. Complicated shapes having very high fibre volume fractions V_f can be produced by this process. This is an open-mould process with low-volume capability. During manufacturing, the prepregs are cut, laid down in desired fibre orientation and then vacuum bagged. Thereafter, the composite is kept inside an autoclave or even along with the mould. After that, the pressure and heat are applied for curing and consolidation of the part to be manufactured:

- Radomes
- Wing structures
- Yacht parts
- Sports goods
- Landing gear door

Merits, demerits and applications: It is an intensive laborious process in which the labour cost is about 50–100 times greater than the pultrusion, filament winding and other high-volume processes. However, it is advantageous for small quantity manufacturing of huge parts. Its major applications are in the manufacturing of the following items.

5.18.2 Wet Lay-Up (or Hand Lay-Up) Process

This process is employed in fabricating the components of marine industry and for other prototype parts. Similar to prepreg lay-up process, this is also an open-mould-type process. In its working, the resin in liquid form is applied on to the mould, and the reinforcement is placed over it. A roller is pressed and rolled over the resin reinforcement mix to impregnate the fibres with resin. Another layer of resin and reinforcement is applied on the first layer. In this way, the stacking of layers is continued until the laminate of suitable thickness is built-up. Different types of fibres, mats and fabric reinforcement can be employed in this process.

Merits, demerits and applications: This is an easier and less expensive, but laborious process. Since the entire manufacturing process is performed manually, therefore, this is also called *hand lay-up process*. Major use of this process may be found in fabrication of the following items:

- Windmill blades
- Storage tanks
- Swimming pools
- Sports boat

5.18.3 Thermoplastic Pultrusion Process

This process is similar to the pultrusion process used for thermoset composites. In it, the thermoplastic prepregs or the fibres are pulled through a die to get the final product. Since the thermoplastic resins are of high viscosity, the processing demands for a greater pulling force.

Merits, demerits and applications: Thermoplastic pultrusion suffers from a drawback that it provides an inferior surface quality than the thermoset pultrusion. Major uses of this process are in the fabrication of the following items:

- Angles and channels section
- Circular tubes
- Rods and squares
- Rectangular bars
- Strips
- Beams
- Flooring
- Handrails
- Ladders
- Light poles
- Electrical enclosures
- Walkways

5.18.4 Comparison of Various Manufacturing Processes

The raw material used, shape and size of the products that can be manufactured, strength, production speed and cost, etc. for various processes are compared in Table 5.7.

TABLE 5.7

Comparison among Various Manufacturing Processes

S. No.	Process	Raw Material	Shape	Size	Strength	Cost	Production Speed
1.	Hand lay-up	Prepreg and fabric with epoxy resin	Simple to complex	Small to large	High	High	Slow
2.	Wet lay-up	Fabric/mat with polyester and epoxy resins	Simple to complex	Medium to large	Medium	Medium	Slow to high
3.	Spray up	Short fibre with catalyzed resin	Simple to complex	Small to medium	Low	Low	Medium to fast
4.	Pultrusion	Continuous fibres, usually with polyester and vinylester resins	Constant cross section	No restriction on length, small to medium size cross section	High (along longitudinal direction)	Low to medium	Fast
5.	Filament winding	Continuous fibres with epoxy and polyester resins	Cylindrical and axisymmetric	Small to large	High	Low to high	Slow to fast
6.	Roll wrapping	Prepregs	Tubular	Small to medium	High	Low to medium	Medium to fast
7.	Compression moulding	Moulded compound, e.g. SMC, BMC	Simple to complex	Small to medium	Medium	Low	Fast
8.	Injection moulding	Pallets (short fibre with thermoplastic)	Complex	Small	Low to medium	Low	Fast
9.	Stamping	Fabric impregnated with thermoplastic tape	Simple to contoured	Medium	Medium	Medium	Fast
10.	SRIM[a]	Fabric or preform with polyisocyanurate resin	Simple to complex	Small to medium	Medium	Low	Fast
11.	RTM[b]	Preform and fabric with vinylester and epoxy	Simple to complex	Small to medium	Medium	Low to medium	Medium

[a] SRIM means structural reaction injection moulding.
[b] RTM means resin transfer moulding.

Review Questions

5.1   What are agglomerated composites? Name some of them. How do they differ from laminated and reinforced composites?

5.2   Write notes on the following and illustrate their applications:

(a) Cermet

(b) Cladded metals

(c) SMC

(d) BMC

5.3   Classify reinforced composites exhaustively. Why are they known as composites?

5.4   Compare between the following:

(a) Hybrid and non-hybrid composites

(b) Flake and particulate composites

(c) Sandwich and off-axis composites

(d) Whisker-reinforced and fibre-reinforced composites

5.5   Justify that a sandwich composite is much lighter than the conventional composite and this lightweightness is an advantageous condition.

5.6   What is a composite? What do the properties of composite materials depend upon?

5.7   Discuss the advantages and limitations of FRP. Name different types of matrices and fibres used in their fabrications.

5.8   Classify fibres from different points of views. How do they affect the quality of composites?

5.9   What is a fibre-reinforced composite? What fibre-reinforcing materials are commonly used?

5.10   Sketch and explain various reinforcing patterns of fibres being used in composites. How is *strand* different from *yarn*?

5.11   Define *Tex* of a fibre. Compare properties of (a) E-glass and S-glass fibres, (b) HM and HS carbon fibres, and (c) Kevlar and boron fibres.

5.12   Enumerate various matrix materials. Discuss the suitability of thermosetting and thermoplast plastics as matrix materials for the composites.

5.13   Derive *rule of mixture* to obtain (a) longitudinal strength of a composite, (b) transverse modulus, (c) shear modulus in LT plane, and (d) Poisson's ratio in LT plane. Write the assumptions made therein.

5.14   How do the strengths and moduli are influenced in U/D composite, cross-plied composite and angle-plied composite?

5.15 Enlist various applications of composites. Also decide the suitable composites for use in the following:

(a) Offshore structures

(b) Desalination plant

(c) Pole in pole vault

(d) Automotive car body

(e) Television body

(f) Body of drilling machine

(g) Railway coach seats

(h) Impeller of centrifugal pump handling drinking water

Elaborate reasons for such selections.

5.16 Draw stress–strain curves for (a) various fibres, (b) various composites, and (c) different matrices.

5.17 Describe GRP composite. Explain its details in the light of fabrication of a helmet. How do glass fibres A, D, S and E differ from each other?

5.18 Explain why is the resin-mixed glass fibres stronger than the resin itself? How can you achieve 3D strengths in a fibrous composite?

5.19 State the various composite manufacturing methods and describe any one of them in detail. Write their major applications also.

5.20 Describe the prepreg lay-up manufacturing process for thermoset composites. Discuss its merits and demerits also.

5.21 Explain the pultrusion process of manufacturing of thermoplastic composites. Write its favourable and unfavourable features. Also write its applications.

Numerical Problems

5.1 Elastic moduli of glass and epoxy are 72 and 3.6 GN/m^2, respectively. A Gl–Ep composite is made up of (a) 10% fibre by volume and (b) 50% fibre by volume. Obtain the fraction of load carried by the epoxy in the aforementioned cases.

[**Answer:** (a) 0.31, (b) 0.05]

5.2 In the sandwich construction of Example 14.1, the densities of GRP and the core are 1.47 and 0.08 g/cm^3, respectively. Optimize the dimensions of GRP skins and the core for maximum stiffness.

[**Answer:** $t_s = 1.22$ mm, $t_c = 89.5$ mm]

5.3 Volume fractions of constituents in a GRP composite were determined by conduct of burnout tests. The following observations were made on electronic digital balance:

Weight of empty crucible = 36.8506 g

Weight of crucible and a piece of composite material = 40.2781 g

Weight of crucible and glass fibres after the plastic is burnt out = 38.9872 g

Calculate the weight and volume fractions of glass fibres and epoxy. Take the densities of fibres and epoxy as 2500 and 1200 kg/m³, respectively.

[**Answer:** $W_f = 62.3\%$, $W_{ep} = 37.7\%$, $V_f = 47\%$, $V_{ep} = 39\%$]

5.4 Calculate density of the composite of problem 3. [**Answer:** 1970 kg/m³]

5.5 A tensile load of 100 N is applied to an aluminium–boron composite of 1 mm² cross-sectional area. The volume of the parallel fibres is 30%. What is the stress in the fibres, when the load axis is (a) parallel to the fibres and (b) perpendicular to the fibres? Young's modulus for aluminium and boron are 71 and 440 GN m-2, respectively.

5.6 A GRP composite 20 mm wide and 6 mm thick is made up of 40% fibre volume fraction. It is subjected to a tensile load of 7.5 kN. The tensile moduli of fibre and plastics are 72 and 300 MPa, respectively. Calculate (a) the ratio of stress produced in fibre to that produced in plastics, (b) the stress produced in plastics and (c) the cross-sectional area of plastics alone without reinforcement. The plastics alone sustain the same load and the same elastic deformation.

[**Answer:** (a) 240, (b) 645 kPa, (c) 11627.9 mm²]

5.7 A GRP laminate is fabricated using 65% glass fibre by volume. The Young's moduli of fibre and epoxy are 72.4 GPa and 13.8 GPa, respectively. The fracture stress of fibre is 3.4 GPa and the tensile strength of epoxy is 62 MPa. Determine (a) longitudinal modulus of composite laminate, (b) tensile strength of laminate and (c) the fraction of load taken by the fibre.

[**Answer:** (a) 51.9 GPa, (b) 2.2 GPa, (c) 91%]

5.8 Calculate density and tensile longitudinal modulus for a U/D composite having (a) glass fibre–polyester resin, $V_f = 50\%$; (b) carbon fibre–epoxy resin, $V_m = 50\%$; and (c) ferro-cement concrete, $V_f = 2\%$. Use materials properties given as follows:

Material	Young's Modulus (GPa)	Density (kg/m³)
Glass fibre	72	2550
Polyester	3	1150
Carbon fibre	390	1900
Epoxy resin	3	1150
Steel	200	7900
Cement concrete	45	2400

[**Answer:** (a) 1850 kg/m³, 37.5 GPa; (b) 1530 kg/m³, 197 GPa; (c) 2510 kg/m³, 48.1 GPa]

5.9 Flat metal sheets of uniform thickness are glued together with epoxy, also of uniform thickness, to make a flaked-type composite. Calculate the ratio of maximum modulus to minimum modulus of composite in terms of Young's modulus of metal E_{me}, Young's modulus of epoxy E_{ep} and volume fraction of metal V_{me}. For which value of V_{me} the largest ratio may be expected? $E_{me} > E_{ep}$.
[**Answer:** $V_{me} = 50\%$]

5.10 Consider data of Kevlar 49–epoxy material system given in Example 14.4, and determine (a) fibre volume fraction of a composite having uniform planer distribution of fibres. The strength of composite is 357.8 MPa and (b) longitudinal modulus of this composite.
[**Answer:** (a) $V_f = 65\%$, (b) 32.9 GPa]

5.11 Determine (a) shear modulus, (b) Poisson's ratio and (c) transverse modulus of a HS graphite–polyester U/D composite having fibre volume fraction 55%. Take required data from Example 14.5.
[**Answer:** (a) 1.55 GPa, (b) 0.35, (c) 9.77 GPa]

5.12 During experiments on a U/D E-glass–Ep composite, the following results are obtained:
Longitudinal strength = 1996 MPa
Poisson's ratio = 0.297
Shear modulus = 2.397 GPa
Transverse modulus = 7.646 GPa
Some properties of Gl–Ep material system are given as follows, and some are misplaced by the investigator. Help him or her to know the misplaced values marked (×) cross in the table.

Material	Tensile Strength σ (MPa)	Longitudinal Modulus E (GPa)	Poisson's Ratio ν	Shear Modulus G (GPa)
E-glass–epoxy	3500×	72	0.25	× 1.2
		×	0.36	

[**Answer:** $E_m = 3.5$ GPa, $G_f = 9.7$ GPa, $V_f = 57\%$, $V_n = 43\%$, $\sigma_m = 2500$ MPa]

5.13 Use data of problem 11, and calculate (a) strength of a cross-plied lamina, (b) longitudinal modulus of composite having 3D random distribution of fibre and (c) strength and modulus of arching fibre composite having rise to span ratio of 0.125.
[**Answer:** (a) 998.5 MPa, (b) 8.34 GPa]

References

1. K.M. Gupta, Chapter 8 Starch based composites for packaging applications. In Dr. S. Pilla (ed.), *Handbook of Bioplastics & Biocomposites Engineering Applications*, Scrivener Publishing LLC, Salem, MA, June 2011, pp. 189–266.
2. A.R. Bunsell, Fibre Reinforcement-Past, Present and Future, in F.L. Matthews (ed.), *Proceeding ICCM 6*, London, U.K., 1987.
3. A.K. Mohanty, M. Misra, and L.T. Drazal, *Natural Fibers, Biopolymers, and Biocomposites*, Taylor & Francis/CRC Press, Boca Raton, FL, 2005, p. 41.

6

Sandwich Composite Materials, and Stitched and Unstitched Laminates

6.1 Introduction

Sandwich composite systems consist of advantages of miscellaneous materials having low density, high strength and high flexural resistance. They also possess the abilities of high-energy absorption and high load damping, sound and vibration insulation and adaptive qualities. Nowadays, these materials are used in innovative research and developments. Sandwich composite materials use a homogeneous or inhomogeneous core of soft foams and other hard materials. Such composite constructions are widely used in aircrafts, marines, spacecrafts, rails, automobiles and structural parts. Different newer materials are being tried for making cores, honeycombs and skin sheets, and also the scope of applications is widening day by day. Sandwich composite (Figure 6.1) is a multilayered material made by bonding the lightweight core material in between the two high-strength fibre-reinforced laminated facings. These are used to fabricate the structural and machine components for large space stations, aircrafts, robot arms, underwater structures, marine uses, etc. In these applications, the sandwich constructions are in the form of beam element, different types of panels, rectangular and circular plates, shallow arch, etc.

6.2 Types of Sandwich Core Materials

Core materials are normally low-strength materials, but their larger higher depth provides the sandwich construction with higher bending stiffness. Various types of homogeneous and non-homogeneous core materials are displayed in Figure 6.2.

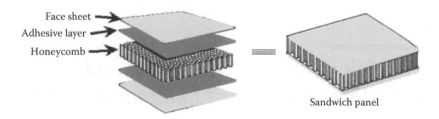

FIGURE 6.1
Sandwich composite structure.

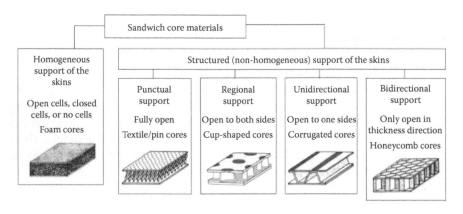

FIGURE 6.2
Various types of cores and their materials.

Structurally, the core serves the following two purposes:

1. It separates the faces and resists deformation in the direction perpendicular to the face plane.
2. It provides certain degree of shear rigidity along the planes that are
   perpendicular to the faces.

A large variety of materials are suitable for use as sandwich cores. Broadly,
they may be put in the following categories:

1. Wood
2. Honeycomb
3. Foam

Among different kinds of woods, the end-grain Balsa wood having elongated
vertical cell structure is a better one. It provides high compressive resistance

to crushing but absorbs a large quantity of resin during lamination. Its density is also high, which is about 100 g/m^3 or more.

6.2.1 Honeycombs

Honeycomb cores (Figure 6.3) are made up of the following main materials:

1. Cardboard honeycomb
2. Aluminium
3. Thermoplastic

Their properties are given in Table 6.1.

6.2.2 Foams

Different kinds of foam materials in current uses are the following:

1. *Cross-linked PVC foams.* This is based on the combination of thermoplastic *PVC* and cross- linked thermoset *polyurea.* Their examples are
   a. Airlite
   b. Klegecell
   c. Divinycell

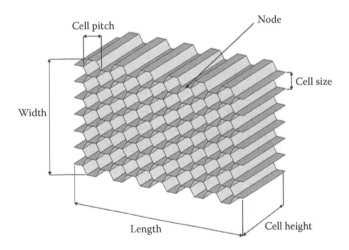

FIGURE 6.3
Honeycomb. (From Schlotter, M., A comparison of core materials for sandwich composite constructions, June 2002.)

TABLE 6.1

Characteristics of Different Honeycomb Materials

Description	Nomex (Cardboard) Paper Honeycomb	Thermoplastic Honeycomb	Aluminium Honeycomb
Basic material	Nomex paper which is based on Kevlar	Aramid, PP, ABS, polyethylene	Aluminium
Density (kg/m³)	80	75	72
Mechanical properties	Very good	Lower	Very good
Compressive modulus (MPa)	250	105	1034
Cost	Expensive	—	Cheaper
Major problem	—	Difficulty in care and skin bonding	Corrosion, delamination
Main applications	Airspace	Interior panelling, floors	Marine application

Source: Schlotter, M., A comparison of core materials for sandwich composite constructions, June 2002.

Main characteristics of these foams are as follows:

  a. Higher strength and stiffness under both static and dynamic situations

  b. Good fatigue resistance

  c. Low absorption of resin and water

  d. Closed cell structure

2. *Styrene acrylonitrile (SAN)–PVC foams.* This is based on thermoplastic SAN and PVC. Its example is corecell.

Main characteristics of these foams are as follows:

  a. Higher impact resistance

  b. Good fatigue resistance

  c. High toughness and good absorption of resin and water

  d. Closed cell structure

3. *Polyurethane (PUR)–polyisocyanurate (PIR) foams.* This is based on the combination of PUR and PIR foams. Their examples are

  a. Rohacell

  b. Airex

Main characteristics of these foams are as follows:

  a. Good compressive strength.

  b. They deteriorate with time, that is, have creeping effect.

  c. Not suitable for stressed panels.

  d. Costly and absorb water.

  e. Are used in aircraft and marine applications.

6.3 Types of Face (Skin) Materials for Sandwich Constructions and Their Characteristics

Face materials in a sandwich construction carry the major applied load. Besides being used in conventional applications, they also serve the special functions in aircraft, transportation and covering fields. They are, therefore, made in special forms such as

1. Profile for proper aerodynamic smoothness
2. Rough non-skidding surface
3. Tough but wear-resistant covering for the floor

Sandwich facing materials, their main use and their specialty are given as follows (Table 6.2).

6.4 Sandwich Composite in Special Applications

Sandwich composites are widely used in many special applications. Depending on the type of applications, they possess different qualities. Accordingly, they are of various grades such as described in the following.

6.4.1 Spacecraft Grade Sandwich Composites

Functional sandwich composites integrate the quite separate objectives of radiation shielding, structural integrity, damage tolerance, thermal

TABLE 6.2

Main Use and Specialty of Different Honeycomb Materials

Facing Materials	Main Use	Specialty
Magnesium alloys	Experimental stage	Very light in weight
Aluminium alloys	Structural, aerospace	Light in weight
Steel alloys	Structural, aerospace	High strength and hard
Cobalt-based alloys	In moderately stressed applications	Work at about 550°C–1000°C
Nickel-based alloys	Heat-resistant sandwich	Work at about 650°C–825°C
Titanium alloys	Moderately high-temperature applications	High strength-to-weight ratio
Glass fibre–reinforced plastics	Airframe applications	Quite low absorption of moisture

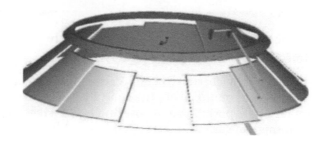

FIGURE 6.4
A cone as used in Ariane for spacecraft. (From Schwingel, D., *J. Acta Astronautica*, 61, 326, 2007.)

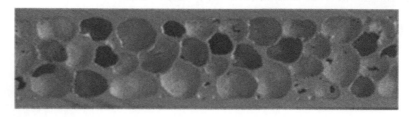

FIGURE 6.5
AFS. (From Schwingel, D., *J. Acta Astronoutica*, 61, 326, 2007.)

insulation and debris/micrometeoroid shielding into a viable structural design. Due to these properties, multifunctional sandwich composite configurations may be used to maximize the efficiency of advanced material usage in future structural spacecraft applications. Typical components of spacecraft structures are tubular truss structures, face sheets for the payload bay door, antenna reflectors, etc. Aluminium foam sandwich (AFS) is used in space components. An example of the component made by AFS is a cone as used in Ariane for spacecraft. It is shown in Figure 6.4. AFS consists of two external aluminium sheets and an internal layer of aluminium foam in between as shown in Figure 6.5 [2].

6.4.2 Marine Grade Sandwich Composites

Low-cost, high-quality sandwich composites are widely used in ships and submarines as primary and secondary load-bearing structures, such as the following:

- Foundations
- Deckhouses
- Hulls

- Propellers
- Mast systems

The use of sandwich composite materials aboard naval vessels is based on the need to increase range, covertness and stability, while reducing weight, fuel consumption and life-cycle costs. Fire resistance is an important issue in marine grade sandwich structures because the core materials tend to degrade rapidly after the degradation of the face sheet when exposed to fire.

6.4.3 Aircraft Grade Sandwich Composites

Sandwich structures have long been recognized as one of the most weight-efficient plate or shell constructions for resisting bending loads. The aerospace industry requires bending stiffness-dominated structures, which have low weight also. Some of the aircraft components made by sandwich composite are shown in Figure 6.6.

A large aircraft structure such as wings and fuselage (Figure 6.7) sections (wing box may be nearly 30 m long and a fuselage section 6 m in diameter and 12 m long) is a complex undertaking. These parts are made of flat or mildly curved monolithic laminates or sandwich panels. A large carbon-fibre fuselage is made up of several subcomponents such as skin panels, stringers, frames and bonds.

It is designed with numerous openings for doors and windows, local reinforcements for point loads and stress concentrations and varying skin thicknesses that are tailored to resist aerodynamic loading, internal pressure or bird strikes. The aircraft grade sandwich composite is suitable for that.

6.4.4 Automobile Grade Sandwich Composites

Sandwich composite is widely used in making automobile parts such as headlight frame and door panel (Figure 6.8). The monologue chassis of a Formula 1 race car is a sandwich structure, made of high-performance carbon-epoxy composite face sheets and an aluminium honeycomb core. High-modulus and high-strength composites, with aerospace-class toughened epoxy resins, are used to obtain the maximum ratio of safety performance to weight ratio.

6.5 Current Fields of Research in Sandwich Composites/Constructions

Although the properties and behaviour of sandwich composite constructions are well explored and many new materials are developed, yet current researches have to produce cheaper sandwich constructions with

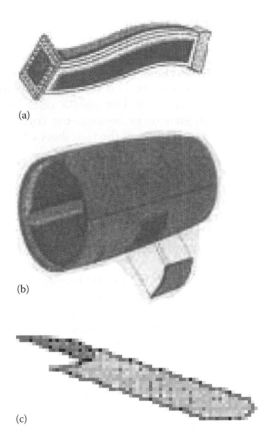

(a)

(b)

(c)

FIGURE 6.6
Aircraft components made by sandwich composite. (a) S-duct, (b) nacelle and (c) leading edge radome.

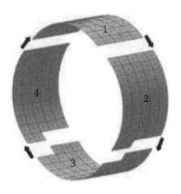

FIGURE 6.7
Aircraft fuselages made by sandwich panels numbered 1, 2, 3 and 4.

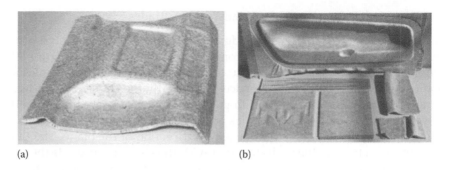

(a)   (b)

FIGURE 6.8
Parts of automobile made by sandwich composite. (a) Head light panel and (b) door panel.

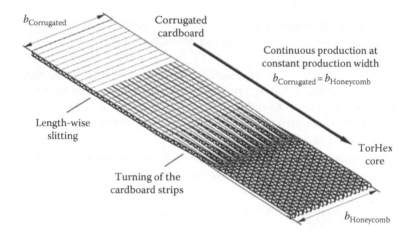

FIGURE 6.9
Folded honeycomb. (From Schlotter, M., A comparison of core materials for sandwich composite constructions, June 2002.)

more improved properties. In this connection, some recent findings are as follows:

- Production of *folded honeycomb* [1] (Figure 6.9)
- Creation of *hybrid foams* to combine the positive properties of existing foams. Toughened PVC and SAN foams may be used to form such hybrid foam.
- Development of partially/fully closed honeycombs that minimize the bonding problems of face skin–core interface.

6.5.1 Sandwich Composites in Wind Turbine Blades

In recent years, the wind energy industry has emerged as the fastest growing market segment for sandwich composites with the blades (Figure 6.10) being the most significant composite component. Sandwich composites are also used in the spinner and the nacelle. Typically the shells use sandwich composites in the leading and trailing edges, whereas the area between the two stiffeners is made up of a thick monolithic construction of primarily unidirectional fibres oriented in the length direction of the blade. The thick monolithic sections of the shells together with the stiffeners provide the stiffness to a rotor blade, whereas the purpose of the sandwich sections is to provide the correct aerodynamic shape and to transfer loads to the monolithic sections [3].

6.5.2 Custom Sandwich Composite for Paddle Surfboard

Custom sandwich composites are basically used for paddle surfboard (Figure 6.11). It consists of the following constitutes [4].

1. *Expanded polystyrene (EPS) foam core*: EPS foam is a very-low-density (as low as 138.4–415.20 kg/m³) lightweight foam. In this application, it is basically used to hold the inner shape of the board while the rest of the strength supplying higher-density foam, and fibreglass is constructed around the core.
2. *Inner fibreglass layer*: This layer of selected weights of fibreglass cloth is the first (from inside out) structural wall that makes up the sandwich.
3. *PVC D-cell wrap*: PVC D-cell is a high-density closed-cell foam (upward of 2767.99 kg/m³) that makes up the main strength of the board. (Regular kite boards are entirely shaped out of this type of foam.)
4. *Outer fibreglass layer*: It is not only the outer skin of the construction but is the second element of sandwich process.

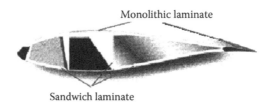

Monolithic laminate

Sandwich laminate

FIGURE 6.10
Cross section of a typical rotor blade. (From Norlin, P., *Reinforced Plast.*, March 2002.)

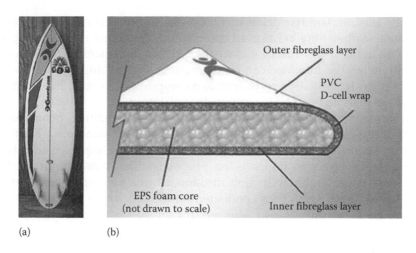

(a)   (b)

FIGURE 6.11
(a) Paddle surfboards. (b) Custom composite sandwich construction.

6.6 Recent Advances

Besides these conventional methods, some advanced techniques are adopted on steel. These are

1. Ion nitriding
2. Plasma carburizing
3. Electron beam hardening
4. Laser hardening

Details on these advanced techniques may be referred in related texts.

6.7 Experimental Studies on Mechanical Behaviour of Stitched and Unstitched Glass/Epoxy Fibre–Reinforced Laminates [5]

Glass/epoxy fibre–reinforced composites have been extensively used in the field of space, aircraft, automotive and sports industries, owing to their high specific strength and stiffness. One of the limitations of these materials is their poor ability to resist impact damage and delamination due to brittleness of the epoxy matrix material. Many investigations have been carried out to study the interlaminar delamination of the thermosetting composites, in particular, carbon fibre–reinforced epoxies and glass fibre–reinforced epoxy. It was found that the interface is weaker. Hence, by stitching the laminate in z-direction, the interface can be strengthened. Stitching can be used to improve the delamination resistance of a composite structure. It is a unique technique that can be used for composite reinforcement.

The research work presented here studied the mechanical behaviour of stitched and unstitched glass/epoxy fibre–reinforced laminates. From the results, various possibilities in structuring the laminate sequence and stitching the fabric layers to improve the properties like interlaminar fracture toughness, considerable reduction in damage area for impact applications were investigated. From tests conducted on the specimens under buckling, delamination and impact loads showed improvements in out-of-plane properties by stitching the laminate in z-direction.

6.8 Literature Review [6–21]

Several research scholars have taken up the study of composite structures, and some of the works pertaining to composite laminates are widely used in industries like defence industries, automobile industries, aerospace and marine.

6.8.1 Tensile and Flexural Testing Review

The influence of stitch bonding on the tensile strength and tensile modulus in plain woven Twaron T-750/vinyl ester composites in the direction of thickness was examined using lock stitch by Mehmet Karahan et al. (2010) [6].

Mouritz (1996) [7] studied the flexural properties of stitched GRP laminates using four-point bending and found out that the flexural properties of GRP laminates were reduced when stitched through the thickness with Kevlar. This reduction was partly caused by the damage introduced during the stitching process, including breakage of glass fibres and formation of polymer-rich regions at the stitch holes.

6.8.2 Mode I Fracture for Double Cantilever Beam and Width-Tapered Double Cantilever Beam

Mouritz et al. (1997) [8] have made extensive reviews of the research works on the effect of through-thickness stitching on the in-plane mechanical properties like tensile, compressive, flexure, interlaminar shear, creep, fracture and fatigue properties for fibre-reinforced polymer composites. Their reviews were compared with one another in order to find out if any similarities or differences could be used as engineering design rules to specify various strength of a laminate. They have also discussed the implications of the findings on the use of stitched composites in lightweight engineering structures.

The effect of fibre orientation on mode I interlaminar fracture toughness of glass/epoxy composites using the double cantilever beam (DCB) test was studied by Shetty et al. (1999) [9]. They found out that the 90°/90° orientation in the neighbourhood of the crack zone exhibits the highest fracture resistance and durability.

Gwo-Chung Tsai and Jun-Wei Chen (2005) [10] emphasized the fact that stitching of graphite/epoxy composites in the through-thickness direction improves the interlaminar fracture toughness. Through their study, they came to a conclusion that the first mode of strain energy release rate of stitched specimens was about three to six times that of unstitched specimens depending on the stiffness of the stitched thread.

On examining the effect of plain stitching by untwisted fibre rovings with respect to the in-plane mechanical properties and mode I interlaminar fracture toughness of glass/polyester composites, Velmurugan and Solaimurugan (2005) [11] arrived at a decision that the traditional FRP laminated composites with 2D fibre architecture have poor interlaminar fracture resistance and suffer extensive damage by delamination cracking when subjected to out-of-plane loading. Hence, they resorted to manual plain stitching by using untwisted fibre rovings. In the plain stitch, thread cross and resin-rich pockets were not formed, and also stitch threads were spread uniformly in the stitches, which resulted in improved interlaminar fracture resistance and in-plane mechanical properties.

Kimberley Dransfield et al. (1994) [12] used the stitching to provide through-thickness reinforcement of CFRP composites as a promising concept. They are of the view that optimal combinations of the stitching and fabrication parameters are to be identified in order to maximize the in-plane mechanical properties.

Solaimurugan and Velmurugan (2008) [13] investigated the influence of in-plane fibre orientation on the mode I interlaminar fracture toughness, G_{IC}, for unstitched and stitched glass/polyester composites. They detected that the influence of fibre orientation on G_{IC} was found to be more in unstitched specimens in comparison with the stitched specimen, whereas in the stitched specimens, stitching improved the G_{IC} and suppressed the influence of fibre orientation. They found out that the fracture toughness increases up to 45° interface angle and later decreases when the angle increases.

A research conducted by Aurodhya Jyoti et al. (2005) [14] proved that the critical energy release rate was practically independent of crack length for unidirectional carbon/epoxy width-tapered double cantilever beam (WTDCB) specimens without adhesive layers. And from various laminates of unidirectional carbon/epoxy, unidirectional E-glass/epoxy specimens with adhesive layers, the self-similar crack growth occurred only in unidirectional carbon/epoxy WTDCB specimens without adhesive layers.

6.8.3 Review for Buckling

Madhusudhana et al. (2007) [15] on investigating delamination buckling of stitched composite laminates state that Kevlar stitched glass/epoxy composite laminate is best for long delamination in retaining its buckling strength. Instead of investigating the effects of fibres or resin on the delamination buckling, they have used Kevlar and Twaron threads for stitching.

6.8.4 Review for Impact

Despite stitching, there might exist cracks on surfaces of the laminates. Longer matrix cracks and isolated matrix cracks in densely stitched composites and moderately stitched composites, respectively, performed by stitches were examined by Tan et al. (2010) [16].

Aymerich et al. (2007) [17] studied the effect of stitching on the impact performance of a graphite/epoxy cross-ply laminates and found that the ability of through-thickness reinforcement can improve the delamination resistance of laminates. In the process, they observed that the stitching does not appear capable of preventing the initiation and spread of delamination but it induces a clear reduction of damage area when stitches bridge delamination to a sufficient length.

Akinori Yoshimura et al. (2008) [18] attempted at experimental and numerical investigations of the damage process in the stitched laminates under the low-velocity impact loading and revealed the characteristic of the improvement on out-of-plane impact resistance due to stitching.

An empirical-based delamination reduction trend (DRT) was developed by Tan et al. (2012) [19] based on an extensive series of low-velocity impact tests using specimens of different laminate thicknesses, stitch densities and stitch thread linear (mass) densities, subjected over a range of impact energy levels.

6.8.5 Research Gap in Existing Available Literature

The following gaps existing in available literature were identified to the best of the scholar's knowledge:

- There is insufficient information available on the comparative study of stitched and unstitched glass/epoxy fibres.
- There is limited open literature for different fibre orientation and stitch density.
- There are no sufficient data for interlaminar fracture and impact strength for stitched laminates.

- So far, no attempt has been made by varying the orientations in WTDCB specimens.

- There is no published research works on different stitched laminates with and without delamination with respect to buckling strength. Finite element analysis in stitched laminate on buckling analysis has not been so far worked on.

- When various studies were compared, it was apparent that many contradictions existed, with available literature with respect to stitching. Some studies reveal that stitching does not affect or may improve slightly the in-plane properties, while others find that the properties are degraded.

6.9 Methodology

The sequence of current research work is given in Figure 6.12. GFRPs made with E-glass (375 gsm) unidirectional fibre with epoxy (LY556) matrix materials along with hardener HY951 were used in the ratio of 35% fibre and 65% matrix, respectively, for empirical study with and without stitching. These specimens were prepared by hand lay-up process. The stitching of dry fabrics was done with polystyrene twisted thread in a sewing machine for lock stitch and modified lock stitch, whereas the plain stitch was made manually. Resin was applied after stitching the eight layers ($0°/45°/-45°/45°/-45°/-45°/45°/0°$) and was impinged strongly inside the fibre in order to wet the laminate completely. Enough pressure was applied to flatten the surface of the laminate and also to remove any excess air inside it, by using a hand roller. The material was allowed to cure. Later, the specimens were cut according to the ASTM standards for tensile, flexural, DCB, WTDCB, buckling and impact analysis. For each test, five numbers of specimens were tested.

The following tests were conducted:

1. For tensile and flexural analyses, unstitch, plain stitch, lock stitch and modified lock stitch were used with the orientation ($0°/45°/-45°/$ m_1 ($45°$)$/m_2$ ($-45°$)$/-45°/45°/0°$), where m_1 and m_2 are midlayers with orientation of $45°/-45°$, respectively.

2. For DCB tests, unstitched and modified lock stitch, midlayer orientation changes as m_1/m_2 of $0°/0°$, $25°/-25°$, $45°/-45°$, $55°/-55°$ and $90°/90°$ were used in a laminate of ($0°/45°/-45°/m_1/m_2/-45°/45°/0°$). Finite element analysis was also done for DCB analysis to validate the experimental results. The fractured surfaces were examined by scanning electron microscope (SEM).

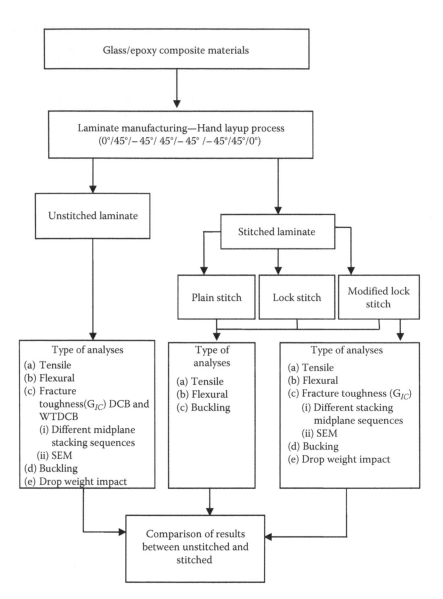

FIGURE 6.12
Flow chart for the present research work. (From Muruganandhan, R., Experimental studies on mechanical behaviour of stitched and unstitched glass/epoxy fibre reinforced laminates, Synopsis in partial fulfilment for the award of the degree of Doctor of Philosophy (Thesis supervisor: Dr. Velamurali), Anna University, Chennai, India, October 2012.)

3. For WTDCB, only unstitched laminates were tested for different midlayer orientations as m_1/m_2 of 0°/0°, 45°/−45°, 90°/90° of laminate (0°/45°/−45°/m_1/m_2/−45°/45°/0°).

4. Unstitched, plain stitch and modified lock stitch were analyzed with delamination and without delamination in between the two midlayer for buckling analysis for the same laminates as mentioned for tensile and flexural analysis. The stitched laminates were also tested by varying the stitch pitch. Finite element analysis was also done for buckling analysis to validate experimental results.

5. For impact analysis, unstitched and modified lock stitch with the same fibre orientations of laminates as given for tensile and flexural analysis were analyzed with respect to damage tolerance.

6.10 Results and Discussion

6.10.1 Effect of Stitching on Tensile and Flexural Properties

The tension test specimens were cut with a dimension of 250 mm × 25 mm and tested as per ASTM D3039 for (0°/45°/−45°/45°/−45°/−45°/45°/0°) fibre-orientated laminate. From the results, in the stitched laminates, there is a decrease in tensile strength. It was also found that unstitched laminates show better stiffness than stitched laminates. Tensile strength was found to be reduced by 19% in lock stitch, 7% in plain stitch and 16% in modified lock stitch in comparison with unstitched laminates. This is due to non-uniformity of matrix and inaccuracy of maintaining needle penetration in equal distances. It was observed that the modified lock stitch offers more stiffness than lock stitch that is due to better tightness in stitch thread. Similar results were found in the literature for the experiments made by Shah Khan et al. (1996) [20].

Three-point flexural bend test specimens having dimension of 65 mm × 13 mm were tested as per ASTM D790-07 on stitched and unstitched laminates of (0°/45°/−45°/45°/−45°/−45°/45°/0°). From the results, the flexural strength of unstitched specimen was found to be 30% greater than plain stitched laminate and 51% higher than modified lock stitched laminate and 69% higher than lock stitched laminate. Among the stitched laminates, plain stitch was found to be of better flexural strength and was 16% higher when compared with modified lock stitch and 31% higher when compared with lock stitch. From the results, it was observed that the flexural stiffness initially increased and then started decreasing. This trend was due to cracking of matrix material, and later fibre had broken suddenly. A similar observation was made by Mouritz (1996) [21].

6.10.2 Effect of Stitching and Midplane Fibre Orientation on Mode I Fracture Toughness

DCB specimens with the dimension of 125 mm × 25 mm were cut from the laminate and tested as per the ASTM standard D5528-01. The test result shows that 45°/–45° midlayer orientation has more resistance to fracture than the other laminates containing different orientated midlayers. The resistance to fracture increases in the order for the laminates (0°/45°/–45°/m_1/m_2/–45°/45°/0°) containing the midlayers m_1/m_2 of 0°/0°, 90°/90°, 25°/–25°, 55°/–55° and 45°/–45°, respectively, which is shown in Figure 6.13a. This is due to the length of the contact area of the fibre with the matrix in a plane of specific dimensions maximum in case of laminates containing the midlayers of the fibre orientation of 45°/–45°. This is also due to the reason

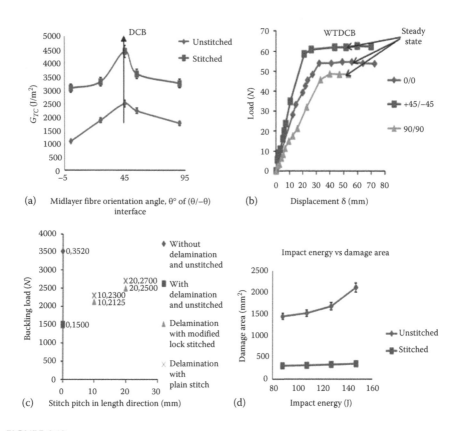

(a) Midlayer fibre orientation angle, θ° of (θ/–θ) interface

(b) Displacement δ (mm)

(c) Stitch pitch in length direction (mm)

(d) Impact energy (J)

FIGURE 6.13
(a) Effect of the midlayer fibre on DCB mode I fracture energy. (b) Effect of midlayer fibre orientation on WTDCB mode I fracture energy. (c) Effect of stitch pitch on buckling load. (d) Effect of impact on unstitched and stitched laminate. (From Muruganandhan, R., Experimental studies on mechanical behaviour of stitched and unstitched glass/epoxy fibre reinforced laminates, Synopsis, in partial fulfilment for the award of the degree of Doctor of Philosophy (Thesis Supervisor, Velamurali), Anna University, Chennai, India, October 2012.)

that the orientation of the fibre in 45°/−45° strengthens in this direction to withstand the maximum shear. Similar trends were found in the stitched laminates. However, the stitched laminates were found to be superior to unstitched laminates with respective fracture toughness, because stitching in z-direction stiffens laminate at the interfaces where they were artificially cracked at the midplane in between midlayers. Fracture toughness calculated experimentally for unstitched and modified lock stitched specimens of the laminate was being validated using finite element analysis. From the literature survey, it is understood that stitching improves the fracture toughness. Many researchers were interested in working using Kevlar and aramid fibre, whereas in this work, twisted polyester thread was used for stitching. Similar midplane orientation results were obtained for WTDCB test also. It was observed that the energy release rate for WTDCB specimens was independent of crack length, as shown in Figure 6.13b.

6.10.3 Scanning Electron Microscopy

The SEM image of unstitched laminate for DCB test observed that no in-plane fibre misalignments were found in the fracture plane surfaces, where as in stitched laminate fibre, misalignment leads to separation of fibres in the roving direction and causes resin-rich region. For unstitched glass/epoxy laminate, the fibre was not pulled out. They were only damaged by mode I loading. So they were able to delaminate easily. Stitching of the laminate was only to prevent delamination in between the layers of the laminate. If the delamination was prevented by stitching, then the lamina fibre at the midplane was pulled out by the stitch thread at the delaminated surface.

6.10.4 Effect of Stitching and Delamination on Buckling Strength

The test specimens were prepared with different aspect ratio (ratio of unsupported length of the specimen a to width of the specimen b) of 0.5, 1, 1.5, 2 and 3, with orientation of (0°/45°/−45°/45°/−45°/−45°/45°/0°) for (1) without delamination, (2) without stitching and (3) with stitching (modified lock stitch and plain stitch considering two different pitches of 10 mm and 20 mm along width direction and both with 5 mm pitch along length direction). The delamination was introduced by inserting 100 μm nylon film at the midplane during the stacking sequence.

In case of unstitched laminate without delamination, when the aspect ratio increases, the buckling load decreases. For example, when the aspect ratio increases 6 times from 0.5 to 3, then the critical buckling load decreases by 10 times. When the delamination was introduced at the midplane in between the midlayers with the ratio of delamination size of 0.65 for the case of unstitched laminates, the critical buckling load decreases in comparison with unstitched without delamination. Uniform reduction was maintained between unstitched without delamination and unstitched with delamination as the aspect ratio increased from 0.5 to 3.

When the pitch length increased along the width direction, it was found to be of better resistance to buckling because of the laminates stitched with less number of stitch penetration causing less damage in the lateral direction. It was noted that for small delamination lengths less than 0.5a, the stitching was found to be less effective in the critical buckling load. It was found to be same as in case of the unstitched laminates. This is due to the damage caused because of stitching in lateral direction that caused reduction in the stiffness of the composites. Below this delamination length, the stitching was found to be not effective, that is, the buckling loads for stitched laminates were less than that of unstitched laminates.

Resin-rich region forms around the stitch threads, due to stitching. Hence, it requires higher buckling load for delamination to buckle locally. The experiments indicate that global buckling mode occurs in all stitched delaminated specimens with different aspect ratios.

The stitching in both plain and modified lock stitchings of the delaminated plate showed that there is increase in the buckling load-carrying capacity in comparison with unstitched delaminated composite. From Figure 6.13c, the effect of stitch density in the laminate could be seen. Stitching with 20 mm pitch by using plain stitch was found to be better with respect to buckling, for a/b ratio varies from 0.3 to 1.5. It is concluded that the plain stitching can withstand buckling strength when compared to modified lock stitching. The experiment results were closely matching with finite element analysis.

6.10.5 Effect of Stitching and Fibre Orientation on Impact Properties

Experiments were conducted for the different stacking sequences of the laminates of size 89 × 89 mm (as per ASTM D7136 for all edges clamped $[0°/45°/-45°/m_1/m_2/-45°/45°/0°]$) containing the midlayers m_1/m_2 of $0°/0°$, $25°/-25°$, $45°/-45°$, $55°/-55°$ and $90°/90°$, respectively, for unstitched composite laminate applied with impact energy of 107.8 J.

It could be noticed that the result of unstitched laminate containing $45°/-45°$ orientation midlayer absorbs more energy, which is due to more resistance to shear. Hence, similar experiments were conducted for the stacking sequence $(0°/45°/-45°/45°/-45°/-45°/45°/0°)$ composite laminate for different impact energy levels like 147, 127.4 and 107.8 J for stitched and unstitched laminates.

From the results at higher energy levels, the curve shows oscillations, which indicate the onset and propagation of damage in the sample. For unstitched samples, it was observed that in-plane damage area was more compared to stitched ones. On the other hand, in case of stitched laminates, damage was within the grid location because stitching arrested the crack growth. Back-surface damage was initiated as a tensile fracture of fabric tow. This facilitated the penetration of the indentor. In case of stitched samples, the spread of back-surface damage was again restricted due to the same reason as explained earlier. Indentation and penetration at the impact location were enhanced more in case of unstitched laminates in comparison with stitched laminates. In all the stitched laminates, the stitch remained intact.

The damage areas in the stitched laminates were always smaller than those in unstitched laminates at the same impact energy level and were shown in Figure 6.13d. The impact damage was found to propagate in the direction of the fibre orientation. This is evident from the experimentally determined backlighting photographic technique. For unstitched laminates, damage spread was conical, with the maximum damage on the back surface. However, for the stitched laminates, the propagation of delamination front was arrested at the stitch location and the damage spreads through the thickness in cylindrical fashion.

In both stitched and unstitched specimens, delamination occurred at each interface and ply cracks connected the delaminations in through-thickness direction. Hence, from the results, it was found that the resistance to damage could be increased by stitching. Stitching also improved interlaminar fracture toughness and reduced energy absorbed during impact. Increase in stitch density decreases the damage area during impact. The stitched laminate increases the delamination strength and reduces impact damage.

6.11 Conclusions

From the work carried out, the following conclusions were arrived at

- Considerable improvement was seen in tensile properties, in plain stitch when compared with other types of stitches, namely, lock stitch and modified lock stitch.
- When unstitch and plain stitches were taken into consideration, tensile strength was found to be degrading less in plain stitch. This is because of resin-rich region.
- For flexural load, unstitch was better when compared to the other stitches, namely, plain stitch, lock stitch and modified lock stitch.
- From the results, the flexural strength of unstitched specimen was found to be 30% greater than plain stitched laminate and 51% higher than modified lock stitched laminate and 69% higher than lock stitched laminate.
- From DCB, the fracture toughness was found to be maximum in the laminate containing midlayer of orientation +45°/−45°.

- Further, fracture toughness can be improved in stitched laminates.
- From WTDCB, the fracture toughness was found to be maximum in the laminate containing midlayer of orientation +45°/−45°. It was observed that fracture toughness was independent of crack length for WTDCB.
- The buckling strength of a laminate can be increased by stitching. This is due to stitch arresting the delamination.
- Buckling strength can be further increased in stitched laminate by increasing stitch pitch. This is due to less penetration in the lateral direction.
- In impact, the damage area is reduced in stitched laminates. The pattern of damage in the thickness direction is conical in unstitched laminates, whereas in stitched laminates, it is cylindrical. More damage occurs at the exit side of the impact that is at the bottom of the laminate in case of unstitched laminates. It was also observed that the orientation angle of the midlayers affects the energy absorption. Angle orientation of 45°/−45° of the midlayer was found to be of more energy absorption for impact loading.

References

1. M. Schlotter, A comparison of core materials for sandwich composite constructions, Design limits and solutions for very large wind turbines. http://wenku.baidu.com/view/59cf8a315a8102d276a22fae.html? June 2002, p. 49.
2. D. Schwingel, H.-W. Seeliger, C. Vecchionacci, D. Alwes, and J. Dittrich, Aluminium foam sandwich structures for space application, *J. Acta Astronautica* 61, 2007, 326–330.
3. P. Norlin, The role of composites blades sandwich in turbine, *Reinforced Plastics*, 46(3), March 2002, 32–34.
4. www.dcboardz.com/surf.html.
5. R. Muruganandhan, Experimental studies on mechanical behaviour of stitched and unstitched glass/epoxy fibre reinforced laminates, Synopsis, in partial fulfilment for the award of the degree of Doctor of Philosophy, Anna University, Chennai, India, October 2012. (Thesis supervisor: Dr. Velamurali.)
6. M. Karahan, N. Karahan, Y. Ulcay, R. Eren, and G. Kaynak, Investigation into the tensile properties of stitched and unstitched woven aramid/vinyl ester composites, *Textile Research Journal*, 80(10), 2010, 880–891.
7. A.P. Mouritz, J. Gallagher, and A.A. Goodwin, Flexural and interlaminar shear properties of stitched GRP laminates following repeated impacts, *Composites Science and Technology*, 57, 1996, 509.
8. A.P. Mouritz, K.H. Leong, and I. Herszberg, A review of the effect of stitching on the in-plane mechanical properties of fibre-reinforced polymer composites, *Composites: Part A*, 28A, 1997, 997–991.

9. M.R. Shetty, K.R. Vijay Kumar, S. Sudhir, P. Raghu, and A.D. Madhuranath, Effect of fibre orientation on mode-I interlaminar fracture toughness of glass epoxy composites, *Journal of Reinforced Plastics and Composites*, 18, 1999, 1–15.

10. G.-C. Tsai and J.-W. Chen, Effect of stitching on mode I strain energy release rate, *Composite Structures*, 69, 2005, 1–9.

11. S. Solaimurugan and R. Velmurugan, Influence of in-plane fibre orientation on mode I interlaminar fracture toughness of stitched glass/polyester composites, *Composites Science and Technology*, 68, 2008, 1742–1752.

12. K. Dransfield, C. Baillie, and Y.-W. Mai, Improving the delamination resistance of CFRP by stitching—A review, *Composites Science and Technology*, 50, 1994, 305–317.

13. R. Velmurugan and S. Solaimurugan, Improvements in mode I interlaminar fracture toughness and inplane mechanical properties of stitched glass/polyester composites, *Composites Science and Technology*, 67, 2007, 61–69.

14. A. Jyoti, R.F. Gibson, and G.M. Newaz, Experimental studies of mode I energy release rate in adhesively bonded width tapered composite DCB specimens, *Composites Science and Technology*, 65, 2005, 9–18.

15. M.R. Parlapalli, K.C. Soh, D.W. Shu, and G. Ma, Experimental investigation of delamination buckling of stitched composite laminates, *Composites: Part A*, 38, 2007, 2024–2033.

16. K.T. Tan, N. Watanabe, and Y. Iwahori, Effect of stitch density and stitch thread thickness on low velocity impact damage of stitched composites, *Composites: Part A*, 41, 2010, 1857–1868.

17. F. Aymerich, C. Pani, and P. Priolo, Damage response of stitched cross-ply laminates under impact loadings, *Engineering Fracture Mechanics*, 74, 2007, 500–514.

18. A. Yoshimura, T. Nakao, S. Yashiro, and N. Takeda, Characterisation of out of-plane impact damage in stitched CFRP laminates, *16th International Conference on Composite Materials*, pp. 1–7, 2007.

19. K.T. Tan, N. Watanabe, Y. Iwahori, and T. Ishikawa, Understanding effectiveness of stitching in suppression of impact damage: An empirical delamination reduction trend for stitched composites, *Composites: Part A*, 43, 2012, 823–832.

20. M.Z. Shah Khan, A.P. Mouritz, Fatigue behaviour of stitched GRP laminates, *Composites Science and Technology*, 56, 1996, 695–701.

21. A.P. Mouritz, Flexural properties of stitched GRP laminates, *Composites: Part A*, 27A, 1996, 525–530.

7

Biocomposite Materials

7.1 Biodegradable Plant Fibre-Reinforced Composite

Environment-friendly, fully biodegradable green composites based on plant natural fibres and resins are increasingly used for various applications as replacements for nondegradable materials. Green composite based on plant fibre is a very promising and sustainable green material without using any toxic chemicals. They contain plant fibre and thermosets or thermoplastics. In comparison to other nondegradable materials, plant fibres are suitable to reinforce plastics due to relative high strength and stiffness, low cost, low density, low CO_2 emission and biodegradability. A modern automobile requires recyclable or biodegradable parts because of the increasing demand for clean environment. A natural and biodegradable composite reinforced by natural fibres provides an important environmental advantage in automobile industry.

Natural plant fibre composite contains biodegradable fibre such as wood fibres, which can be mixed with the polymers to form a green composite product. Plant fibre composite consists of plant fibres as fillers and reinforcements for polymeric matrix, that is, polypropylene (PP) and polyvinyl chloride (PVC). Such composites are made by using wood flour (a waste from sawmills) or wood fibres obtained from waste wood products such as packaging pallets, old furnitures and construction wood scraps. These are also used as inexpensive filler for PP and PVC, resins, epoxies, phenolics and soybean oil–based resins, which are forest products. Green composites made from biodegradable agricultural resources are widely used in automotive applications.

7.2 Advantages and Disadvantages of Plant Fibre Composite

Plant fibres have a number of advantages and disadvantages over traditional glass fibres. These are the following:

- Ecological character
- Biodegradability
- High specific properties
- Nonabrasive nature
- Low density
- Safe fibre handling
- High possible filling levels
- Low energy consumption

These advantages are very important factors for their acceptance in automotive industry.

Natural fibres have some drawbacks also, which are the following:

- Low thermal stability.
- Tendency to form aggregates during processing.
- Low resistance to moisture and seasonal quality variations (even between individual plants in the same cultivation) greatly reduces the potential of plant fibres to be used as reinforcement for polymers.
- The high moisture absorption of plant fibres leads to swelling and presence of voids at the interface (porous products), which results in poor mechanical properties and reduces dimensional instability of composites.
- Treatment of plant fibres with hydrophobic chemicals or modification with vinyl monomers can reduce the moisture gain.
- Poor compatibility exhibited between the fibres and the polymeric matrices, which results in nonuniform dispersion of fibres within the matrix and poor mechanical properties.

7.3 Different Types of Plant Fibres for Green Composite

Plant fibres are classified according to the part of the plant from which they are obtained. In general, they are subdivided based on their origins from the plants. All plant fibres are composed of cellulose. Longer plant-based fibres

TABLE 7.1

List of Important Plant Fibres and Their Related Details

Plant Fibre	Origin	Species	Relative Density	Young's Modulus (GPa)
Cotton	Seed	*Gossypium* species	1.5	5.5–12.6
Jute	Stem	*Corchorus capsularis*	1.3	26.5
Sisal	Leaf	*Agave sisalana*	1.5	9.42–22.0
Bamco	Glass	(>1,250 species)	0.8	48.88
Pineapple	Leaf	*Ananas comosus*	—	34.5–82.5
Soft wood	Stem	(>10,000 species)	1.5	40.0
Flax	Stem	*Linum usitatissimum*	1.5	27.6
Coir	Fruit	*Cocos nucifera*	1.2	4.0–6.0
Ramie	Stem	*Boehmeria nivea*	1.5	61.4–128

Source: Ashori, A., *Bioresour. Technol.*, 2007.

such as abaca, bamboo, flax, henequen, hemp, jute, kenaf, pineapple, ramie and sisal have good mechanical properties. Important fibres and their origin, species, density and Young's modulus are shown in Table 7.1 [1].

7.4 Contribution of Plant Fibre–Based Green Composite for Various Applications

Material scientists are always on the lookout for new materials and improved processes for better products. Plant fibre composites are being used in a large number of applications in automotives, constructions, marine, electronic and aerospace as shown in Figure 7.1. For the first time, Daimler–Benz gave the

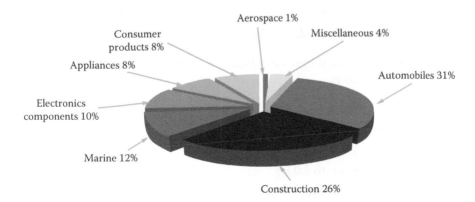

FIGURE 7.1

Applications of plant fibre composites in various fields. (From http://www.edmunds.com/advice/alternativefuels/articles/105341/article.html.)

(a)

(b)

FIGURE 7.2
Application of plant fibre–based green composite. (a) 50 Mercedes-Benz E-Class components; (b) underfloor panel. (From Ashori, A., *Bioresour. Technol.*, 2007.)

idea of replacing glass fibres with plant fibres in automotive components in 1991. Mercedes used the jute-based door panels in its E-class vehicles in 1996. In September 2000, Daimler Chrysler began using plant fibres for their vehicle production. The advantages of using plant fibres in the automotive industry include a weight saving of between 10% and 30% and corresponding cost savings. Plant fibre–based green composites are used for various automobile parts (Figure 7.2) such as the following:

- Dashboards
- Door panels
- Parcel shelves
- Seat cushions
- Backrests
- Cabin lining
- Thermoacoustic insulation purposes
- Package shelves (the space behind the rear seats of sedans)
- Car disc brakes to replace asbestos fibres

Typical amounts of plant fibres used for different applications in automotive industry are as follows:

- Front door linens: 1.2–1.8 kg
- Rear door linens: 0.8–1.5 kg
- Boot linens: 1.5–2.5 kg
- Parcel shelves: up to 2.0 kg
- Seat backs: 1.6–2.0 kg
- Sunroof sliders: up to 0.4 kg
- Headliners: average 2.5 kg

From 1996 till 2003, the use of plant fibre composites in German automotive industry increased from 4,000 tonnes to 18,000 tonnes/year. For Europe, it amounts to almost 70,000 tonnes of these new plant fibre materials [2].

Various automobile manufacturers using natural plant fibre composite into their cars are as follows:

- The BMW Group incorporates the flax and sisal in the interior door linings, panels and seatback cushion.
- Toyota using kenaf into the body structure of Toyota's i-foot and i-unit concept vehicles.
- At General Motors, a kenaf and flax mixture is used for package trays and door panel inserts for Saturn L300s and Opel Vectras, while wood fibre is being used in seatbacks for the Cadillac DeVille.
- Honda is using wood fibre in the cargo area floor for the Pilot SUV.
- Ford mounts Goodyear tyres that are made with corn on its fuel-sipping Fiestas in Europe. The sliding door inserts for the Ford Freestar are made with wood fibre.

7.5 Starch-Based Composites

In the following sections, environmental problems of synthetic polymeric composites are explained. The importance of starch-based composites is discussed in favour of the clean and green environment. Different kinds of starches suitable as matrix components of composites are highlighted along with their characteristics and mechanical properties. Vivid kinds of biopolymers and reinforcing agents such as natural fibres are described along with their suitability and usefulness. The concept of various kinds of composites, namely, flaked composite, fibre composite, sandwich composite and hybrid composites, is included. Starch is discussed as a green material along with its history, characteristics and structure. Different sources of starch and various kinds of modified starches are explained. Processing of starch before using as matrix in composite, and methods of improving the properties of starch are also described. Blending with synthetic degradable polymers and biopolymers, chemical derivatives, biopolymers/biodegradable polymers for use as matrix of the composite are incorporated. Biodegradable thermoplastic polymer, polylactic acid, starch as a source of biopolymer (agropolymer) are also explained. Finally, the classification of starch-based biocomposites are enlisted.

7.6 Starch as a Green Material [5]

Starch is used as a starting material for a wide range of green biomaterials. Different routes are used to modify starch to improve the product properties and to extend the application range. Seventy-five percent of all organic material on earth is present in the form of polysaccharides. An important polysaccharide is starch. Plants synthesize and store starch in their structure as an energy reserve. It is generally deposited in the form of small granules or cells with diameters between 1–100 μm. Starch is found in seeds (i.e. corn, maize, wheat, rice, sorghum, barley or peas) and in tubers or roots (i.e. potato or cassava) of the plants. Most of the starch produced worldwide is derived from corn. The worldwide production of starch in 2008 is estimated to be around 66 million tonnes. Starch is generally extracted from the plant by wet milling processes. Starch is a polymer consisting of two types of anhydroglucose (AHG) polymers, amylose and amylopectin. Amylose is essentially a linear polymer in which AHG units are predominantly connected through α-D-(1,4) glucosidic bonds. Amylopectin is a branched polymer, containing periodic branches linked with the backbones through α-D-(1,6) glucosidic bonds. The content of amylose and amylopectin in starch varies and largely depends on the starch source. Typically, the amylose content is between 18% and 28%.

Utilization of polymer products at the end of their service life is an environmental problem. The traditional way of treating them, namely, burning, recycling, waste disposal and pyrolysis, cannot improve the ecological situation. Environmental pollution caused by the accumulation of nondestructible solid waste is a matter of serious concern today. It has created stir and has attracted the attention of researchers to develop bio-based biodegradable polymers and composite materials. The biodegradable materials are the appropriate answers to this menace. Such materials are composed quickly under the action of the natural environment such as soil microorganism, light, water and other factors. In developing such materials, one has to meet mutually contradictory environments, namely, high parameters of mechanical properties on the one hand and the accelerated biodegradability on the other hand. In this connection, the material based on biodegradable polymer obtained from renewable vegetable as raw materials is of increasing interest. Special among them is starch-based plastics. Starch is a typical natural polymer. It is biodegradable and after plasticization can be processed into items on processing equipment.

Basic drawback of materials based on starch is its low mechanical properties. A promising method to improve the mechanical properties of starch is to introduce fibrous or lamellar particles of filler. By doing so, the effect of reinforcement (strengthening) can be reached not only because of the considerably higher values of strength and rigidity of the filler but also because of the particle geometry (the characteristic ratio of their sizes, known as the aspect ratio). Good results are obtained on using cellulose microfibres as the filler. Therefore, during the past years, an interest in starch-based composites containing natural fibres, layered silicate, etc., as filler has increased. These composites consist of a completely biodegradable matrix and ecologically safe filler.

7.7 Starch: History, Characteristics and Structure [5]

Starch, also known as amylum, is an important carbohydrate that remains present in human diet. It is composed of a large number of glucose units (Figure 7.3). These units are joined together by glycosidic bonds. Its molecular formula is $(C_6H_{10}O_5)_n$. All green plants produce it as a store of energy. Starch molecules arrange themselves in the plant in semicrystalline granules. Each plant species has a unique starch granular size. For example, the rice starch is relatively small (about 2 µm), while potato starches have larger granules (up to 100 µm). Although in absolute mass only about one quarter of the starch granules in plants consists of amylose, there are about 150 times more amylose molecules than amylopectin molecules. Amylose is a much smaller molecule than amylopectin.

The word starch is derived from *structure*, an English word which means *stiffness*. It is named as *"amylum"* in Latin and *"amulon"* in Greek language. The history of usage of starch is very old. Egyptians were using wheat starch as paste for stiffening cloth and weaving of linen. Romans used wheat starch pastes in cosmetic creams for powdering the hair and thickening

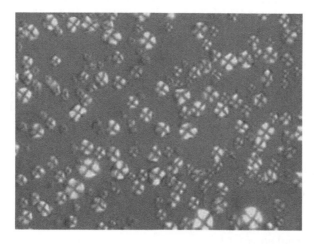

FIGURE 7.3
Starch, 800× magnified, under polarized light. (From Starch-Wikipedia, the free Encyclopedia, http://en.wikipedia.org/wiki/Starch.)

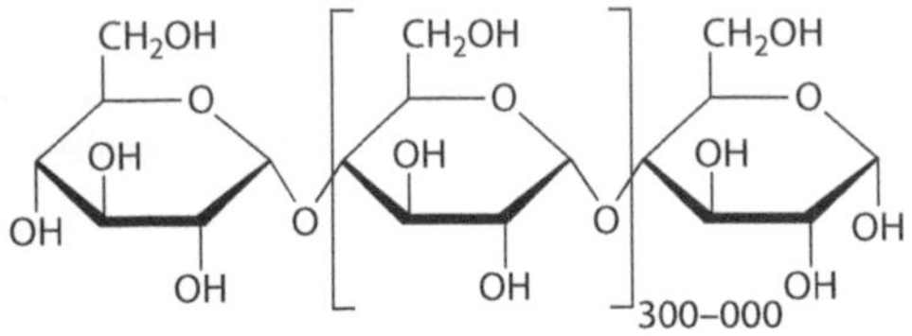

FIGURE 7.4
Structure of the amylase molecule.

sources. Persians used wheat starch to make dishes. Indians used it to make wheat halva. The Chinese used rice starch for surface treatment of paper. Starch in pure form is a powder of white appearance. It is odourless and taste-less and is insoluble in alcohol and cold water. It is heavier than water having the density of 1500 kg/m³. It does not possess a sharp melting point, rather decomposes over a temperature range. It autoignites at 410°C. Starch is com-posed of two different kinds of molecules, namely, linear and helical amylase structure and branched amylopectin structure. These molecules vary from plant to plant. Generally, 20%–25% amylase and 75%–80% amylopectin are found in starch. The structure of an amylase molecule is shown in Figure 7.4.

Starch becomes soluble in water when heated. The granules swell and burst, the semicrystalline structure is lost and the smaller amylose molecules start leaching out of the granule, forming a network that holds water and increases the mixture's viscosity. This process is called starch gelatinization. During cooking, the starch becomes a paste and increases further in vis-cosity. During cooling or prolonged storage of the paste, the semicrystalline structure partially recovers and the starch paste thickens, expelling water. This is mainly caused by the retrogradation of the amylose. This process is responsible for the hardening of bread or staling and for the water layer on top of a starch gel (syneresis).

Some cultivated plant varieties have pure amylopectin starch without amylose, known as waxy starches. The most used is waxy maize; others are glutinous rice and waxy potato starch. Waxy starches have less retrograda-tion, resulting in a more stable paste. High-amylose starch, amylomaize, is cultivated for the use of its gel strength. Major source of starch is found in foods such as wheat, rice, corn (maize), potatoes and cassava.

7.7.1 Different Sources of Starch and Modified Starches

Various sources of starch that are used for making biocomposites are the following:

- Commercial corn
- Sugary mutant corn
- Buckwheat

- Alderman pea
- Steadfast pea
- Water chestnut
- Waxy corn
- Sweet potato
- White potato
- Mung bean
- Colocasia
- Sorghum
- Oat
- Arracacha
- Rye
- Kudzu
- Yarns
- Wheat
- Arrowroot
- Tapioca
- Cannas
- Oca
- Sago
- Chestnuts
- Lentils
- Barely
- Banana
- Millet
- Breadfruit
- Kataburi
- Taro
- Malaga
- Favas

Starches are modified to make them suitable for specific applications. They are generally chemically treated to function properly during processing and storage. The treatments involve high heat, cooling, freezing, high shear and low pH. Various modified starches are listed as follows:

- Hydroxypropyl starch
- Cationic starch

- Carboxymethylated starch
- Alkaline treated starch
- Acid treated starch
- Bleached starch
- Oxidized starch

These starches find use in different applications such as food, food additives, paper making, biofuel, corrugated board adhesives and clothing/laundry. They are also used in the textile industry; in printing industry; in biodegradable bioplastic production, for example, polylactic acid (PLA); as body powder and in oil exploration to adjust the viscosity of drilling fluid.

Foamed starch: Starch can be blown by environmentally friendly means into a foamed material using water steam. Foamed starch is antistatic, insulating and shock absorbing, therefore constituting a good replacement for polystyrene foam.

7.7.2 Processing of Starch before Using as Matrix in Composite

In making the starch suitable for industrial use, first of all, the starch is extracted from roots, tubers and seeds of the plants. It is then refined by conducting the processes of wet grinding, washing, sieving and drying. The following refined starches are mainly used commercially:

- Wheat starch
- Tapioca
- Rice starch
- Mung bean starch
- Potato starch
- Cornstarch
- Sweet potato starch
- Sago starch

Untreated starches are heated for thickenings (i.e. gelatinizing). Starches are also precoated and thickened in cold water instantly. The starch is used for sizing and coating of paper. It is preferable because of its potential to reduce total cost of composite. Proper cooking and complete hydration of starch molecules are essentially desired for all applications. Potato starch is very easy to cook. Complete dispersion of starch granule depends on cooking temperature and shear ability. High shear is a desired property. Modified starches such as acid modified, oxidized and ethylated are used in the textile industry. These starches are used to treat cloth as it comes from the loom. The starch removes the impurities of various applied dyes, chemicals, softeners, etc. It is also used for desizing.

7.7.3 Improving the Properties of Starch [5]

To improve the properties of starch, various physical and chemical modifications such as blending, derivation and graft copolymerization are done. It can be done in the following ways.

7.7.3.1 Blending with Synthetic Degradable Polymers

To prepare completely biodegradable starch-based composites, the components to blend with starch are aliphatic polyesters, polyvinyl alcohol (PVA) and biopolymers. The commonly used polyesters are poly(β-hydroxyalkanoates) (PHAs) obtained by microbial synthesis and polylactide or poly(ε-caprolactone) (PCL), derived from chemical polymerization. The goal of blending completely degradable polyester with low-cost starch is to improve its cost competitiveness while maintaining other properties at an acceptable level.

PLA is one of the most important biodegradable polyesters with many excellent properties. It possesses good biocompatibility and processability as well as high strength and modulus. However, PLA is very brittle under tension and bending loads and develops serious physical ageing during application. Moreover, PLA is a much more expensive material than the industrial polymers. To improve the compatibility between PLA and starch, suitable compatibilizer is added. Besides, gelatinization of starch is also a good method to enhance the interfacial affinity. Starch is gelatinized to disintegrate granules and overcome the strong interaction of starch molecules in the presence of water and other plasticizers, which leads to well dispersion. The glass transition temperature and mechanical properties of TPS/PLA blend depend on its composition and the content of plasticizer.

PCL is another important member of synthetic biodegradable polymer family. It is linear, hydrophobic, partially crystalline polyester and can be slowly degraded by microbes. Blends between starch and PCL have been well used. The weakness of pure starch materials including low resilience, high moisture sensitivity and high shrinkage is overcome by adding PCL to starch matrix even at low PCL concentration. Blending with PCL, the impact resistance and the dimensional stability of native starch are improved significantly. PCL/starch blends can be further reinforced with fibre and nanoclay, respectively. Moreover, the other properties of the blends such as hydrolytic stability, degradation rate and compatibilization between PCL and starch are also improved.

PVA is a synthetic water-soluble and biodegradable polymer. PVA has excellent mechanical properties and compatibility with starch. PVA/starch blend is assumed to be biodegradable since both components are biodegradable in various microbial environments.

7.7.3.2 Blending with Biopolymers

- Natural polymers such as chitosan and cellulose and their derivatives are inherently biodegradable and exhibit unique properties. Starch and chitosan are abundant naturally occurring polysaccharide. Both of them are cheap, renewable, nontoxic and biodegradable. The starch/chitosan blend exhibits good film-forming property, which is attributed to the inter- and intramolecular hydrogen bonding that formed between amino groups and hydroxyl groups on the backbone of two components. The mechanical properties, water barrier properties and miscibility of biodegradable blend films are affected by the ratio of starch and chitosan.

7.7.3.3 Chemical Derivatives

- One problem for starch-based blends is that starch and many polymers are nonmiscible, which leads to the mechanical properties of the starch/polymer blends generally becoming poor. Thus, chemical strategies are taken into consideration. Chemical modifications of starch are generally carried out via the reaction with hydroxyl groups in the starch molecule. The derivatives have physicochemical properties that differ significantly from the parent starch but the biodegradability is still maintained. Consequently, substituting the hydroxyl groups with some groups or chains is an effective means to prepare starch-based materials for various needs.

7.8 Biopolymers/Biodegradable Polymers for Use as Matrix of the Composite [5]

Biopolymers are obtained in different ways, namely, (1) from renewable resources, (2) by synthesizing microbially and (3) by synthesizing from petroleum-based chemicals. The biopolymers based on renewable resources are the following:

1. Agropolymers such as starch and cellulose plastics
2. PHA, for example, polyhydroxybutyrate (PHB)
3. Polylactides

TABLE 7.2

Properties of Some Traditional Polymers and Biodegradable Polymers

S. No.	Property	PBS[a]	PS	LDPE	PP	PLA (Eco-PLA)	PHB[b] (P226[c])	PHBV	PCL (Tone 787)
1	Tensile stress at break (MPa)	57	35–64	8–10	34	45	24–27	25	41
2	Tensile modulus (MPa)	—	2800–3500	100–200	—	2800	1700–2000	1000	386
3	Elongation at break (%)	700	1–2.5	150–600	12	3	6–9	25	900
4	Density (g/cm³)	1.26	1.04–1.09	0.92	0.90	1.21	1.25	1.25	1.145
5	Melting point (°C)	115	—	124	164	177–180	168–172	135	60

Source: Biomer, Germany, http://www.biomer.de/.
[a] Bionolle 1001b film grade.
[b] Plasticizes PHB.
[c] From technical data sheet, Showa Highpolymer Co. Ltd.

Biodegradable polymers may be used to constitute the matrix part of composite materials. But they generally have high cost and performance limitations. This deficiency, however, may be overcome by using the biodegradable films such as starch (Table 7.2). This is an attractive way to get cost-effective bio-based polymer. Such matrix material will be less expensive, biodegradable and with ample availability. Due to the hygroscopic nature of starch and lack of affinity, the adhesion between starch and hydrophobic biopolymers can be increased by preparing the multilayered biocomposites.

Different types of biodegradable polymers are as follows:

1. Petroleum/fossil fuel based such as
   a. Aliphatic polyesters
   b. Aliphatic–aromatic polyesters
   c. Poly(ester amide)
   d. Poly(alkyene succinates)
   e. Poly(vinyl alcohol)
2. Renewable resource based such as
   a. Polylactides (PLAs)
   b. Cellulose esters

    c. Starch plastics

    d. PHAs

  3. Mixed resource based: renewable resources + petroleum resources such as blendings of

    a. Two/more biodegradable polymers(e.g. starch plastic + PLA)

    b. One biodegradable + one fossil fuel–made polymer(e.g. starch plastic + polyethene)

    c. Epoxidized soya bean oil + petro-based epoxy resin

7.8.1 Biodegradable Thermoplastic Polymer: Polylactic Acid

Polylactic acid is a class of crystalline biodegradable thermoplastic polymer with relatively high melting point and excellent mechanical properties. Recently, it has been highlighted because of its availability from renewable resources such as corn and sugar beets. PLA is synthesized by the condensation polymerization of D- or L-lactic acid or ring opening polymerization of the corresponding lactide. Under specific environmental conditions, pure PLA can degrade to carbon dioxide, water and methane over a period of several months to 2 years, a distinct advantage compared to other petroleum plastics that need much longer periods. The final properties of PLA strictly depend on its molecular weight and crystallinity. PLA has been extensively studied as a biomaterial in medicine, but only recently, it has been used as a polymer matrix in composites.

7.9 Starch as a Source of Biopolymer (Agropolymer) [5]

Potato: Potato (Figure 7.5) is a starchy, tuberous crop of the Solanaceae family. It is the world's fourth-largest food crop, following rice, wheat and maize. Potatoes yield abundantly with little effort and adapt readily to diverse climates as long as the climate is cool and moist enough for the plants to gather sufficient water from the soil to form the starchy tubers. In terms of nutrition, the potato is best known for its carbohydrate content (approximately 26 g in a medium potato). The predominant form of this carbohydrate is starch. Its starch content along with other nutritional values is given in Table 7.3.

Sweet potato: The sweet potato (Figure 7.5) is a dicotyledonous plant that belongs to the family of Convolvulaceae. Its large, starchy, sweet tasting tuberous roots are an important root vegetable. Its starch content along with other nutritional values is given in Table 7.3.

Rice: Rice (Figure 7.5) is the seed of the monocot plant. As a cereal grain, it is the most important staple food for a large part of the world's human

FIGURE 7.5
Starch as a source of biopolymer (agropolymer). (From Starch-Wikipedia, the free Encyclopedia, http://en.wikipedia.org/wiki/Starch.)

TABLE 7.3

Starch Content and Nutritional Values of Biopolymers per 100 g

Item	Potato	Sweet Potato	Rice	Wheat	Maize	Banana
Energy (kJ)	321	360	1527	1506	360	371
Carbohydrate (g)	19	20.1	79	51.8	19.02	22.84
Starch (g)	15	12.7	75.8	71.9	79.5	—
Sugar (g)	—	12.7	0.12	—	3.22	12.23
Fat (g)	2.2	0.1	0.5	0.5	4.6	0.016–0.4
Dietary fibre (g)	0.1–2	3.0	1.3	1.3	2.7	2.6
Protein (g)	75	1.6	7.12	7.12	10.2	1.09
Water	1.8 mg (14%)	—	11.62 g	11 g	—	68.6 78.1 g

Source: Starch-Wikipedia, the free Encyclopedia, http://en.wikipedia.org/wiki/Starch, nutritiondata.com and USDA Nutrient database.

Note: Percentages are relative to US recommendations for adults.

population. It is the grain with the second highest worldwide production, after maize (corn). Its starch content along with other nutritional values is given in Table 7.3.

Wheat: Wheat (Figure 7.5) is a grass. In 2007, the world production of wheat was 607 million tonnes, making it the third most-produced cereal after maize (784 million tonnes) and rice (651 million tonnes). Much of the carbohydrate fraction of wheat is starch. Wheat starch is an important commercial product of wheat. The principal parts of wheat flour are gluten and starch. These can be separated in a kind of home experiment, by mixing flour and water to form a small ball of dough and kneading it gently while rinsing it in a bowl of water. The starch falls out of the dough and sinks to the bottom of the bowl, leaving behind a ball of gluten. Its starch content along with other nutritional values is given in Table 7.3.

Maize: Maize is also known as corn. Maize is one of the most widely grown crops. There are many maize varieties. Sweet corn is usually shorter than field corn varieties. Many forms of maize (Figure 7.5) are used for food. They are also classified as the following subspecies depending upon the amount of starch each had:

- Flour corn
- Waxy corn
- Dent corn
- Pod corn
- Sweet corn
- Popcorn
- Amylomaize
- Flint corn
- Striped maize

Maize is a major source of starch. Starch from maize can also be made into other chemical products. The corn steep liquor, a plentiful watery by-product of maize wet milling process, is widely used in the biochemical industry and research as a culture medium to grow many kinds of microorganisms. Maize (corn) contains about 70% starch and other components being protein, fibres and fat. The basis of the maize milling process is the separation of the maize kernel into its different parts. Maize starch is produced by the wet milling process, which involves grinding of softened maize and separation of corn oil seeds (germs), gluten (proteins), fibres (husk) and pure starch. Its starch content along with other nutritional values is given in Table 7.3.

Cassava: Cassava (Figure 7.6), also called yuca or manioc, is a woody shrub of the spurge family. It is extensively cultivated as an annual crop in tropical and subtropical regions for its starchy tuberous root, a major source of

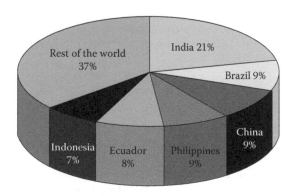

FIGURE 7.6
Distribution of the world banana production. (From NNCTAD from FAO statistics.)

carbohydrates. Cassava is the third largest source of carbohydrates for meals in the world. It is classified as *sweet* or *bitter* depending on the level of toxic cyanogenic glucosides. Commercial varieties can be 5–10 cm in diameter at the top and around 15–30 cm long. Cassava roots are very rich in starch and contain significant amounts of calcium (50 mg/100 g), phosphorus (40 mg/100 g) and vitamin C (25 mg/100 g). However, they are poor in protein and other nutrients. Its tuberous roots have innumerable industrial uses also, particularly for starch extraction. Cassava starch has very good properties that are highly desirable for the paper manufacturer. Cassava starch possesses a strong film, clear paste, good water holding properties and stable viscosity. Properties of the starch used are abrasion resistance, flexibility and ability to form a bond to the fibre, to penetrate the fibre bundle to some extent and to have enough water holding capacity so that the fibre itself does not rob the size of its hydration. Starch is a popular base for adhesives, particularly those designed to bond paper in some form to itself or to other materials such as glass, mineral wool and clay. Starch can also be used as a binder or adhesive for nonpaper substances such as charcoal in charcoal briquettes, mineral wool in ceiling tiles and ceramics before firing. The starches most commonly used for the manufacture of adhesive pastes are maize, potato and cassava; of these, cassava starch appears more suitable in several respects.

Cassava starch adhesives are more viscous and smoother working. They are fluid, stable glues of neutral pH that can be easily prepared and can be combined with many synthetic resin emulsions. Corn and rice starches take a much longer time to prepare and a higher temperature to reach the same level of conversion. For top-quality work, cassava starch is thought to be ideal, because it is slightly stronger than a potato starch adhesive while being odourless and tasteless and excellent as an adhesive for postage stamps, envelope flaps and labels. Certain potato pastes have bitter tasting properties, while cereal starches exhibit a cereal flavour.

Banana: The banana plant has long been a source of fibre for high-quality textiles. Harvested shoots are first boiled in lye to prepare fibres for yarn making. These banana shoots produce fibres of varying degrees of softness, yielding yarns and textiles with differing qualities for specific uses. For example, the outermost fibres of the shoots are the coarsest and are suitable for tablecloths, while the softest innermost fibres are desirable for cloth-making process that requires many steps, all performed by hand. Banana fibre is used in the production of banana paper. Banana paper is used in two different senses: to refer to a paper made from the bark of the banana plant, mainly used for artistic purposes, or paper made from banana fibre, obtained with an industrialized process from the stem and the nonusable fruits. The paper itself can be either handmade or made by industrial processes. The average distribution of the world banana production in the 2003–2007 period is shown in Figure 7.6. There is abundance of starch available in agropolymer crops. Worldwide production of various starch yielding agropolymer crops in million metric tonnes is displayed in Table 7.4.

TABLE 7.4

Worldwide Production of Various Starch Yielding Agropolymer Crops in Million Metric Tonne

Producer Countries	Potato in 2006	Rice in 2007	Wheat in 2008	Cassava
World total	315	—	690	202.58
People's Republic of China	70	187	112	4.20
Russia	39	—	64	—
India	24	144	79	6.7
United States	20	—	68	—
Indonesia	—	57	—	19.26

Source: UN Food & Agriculture Organization (FAO)/fao.org from FAO statistics average on the 2003–2007 period.

FIGURE 7.7
Starch as a source of biopolymer (agropolymer). (Starch-Wikipedia, the free Encyclopedia, http://en.wikipedia.org/wiki/Starch.)

Barley: Barley (Figure 7.7) has a wider ecological range than any other cereal because it is more adaptable than other cereals. Barley can be grown on soils unsuitable for wheat and at altitudes unsuitable for wheat or oats. Because of its salt and drought tolerance, it can be grown near desert areas also.

Buckwheat: Buckwheat (Figure 7.7) refers to variety of plants in the dicot family. The name *buckwheat* or *beech wheat* comes from its triangular seeds, which resemble the much larger seeds of the beech nut from the beech tree and the fact that it is used like wheat. The starch content in seeds of buckwheat is as follows:

71%–78% in groats and 70%–91% in different types of flour. Starch is 25% amylose and 75% amylopectin. Depending on hydrothermal treatment, buckwheat groats contain 7%–37% of resistant starch.

Rye: Rye (Figure 7.7) is a cereal crop that is grown extensively for its grains. Scientifically known as *Secale cereale*, it belongs to the wheat tribe *Triticeae*. Rye bears a lot of resemblance to wheat and produces kernels in the same manner as wheat. However, the kernels of rye are much smaller as compared to those of wheat. Rye contains many healthy nutrients like dietary fibres and proteins.

Taro: Taro (Figure 7.7) is a tropical plant grown primarily as a vegetable food for its edible corm and secondarily as a leaf vegetable. Taro is loosely called elephant ear. Taro leaves are rich in vitamins and minerals. Taro corms are very high in starch and are a good source of dietary fibre. In North India, it is called *arbi*.

7.10 Fibres

Fibres are classified in different ways such as the following:

1. Natural and synthetic fibres
2. Continuous and short fibres
3. Organic and inorganic fibres

Natural fibres such as jute, hump, silk, felt, cotton and flax are obtained from natural sources such as plants, animals and minerals. Synthetic fibres are produced in industries. They are cheaper and more uniform in cross section than the natural fibres. Their diameters vary between 10 and 100 μm. Biofibres such as carbon and graphite fibres are light in weight, flexible, elastic and heat sensitive. Inorganic glass, tungsten and ceramic fibres have high strength, low fatigue resistance and good heat resistant. The strength of composites increases when it is made of long continuous fibres. A smaller diameter of fibres also enhances the overall strength of composite.

7.10.1 Natural Fibres [5]

Most natural leaf, stalk (bast) and seed fibres can be used in filling or reinforcing thermoplastics. Typically, the greater the aspect ratio (length/diameter) of fibre, the greater the improvement in properties over pure thermoplastic. Bast fibres are typically the best for improvements in tensile and bending strength and modulus. For toughness, the coarse fibres such as sisal and coir (coconut husk fibre) are best. In addition to fibres, fines from processing of wood, coir, agave, hemp and jute as well as rice and nut hulls, cereal straws (oat, rye, wheat, etc.) and corn cobs can be used to improve dimensional stability and stiffness of thermoplastics. Some examples of natural fibres and their category are listed as follows:

Leaf: pineapple

Bast: kenaf, hemp, jute, ramie, flax, sugarcane, banana, wood, bamboo, sisal

Seed: cotton, coconut, milkweed, rice hulls

Ramie fibres: Ramie is a flowering plant of the nettle family. The true ramie (or china grass) also called Chinese plant or white ramie is the Chinese cultivated plant. A second type, known as green ramie or rhea, has smaller leaves which are green on the underside. Ramie plant is shown in Figure 7.8. Ramie is one of the oldest fibre crops. It is a bast fibre, and the part used is the bark (phloem) of the vegetative stalks. Unlike other bast crops, ramie requires chemical processing to degum the fibre. The extraction of the fibre occurs in three stages.

Ramie is one of the strongest natural fibres. It exhibits greater strength when wet. It is not as durable as other fibres and so is usually used as a blend with other fibres such as *cotton* or *wool*. It is similar to flax in absorbency, density and microscopic appearance. Because of its high molecular crystallinity, ramie is stiff and brittle and breaks if folded repeatedly in the same place. It lacks resiliency and is low in elasticity and elongation potential. Ramie fibre has a moisture content of 8.0 wt% and fracture strain as 0.25%.

Sisal fibres. Sisal (i.e. *Agave sisalana*) is a plant of the agave family. Sisal fibres are made of the leaves of the plant. The fibre is usually obtained by machine decortications in which the leaf is crushed between rollers. The resulting pulp is scraped from the fibre, and the fibre is washed and then dried by mechanical or natural means. The lustrous fibre strands are usually creamy

FIGURE 7.8
Ramie plant.

FIGURE 7.9
(a) Sisal plant. (From Wigglesworth & Co. Limited, London, U.K.) (b) Sisal fibre. (From http://www.matbase.com/material/fibres/natural/sisal/properties.) (c) Banana tree. (From http://askpari.files.wordpress.com/2009/06/100_4807_banana_tree.jpg.) (d) Banana fibre. (From http://ropeinternational.com/images/uploaded_images/banana%20silky%20fibre.jpg.)

white, with an average length and diameter of 100–125 cm and 0.2–0.4 cm, respectively. Sisal plant and fibres are shown in Figure 7.9a and b.

Sisal fibre is fairly coarse and inflexible. It is valued for its strength, durability, ability to stretch, affinity for certain dyestuffs and resistance to deterioration in saltwater. Sisal ropes are employed for marine, agricultural, shipping and general industrial uses. The water absorption capacity of sisal fibre is 5.8%–6.1% and elongation at break is 4.3% [7]. Other mechanical and physical properties are given in Table 7.5.

Banana fibres: Banana fibre products are popular for their household utility. These utility items are like laundry basket, office waste paper basket and fruit or egg trays. Banana fibre products also serve as house deco. They can

TABLE 7.5

Characteristic Values for the Density, Diameter and Mechanical Properties of
Vegetable Fibres

Fibre	Density (g/cm³)	Diameter (µm)	Tensile Strength (MPa)	Young's Modulus (GPa)	Elongation at Break (%)
Ramie	1.55	—	400–938	61.4–128	1.2–3.8
Sisal	1.45	50–200	468–700	9.4–22	3–7
Banana	—	—	540–600	—	2.82–3
Coir	1.15–1.46	100–460	131–220	4–6	15–40
Flax	1.5	40–600	345–1500	27.6	2.7–3.2
Jute	1.3–1.49	25–200	393–800	13–26.5	1.16–1.5
Hemp	1.47	25–500	690	70	1.6
Cotton	1.5–1.6	12–38	287–800	5.5–12.6	7–8
Kenaf	—	—	930	53	1.6
Oil palm EFB	0.7–1.55	150–500	248	3.2	25

Source: Mohanty, A.K. et al., *Natural Fibers, Biopolymers and Biocomposites*, Taylor & Francis,
  CRC Press, Boca Raton, FL, 2005, p. 41, Table 2.1 and modified.

also be used as paper, for both industrial grade and stationery products. The
paper is made of 20% natural banana fibre and 80% postconsumer waste.
By using this method instead of wood fibre, the burden of deforestation is
assuaged. A banana tree and banana fibres are shown in Figure 7.9c and d.
Other mechanical and physical properties are given in Table 7.5.

Coir fibre: Coir fibre is produced in India, Sri Lanka and Thailand. Fibres
are gained from coconut husks. Approximately 40%–50% of a matured husk
consists of fibres. After harvesting, the retting process takes place for gaining white fibres. This fruit fibre is contained in the husk of coconuts. Fibre
length ranges from 100 to 300 m. Coir fibres are light in weight, strong and
elastic and have a low light resistance. They have a high durability (because
of the fibre composition, 35%–45% cellulose, 40%–45% lignin and 2.7%–4%
pectins and 0.15%–0.25% hemicelluloses). A coir tree and coir fibres are
shown in Figure 7.10a and b, respectively. Thermal conductivity of coir fibre
is 0.047 W/mK and value of elongation at break is 15%–17.3% [7]. Other
mechanical and physical properties are given in Table 7.5.

Flax fibres: Flax is obtained from flax fibre plant. Its botanical name is *Linum
usitatissimum*. It is commonly known as patsan, which is a substitute of *jute*.
The flax fibre is strong and wiry, longer and finer in nature. It lies in the category of bast (or soft) fibres. Flax fibre has a length of 75–120 cm and is valued
for its strength, lustre, durability and moisture absorbency. It absorbs little
dirt, is free of bacteria, does not cause fluffs and has good resistance against
bases. The flax tree and fibres are shown in Figure 7.10 c and d. The thermal
conductivity of flax fibre is 0.055 W/mK and the value of elongation at break is
1.5%–4%. The water absorption capacity is 8%–10% [7]. Other mechanical and
physical properties are given in Table 7.5.

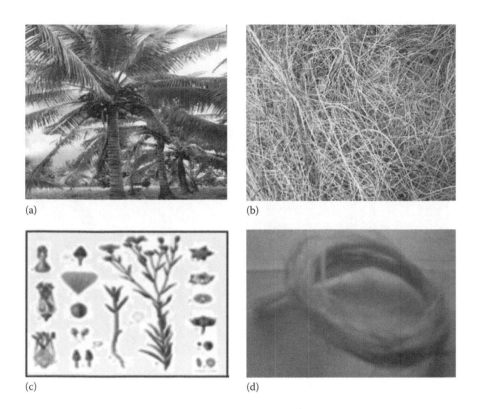

FIGURE 7.10
(a) Coir tree. (From http://fida.da.gov.ph/Coco%20tree.jpg.) (b) Coir fibre. (From http://product-image.tradeindia.com/00281409/b/0/Coir-Fibre.jpg.) (c) Flax fibre plant. (d) Natural flax fibre. (From Kumar, A., Dual matrix -single fibre hybrid composite, MTech thesis, Applied Mechanics Department, Moti Lal Nehru National Institute of Technology, Allahabad, India, under the supervision of Dr. K.M. Gupta, 2009.)

Jute fibres: It is a bast fibre and one of the cheapest natural fibres. It possesses a poor resistance against moisture, brittles under influence of light and absorbs paint easily. Jute consists of very short elementary fibres (length 0.7–6 mm), which are stuck together by lignin to form long brittle fibres (length of 300–400 mm). The jute tree and fibres are shown in Figure 7.11a and b. For jute fibre, the value of elongation is 0.8%–2% and the water absorption capacity is 2%–35% [7]. Other mechanical and physical properties are given in Table 7.5.

Hemp fibres: It is a bast fibre and is yellow-brown in colour. It resembles flax in appearance, but is coarser and harsher. It is strong, lightweight and has very little elongation. The hemp tree and fibres are shown in Figure 7.11c and d, respectively. The thermal conductivity of hemp fibre is 0.048 W/mK and the value of elongation at break is 1%–6%. The water absorption capacity is 8%–30% [7]. Other mechanical and physical properties are given in Table 7.5.

(a)   (b)

(c)   (d)

FIGURE 7.11
(a) Jute fibre plant. (b) Natural jute fibre. (From Srivastava, A., Mechanical characterization of jute epoxy hybrid composite, MTech thesis, Department of Applied Mechanics, Moti Lal Nehru National Institute of Technology, Allahabad, India, Submitted by under supervision of Dr. K.M. Gupta, 2009.) (c) Hemp in field. (From http://keetsa.com/blog/wp-content/uploads/2008/09/hemp_field2.jpg.) (d) Hemp fibre. (From http://www.hempsa.co.za/images/Fibre/HempFibreRaw.jpg.)

Cotton fibres. It is a seed fibre. The length of its fibres is 8–50 mm, diameter 12–20 μm and width 16–40 μm. Cotton fabrics are usually available in 1.2–1.4 m width. Well-known varieties are furniture fabric, velours, gobelin and velvet. A cotton plant and cotton fibres are shown in Figure 7.12a and b, respectively. For cotton fibre, the value of elongation at break is 3%–8%. And the water absorption capacity is 20%–100% [7]. Other mechanical and physical properties are given in Table 7.5.

Palmyra fibres: Palmyra fibre is also known as toddy, wine siwalana, lontar, or taluuria baha palm. It is obtained from palm (toddy) plant. It is a bast fibre. Its botanical name is *Borassus flabellifer.* Palmyra fibres are produced from the leaf sheath (petioles) of palmyra tree. The fibre is strong and wiry, shorter and finer in nature. It lies in the category of hard fibres. Palmyra tree and palmyra fibre are shown in Figure 7.12c and d. For palmyra fibre, the value of elongation at break is 3%–8%. Its specific strength is 70–270 MPa. Other mechanical and physical properties are given in Table 7.5.

(a) (c)

(b) (d)

FIGURE 7.12
(a) Cotton plant. (From http://www.djc.com/blogs/BuildingGreen/wp-content/uploads/2009/03/cotton-plant.jpg.) (b) Cotton fibre. (From http://www.diplomatie.gouv.fr/en/IMG/jpg/trans2_150.jpg.) (c) Palmyra tree. (d) Palmyra fibre. (From Gupta, K.M. and Srivastava, A., Tensile characterization of individual palmyra fibres, *International Conference on Recent Advances in Composite Materials (ICRACM-2007)*, Srivastava, V.K. et al. (eds.), Air Force Office of Scientific Research-Asian Office of Aerospace Research & Development and the Department of the Navy, Allied Publishers private ltd, New Delhi, India, February 2007, pp. 424–429.)

7.11 Classification of Starch-Based Biocomposites

The starch-based biocomposites may be classified in many ways such as the following:

1. On the basis of the nature of reinforcing material used:
   a. Particulate composite
   b. Fibre composite
   c. Flake composite
   d. Sandwich composite

    e. Hybrid composite

    f. Whisker-reinforced composite

2. Natural fibre–reinforced biocomposites [5–20] such as

    a. PLA–cellulose composite

    b. PLA–kenaf composite

    c. PLA–flax composite

    d. PLA–hemp composite

    e. PLA–bamboo composite

    f. PLA–jute composite

    g. PLA–wood fibre composite

    h. PLA–cotton fibre composite

    i. PLA–silkworm silk fibre composite for tissue engineering applications

    j. PLA–agriculture waste fibre composite (water bamboo husk, rice straw)

    k. PLA–rice straw

    l. PHBV–rice straw

    m. PLA nanocomposites

    n. PLA–chicken feather fibre composite

3. Synthetic fibre–reinforced biocomposites such as

    a. PLA–rayon composite

    b. PLA–PP composite

4. On the basis of the type of aspect ratio of fibres used:

    a. Short-fibre biocomposite

    b. Long-fibre biocomposite

5. Based on the arrangement of fibre lay:

    a. Unidirectional (U/D) composite

    b. Bidirectional or cross plied composite

    c. Angle-plied composite

    d. Off-axis composite

    e. Randomly oriented composite

6. On the basis of type of surface modification of fibre biocomposites:

    a. Modification by silane treatment

    b. Modification by alkali–silane treatment

    c. Modification by sodium hydroxide

   d. Modification by hybridization

   e. Modification by steam explosion

  7. On the basis of method of manufacturing/fabrication:

   a. Hot press composite

   b. Film stacking composite

   c. Injection moulded composite

   d. Microbraining technique composite

   e. Melting compound method

   f. Coupling agent

  8. Starch-based nanocomposites such as

   a. Starch nanoclay composite

   b. MMT–potato starch nanocomposite

   c. Cellulose–starch nanocomposite

   d. Sweet potato–OMMT nanocomposite

  9. Starch-based flexible films and coatings such as

   a. Chitosan–starch foam

   b. Cassava starch foam

References

1. M.Q. Zhang, M.Z. Rong, and X. Lu, Fully biodegradable natural fibre composites from renewable resources: All-plant fibre composites. *Composites Science and Technology*, 65, 2005, 2514–2525.
2. A. Ashori, Wood–plastic composites as promising green-composites for automotive industries, *Bioresource Technology*, 99, 4661–4667, 2008.
3. http://www.edmunds.com/advice/alternativefuels/articles/105341/article.html.
4. Biomer, Germany, http://www.biomer.de/.
5. K.M. Gupta, Chapter 8 Starch based composites for packaging applications, in *Handbook of Bioplastics and Biocomposites Engineering Applications*, Dr. Srikanth Pilla (ed.), Scrivener Publishing LLC, Salem, MA, June 2011, pp. 189–266.
6. Starch-Wikipedia, the free Encyclopedia, http://en.wikipedia.org/wiki/Starch.
7. http://www.matbase.com/material/fibres/natural/.
8. A.K. Mohanty, M. Misra, and L.T. Drazal, *Natural Fibers, Biopolymers, and Biocomposites*, Taylor & Francis, CRC Press, Boca Raton, FL, 2005.
9. Wigglesworth & Co. Limited, London, U.K.
10. http://www.matbase.com/material/fibres/natural/sisal/properties.
11. http://askpari.files.wordpress.com/2009/06/100_4807_banana_tree.jpg.

12. http://ropeinternational.com/images/uploaded_images/banana%20 silky%20fibre.jpg).
13. http://fida.da.gov.ph/Coco%20tree.jpg.
14. http://product-image.tradeindia.com/00281409/b/0/Coir-Fibre.jpg.
15. A. Kumar (under the supervision of Dr. K.M. Gupta), Dual matrix -single fibre hybrid composite, MTech thesis, Applied Mechanics Department, Moti Lal Nehru National Institute of Technology, Allahabad, India, 2009.
16. A. Srivastava (under the supervision of Dr. K.M. Gupta), Mechanical character-ization of jute epoxy hybrid composite, MTech thesis, Department of Applied Mechanics, Moti Lal Nehru National Institute of Technology, Allahabad, India, 2009.
17. http://keetsa.com/blog/wp-content/uploads/2008/09/hemp_field2.jpg.
18. http://www.hempsa.co.za/images/Fibre/HempFibreRaw.jpg.
19. http://www.djc.com/blogs/BuildingGreen/wp-content/uploads/2009/03/cotton-plant.jpg.
20. http://www.diplomatie.gouv.fr/en/IMG/jpg/trans2_150.jpg.
21. K.M. Gupta and A. Srivastava, Tensile characterization of individual pal-myra fibres, *International Conference on Recent Advances in Composite Materials (ICRACM-2007)*, V.K. Srivastava et al. (eds.), Air Force Office of Scientific Research-Asian Office of Aerospace Research & Development, U.S.A, and the Department of the Navy, U.S.A; Allied Publishers private ltd, New Delhi, India, February 2007, pp. 424–429.

8

Special Kinds of Composites

8.1 Composites for Marine Applications

8.1.1 Introduction of Composites to Marine Industry

There have been numerous applications of polymer engineering composites in marine industry to compete with and complement conventional materials. Typical applications include boat hulls, sonar domes, masts, tanks, diving equipment, decks and oil platforms. These applications are due to the unique characteristics associated with polymer composites that are

- High strength and stiffness
- Low weight
- Resistance to aggressive marine, water and high UV environments
- Monolithic and seamless construction without leakage and assembly problems
- Durability and ease of maintenance and repair
- High energy absorption
- Design flexibility for specific requirements
- Good dielectric, absence of magnetic properties and low thermal conductivity

The main material involved in the marine industry is glass fibre where it is extensively used. Prior to 1980, glass fibre–reinforced plastic (GFRP) was used on small yachts and boats. The industry did not fully trust the polyester resins and orthophthalic resin gel coat materials due to the porous nature. Modern epoxy and the more expensive polyurethane resins have proved more resistant to water ingress and thus more suitable to larger structures such as ships to which the greatest developments have been made. Additionally, glass fibre has been combined with carbon and Kevlar fibres to provide lighter, stronger and tougher materials.

8.1.2 Current Uses of Composites in Marine Applications

Major uses of composites in marine applications are the following and shown in Figure 8.1a through e:

- Minesweeper
- Small boats
- Fishing boats
- Larger passenger and cargo vessels
- Racing yachts
- Small sailing boats and sailing skiffs
- Surfboards and wind surfers
- Canoes and kayaks

Minesweeper or minesweep vessels are one of the largest glass fibre–reinforced polymer composite boats fabricated. In addition to generic advantages, a nonmagnetic characteristic is the unique feature of these materials compared to metals. These military vessels are immune to the threat from

(a)   (b)

(c)   (d)   (e)

FIGURE 8.1
(a) Minesweeper. (b–e) Racing Yacht.

magnetic mines. Their lightweight and corrosion resistance in marine environment make them attractive and competitive for use in military minesweeper applications. Navies in many different countries have built composite mine hunters and have successfully accomplished their objectives. Other naval applications of fibre-reinforced polymers (FRPs) include construction of landing craft, fast patrol boats and submersibles.

Racing yachts: Racing yachts use composites more extensively than any other marine structure. The materials used are not typical of marine construction because of special requirements. Minimal weight and maximal stiffness are crucially important in their design so that they can sail with maximal speed and resistance to the impact of waves and other elements in marine environments. Instead of conventional glass fibre–reinforced polyester composites, racing yachts are made from advanced polymer composites using aerospace materials. Carbon fibre–reinforced epoxy composites are usually used in the hull cored with aluminium honeycomb, frames, keels, masts, poles and boom, carbon winch drums and shafting. The use of advanced polymer composites can contribute to improved performance and minimize the danger of sailing drawbacks and failure in the different international sailing conditions.

8.1.3 Desired Requirements of Composites in Ships and Marine Structures

The short-term goal of testing composites for the shipbuilding industry is to design a lightweight, robust, low-cost seaworthy vessel that won't capsize or corrode prematurely increasing maintenance costs. The long-term objective is to gain data that will become instrumental in refining and improving ship designs. Replacing metals with composite materials has numerous benefits, and the marine industry has merely skimmed the surface. Over time, rigorous testing will prove composites are superior materials for many marine applications.

Corrosion resistance of marine composites: Another reason that composites are being used in shipbuilding and various marine applications is its noncorrosive nature. Unlike metals that corrode and decay, composites last for many years. Consequently, composites are ideal for combating nonambient environments like extreme temperatures and seawater. And with some marine apparatus like propeller shafts, buoys and light stations that are designed to remain in the water at all times, use of composite materials ensures long life.

Adhesive testing: Similarly, because composite structures and components are often bonded together with adhesives, the bonded joints must be tested as well. Static and cyclic tests measure and detect bond strength, debonding modes, peel failure and fatigue life.

Damage tolerance: Testing also helps shed light on damage tolerance. Test results establish if one extreme event will cause catastrophic damage or if it will take many years of cumulative damage use to render a ship unsafe.

8.1.4 Resins in Marine Applications

Polyester: Polyester resins are the simplest, most economical resin systems that are easiest to use and show good chemical resistance. Unsaturated polyesters (UPEs) consist of unsaturated material, such as maleic anhydride or fumaric acid, that is dissolved in a reactive monomer, such as styrene. Polyester resins have long been considered the least toxic thermoset to personnel. The two basic polyester resins used in the marine industry are orthophthalic and isophthalic. The ortho resins were the original group of polyesters developed and are still in widespread use. They have somewhat limited thermal stability, chemical resistance and processability characteristics. The iso resins generally have better mechanical properties and show better chemical resistance. Their increased resistance to water permeation has prompted many builders to use this resin as a gel coat or barrier coat in marine laminates. The rigidity of polyester resins can be lessened by increasing the ratio of saturated to unsaturated acids. Flexible resins may be advantageous for increased impact resistance; however, this comes at the expense of overall hull girder stiffness. Curing of polyester without addition of heat is accomplished by adding accelerator along with the catalyst. Gel times can be carefully controlled by modifying formulations to match ambient temperature conditions and laminate thickness.

Vinyl ester (VE): VE resins are unsaturated resins prepared by the reaction of a monofunctional unsaturated acid, such as methacrylic or acrylic, with a bisphenol diepoxide. The resulting polymer is mixed with an unsaturated monomer, such as styrene. The handling and performance characteristics of VEs are similar to polyesters. Some advantages of the VEs, which may justify their higher cost, include superior corrosion resistance, hydrolytic stability and excellent physical properties, such as impact and fatigue resistance. It has been shown that a 20–60 mil layer with a VE resin matrix can provide an excellent permeation barrier to resist blistering in marine laminates.

Epoxy: Epoxy resins are a broad family of materials that contain a reactive functional group in their molecular structure. Epoxy resins show the best performance characteristics of all the resins used in the marine industry. Aerospace applications use epoxy almost exclusively, except when high-temperature performance is critical. The high cost of epoxies and handling difficulties have limited their use for large marine structures.

Thermoplastics: Thermoplastics have 1D or 2D molecular structures, as opposed to 3D structures for thermosets. The thermoplastics generally come in the form of moulding compounds that soften at high temperatures. Polyethylene, polystyrene, polypropylene, polyamides and nylon are examples of thermoplastics. Their use in the marine industry has generally been limited to small boats and recreational items. Some attractive features include no exotherm upon cure, which has plagued filament winding of extremely thick sections with thermosets, and enhanced damage tolerance.

8.1.5 Core Materials in Marine Applications

Balsa: End-grain balsa's closed-cell structure consists of elongated, prismatic cells with a length (grain direction) that is approximately 16 times the diameter (see Figure 8.2). In densities between 0.1 and 0.25 g/cm³, the material exhibits excellent stiffness and bond strength. Stiffness and strength characteristics are much like aerospace honeycomb cores, although the static strength of balsa panels is higher than the PVC foams and impact energy absorption is lower. Local impact resistance is very good because stress is efficiently transmitted between sandwich skins. End-grain balsa is available in sheet form for flat panel construction or in a scrim-backed block arrangement that conforms to complex curves.

Thermoset foams: Foamed plastics such as cellular cellulose acetate (CCA), polystyrene and polyurethane are very light and resist water, fungi and decay. These materials have very low mechanical properties and polystyrene will be attacked by polyester resin. Use is generally limited to buoyancy rather than structural applications. Polyurethane is often foamed in place when used as a buoyancy material.

Syntactic foams: Syntactic foams are made by mixing hollow microspheres of glass, epoxy and phenolic into fluid resin with additives and curing agents to form a mouldable, curable, lightweight fluid mass.

Cross-linked PVC foams: Polyvinyl foam cores are manufactured by combining a polyvinyl copolymer with stabilizers, plasticizers, cross-linking compounds and blowing agents. The mixture is heated under pressure to initiate the cross-linking reaction and then submerged in hot water tanks to expand to the desired density. The resulting material is thermoplastic, enabling the

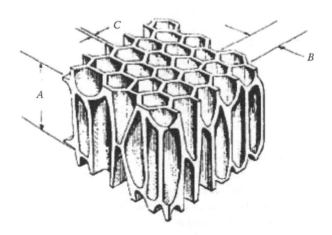

FIGURE 8.2
Balsa cell geometry with A = average cell length = 0.635 mm; B = average cell diameter = 0.35 mm; and C = average cell wall thickness = 0.0015 mm. (From Baltek Corporation, Northvale, NJ.)

material to conform to compound curves of a hull. PVC foams have almost exclusively replaced urethane foams as a structural core material, except in configurations where the foam is *blown* in place.

Linear PVC foam: Unique mechanical properties of linear PVC foam core are a result of a nonconnected molecular structure, which allows significant displacements before failure. In comparison to the cross-linked (nonlinear) PVCs, static properties will be less favourable and impact will be better.

Honeycomb: Constituent materials include aluminium, phenolic resin–impregnated fibreglass, polypropylene and aramid fibre phenolic–treated paper. Physical properties vary linearly with density. Although the fabrication of extremely lightweight panels is possible with honeycomb cores, applications in a marine environment are limited due to the difficulty of bonding to complex face geometries and the potential for significant water absorption.

8.2 Fire-Resistant Composites

8.2.1 Flammability Parameters

The fire behaviour of a material is commonly characterized using two parameters, the heat release rate (HRR) and the specific optical smoke density (D_s). The HRR is the rate at which heat energy is evolved by a material when burned and is used to measure how large and how quickly a fire environment grows. Recently, advances in fire research and fire dynamics have emphasized the importance of the HRR as the primary fire hazard indicator of a material. The rate of heat release, especially the peak HRR, is the primary characteristic determining the size, growth and suppression requirements of a fire environment. To quantify the heat release properties of any material, the Ohio State University (OSU) rate of heat release test is utilized in accordance with ASTM test method E906. The HRR is expressed in terms of power per unit area (kW/m^2) and reaches maximum value when a material is burning most intensely. In contrast, the heat release of a burning material is the amount of heat energy evolved and is expressed in terms of energy per unit area ($kW\ min/m^2$).

The Ds is a dimensionless measure of the amount of smoke produced per unit area when a material is exposed to both flaming and radiant heat sources. To determine this smoke-generating characteristic, the National Bureau of Standards (NBS) smoke test is employed in which a material is burned in the NBS smoke chamber. The maximum value of Ds that occurs during the first 4 min of the test is the most important parameter measured.

8.2.2 Aluminosilicate–Carbon Composites: A Geopolymer Fire-Resistant Composite [1]

Flammability of organic matrices limits the use of fibre-reinforced composites in applications where fire is an important design parameter, such as the interior of aircraft. Many of the commonly used organic matrix materials soften and ignite at 400°C–600°C. This is unacceptable in structures where egress is restricted. Composites made using inorganic matrices can be utilized when high use temperatures are expected.

Geopolymer composites have fire properties that are superior to all organic matrix composites currently available. The fire properties are weight loss, time to ignition, peak HRR, 300 s average HRR, total heat release and smoke production. Geopolymer's fire properties to organic matrix materials that are currently used in aircraft interiors are better. In addition to these properties, perhaps the best indicator of fire resistance is the predicted time to flashover value. Flashover occurs in closed compartments when flammable gases from material combustion are finally heated to a point where they ignite. This event marks the end of human survivability in post-crash scenarios. The geopolymer composite had an infinite time to flashover. This is the logical result for an inorganic matrix composite. There is no flammable material in the composite; hence, there can never be a flashover.

Composites made with 3k plain weave carbon fabric, and geopolymer had a tensile strength of 327 MPa and a flexural strength of 245 MPa. Both of these values are comparable to the strengths from similar organic matrix composites. Geopolymer composite samples retained 63% of their flexural load-carrying ability after 1 h of exposure at 800°C. In shear, geopolymer samples had a strength of 14 MPa. This strength decreases to a value of 4.6 MPa for samples heated to 1000°C for 1 h. Under fatigue loading, composites made with geopolymer matrix can sustain about 10 million cycles at a stress.

8.2.3 Fire Resistance of Inorganic Sawdust Biocomposite [2]

Wood plastic composites (WPCs) or biocomposites are relatively new categories of materials that cover a broad range of composite materials, utilizing an organic resin binder (matrix) and fillers composed of cellulose materials. The new and rapidly developing biocomposite materials are high-technology products, which have one unique advantage—the wood filler can include sawdust and scrap wood products.

Consequently, no additional wood resources are needed to manufacture biocomposites. Waste products that would traditionally cost money for proper disposal now become a beneficial resource, allowing recycling to be both profitable and environmentally conscious. The use of biocomposites and WPCs has increased rapidly all over the world, with the end users for these composites in the construction, motor vehicle and furniture industries.

One of the primary problems related to the use of biocomposites is the flammability of the two main components (organic binder and cellulose-based filler). If a flame retardant were added, this would require the adhesion between fibre and matrix not be disturbed by the retardant. The challenge is to develop a fire-resistant biocomposite that will maintain its level of mechanical performance. In lieu of organic matrix compounds, inorganic matrices can be utilized to improve the fire resistance. Inorganic-based wood composites are those that consist of a mineral mix as the binder system. Such inorganic binder systems include gypsum and Portland cement, both of which are highly resistant to fire and insects. The main disadvantage with these systems is that the maximum amount of sawdust or fibres that can be incorporated is low. This drawback stems from the inherently high viscosity of the inorganic resin, which reduces the ability of the sawdust particles to become fully saturated with resin during the mixing process. In addition, chemical sizings placed on reinforcing fibres are specifically engineered to improve the bond between fibre and organic resin. Since the chemical makeup of organic and inorganic resins differs considerably, these sizings often interfere with bonding between fibres and inorganic resin. As a result, less fibre can be successfully incorporated in the composite.

The primary objective of the research reported in this paper was to fabricate fire-resistant biocomposite sandwich plates by combining the aforementioned potassium aluminosilicate matrix with waste sawdust. Earlier research by the authors found that the compressive and flexural strengths of this biocomposite were approximately 39 and 1.79 MPa, respectively, while the increase of 265% in flexural capacity was achievable with one row of carbon fibre reinforcement on the tension face of a biocomposite beam. Using similar formulations of the biocomposite used to establish these mechanical properties, the authors investigated the fire response of the material in this paper. Small plates of this biocomposite material were reinforced with glass and carbon fibres to examine the effect of reinforcement type on HRR and smoke emission of the resulting sandwich plate.

8.2.4 Potassium Aluminosilicate Matrix [2]

One relatively new type of inorganic matrix is potassium aluminosilicate, an environmentally friendly compound made from naturally occurring materials. This matrix is a two-part system consisting of an alumina liquid and a silica powder that cures at a reasonably low temperature of 150°C. In addition, hardeners can be added to facilitate room-temperature curing. The Federal Aviation Administration has investigated the feasibility of using this matrix in commercial aircraft due to its ability to resist temperatures of up to 1000°C without generating smoke and its ability to enable carbon composites to withstand temperatures of 800°C and maintain 63% of their original flexural strength. Potassium aluminosilicate matrices are compatible with many common building materials including clay brick, masonry, concrete, steel, titanium, balsa, oak, pine and particleboard. Processing requirements and mechanical properties of carbon/carbon composites; ceramic matrix composites made with silicon carbide, silicon nitride and alumina fibres; and carbon/potassium aluminosilicate composites indicated that the carbon/ potassium aluminosilicate composites have mechanical properties that are better than most fire-resistant composites.

8.2.5 Fire-Resistant Ecocomposites Using Polyhedral Oligomeric Silsesquioxane Fire Retardants [3]

Polymer matrix composites (PMCs) are used in military, automotive and civil infrastructure applications because of their overall good thermal, mechanical and electrical properties; low weight; and low cost compared to conventional materials. VE resins are often preferred resin materials because they offer ease of processing and lower cost over epoxy resins while having superior properties relative to UPE resins. Manufacture of these composites, however, generates hazardous air pollutants (HAPs), including styrene and methyl methacrylate, during all stages of composite manufacture and during the lifetime of the part.

PMCs are also highly flammable and can produce toxic gases during combustion. Polymers and PMC materials soften and ignite at much lower temperatures than conventional metals they are often designed to replace. When they do ignite, polymers are very flammable and produce toxic gases, soot and smoke, which are a health and environmental impediment to their full implementation in numerous engineering designs. Fire retardants, including halogenated organics (brominated or chlorinated), phosphorous-based chemicals, nitrogen-based chemicals and inorganic hydroxides and oxides, are added to polymers and PMCs to reduce flammability and arrest flame spread and the production of gaseous and combustible vapours and contaminants.

The class of fire retardant used depends upon the choice of polymer resin. Bromination is the most common means to render resins fire retardant. Brominated diglycidyl ether of bisphenol A is used to prepare fire-retardant VE and epoxy resins, while tetrabromophthalic anhydride and dibromoneopentyl glycol are used in the chemical synthesis of UPE resins. Yet halogenated chemicals produce toxic smoke when exposed to high temperatures and fire. Phosphorous-based chemicals, including triethyl phosphate and dimethyl methylphosphonate, are used in epoxy resins, but not generally in VE or UPE resins. Furthermore, phosphorous additives plasticize the resin, thereby reducing modulus glass transition Tg and strength. There are also environmental concerns associated with these phosphorous additives. Inorganic oxides, including antimony trioxide, do not improve fire retardancy on their own. These additives act as synergists only, improving the fire retardancy effects of bromine or phosphorous fire retardants.

Metal hydrates, such as magnesium hydroxide and aluminium trihydrate, require high loadings to effectively reduce polymer flammability.

Polyhedral oligomeric silsesquioxanes (POSSs) are new-generation materials that can be used as flame retardants in thermosetting resins. These materials that are chemically modified particles of silica have dimensions at the nanometre scale. Their inorganic (silicon based) structure provides thermal stability and fire resistance, whereas their organic (carbon based) structure provides compatibility and/or reactivity with the resins. This covalent bonding allows the POSS to react into the thermoset network and reinforce the polymer on the molecular level. These chains act as nanoscale reinforcement that provides exceptional gains in heat resistance. POSS has been used as an additive in heat-resistant paints and coatings, mechanical property modifiers, cross-linking agents and viscosity modifiers. Because POSS exists on the nanoscale, the motions of the chain can be controlled and the processing and mouldability of the resin are theoretically maintained. POSS is easily incorporated into common plastics via copolymerization or blending and hence requires little or no alteration to existing manufacturing processes. In addition, POSS molecules are odourless, release no volatile organic compounds, and are therefore environmentally friendly even when ignited.

Fabrication and use of UPE and VE composites produces significant amounts of volatile emissions during mixing, moulding and fielding. A potential solution for manufacturing National Emission Standards for Hazardous Air Pollutant (NESHAP)-compliant VE resins with polymer properties and performance similar to that of commercially available VE resins has been developed. Fatty acid vinyl esters (FAVEs) use fatty acid monomers, such as methacrylated lauric acid (MLau), as a reactive diluent to replace all but 10%–25% of the resin styrene content in VE resins. The FAVE resins have similar resin viscosity and polymer properties relative to commercial VE resins. Fatty acid monomers are excellent alternatives to styrene due to their low cost and extremely low volatility. In addition, fatty acids are renewable resources because they are derived from plant oils. Therefore, not only would the use of fatty acid monomers reduce HAP emissions in liquid moulding resins, thereby reducing health and environmental risks, but also would promote global sustainability.

8.2.6 Fire-Resistant Behaviour of Bottom Ash–Based Cementitious Coating–Applied Concrete Tunnel Lining [4]

Underground structures such as tunnels are situated in confined locations where accidental fires can result in severe human casualties and structural damage. In addition, given the high relative humidity in tunnels (e.g. relative humidity of 75% or greater) when compared to those of air-exposed buildings in general (e.g. relative humidity of 50%), the risk of occurrence of explosive thermal spalling in tunnels during a fire increases due to the availability of moisture-entrained concrete pores. Hydrocarbon-type tunnel fires are also more severe than cellulose-type building fires because of the described confined conditions. Therefore, tunnels are subject to very stringent and strictly enforced fire safety requirements and preventive measures, which are gaining great importance in tunnel design. In order to prevent problems such as those mentioned earlier, clients typically request that tunnel linings be fire resistant and that concrete linings include fire protection coatings.

In light of these unfavourable conditions surrounding tunnel structures, since the early 1990s, concrete lining materials have exhibited delamination and spalling due to cracks and water leakage requiring repairs. In particular, major economic and social losses are caused by fires in road and railway tunnels. More specifically, loss of concrete strength due to fires in enclosed spaces such as tunnels can not only cause structural collapse but may also bring about long-term blockage of traffic infrastructure. Therefore, various types of fire-resistant materials for tunnel usages have been and are being developed. There are four main methods of fire protection for tunnels. The first involves increasing a structure's fire resistance by improving the tunnel's main material: for instance, shotcrete is manufactured to be fire resistant. The second method involves treating the structure's surface with a coating material that is able to prolong heat transfer, while the third fire-prevention method involves spraying a secondary lining onto the surface material. The final method involves installing fire-resistant precast panels or boards.

Recently, many researchers have focused their studies on developing fire protection coating materials. In many cases, lightweight and porous aggregates such as shell sand are adopted for insulation. However, when using these lightweight aggregates for fire protection coatings, the compressive and tensile strengths of the resulting coating materials tend to be low. The average compressive strength of fire protection coating materials currently sold in the market is under 10 MPa. On the other hand, the surface pressure from passing traffic in tunnels is estimated to be in the range of 25–600 Pa in general. In the case of high-speed railway tunnels, the surface pressures and vibrations are much higher. The high pressures and vibrations induced within tunnels could cause spalling and fatigue failure due to negative pressure being applied to the concrete lining. On the other hand, poly-fibres are adopted in concrete for releasing internal evaporative pressure caused by high thermal loading. Fibres are effective in providing channels for the release of internal pressure via melting at high temperatures and connecting internal pores. However, since the addition of fibres makes the concrete more porous, the strength of the concrete is likely to be degraded.

8.2.7 Fire-Resistant Polypropylene Fibre–Reinforced Cement Composites [5]

Large-scale fires often occur in tunnels and underground spaces worldwide, resulting in disastrous economic and social consequences. When a fire occurs in an enclosed space, such as a tunnel, the strength of the concrete is weakened, resulting in spalling as well as the total breakdown of the infra-structure system. The massive fires in the Mont Blanc and Tauern tunnels in Italy and Austria, respectively, resulted in many human casualties and huge repair costs due to structural damage. Therefore, there is an increas-ing awareness and interest in the safety aspects of tunnel fires. In European countries, the serious impact of these fires has been acknowledged and related research is being performed.

Concrete is widely used as the main material for tunnels and underground structures because of its excellent fire-resistant properties. However, during a fire, it is subjected to high temperatures that cause spalling. As a result, the heat is transferred to the inside of the structures, which weakens and occa-sionally destroys them. Such spalling occurs easily in high-strength concrete with a dense internal structure and a low ratio of water to cement. Spalling can also be caused by the condition of the submaterials, the functional ratio within the concrete and certain types of pressure. The type of transportation vehicle that causes the fire also plays a role in the amount of spalling.

Research and development of products with polypropylene fibre has been undertaken to reduce the amount of spalling in concrete. The spalling resis-tance of concrete is affected by the type and length of the polypropylene fibres. Multifilament polypropylene fibres are more effective at spalling resistance than fibrillated bundle polypropylene fibres. This is because the diameter of multifilament polypropylene fibres is less than that of fibrillated bundle polypropylene fibres, so the former has more fibres per unit dimen-sion than the latter, forming a matrix with excellent air permeability.

Polypropylene fibre mixtures have several important properties. Among these is a distinctive decrease in the void pressure within concrete exposed to high temperatures. As the volume fraction of polypropylene fibre increases, the internal pressure of the concrete decreases. When concrete with poly-propylene fibres is exposed to temperatures of 400°C, more cracks form compared to concrete without polypropylene fibres. The cracks are smaller than 1 μm and are formed in densely packed regions between thin and thick aggregates. In contrast, a relatively small number of larger 10 μm cracks form in concrete without polypropylene fibres because the fibre bed creates cracks during the transitional expansion–contraction phase when the aggregates expand due to heat and the concrete contracts due to dehydration of the residual water in the cement paste.

When the fibres dissolve with water, the concrete expands, creating a tensile strength within its matrix and hence generating cracks. The air per-meability of concrete mixed with polypropylene fibres increases with the internal temperature of the concrete due to two changing factors within the concrete microstructures. The first factor is the size of the void. The size of the void increases at temperatures between 80°C and 300°C, and the air per-meability is dependent on the crack size especially at temperatures above 400°C. Accordingly, the air permeability of concrete mixed with polypropyl-ene fibres increases radically when the temperature is below 200°C, which is the starting point of fibre dissolution for a mixture ratio of 3 kg/m^3. However, once the fibres are dissolved, the degree of increase in the air permeability coefficient radically decreases.

Various studies have attempted to reduce the amount of concrete spalling and enhance the fire resistance when fires occur in tunnels or underground structures. Fire-resistant spray materials have been used for steel structure materials and are increasingly applied as fire-resistant materials in tunnels and underground structures. Although the main function of such spray materials is to ensure an effective level of fire resistance, the strength of the materials and the bond strength with concrete are low, resulting in falling off and exfoliation problems. These are exacerbated if the materials are exposed to an environment with large vibrations and internal air pressures, such as that found inside road and railway tunnels.

The objective of this study [5] was to prevent fire-resistant spray materials from falling off or exfoliating from the tunnel or underground concrete structure over the long term and to determine an appropriate mixing ratio of fire-resistant wet-mixed high-strength polypropylene fibre–reinforced sprayed polymer cement composite that enables fast and efficient construction.

8.3 Eco-Friendly Fireproof High-Strength Polymer Cementitious Composites [6]

Tunnels and their associated underground structures are steadily growing in size. Because of this, the use of high-strength concrete, which improves the physical and mechanical performance of concrete, has been increasing. In addition, traffic volume has increased greatly due to population increase and industrial development, leading to an increased risk of fires occurring as a result of traffic accidents. A tunnel and its underground structure are an enclosed space, rendering extinguishing a fire and evacuating people very difficult. Aside from the loss of life, potential long-term disruption of the traffic network is another issue.

The best countermeasure technology to prevent damage to a tunnel and its underground structure in a fire, to enable safe evacuation of people and to recover traffic flow after a fire is fireproofing. In Europe and Japan, fire-proof countermeasures have recently been applied, depending on the risk of fire in the tunnel or underground structure. Research on fireproof materials and construction methods has been advancing. The most general way to fireproof a structure is to apply a fireproof covering, such as fireproof panels attached with fireproof mortar. However, these fireproof materials use cement as the main component, which is not in accord with the global desire to reduce carbon dioxide emissions. Carbon dioxide is the main cause of global warming. About 1 tonnes of carbon dioxide is created from the production of 1 tonnes of cement; hence, there is great interest in reducing the cement content of fireproof coatings.

Previous investigations have focused on developing more environmentally friendly mortars and concretes by using industrial by-products, for example, pozzolans such as fly ash or blast furnace slag, instead of cement. In this study [6], the formulation of an eco-friendly fireproof high-strength polymer cementitious composite suitable for various concrete structures has been optimized. In this study, the use of cement was minimized by adding blast furnace slag (an industrial by-product), porcelain (which forms a dense structure at high temperature) and polypropylene fibre.

8.3.1 Materials and Formulation

Type I cement was used in this study. The fine aggregates were quartz sand and expanded perlite having a specific gravity 0.15. Expanded perlite is an aggregate with 90% porosity and low weight (density of 0.05–0.30 g/cm^3) and provides thermal insulation (thermal conductivity of 0.03–0.05 kcal/h m K). In addition, since the material is a nonflammable inorganic substance, there is no risk of poisonous gas creation during a fire. Polypropylene fibre reduces mortar cracking and explosive spalling by melting during a fire, thereby reducing the vapour pressure inside the mortar. Fibrillated bundle polypropylene fibres, 35 μm in diameter and 6 mm long, were used in this study. The blast furnace slag used was an industrial by-product of the manufacture of pig iron; about 300 kg is produced per ton of pig iron. Pulverized furnace slag is ground granulated blast furnace slag and has angular particles. Sufficient quantities of OH$^-$ or SO$_4^{-2}$ ions are required for ground granulated blast furnace slag to be hardened inside a cement composite because of its latent hydraulic activity. The main components of the ground granulated blast furnace slag used in this study were CaO, SiO2 and Al$_2$O$_3$. It had hydraulic activity like cement due to stimulating material. It caused a pozzolan reaction that created the insoluble hardening material when it reacted with the Ca(OH)$_2$ created during the hydration reaction. The blast furnace slag was a pozzolan admixture having similar chemical elements as cement and could be regarded as a more environmentally friendly material since an industrial by-product was being used as a substitute for cement. Accordingly, blast furnace slag was used to develop an environmentally friendly fireproof material having a reduced carbon footprint.

An activator is necessary to function as a catalyst to induce the hydration of blast furnace slag. The strong base NaOH was used in initial experiments with activated slag, but it was difficult to handle due to its deliquescence. Sodium silicate (Na$_2$O–nSiO$_2$), known as water glass, was considered as an alternative activator; however, there are health hazards associated with SiO$_2$. Hence, the safer powdered anhydrous sodium metasilicate was as an activator. A ratio of 1.0:0.2 of blast furnace slag to meta sodium silicate was used. When heated above 1000°C, concrete and mortar made mainly from cement experience a great reduction in strength, that is, its residual compressive strength decreases by more than 80%. A cement composite applied to an underground structure and tunnel must have high residual strength to prevent extensive damage during a fire. Porcelain was used in this study to improve the residual strength of cement composites exposed to high temperature. Porcelain exposed to high temperature over 1000°C acquires a vitreous structure by combining with silicate ions and melted alkali oxides such as K_2O and Na_2O. Porcelain in its original form appears as a white clay. But it must be added as a powder in a cement composite. Therefore, the porcelain clay was fully dried in a drying oven at 110°C and then crushed into a powder using a pulverizer. The powder that passed through a number 16 sieve (1.18 mm) was accepted for further use.

The polymer was used as an admixture to ensure a sufficient amount of air in the concrete and to improve the bond strength between the cement paste and the aggregate. The polymer was in the form of small spherical particles, 0.5–5.0 μm in diameter, and coated with surfactant. All particles remained in suspension because of the adsorbed surfactant. Using the polymer suspension improved the workability of the unhardened cementitious composite and enabled a reduced water–cement ratio, and this increased the strength. During the hydration reaction, latex particles formed a film and reduced the permeability by filling pores and attaching the hydration products to the aggregate surface. In addition, the latex film increased the tensile strength.

8.3.2 Mix Proportion

Previous work established the optimum formulation for a high-strength fireproof polymer cement mortar and was taken as the basic formulation in the work reported here. This formulation had a compressive strength greater than 40 MPa at 28 days. Light aggregate perlite and polypropylene fibre were used to improve the resistance to high temperature. Finally, the polymer was used to improve the tensile and flexural strengths, which were key mechanical characteristics of the cement composite. The experimental variables were the substitution ratio of blast furnace slag, the addition ratio of porcelain and the volume fraction of polypropylene fibre. Substitution levels of 0, 50 and 100 wt.% of blast furnace slag for cement were set to evaluate the mechanical characteristics. The porcelain was added at 0, 15 and 30 wt.% to the cement to assess the improvement in residual strength following exposure to fire.

8.3.3 Results and Discussion

The compressive strength test results for the eco-friendly fireproof high-strength polymer cementitious composites indicated that each mixture except the ones with 50% of the cement replaced with blast furnace slag gave compressive strengths exceeding 40 MPa. The mixtures with 100% of blast furnace slag showed higher compressive strength than the mixture with cement. The improvement in strength with ageing was attributed to latent hydraulic activity, not only from the dense structure of the blast furnace slag caused by the fineness of the material. However, the mixtures containing 50% substituted blast furnace slag showed lower compressive strengths than the other mixtures. This was because less sodium silicate was present in these mixtures compared to those with 100% blast furnace slag. The sodium silicate leached the alkali elements from the blast furnace slag, and the material was strengthened as C–S–H bonds were formed with Si ions. However, when only 50% blast furnace slag was used, the resulting material did not have sufficient strength due to the lower concentration of SiO_4.

As the ratio of porcelain increased, so did the strength. The compressive strength was reduced as the volume fraction of the polypropylene fibre increased. This may be due to the difference in density between the polypropylene fibre and the cement matrix. The particle size of porcelain was very small (4–6 μm), and it filled the pores inside the structure of the matrix, densifying it and thereby increasing the compressive strength. However, interfacial pores were created between the cement matrix and fibre, and more pores were formed as the volume fraction of polypropylene fibre increased. A less dense cement matrix structure decreases the compressive strength.

8.3.4 Compressive Strength after Fire Test

The compressive strengths of the composites were measured after exposure to high temperature. The results show that as the ratio of porcelain increased, the residual compressive strength after the fire test increased. The effect was greatest at 42% of the residual ratio in a mixture with no blast furnace slag (100% cement), with 0.1% polypropylene fibre and 15% porcelain added (F1C100P15). The main reason for the low strength of the cement composites exposed to high temperature was the creation of microcracks due to reduced adhesion between the aggregate and cement paste. These microcracks formed because of decomposition of $Ca(OH)_2$ inside the cement paste. The added porcelain crystallized into a vitreous form within the cement matrix and increased the density. We believe that using perlite helped to overcome the reduction of adhesion between the cement matrix and aggregate and thereby improved the residual compressive strength.

8.4 Composite Materials in Alternative Energy Sources

8.4.1 Introduction

The blades are the most important components in a wind turbine. They have to be made of such material designed such as to capture the maximum energy from the wind flow. Blades of horizontal axis are made of composite materials. Composite materials satisfy various design constrains such as low weight and high stiffness while providing reasonable resistance to the static and fatigue loading. With growing needs of wind energy, wind turbines are the composite industry's fastest-growing application. Ultralight, strong, highly resistant and durable composites are ideal for producing lightweight blades with tremendous performance.

8.4.2 Requirements for Wind Turbine Blade Materials

When designing a wind turbine, one desires to attain the maximum possible power output under specified atmospheric conditions. From the technical view point, this depends on the shape of the blade. The change of the shape of blades is one of the methods to modify their stiffness and stability, and the other method is to modify the composite material of which the blades are made of. The problem of determining the proper shape of blade and determining the most appropriate composite material depends on the aerodynamic load and a number of constraints and objectives to be satisfied.

Minimization of vibrations is essential for successful design of blade structure. However, when minimizing vibrations of the blade, the natural frequency of the blade must be separated from the harmonic vibration associated with rotor rotation. Such approach prevents the occurrence of resonance, which under high amplitude of vibration could lead to destruction of the structure. A major problem influencing the design and operation of wind turbine is fatigue. The lifetime of most components are gradually reduced by the high number of revolutions that occurs in relatively low stress magnitudes. Turbine blades are the components which exhibit the largest proportion of fatigue failure. The components of turbine are changing as the technology improves and evolves. There is a trend towards lighter weight systems. Lightweight, low-cost materials are especially important in blades and towers for several reasons.

References

1. J.A. Hammella, P.N. Balaguru, and R.E. Lyon, Strength retention of fire resistant aluminosilicate–carbon composites under wet–dry conditions, *Composites: Part B*, 31, 2000, 107–111.
2. J. Giancaspro, C. Papakonstantinou, and P. Balaguru, Fire resistance of inorganic sawdust biocomposite, *Composites Science and Technology*, 68, 2008, 1895–1902.
3. T.E. Glodek, S.E. Boyd, I.M. McAninch, and J.J. LaScala, Properties and performance of fire resistant eco-composites using polyhedral oligomeric silsesquioxane (POSS) fire retardants, *Composites Science and Technology*, 68, 2008, 2994–3001.

4. J.-H. Jay Kim, Y.M. Lim, J.P. Won, and H.G. Park, Fire resistant behavior of newly developed bottom-ash-based cementitious coating applied concrete tunnel lining under RABT fire loading, *Construction and Building Materials*, 24, 2010, 1984–1994.
5. J.-P. Won, S.-W. Choi, S.-W. Lee, C.-II Jang, and S.-J. Lee, Mix proportion and properties of fire-resistant wet-mixed high-strength polypropylene fiber-reinforced sprayed polymer cement composites, *Composite Structures*, 92, 2010, 2166–2172.
6. J.-P. Wona, H.-B. Kang, S.-J. Lee, J.-W. Kang, Eco-friendly fireproof high-strength polymer cementitious composites, *Construction and Building Materials*, 30, 2012, 406–412.

9

Biomimetics and Biomimetic Materials

9.1 Introduction

Biomimetics is a new field of materials science. It opens a new era of future of engineering. Biomimetic materials are novel materials having unique characteristics. They possess nature graded property and are being tried in almost all fields of engineering. Due to their versatility of behaviour, their concepts are now used in aircraft construction, robot construction and industrial bearings. The study and simulation of biological systems with desired properties is popularly known as biomimetics. Such an approach involves the transformation of the principles discovered in nature into man-made technology. Biomimetics, already popular in the fields of materials science and engineering, is being applied in diverse areas ranging from micro-/nanoelectronics to structural engineering and in tribology.

Design solutions can draw inspiration from many sources, including the anatomy, physiology and behaviour of living systems. Industry also imitates nature. It is well known that the Wright brothers were bird watchers, and their airplane wing design was modelled after birds. In recent times, scientists have begun to take more ideas from nature with the explosion in biotechnological progress. Biomimetics is currently being used to explore a variety of design projects, including the development of different biomaterials (most notably spider silk) as well as robots based on animal models. In the 1960s, Jack Steele, a researcher of the United States, used the word bionics, referring to copying nature or taking ideas from nature. Biomimicry refers to the study of nature's most successful developments and then imitating these designs and processes to solve human problems [1]. Biomimetic materials are classified under the following ways:

1. Trends in the development of hydrobiomimetic-inspired biomimetic materials and their novel applications

2. Reptile-inspired biomimetic materials

3. Development of insect-inspired biomimetic materials and their novel applications

4. Trends in the development of plant-inspired biomimetic materials and their novel applications

5. Trends in the development of bird-inspired biomimetic materials and their novel applications

6. Miscellaneous biomimetic principles and inspirations

9.2 Biomaterials

One major application of biomimetics is in the field of biomaterials, which involves mimicking or synthesizing the natural materials and applying this to practical design. There are many examples of materials in nature that exhibit unique useful properties. One of the major advantages of biomaterials is that they are normally biodegradable. In addition, the extreme temperatures and hazardous chemicals often used in man-made construction are usually unnecessary with natural alternatives.

S. No.	Nature's Design	Inspirations in the Works
1.	Abalone and conch shell nacre (mother of pearl coating)	Consists of alternating layers of hard and soft material so cracks in the hard part are absorbed by the soft. Therefore, this structure self-assembles and self-repairs. Inspiration: The bodies of cars or anything that needs to be lightweight but fracture resistant.
2.	Bat and marine mammal navigation	Recent research suggests that animals use a combination of magnetism, the sun, stars and sight to navigate. Inspiration: Sonar, and a walking cane has been created using fruit bat sonar techniques to aid the blind.
3.	Birds	Birds are perfect flying machines. In a bird's body, however, the air circulation works just like a cooling system. It is therefore impossible to hit a bird with a heat-seeking missile as one can with a plane. Inspiration: Aircrafts which are less heat generating. Stealth airplanes, model for high-speed trains, etc.
4.	Camouflaging chameleons	Chameleons are able to change and camouflage with their environment using three types of colour organs which allow it to create combinations of pigment and iridescence. Inspiration: Changeable clothing or furnishing.
5.	Chlorophyll and enzymes	Photozymes – chlorophyll-like molecules that attract, conduct and absorb light energy. Like enzymes, they use available energy (sun) to aid needed chemical reactions. Inspiration: When added to water, photozymes can break down pollutants such as PCBs into harmless compounds using the sun's energy.

(Continued)

S. No.	Nature's Design	Inspirations in the Works
6.	Cockroach cuticle	A springy protein known as reslin found in cockroaches does not swell on contact with organic solvents. Inspiration: Protective gloves or tubing would greatly aid out handling of fuels and other hydrocarbon-based chemicals. Fast running roots.
7.	Crocodile skin	Able to deflect spears, arrows and sometimes even bullets. Inspiration: Protective clothing or coverings for cars.
8.	Dolphin and shark skin; Narwhal tusk	This material deforms slightly to shrug off water/air pressure. Inspiration: Airplane or submarine hulls. Speedo is already looking into a racing wet suit/swimsuit design.
9.	Elephant's trunk	The elephant is able to move its trunk in any direction it wants and can perform tasks requiring the greatest care and sensitivity. Inspiration: Robotic arm.
10.	Gecko toes	Geckos are able to walk up walls and ceilings even of smooth glass using molecular van der Waals forces. Inspiration: Applications for nonmarking adhesives, closures and of course rock-climbing gloves.
11.	Hummingbirds	Able to fly 35 mph and travel 2000 miles per year. They have to make long overwater flights on very little fuel (600 miles on 2.1 g!) and the process by which they fuel up pollinates, in other words, contributes. Inspiration: More efficient and biologically based fuel systems and flight technology could be discovered.
12.	Honeycomb	The honeycomb is made by bees in total darkness and it consists of a perfect hexagonal cellular structure that offers an optimal packing shape. Inspiration: Sinosteel International Plaza and honeycomb sandwich panels for aerospace.
13.	Leaves	The centre of photosynthesis in plants; they are designed to efficiently capture and process the sun's energy depending on the surrounding conditions. Inspiration: 'Pentads are solar batteries that mimic the leaf's reaction centre. Molecular in size, they could one day be used to split water into clean-burning hydrogen gas and oxygen. Or they could be used as computer switching devices that shuttle light instead of electrons. Or they could be the light-activated' 'power'.
14.	Lotus flower	Lotus flowers have always inspired awe for their ability to emerge from the mud a pristine beauty. Microanalysis revealed that the petal surfaces had tiny mountains and valleys that resulted in waterdrops picking up all dirt when it rained. Inspiration: These self-cleaning surfaces have already stimulated a German paint company ISPO to create Lotusan products.
15.	Mosquito and snake fangs	Able to penetrate skin and inject material into the bloodstream. Inspiration: The engineering inspiration for the hypodermic needle.
16.	Moth's eye	Antireflecting coatings. Inspiration: TV screens; the AQUOS XL series.

(Continued)

S. No.	Nature's Design	Inspirations in the Works
17.	Penguin insulation	Penguins live in the Antarctic and despite cold temperatures and major loss of body fat during nesting seasons, they are able to stay warm using air pocket and feather networks. Inspiration: For cold weather clothing.
18.	Spider	Spider web is a structure built of a 1D fibre. The fibre is very strong and continuous and is insoluble in water. Inspiration: Spider silk.
19.	Snakes	Sensing system, balancing system and wear-resistant surfaces. Inspiration: Mini-VIPeR robot for missile detection.
20.	Scorpions	For harsh desert conditions. Inspiration: Scorpion robots for military applications.
21.	Termites	Termites regulate temperature, humidity and airflow in their mounds. Inspiration: Termite mound inspired for air conditioning.
22.	Tree frog	Frog's feet have microscopic bumps on their toes that create adhesion to the surface they are on while also channelling away excess moisture from their wet environment. Inspiration: Treads on automobile tyres.
23.	Vulture wings	It is aerodynamically designed to allow for lift, gliding and changing of direction and altitude. Inspiration: Wright brothers learned a lot about airplane design from birds.
24.	Wood	Wood absorbs the energy from low-velocity impacts; it is highly effective at restricting damage to one specific location. Inspiration: Bulletproof clothing.

9.3 Spider Silk [1]

Spider silk is one of the most sought after biomaterials. This material produced by special glands in a spider's body has the advantage of being both light and flexible. It is roughly three times stronger than steel. The tensile strength of the radial threads of spider silk is 1154 MPa, while that of steel is 400 MPa only. For a flying insect to be caught, the spider's web must slow its motion to a halt by absorbing kinetic energy. The force required to stop the insect's motion is inversely proportional to the distance over which the motion must be stopped. The longer the distance over which the insect is slowed down, the smaller the force necessary to stop it, reducing the potential for damage to the web.

The incredible properties of spider silk are due to its unique molecular structure (Figure 9.1). X-ray diffraction studies have shown that the silk is composed of long amino acid chains that form protein crystals. The majority of silks also contain β-pleated sheet crystals that form from tandemly repeated amino acid sequences rich in small amino acid residues. The resulting beta-sheet crystals cross-link the fibroins into a polymer network with great stiffness, strength and toughness. This crystalline component is embedded in a rubbery component that permits extensibility, composed of amorphous

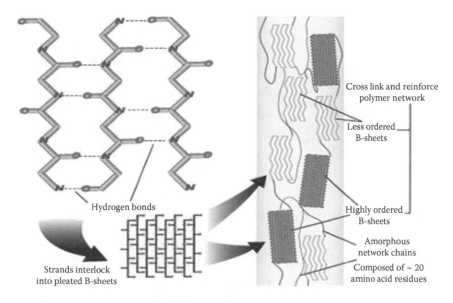

FIGURE 9.1
The structure of a strand of silk. (From Kennedy, S., *Sci. Creative Quart.*, (6), August 2004.)

network chains. It is this extensibility and tensile strength combined with its light weight that enables webs to prevent damage from wind and their anchoring points from being pulled off.

9.3.1 Likely Applications of Spider Silk [1]

Despite the high demand for spider silk as a building material, the difficulties surrounding its harvest have precluded large-scale production. A new biotechnology firm (Nexia Biotechnologies, Montreal, Quebec, Canada) has successfully expressed the silk genes of two spider species in the milk of a transgenic goat. This technology could have applications in the field of medicine as a new form of strong, tough artificial tendons, ligaments and limbs. Spider silk could also be used to help tissue repair and wound healing and to create superthin, biodegradable sutures for eye or neurosurgery, as well as being used as a substitute for Kevlar fibre.

9.4 Biomimetic Robot: Chemistry, Life and Applications [1]

A second application of biomimetics is the field of robotics. Animal models are being used as the inspiration for different types of robots. Researchers closely study the mechanics of various animals and then apply these

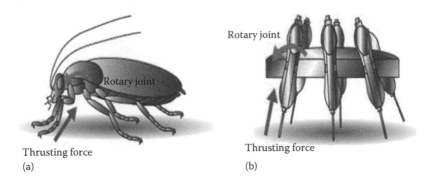

Rotary joint

Rotary joint

Thrusting force   Thrusting force
(a)   (b)

FIGURE 9.2
The (a) cockroach leg is a prime candidate for biomimicry for (b) a robot. (From Kennedy, S., *Sci. Creative Quart.*, (6), August 2004.)

observations to robot design. The goal is to develop a new class of biologically inspired robots with greater performance in unstructured environments. One example is to mimic the leg and joint structure of animals for
use in robot mobility such as modelling the joint and leg structure of the
cockroach for the development of a hexapedal running robot.

Researchers are using biomimicry of the cockroach, one of nature's most
successful species, to design and build sprawl-legged robots that can move
very quickly (Figure 9.2). In addition, these robots are very good at manoeu-
vring in changing terrain and can continue forward motion when encountering hip-height obstacles or uphill and downhill slopes of up to 24°. These
types of small, fast robots could potentially be used for military reconnais-
sance, bomb defusion and demining expeditions.

9.5 Shark Skin Effect [2]

Friction between a solid surface and a fluid can also be considered a tribological phenomenon. Inspiration from aquatic animals' surface material and
texture would benefit the design of surfaces that could increase efficiency in
cases such as underwater navigation. One such example is the *shark skin effect*.
Sharks' skin has grooved scales on its entire body. The scales are directed
almost parallel to the longitudinal body axis of the shark. The presence of
this nonsmooth surface texture on the shark skin effectively reduces the drag
by 5%–10% (Figure 9.3a and b). The benefit of this property has been utilized
in the following applications:

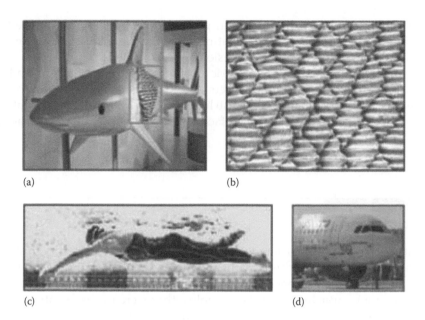

FIGURE 9.3
(a) Shark; (b) riblets and grooves found on shark skin; (c) a swimming cloth with similar surface as a shark skin could reduce frictional resistance; (d) an aircraft coated with a plastic film that has similar microscopic texture as found on a shark skin. (From Singh, R.A. et al., *Lubr. Technol.*, 65, 41, 2009.)

- Swimsuits [2] with biomimetically designed surfaces that mimic the nonsmooth surface texture on the shark skin have proved to be faster than conventional suits, as they reduce drag along key areas of the body (Figure 9.3c).

- Surface texture such as those on sharks' skin also helps to reduce the friction between a solid surface and air. A transparent plastic film with similar microscopic texture (ribs parallel to the direction of flow) reduces aircraft drag by about 8% and is effective in saving fuel by about 1.5%. The commercial aircraft Cathay Pacific Airbus 340 [2] already has been fitted with ribbed structures on its body surface (Figure 9.3d).

9.6 Snake Scales [2]

Studying the frictional surfaces of snake scales would benefit when designing the surfaces with anisotropic frictional characteristics. Snakes have friction-modifying structures consisting of ordered double-ridge microfibrillar array. The double-ridge microfibrillar geometry provides significant reduction in adhesive forces, thereby creating ideal conditions for sliding in forward direction with minimum adhesion.

Meanwhile, the highly asymmetric profile of the microfibrillar ending induces frictional anisotropy, as it acts as a locking mechanism prohibiting the backward motion. Snake skin also has micropores that deliver an antiadhesive lipid mixture, which further facilitates easy motion owing to boundary lubrication.

9.7 Gecko Effect [2]

Creatures such as beetles, flies, spiders and lizards have the ability to attach themselves to surfaces without falling off, even when the surfaces are vertically inclined. The presence of micro-/nanostructures—small hairs called *setae* on their attachment pads—enables these creatures to attach and detach easily over any surface. As the mass of the creature increases, the radii of the terminal attachment structures decrease while the density of the structures increase (Figure 9.4a through d). The gecko is the largest animal that has this kind of dry attachment system and, therefore, is the main interest for scientific research (Figure 9.4e and f). In functional terms, the tiny hairs found on gecko feet are able to conform to the shape of surface irregularities to which the gecko is adhering. By mimicking the shape and geometry of gecko setae, synthetic adhesives have been made from polymeric materials:

- An example is gecko tape, which can be used for several detachment – attachment cycles before degradation of its adhesive property. This tape has arrays of flexible polyimide pillars fabricated using electron beam lithography and dry etching in oxygen.
- To create a gecko adhesive, the pillars must be sufficiently flexible and placed on a soft, flexible substrate so that the individual tips can act in unison and attach to uneven surfaces all at the same time. To demonstrate the effectiveness of the gecko tape as a dry adhesive, a Spider-Man toy was attached to a glass plate through the microfabricated gecko tape. Polyurethane elastomer microfibre array with flat spatulate tips was fabricated by moulding a master template using deep reactive ion etching, and the notching effect also can act as effective adhesive surfaces. These surfaces, when tested against a smooth glass hemisphere for their adhesive property, show three times higher adhesion than the flat polymeric surfaces.

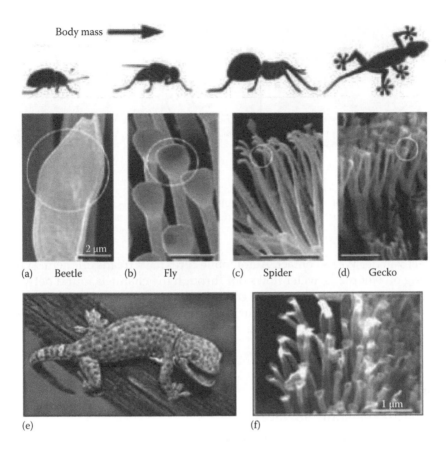

FIGURE 9.4
(a–d) Terminal elements found on the attachment pads of various insects and gecko. As the size (mass) of the creature increases, the radius of the terminal attachment structures decreases, while the density of the structures increases. (e) Gecko; (f) SEM image of gecko setae. (From Singh, R.A. et al., *Lubr. Technol.*, 65, 41, 2009.)

9.8 Tread Effect

Another of man's creations analogous to a principle found in nature, namely, the distinctive patterns on a tree frog's foot, is the design of treads on automobile tyres. Tree frogs such as *Amolops* sp. possess large disklike pads at the tip of their toes that assist them in attaching to surfaces such as leaves (Figure 9.5a through c). The pads consist of flat-topped cuboidal columnar cells, which are separated from each other by canal-like spaces. During climbing, water gets squeezed out from the contact through the channels between the foot and the surfaces on which they climb, making a

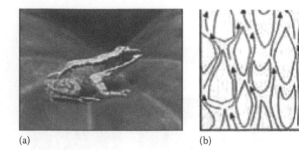

(a)   (b)   (c)

FIGURE 9.5
(a) A tree frog, (b) schematic of flat-topped cuboidal columnar cells separated by canal-like spaces in tree frogs and (c) a close-up view of treads on a car tyre. (From Singh, R.A. et al., *Lubr. Technol.*, 65, 41, 2009.)

perfect van der Waals contact. Automobile tyres have treads on them. While driving on wet roads, water flows out through the channels found between the treads, giving rise to intimate contact between the treads and the road, thereby creating sufficient grip during motion. This effect is known as the *tyre tread effect* [2].

9.9 Wear-Resistant Surfaces

9.9.1 Wear Resistance of Sandfish in Desert [2]

Man has begun to learn from nature to design materials and textures that have superior wear-resistant characteristics. For example, a sandfish, living in the Sahara desert, moves rapidly over the desert sand and the scales on its body have excellent sand erosion wear resistance. Erosion experiments conducted using sand on the sandfish's scales, glass and soft steel for 10 h showed that the wear trace on the sandfish's skin was the smallest, which suggests that its wear resistance is comparatively much higher than that of glass and soft steel. The biomaterial comprising the sandfish's scales and their surface texture together contribute towards its high wear resistance.

9.9.2 Erosion Resistance of Mollusc (Conch) Shells on Sandy Beach [2]

Mollusc (conch) shells experience water–sand erosion on sandy beaches. Their antiwear mechanism arises from the combination of their biotissues and their unique shape, which together prevent abrasion. The biomaterial of mollusc shells is a bioceramic composite that has a complex microstructure, the result of a billion years of evolution. Understanding the formation and microstructure of these bioceramic composites will help in the design of ceramics with superior mechanical properties such as toughness. Current methods used to fabricate ceramics still are unable to control parameters such as crystal density, orientation and morphological uniformity to the degree of perfection nature has achieved in mollusc shells. Studies on the microstructure of bivalve shells are expected to provide guidelines in developing biomimetic composite materials with better tribological properties.

9.9.3 Pangolin Scales [2]

A pangolin is a soil-burrowing animal with a layer of scales covering its body. As the pangolin burrows into soil, its scales are subjected to wear. A study on the chemical constitution of pangolin scales revealed the presence of 18 amino acids. The protein in the scales mainly consists of α-keratin and β-keratin. The specific elongation of pangolin scales is about 15% due to the presence of the proteins. Therefore, the plasticity of the pangolin scales is low, resulting in antiabrasive characteristics. Studies on biomaterials that exhibit remarkable friction and wear characteristics would provide insights towards enhancing the performance of soil-engaging engineering components such as those in agricultural machinery and earth moving machinery.

9.10 Lotus (or Self-Cleaning) Effect [2]

A number of plants have water-repellent leaves, which exhibit superhydrophobic property. Lotus (*Nelumbo nucifera*) is the most popular example (water contact angle ~162°). The unique ability of lotus leaf surface to avoid wetting, popularly known as the *lotus effect*, is mainly due to the presence of microscale protuberances covered with waxy nanocrystals on their surface. *Colocasia* (*Colocasia esculenta*) is another example of a plant whose leaves are superhydrophobic in nature (water contact angle ~164°), due to the presence of micron-sized protuberances and wax crystals (Figure 9.6a through f).

Both the protuberances and the wax crystals make the surfaces of lotus and *Colocasia* leaves superhydrophobic in nature, which means water droplets easily roll over leaf surfaces taking contaminants and dust particles with them. This phenomenon is popularly known as the *self-cleaning effect*.

The self-cleaning property is highly important for water plants. In their habitats, these plants observe the presence of free water which supports pathogenic organisms. These plants protect themselves from waterborne infections by hindering any adhesion of water necessary for the germination of pathogens. Another reason for water repellency is the fact that CO_2 diffuses 10% times slower in water than in air. The presence of water-repellent surface ensures that sufficient intake of CO_2 is observed for photosynthesis.

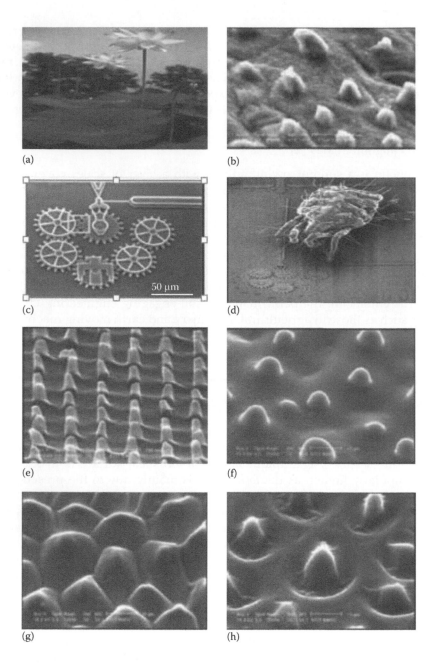

FIGURE 9.6
(a) Lotus (*N. nucifera*), (b) protuberances on the surface of a lotus leaf, (c) a microelectromechanical system that has six gear chains, (d) the same MEMS device in comparison with the size of a dust mite, (e) nanopatterns that mimic the protuberances of a lotus leaf, (f) lotus-like (fresh) surface, (g) *Colocasia*-like (fresh) surface and (h) *Colocasia*-like (dry) surface. (From Singh, R.A. et al., *Lubr. Technol.*, 65, 41, 2009.)

9.10.1 Reducing Adhesion and Friction [2]

By studying the surface morphologies of water-repellent leaves, tribologists design and create hydrophobic surfaces to reduce adhesion between surfaces at small scales, which arise due to water condensation. Scientists and engineers are aware of the fact that surface forces such as adhesion and friction significantly oppose easy motion between tiny elements in miniaturized devices such as micro-/nanoelectromechanical systems (MEMs/NEMs).

Minimizing the surface forces such as adhesion and friction, and also the occurrence of wear in miniaturized devices, is a real challenge, as the size of these devices is extremely small (usually their sizes are smaller than insects such as dust mites). Among the various attractive forces that contribute to adhesion at small scales, the capillary force that arises due to the condensation of water from the environment is the strongest. Hence, there arises a need to modify surfaces at nano-/microscale in order to achieve increased hydrophobicity that would drastically reduce adhesion due to capillary force, which in turn would also reduce friction.

9.11 Biomimetic Human Joints [2]

The design of lubricated bearings also can benefit greatly from the knowledge of mechanisms at work in biological joints. Biological joints are much different than the conventional industrial bearings. Friction, wear and lubrication of surfaces greatly affect the reliability and efficiency of the various joints in the human anatomy. When these joints fail, such as with severe osteoarthritis, the joints may require surgical repair or replacement. However, the materials (metals, ceramics and polymers) used in most artificial joints are much different than the actual biomaterials in a human joint and are more closely related to industrial bearings. A typical joint usually has a softer layer of cartilage and other materials separating the hard bone. The natural joints use soft material much more liberally than hard material, and perhaps it is to their advantage. More recent approaches use softer materials that are more similar to the original tissue such as hydrogels and actual living cells to replace joint surfaces.

A large number of works have numerically modelled the elastohydrodynamic lubrication of natural joints. Even then, the fundamental mechanical models used are essentially the same as are used to consider the industrial hydrodynamic bearings in rolling element bearings. Recently, numerical modelling has been used to design prototype surfaces designed to be soft and deformable like biological joints. These biomimetic self-adapting surfaces change their surface profiles at the nano- and microscale to improve performance. The mechanism is similar to that seen in gas foil bearings.

9.12 Development of Hydrobiomimetic-Inspired Biomimetic Materials and Their Novel Applications

The term biomimetic is derived from the Greek words bios, that is, life, and mimesis, that is, to imitate. It is the science of mimicking nature in the development of newer materials and products. Biomimetics is the study and development of synthetic systems that mimic the formation, function and structure of biologically produced substances and materials. Biomimetics is aimed at exploiting the natural structures and functionalities for use in technological applications. Biomimetic materials are a novel class of materials having unique characteristics. They are used as micro–robot fish, biomimetic fin, biomimetic drag-reducing surfaces, grid shell roof, armour resistant to ballistic impact, pulse-jet propulsion, etc. The scope of biomimetic materials extends to the fields of electronic signals, synthesis of functional materials, biomineralization (nature's way of making materials), information processing by biological systems, etc. In the past few decades, materials scientists have shown increasing interest in studying a whole variety of biological materials including hard and soft tissues and to use discovered concepts to engineer new materials with unique combinations of properties. This section aims at elaborating the development of such biomimetic materials by compiling the ongoing researches. In this regard, the research developments of some newer materials by other investigators have been presented here. Brief details of the development of micro–robot fish, biomimetic drag-reducing surfaces, seashells and sonar technology are discussed. In these elaborations, it is shown that these biomimetic materials can be effectively used in a large variety of application in near future.

9.12.1 Analogy between Biological Examples and Biomimetic Materials

Biomimetics is the science of mimicking nature in the development of new materials and products. Mankind has learnt a lot from nature, and designing of materials is no exception. Living organisms provide inspiration for innovations in different fields and for entirely different reasons. Energy is stored in chemical form by plants with almost 100% efficiency. Animal muscle is an efficient mechanical motor capable of an exquisite degree of control. Transmission of information in the nervous system is more complex than in the largest telephone exchanges. And the problem-solving capabilities of a human brain greatly exceed those of the most powerful supercomputers. Nature's designs of materials have been some of the best known designs; hence, *biomimetic* has emerged as an important strategy for material design. The mimicking of natural materials can be of multiple types including chemical mimicking and physical mimicking.

Various forms of bioinspiration and related examples are listed as follows for a ready reference:

Biological Example	Type of Analogy	Biomimetic Materials
Some fishes	Biochemical (observation of biochemical mechanisms)	Shape memory alloys (SMAs), ionic exchange polymer metal composites (IPMCs), or piezoelectric lead–zirconate–titanate (PLZT) actuators
Shells	Morphostructural	China clay, porcelain, bricks, tiles, glasses, etc.
Superhydrophobic surfaces (e.g. shark skin)	Functional	Nylon, elastane, polyester, etc.

9.13 Design of Micro–Robot Fish Using Biomimetic Fin [3]

A flexible biomimetic fin–propelled micro–robot fish prototype (Figure 9.7a and b) is presented [3]. In it, the fish muscle and the musculature of squid/cuttlefish fin are analyzed. Since the latter one is easier to be realized in engineering, it is emulated by biomimetic fin. SMA wire is selected as the most suitable actuator of it. Elastic energy storage and exchange mechanism is incorporated into the biomimetic fin for efficiency improvement. Fish swimming mechanism is reviewed as the foundation of the robot fish. A radiofrequency-controlled micro–robot fish propelled by biomimetic fin was built. Experimental results show that the micro–robot fish can swim straight and turn at different frequencies. The maximum swimming speed and the minimum turning radius reached 112 mm/s and 136 mm, respectively. Fishtail-like propeller is considered as an alternative apparatus for thrust to commonly used rotator of underwater vehicle.

As the propulsive efficiency of small underwater vehicles is usually below 40%, fishtail-like propeller is superior to rotator used in small underwater vehicles. Micro–robot fish have great advantages in exploring complex, narrow underwater environments. They may even be used in blood vessels for microsurgery. Many micro–robot fish have been developed in recent years. Most reported micro–robot fish are driven by SMAs, IPMCs, or piezoelectric (PZT) actuators. The difficulties of developing a micro–robot fish mainly lie in the selection of actuators, sealing, compactness, wireless controlling, reducing noise and improving propulsive efficiency. The micro–robot fish has no traditional components such as gears and bearings, and it can swim noiseless and flexibly. The biomimetic fin consists of elastic substrate, skin and transverse muscle-like SMA wires. Further research on the structure, material and control strategy would be made to improve the performance of biomimetic fin.

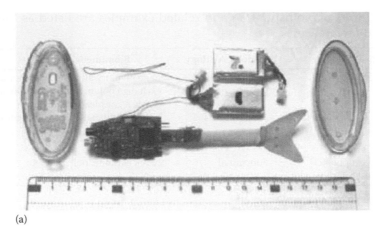

(a)

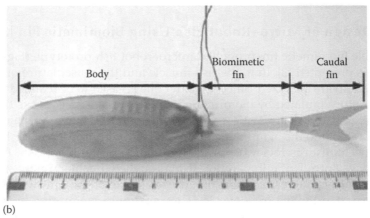

(b)

FIGURE 9.7

Micro–robot fish prototype. (a) Before assembly (without skin). (b) After assembly. (From Wang, Z. et al., *Sensor. Actuat. A*, 144, 354, 2008.)

9.14 Shark Skin–Inspired Biomimetic Drag-Reducing Surfaces [4]

In Olympic swimming competitions, 1/100th of a second can make the difference between winning and losing. Because the resistive drag opposing the motion of swimmers' bodies is of great importance, many swimmers choose newly designed swimsuits that reduce the drag. These tightly fitting suits, covering a rather large area of the body, are made out of a fabric which was designed to mimic the properties of a shark (Figure 9.8a) and a shark's skin by superimposing vertical resin stripes. Scanning electron microscope

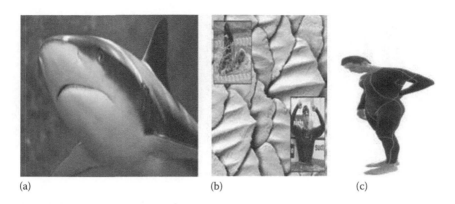

(a)   (b)   (c)

FIGURE 9.8
(a) Shark, (b) Scanning electron microscope image of shark skin and (c) swimsuits with the same properties as shark skin. (From Han, X. et al., *Chin. Sci. Bull.*, 53(10), 1587, 2008.)

studies (Figure 9.8b) have revealed that tiny *teeth* (riblets) cover the surface of a sharks' skin that produce vertical vortices or spirals of water, keeping the water closer to the shark's body and thus reducing drag. This phenomenon is known as the riblet effect, and research into shark skin is ongoing at NASA Langley Research Centre. Swimsuits made with new fibres and weaving techniques are produced to cling tightly to the swimmer's body and reduce drag as much as possible. Research has shown that such garments can reduce drag by 8% over ordinary swimsuits. The swimsuit (Figure 9.8c) is made of a knitted superstretch nylon/elastane/polyester fabric (flexskin) that has V-shaped ridges and a denticle surface print.

On the investigation of biomimetic drag-reducing surface, the results of the resistance measurements in a water tunnel according to the flat-plate sample pieces have shown that the biomimetic shark skin coating fabricated by the bioreplicated forming method has significant drag reduction effect and that the drag reduction efficiency reached 8.25% in the test conditions. The low-resistance scarfskins of some underwater animals (such as sharks and tunas, swimming speed up to 60 km/h) have provided profuse configuration resources for the investigation of the biomimetic drag-reducing surfaces. The biomachining technology may breach the bottlenecks of the existing forming methods both in techniques and efficiency by bringing directly the complex biological body shapes to the fabrication of the functional configurations.

9.15 Seashell-Inspired Design of Grid Shell Roof Covering [5]

Seashells (Figure 9.9a) are natural ceramics similar to our bones and teeth. Conventionally, ceramics are defined as compounds between metallic and nonmetallic elements. The term *ceramic* comes from the Greek word

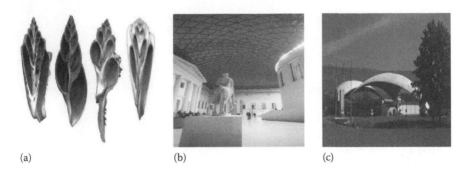

(a)   (b)   (c)

FIGURE 9.9
(a) Cut cross sections of a selection of seashells found in nature, (b) grid shell roof covering the Great Court of the British Museum London and (c) concrete shell structure designed by Heinz Isler. (From Akella, K., *Resonance*, 2012.)

keramikos, which means burnt stuff. These materials achieve their desired state through a high-temperature heat treatment process called firing. Traditional ceramics such as china clay, porcelain, bricks, tiles and glasses are made using clay as the raw material. With advancement in understanding of these materials, the use of ceramics has surpassed traditional applications. Popular modern ceramics such as aluminium oxide, magnesium oxide and silicon carbide are used in a wide array of application areas such as electronics, automobiles, aerospace and defence. However, applications of ceramics are limited due to their brittle failure characteristics with low energy absorption during failure. Natural ceramics such as bones, teeth and seashells are different from conventional ceramics. These materials are called ceramics as they are compounds of metallic and nonmetallic elements (such as calcium carbonate), but are different since they are not fired like human-made ceramics.

Natural ceramics, especially seashells with pearly layers, also known as nacreous layers, have a distinctly different behaviour from the brittle behaviour of conventional ceramics (Figure 9.9a). They have a layered brick-and-mortar architecture with calcium carbonate platelets as the bricks and protein as the mortar. Behavioural aspects of natural ceramics such as reduced effect of flaws on tensile strength and higher toughness make them desirable for applications such as armour resistant to ballistic impact. Engineers want to exploit the advantages of ceramics but are struggling to work around its disadvantages. Ceramics are stronger than metals under heavy compressive loads acting in the vicinity of areas subjected to impact. Their much lower density makes them attractive lightweight alternatives to metallic armour. But processing of ceramics requires high temperature and pressure. Hence, they are more expensive than metals. Processing limitations also make manufacture of large and thick parts infeasible. Another drawback of ceramics is their brittle behaviour due to which armour suffers heavy degradation on impact. It therefore has lower multiple hit–resisting capability than metallic armour.

Armour designers desire ductile ceramics. Attempts were made to achieve this goal by embedding ceramic fibres inside the bulk ceramics. Such materials are popularly known as ceramic matrix composites (CMCs). However, CMCs are difficult to process and much more expensive than conventional ceramics. Their application is hence restricted to niche areas such as the nosecone of a space reentry vehicle, to improve the design of armour at reasonable cost. Studying the microstructure of shells and methods of replication therefore offers a promising opportunity. It opens up a possibility of attaining ductility in ceramics at lower cost. Exploring this possibility might lead us to light, cost-effective and efficient armour. Biomimesis has been successfully applied to architecture (Figure 9.9b and c). A shell is a 3D curved structure which resists load through its inherent curvature. Loads applied to a shell surface can be accommodated through two actions, through bending and through stretching. Bending forces are carried through moments and shear forces and stretching forces through axial thrusts and tensions. The latter of these phenomena is called membrane action. Membrane action is much more efficient than bending, and therefore a shell will work by membrane action if it can; however, to determine whether a shell can or not depends greatly upon its geometry and support conditions. Membrane action also relies on bending stiffness to prevent buckling.

9.16 Dolphin Sound Wave–Inspired Sonar Technology [6]

From a special organ known as the melon (Figure 9.10a) in its head, a dolphin can sometimes produce as many as 1200 clicks a second. Simply by moving its head, this creature is able to transmit the waves in the direction it wishes. When the sound waves (Figure 9.10b) strike an object, they are reflected and return to the dolphin. The echoes reflected from the object pass through the dolphin's lower jaw to the middle ear and from there to the brain. Thanks to the enormous speed at which these data are interpreted, very accurate and sensitive information is obtained. The echoes let the dolphin determine the direction of movement and speed and size of the object that reflects them.

The dolphin sonar is so sensitive that it can even identify one single fish from among an entire shoal. It can also distinguish between two separate metal coins, 3 km away in the pitch dark. In the present day, the instrument known as sonar (Figure 9.10c) is used to identify targets and their directions for ships and submarines. Sonar works on exactly the same principle as that employed by the dolphin.

FIGURE 9.10
(a) Special organ, melon, in dolphin head, (b) dolphin 1 and its sound waves 2, and (c) sonar device (koala robot). (From Yahya, H., *Biomimetics: Technology Imitates Nature*, Global Publishing, Istanbul, Turkey, 2006.)

9.17 Reptile-Inspired Biomimetic Materials and Their Novel Applications

Research in the field of biological and biomimetic materials constitutes a case study of how traditional research boundaries are becoming increasingly obsolete. Positioned at the intersection of life and physical sciences, it is becoming more and more evident that the future development in this area will require extensive interaction between materials and life scientists. Nonetheless, the gap with nature is gradually closing. Researchers are using electron- and atomic force microscopes, microtomography and high-speed computers to peer ever deeper into nature's microscale and nanoscale secrets and a growing array of advanced materials to mimic them more accurately than ever before. And even before biomimetic matures into a commercial industry, it has itself developed into a powerful new tool for understanding life. For all nature's sophistication, many of its clever devices are made from simple materials like keratin, calcium carbonate and silica, which nature manipulates into structures of fantastic complexity, strength and toughness. The natural structure provides a clue to what is useful in a mechanism. Lessons from the lizard may enhance the water-collection technology.

In the past few decades, materials scientists have shown increasing interest in studying the whole variety of biological materials including hard and soft tissues and to use discovered concepts to engineer new materials with unique combinations of properties. This section aims at elaborating the development of such biomimetic materials by compiling the ongoing researches. In this regard, the research developments of some newer materials by other investigators have been presented here. Brief details of the development of gecko feet inspired StickyBot, Mini-VIPeR model robot, clothes that change colour, snake-imitating robot and robot scorpion are discussed. In these elaborations, it is shown that these biomimetic materials can be effectively used in a large variety of application in near future.

9.17.1 Analogy of Biology and Materials

Biomimetics is the word most frequently used in scientific and engineering literature that is meant to indicate the process of understanding and applying (to human designs) biological principles that underlie the function of biological entities at all levels of organization. Among the many fields of study of biomimetic, one area is the mobile robot. Biomimetics and biomimicry are both aimed at solving problems by first examining and then imitating or drawing inspiration from models in nature. Biomimetics is the term used to describe the substances, equipment, mechanisms and systems by which humans imitate natural systems and designs, especially in the fields of defence, nanotechnology, robot technology and artificial intelligence. Designs in nature ensure the greatest productivity for the least amount of materials and energy. They are able to repair themselves, are environmentally friendly, and are wholly recyclable. They operate silently, are pleasing in aesthetic appearance, and offer long lives and durability. Biomimetic materials are inspired by biology from molecules to materials and from materials to machines. Some examples of mimicking in nature are adhesives that mimic gecko fingers, heat-sensing system that mimic viper, colour-changing clothes that mimic chameleons and constant state of balance inspired by snake. They are presented as examples of next-generation bioinspired materials. Biomimetic articulated robots are robots that imitate living creatures and have many modules. Various forms of bioinspiration and related examples are listed in the following for a ready reference:

Biological Example	Type of Analogy	Biomimetic Materials
Gecko	Adhesion	Dry adhesives
Snakes	Sensing system and balancing system	SMAs, piezoelectric materials and electroactive polymers
Chameleons	Colour changing	Choleric liquid crystals (CLCs)
Scorpion	Behavioural	Central pattern generators

9.18 Gecko Feet–Inspired Biomimetic Products and Materials [7]

Small lizards are able to run very fast up the walls and walk around cling-
ing to the ceiling very comfortably. Until recently, we could not understand
as to how it could be possible for any vertebrate animal to climb up walls
like the cartoon and film hero Spider-Man. Now, years of research have
finally uncovered the secret of their extraordinary ability. Little steps by
the gecko have led to enormous discoveries with tremendous implications,
particularly for robot designers. A few of them can be summarized as
follows:

- Researchers in California believe that the lizard's *sticky* toes can help
  in developing a dry and self-cleaning adhesive.
- Geckos' feet (Figure 9.11a) generate an adhesive force 600 times
  greater than that of friction. Gecko-like robots could climb up the
  walls of burning buildings to rescue those inside.
- Dry adhesives could be of great benefits in smaller devices, such as
  in medical applications and computer architecture.
- Their legs act like springs, responding automatically when they
  touch a surface. This is a particularly appropriate feature for robots,
  which have no brain. Geckos' feet never lose their effectiveness, no
  matter how much they are used; they are self-cleaning and they also
  work in a vacuum or underwater.
- A dry adhesive could help hold slick body parts in place during
  nanosurgery.
- Such an adhesive could keep car tyres stuck to the road.

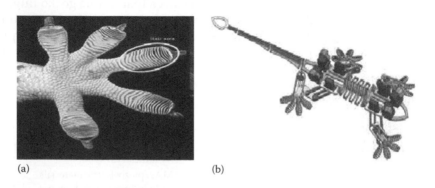

(a)   (b)

FIGURE 9.11
(a) Gecko feet and (b) StickyBot: gecko-inspired wall climbing robot. (From Cho, K-J. et al., *Int. J. Precis. Eng. Manuf.*, 10(3), 171, 2009.)

- Gecko-like robots could be used to repair cracks in ships, bridges and piers and in the regular maintenance of satellites.
- Robots modelled after the geckos' feet could be used to wash windows, clean floors and ceilings. Not only will they be able to climb up flat vertical surfaces, but they will overcome any obstacles they meet on the way.

9.18.1 Hydrophobic Spin with Hidden Capillaries

The water is spreading out incredibly fast as drops fell onto the lizard's back and vanished. Its skin is far more hydrophobic. There may be hidden capillaries, channelling the water into the mouth. A subsequent examination of the thorny lizard's skin with an instrument called a micro-CT scanner confirmed the presence of tiny capillaries between the scales evidently designed to guide water towards the lizard's mouth. With this in mind, Cutkosky endowed his robot with seven-segmented toes that drag and release just like the lizard's and a gecko-like stride that snugs it to the wall. He also crafted StickyBot's legs and feet with a process, which combines a range of metals, polymers and fabrics to create the same smooth gradation from stiff to flexible that is present in the lizard's limbs and absent in most man-made materials. StickyBot (Figure 9.11b) is a four-legged robot capable of climbing smooth surfaces. He subsequently embedded a branching polyester cloth *tendon* in his robot's limbs to distribute its load in the same way evenly across the entire surface of its toes. StickyBot now walks up vertical surfaces of glass, plastic and glazed ceramic tile, though it will be some time before it can keep up with a gecko. For the moment, it can walk only on smooth surfaces at a mere 4 cm/s, a fraction of the speed of its biological role model. The dry adhesive on StickyBot's toes isn't self-cleaning like the lizard's, so it rapidly clogs with dirt. 'There are a lot of things about the gecko that we simply had to ignore'.

9.19 Viper as a Model in Its Defence

Dr. John Pearce, of the University of Texas Electrical and Computer Engineering Department, has studied crotalines, better known as pit vipers. His research focused on the pit organs of these snakes. In front of the snake's eye (Figure 9.12a) is a tiny nerve-rich depression, called the pit, which is used in locating warm-blooded prey. It contains a sophisticated heat-sensing system–so sensitive that the snake can detect a mouse several metres away in pitch darkness. The researchers stated that when they unravel the secrets of the pit viper's search-and-destroy mechanism, the methods the snake employs can be adapted more widely to protect the country from enemy missiles.

(a)   (b)

FIGURE 9.12
(a) Snake and (b) the Mini-VIPeR model. (From Lappin, Y., Elbit unveils new defense products at Latrun Conference, Jerusalem, Israel, July 2010.)

The snake's pit is a thin membrane rich in blood vessels and nerve bundles. The membrane is so sensitive and the variations in responses so minute that to catch and study these signals has proved exceedingly difficult. To understand the functioning of the pit organ, it is necessary to work with delicate measurements and photomicrographs. The Mini-VIPeR model robot in Figure 9.12b weighs around 3.5 kg and is equipped with an array of sensors. Most of these conventional sensors are strain sensors, thermal sensors and optical sensors. More advanced actuation concepts are typically employed using active materials such as SMAs, piezoelectric materials and electroactive polymers. Small enough to move through tunnels and narrow alleys, it can be thrown into a building through a window and automatically begins scanning its environment. The robot is designed to protect infantry soldiers from explosives, booby traps and hostile forces lying in ambush.

9.20 Chameleon-Inspired Colour-Changing Clothes [9]

The impressive ability that chameleons (Figure 9.13a and b) have to change colours to match their surroundings is both astonishing and aesthetically pleasing. The chameleon can camouflage itself at a speed that quite amazes people. With great expertise, the chameleon uses its cells called chromatophores, which contain basic yellow and red pigments, the reflective layer reflecting blue and white light and the melanophores containing the black to dark brown pigment melanin, which darkens its colour. The technology in colour-changing clothes (Figure 9.13c) and the chameleon's ability to change colour may appear similar but are in fact very different. Even if this technology can change colour, it still entirely lacks the chameleon's camouflage

(a)   (b)

(c)

FIGURE 9.13
(a and b) Chameleon's body with a system that lets it change colour to match its surroundings and (c) baby clothes that change colour with temperature. (From Baby clothes that change color with temperature, http://elementalproject.org/839.)

ability that lets it match its surroundings in moments. For instance, place a chameleon into a bright yellow environment, and it quickly turns yellow. In addition, the chameleon can match not only one single colour but a mixture of hues.

The secret behind this lies in the way pigment-containing cells under the camouflage's skin expand or contract to match their surroundings. God has created the chameleon's body with a system that lets it change colour to match its surroundings, endowing it with a considerable advantage. Chameleons inspired for making clothes, bags and shoes that are able to change colours the same way as the chameleon does. Researchers envision clothing made from the newly developed fibre, which can reflect all the light that hits it, and equipped with a tiny battery pack. This technology will allow the clothing to change colours and patterns in seconds by means of a switch on the pack. Yet this technology is still very expensive. Scientists have designed CLCs to alter the visible colour of an object to create the thermal and visual camouflage in fabrics. The colour of CLCs can be changed with temperature-sensitive thermocouples [9]. The heating–cooling ability of thermocouples can be used to adjust the colour of the liquid crystals to match the object's background colour, providing camouflage or adaptive concealment.

9.21 Snake-Imitating Robot to Overcome the Problem of Balance

For those engaged in robotics, one of the problems they encounter most frequently is of maintaining equilibrium. Even robots equipped with the latest technology can lose their balance when walking. Robot experts attempt to build a balance-establishing learning that the snake (Figure 9.14a) never loses its balance. Unlike other vertebrates, snakes lack a hard spine and limbs and have been created in such a way as to enter cracks and crevices. They can expand and contract the diameter of their bodies and can cling to branches and glide over rocks. Snakes' properties inspired for a new robotic, interplanetary probe developed by NASA's Ames Research Centre, which they called the *snakebot* (Figure 9.14b). This robot thus was designed to be in a constant state of balance, without ever getting caught up by obstacles.

(a)

(b)

FIGURE 9.14
(a) Snake and (b) Snake-bot.

9.22 Robot Scorpion Can Work in Harsh Desert Conditions

Sand or other abrasive particles have a way of eroding anything they encounter. Scorpions (Figure 9.15a) have been able to survive harsh desert conditions ever since their creation. They live their entire life under blowing sand, yet they never appear eroded. As a result, items such as helicopter rotor blades, airplane propellers and rocket motor nozzles and pipes regularly wear out and need to be replaced. A group of scientists recently set out to discover their secret, so it could be applied to man-made materials. In the United States, the Defense Advanced Research Projects Agency (DARPA) is working to develop a robot scorpion (Figure 9.15b) [10]. The reason the project selected a scorpion as its model is that the robot was to operate in the desert. The scientists subsequently applied what they observed in the scorpions' exoskeletons to man-made surfaces.

They determined that the effects of erosion on steel surfaces could be significantly reduced, if that steel contained a series of small grooves set at a 30° angle to the flow of abrasive particles. But another reason why DARPA selected a scorpion was that along with being able to move over tough terrain very easily, its reflexes are much simpler than those of mammals and can be imitated. Before developing their robot, the researchers spent a long time observing the movements of live scorpions using high-speed cameras and analyzed the video data. Later, the coordination and organization of the scorpion's legs were used as a starting point for the model's creation. The robot is controlled using a biomimetic approach of ambulation control. The approach is based on two biological control primitives, central pattern generators and reflexes. Using this approach, omnidirectional walking and smooth gait transitions can be achieved. Additionally, the posture of the robot can be changed while walking. The robot was

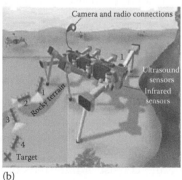

(a)   (b)

FIGURE 9.15

(a) Scorpion and (b) scorpion robot. (From Coxworth, B., Wear-resistant surfaces inspired by scorpions, 2012.)

successfully tested in rough terrain with obstacles as high as the robot's body and in different terrains such as sand, grass, concrete and rock piles.

9.23 Development of Insect-Inspired Biomimetic Materials and Their Novel Applications

Biomimetics refers to human-made processes, materials, devices or systems that imitate nature. The word biomimetics means 'the study of the formation, structure, or function of biologically produced substances and materials'. The field of biomimetics is highly interdisciplinary. It involves the understanding of biological functions, structures and principles of various objects found in nature by biologists, physicists, chemists and material scientists and the biologically inspired design and fabrication of various materials and devices of commercial interest by engineers, materials scientists, chemists, biologists and others. Biomimetics allows biologically inspired design, adaptation, or derivation from nature. Biomimetic-inspired materials and surfaces are eco-friendly or green, which have generated significant interest and are helping to shape green science and technology. The understanding of the functions provided by objects and processes found in nature can guide us to design and produce nanomaterials and nanodevices. This section aims at elaborating the development of such biomimetic materials, which are inspired by insects. In this regard, the research developments of some newer materials by other investigators have been presented here. Brief details of the development of moth eye–inspired biomimetic materials; termite-inspired biomimetic materials; *Stenocara*, a fully fledged water capturing unit; mosquito bite–inspired biomimetic materials and honeycomb are discussed. In these elaborations, it is shown that these biomimetic materials can be effectively used in a large variety of application in near future.

9.24 Moth Eye–Inspired Biomimetic Materials

The U.K. National Physics Laboratory [11] made clear that a periodical array of convexoconcave structures (moth eye structures) had the effect of gradually changing refractive index in the direction normal to the surface, resulting in a nonreflective surface. The eyes of moths are antireflective to visible light and consist of hundreds of hexagonally organized nanoscopic pillars, each approximately 200 nm in diameter and height, which result in a very low reflectance for visible light. These nanostructures' optical surfaces make the eye surface nearly antireflective in any direction. The surface of a moth's compound eyes (Figure 9.16a and b is covered with an array of cones).

FIGURE 9.16
(a and b) Moth's eye. (From Rigg, J., Sharp announces first TVs with Moth-Eye technology: The AQUOS XL series, October 2012.) (c) TV screens the AQUOS XL series. (From Hadhazy, A., *Christ. Sci. Monit.*, 25 May 2010.)

As they create a refractive index gradient, the eye does not reflect light. The antireflective film, with its moth eye structure, received attention in the development of optical materials. Moth's eye inspires glare-free TV screens (Figure 9.16c) and antireflective coatings in moth eye studies that have led to brighter screens for cellular phones. The tiny bumps on the eyes of a moth make them nonreflective. The same technology is being used to develop glare-free screens for cell phones and televisions. Moth's eyes don't glimmer in the light, helping these insects stay hidden from predators.

9.25 Termite-Inspired Biomimetic Materials

Termite mound (Figure 9.17a) inspired for air conditioning. The Eastgate Centre in Harare, Zimbabwe (Figure 9.17b), typifies the best of green architecture and ecologically sensitive adaptation. The country's largest office and

 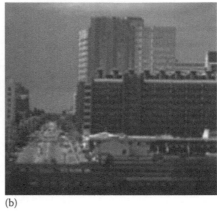

(a)   (b)

FIGURE 9.17
(a) Termite mound and (b) green building in Zimbabwe modelled after termite mounds. (From Turner, J.S. and Soar, R.C., Beyond biomimicry: What termites can tell us about realizing the living building, *First International Conference on Industrialized, Intelligent Construction (I3CON)*, Loughborough University, Leicester, U.K., 14–16 May 2008.)

shopping complex is an architectural marvel in its use of biomimicry principles. The midrise building, designed by architect Mick Pearce in conjunction with engineers at Arup Associates, has no conventional air conditioning or heating yet stays regulated year-round with dramatically less energy consumption using design methods inspired by indigenous Zimbabwean masonry and the self-cooling mounds of African termites! Termites and the structures they build have been used as exemplars of biomimetic designs for climate control in buildings, like Zimbabwe's Eastgate Centre, and various other *termite-inspired* buildings. Architects in Zimbabwe are studying how termites regulate temperature, humidity and airflow in their mounds in order to build more comfortable buildings [13].

Termites in Zimbabwe build gigantic mounds inside of which they farm a fungus that is their primary food source. The fungus must be kept at exactly 87°F, while the temperatures outside range from 35°F at night to 104°F during the day. The termites achieve this remarkable feat by constantly opening and closing a series of heating and cooling vents throughout the mound over the course of the day. With a system of carefully adjusted convection currents, air is sucked in at the lower part of the mound, down into enclosures with muddy walls, and up through a channel to the peak of the termite mound. The industrious termites constantly dig new vents and plug up old ones in order to regulate the temperature.

The Eastgate Centre, largely made of concrete, has a ventilation system which operates in a similar way. Outside air that is drawn in is either warmed or cooled by the building mass depending on which is hotter, the building concrete or the air. It is then vented into the building's floors and offices before exiting via chimneys at the top. The complex also consists of two buildings side by side that are separated by an open space that is covered by glass and open to the local breezes. Air is continuously drawn from this open space by fans on the first floor. It is then pushed up the vertical supply sections of ducts that are located in the central spine of each of the two buildings. The fresh air replaces stale air that rises and exits through exhaust ports in the ceilings of each floor. Ultimately, it enters the exhaust section of the vertical ducts before it is flushed out of the building through chimneys. The Eastgate Centre uses less than 10% of the energy of a conventional building its size. These efficiencies translate directly to the bottom line: Eastgate's owners have saved $3.5 million alone because of an air-conditioning system that did not have to be implemented. Outside of being eco-efficient and better for the environment, these savings also trickle down to the tenants whose rents are 20% lower than those of occupants in the surrounding buildings.

9.26 *Stenocara*: A Water Capturing Insect

In the desert, where few living things are to be found, some species possess the most astonishing designs. One of these is the tenebrinoid beetle *Stenocara* (Figure 9.18a), which lives in the Namib Desert, in Southern Africa. A report in the 1 November 2001, edition of *Nature* describes how this beetle collects the water so vital to its survival. *Stenocara's* water capture system basically depends on a special feature of its back, whose surface is covered with tiny bumps. The surface of the regions between these bumps is wax coated, though the peaks of the bumps are wax-free. The exoskeleton (Figure 9.18b) exhibits rounded nodules that are densely packed in a hexagonal pattern. The matrix is composed of larger bumps that help to attract water, while the in-between nodules are waxy and hydrophobic in nature–having characteristics similar to a Teflon surface. The varying matrix helps to attract water and repel it at the same time. The specific form of the bumps and nodules helps to move the water from one point to the other. This allows the beetle to

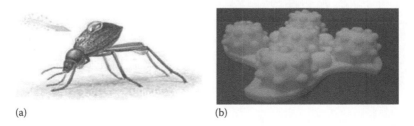

(a)   (b)

FIGURE 9.18
(a) *Stenocara* and (b) *Stenocara* beetle's exoskeleton.

collect in a more productive manner. *Stenocara* extracts from the air the water vapour that occurs only rarely in its desert environment.

What is remarkable is how it separates out the water from the desert air, where tiny water droplets evaporate very quickly due to heat and wind. Such droplets, weighing almost nothing, are borne along parallel to the ground by the wind. The beetle, behaving as if it knew this, tilts its body forwards into the wind. Thanks to its unique design, droplets form on the wings and roll down the beetle's surface to its mouthparts. The mechanism by which water is extracted from the air and formed into large droplets has so far not been explained, despite its biomimetic potential. Examining the features of this beetle's back under an electron microscope, scientists established that it's a perfect model for water-trapping tent and building coverings or water condensers and engines. Designs of such a complex nature cannot come about just by themselves or through natural events. Also, it's impossible for a tiny beetle to have *invented* any system of such extraordinary design. Just *Stenocara* alone is sufficient to prove that our Creator designed everything that exists.

9.27 Mosquito Bite–Inspired Biomimetic Materials [14]

Japanese medical researchers are reducing the pain of an injection by using hypodermic needles edged with tiny serrations, like those on a mosquito's (Figure 9.19a) proboscis, minimizing nerve stimulation. Puncturing of the human skin with a hollow needle is perhaps the most common invasive medical procedure to inject fluids into or extract fluids from the human

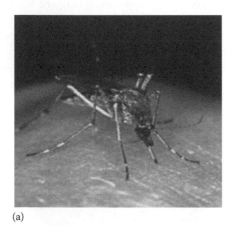
(a)

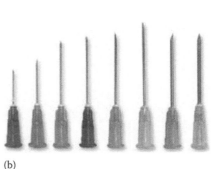
(b)

FIGURE 9.19
(a) Mosquito and (b) mosquito bite–inspired near-painless hypodermic needles. (From Ramasubramanian, M.K. et al., *Bioinspir. Biomim.*, 3, 046001, 2008.)

body. Regardless of the size of the needle, unless an anaesthetic is used, it is at least uncomfortable if not painful.

During the past few years, considerable effort has been put into developing microneedles (Figure 9.19b) to replace the traditional hypodermic needle. Microneedles are sharp hypodermic needles that are intended to be minimally invasive while accomplishing the task of penetrating the skin's outermost noninnervated layer to draw small volumes of blood or inject small quantities of therapeutic agents into the capillary-rich dermis layer, painlessly [15]. It has been reported that a penetration up to 1.5 mm can be painless. Typical sizes of these microneedles are 40–100 μm in diameter with submicron in tip radii. For comparison, the smallest hypodermic needle typically used (30 gauge) is about 320 μm in outer diameter and 160 μm in inner diameter. Some microneedles are hollow, while some are solid with the therapeutic agents coated on the surface that dissolves once the needle is in the bloodstream. The mechanics of a fascicle insertion into the skin by a mosquito of the type *Aedes aegypti* has been studied experimentally using high-speed video (HSV) imaging and analytically using a mathematical model.

The fascicle is a polymeric microneedle composed of a ductile material, chitin. It has been proposed that the mosquito applies a nonconservative follower force component in addition to the Euler compressive load in order to prevent buckling and penetrate the skin. In addition, the protective sheath surrounding the fascicle (labium) provides lateral support during insertion. Despite the advances in microneedle fabrication and experiments with infusion into the skin, limited work has been done to understand the mechanics of penetration into the skin. Yang and Zahn [16] claim that a mosquito uses vibratory cutting at a frequency of 200–400 Hz, but no reference is provided. They state that current designs of needles do not tolerate forces associated with insertion and intact removal and are typically either too fragile or too ductile. Vibratory techniques are proposed as a method for inserting the needle into the skin. A silicon microneedle (100 μm diameter and 6 mm long) was used in their experiments. They show that there is a greater than 70% decrease in insertion force when the needle is vibrated compared to quasistatic insertion. The vibration applied was parallel to the skin surface and had an amplitude of 0.6 mm, vibrating in the kHz range. The static peak insertion force was found to be 300 mN; upon inserting the needle with vibration, the force drops to 50 mN.

9.28 Honeycomb

The honeycomb (Figure 9.20a) is made by bees in total darkness, and it consists of a perfect hexagonal cellular structure that offers an optimal packing shape. For honeybees, the geometry meets their need for making a structure that provides the maximum amount of stable containment (honey, larvae)

FIGURE 9.20
(a) Honeycomb. (From www.reflectionsintheword.org.) (b) Honeycomb sandwich panels for aerospace. (From www.inhabitat.com.) (c) Sinosteel International Plaza. (From www.hexcel. com.)

using the minimum amount of material. For the same reasons, the honeycomb is an ideal structure for the construction of control surfaces of an aircraft (Figure 9.20b), and it can be found in the wing, elevators, tail, the flan and many other parts that need strength and large dimensions while maintaining low weight. The construction of honeycombs offers many great important advantages, including stability. As the bees in the hive give directions to one another in the so-called waggle dance, they set up vibrations that, in a structure of such small dimensions, can be equated to an earthquake. The walls of the comb absorb these potentially damaging vibrations. *Nature* magazine stated that architects could use this superior structure in designing earthquake-proof buildings. Included in the report was the following statement by Jurgen Tautz of the University of Wurzburg, Germany:

> Vibrations in honeybee nests are like miniature earthquakes generated by the bees, so it's very interesting to see how the structure responds to it... Understanding the phase reversal could help architects predict which parts of a building will be especially vulnerable to earthquakes... They could then strengthen these areas, or even introduce weak spots into non-critical areas of buildings to absorb harmful vibrations.

As this all shows, the combs that bees construct with such flawless precision expertise are marvels of design. This structure within the comb thus paves the way for architects and scientists, giving them new ideas. Be it their biomimetic form, their integral strength, or their beautiful visual texture, lately we can't get enough of hexagonal honeycomb structures. The latest to catch our eye is the stunning Sinosteel International Plaza (Figure 9.20c) by Beijing-based MAD architects. More than just a striking façade, the building's hexagonal curtain is based upon climate modelling and serves to regulate the structure's temperature and daylight by varying the size of each cell's window.

9.29 Conclusions

The biomaterials will be very much useful in biomimeting robots, swimsuits, tyre tread design, wear-resistant surfaces, self-cleaning systems, etc. The study of biomimetics will also help tribologists in designing systems with reduced adhesion and friction. The biomimetic MEMs and NEMs and biomimetic tribological surfaces are also the likely outcome. Finally, the human joints can also be biomimetized. Thus, the already popular in materials science and engineering, biomimetics is now emerging in tribology. *Its very fine application is found in the outer surface of the Cathay Pacific Airbus 340, modelled after the grooved scales of shark skin, reduces drag and improves fuel efficiency.*

From the aforementioned studies, the following conclusions have been drawn:

1. The micro–robot fish has no traditional components such as gears and bearings, and it can swim noiseless and flexibly. The embedded SMA wire actuated biomimetic fin imitates the structure of squid/cuttlefish fin.
2. The biomimetic drag-reducing surface is used for energy saving. New type biomimetic drag-reducing surface which is light in weight and powerful in function is valuable, especially to the swimsuits that reduce the drag.

3. Seashells' 3D curved structures have been successfully applied to architectural structures. Studying the microstructure of seashells opens up a possibility of attaining ductility in ceramics at lower cost. Exploring this possibility might lead us to light, cost-effective and efficient armour.

4. The dolphin sonar is so sensitive that it can even identify one single fish from among an entire shoal. Similarly, a sonar device is used for identifying targets and their directions for ships and submarines.

5. Gecko feet inspires in developing a dry and self-cleaning adhesive. A dry adhesive could help hold slick body parts in place during nanosurgery. StickyBot walks up vertical surfaces of glass, plastic and glazed ceramic tile and is also used to repair cracks in ships, bridges and piers and in the regular maintenance of satellites.

6. Inspired from the viper, a mini robot is designed to protect infantry soldiers from explosives, booby traps and hostile forces lying in ambush. Viper's search-and-destroy mechanism, the methods the snake employs, can be adapted more widely to protect the country from enemy missiles

7. Smart and intelligent textiles are an important developing area in science and technology because of their major commercial viability and public interest. Chameleons inspired for making clothes, bags and shoes able to change colours.

8. Snakes' properties inspired for a new robotic interplanetary probe, which has a constant state of balance, without ever getting caught up by obstacles. The balancing problem can be overcome.

9. By mimicking the scorpion, a robot is made to withstand harsh desert conditions. The robot can be asked to go to a specific region and, with a camera in its tail, send back to base images of the location.

10. Moth's eye inspires glare-free TV screens and studies of the antireflective coatings in moth eyes have led to brighter screens for cellular phones.

11. Air-conditioning system is inspired by termite mounds. Termite structures have been used for climate control in buildings. Office and shopping complexes are an architectural marvel in its use of biomimicry principles.

12. *Stenocara* inspired for water capture system. Scientists established that it's a perfect model for water-trapping tent and building coverings or water condensers and engines.

13. Mosquito bite inspires for developing near-painless microneedles to replace the traditional hypodermic needle. Vibratory techniques are proposed as a method for inserting the needle into the skin.

14. The combs that bees construct with such flawless precision expertise are marvels of design. Be it their biomimetic form, their integral
strength, or their beautiful visual texture, we get hexagonal honeycomb structures.

References

1. S. Kennedy, Biomimicry/biomimetics: General principles and practical examples, *The Science Creative Quarterly*, Issue 6, August 2004.
2. R.A. Singh, E.-S. Yoon, and R.L. Jackson, Science of imitating nature, *Tribology and Lubrication Technology*, 65, February 2009, 41–47, www.stle.org.
3. Z. Wang, G. Hang, J. Li, Y. Wang, and K. Xiao, A micro-robot fish with embedded SMA wire actuated flexible biomimetic fin, *Sensors and Actuators A*, 144, 2008, 354–360.
4. X. Han, D.Y. Zhang, X. Li, and Y.Y. Li, Bio-replicated forming of the biomimetic drag-reducing surfaces in large area based on shark skin, *Chinese Science Bulletin*, 53(10), May 2008, 1587–1592.
5. K. Akella, Biomimetic designs inspired by seashells, *Resonance*, 573–591, June 2012.
6. H. Yahya, *Biomimetics: Technology Imitates Nature*, Global Publishing, Istanbul, Turkey, March 2006.
7. K.-J. Cho, J.-S. Koh, S. Kim, W.-S. Chu, Y. Hong, and S.-H. Ahn, Review of manufacturing processes for soft biomimetic robots, *International Journal of Precision Engineering and Manufacturing*, 10(3), July 2009, 171–181.
8. Y. Lappin, Elbit unveils new defense products, in *Latrun Conference*, The Jerusalem Post, Jerusalem, Israel, July 2010.
9. Brian, Baby clothes that change color with temperature, Posted in: Stories, Elemental/Exposing the Positive, October 14, 2010. http://elementalproject.org/839.
10. B. Coxworth, Wear-resistant surfaces inspired by scorpions, *American Chemical Society Journal Langmuir*, 26 January 2012.
11. J. Rigg, Sharp announces first TVs with Moth-Eye technology: The AQUOS XL series, October, 2012.
12. A. Hadhazy, How a moth's eye inspires glare-free TV screens, *The Christian Science Monitor*, 25 May 2010.
13. J.S. Turner and R.C. Soar, Beyond biomimicry: What termites can tell us about realizing the living building, in *First International Conference on Industrialized, Intelligent Construction (I3CON)*, Loughborough University, Leicester, U.K., 14–16 May 2008.
14. M.K. Ramasubramanian, O.M. Barham, and V. Swaminathan, Mechanics of a mosquito bite with applications to microneedle design, *Bioinspiration & Biomimetics*, 3, 2008, 046001.
15. Sportstechreview, Mosquito designed surgical needle, September 6, 2010.
16. M. Yang and J. Zahn, Microneedle insertion force reduction using vibratory actuation, *Biomedical Microdevices*, 6, 2004, 177–182.

10

Superhard Materials

10.1 Introduction

Hardness is a *surface property*. It is defined as the resistance of a material against permanent deformation of the surface in the form of scratch, cutting, indentation or mechanical wear. Diamond is the hardest known material.

10.1.1 Need of Hardness Test

The need of hardness test arises from the fact that in numerous engineering applications, two components in contact are made to slide or roll over each other. In due course, their surfaces are scratched and they may fail due to mechanical wear. This results in not only a quick replacement of both parts but also incurs a big loss in terms of money. For example, piston rings of an IC engine remain in sliding contact with the cylinder body when the piston reciprocates within the cylinder. If proper care is not taken in selection of materials for them, the piston rings and cylinder will wear soon. In this case, the replacement or repairing of cylinder block will involve much time, trouble and money. Therefore, the materials of piston rings and cylinder block should be taken such that the wear is least on the cylinder. Thus, in case of repairing, comparatively cheaper piston rings can be easily replaced. This envisages that material of cylinder block should be harder than the material of piston rings so that the cylinder wears the least. This can be ascertained by conduct of a hardness test. That is why it is essential to know as to how this test can be conducted.

10.1.2 Different Types of Hardness Tests

In various hardness tests, the indentors are used to introduce indentation on the surface. The shape of indentors may be a spherical ball, a cone or a pyramid. Various hardness test methods are as follows:

1. Mohs hardness test with scale range 0–10
2. Brinell hardness test with scale range 0–3000
3. Rockwell hardness test with scale range 0–1000

4. Vickers hardness test with scale range 0–3000

5. Knoop hardness test for hardness of microscopic areas

6. Shore hardness test for hardness of rubber, plastics, paper, etc.

7. Barcol hardness test for checking the degree of cure of plastics and composites

8. Jominy hardness test for end-quenched metals

9. Rebound hardness test or Shore Scleroscope method

10.2 Brinell Hardness Test

10.2.1 Test Setup, Specifications of Hardness Testing Machines and Indentors

A hardness test can be conducted on Brinell testing machine, Rockwell testing machine or Vickers testing machine. The specimen taken for them may be a cylinder, cube, thick or thin metallic sheet. A Brinell-cum-Rockwell hardness testing machine along with the specimen is shown in Figure 10.1. Its various components are shown therein whose details are self-explanatory.

Its specifications are as follows:

- Ability to determine hardness up to 500 BHN (Brinell hardness number)
- Diameter of ball (as indentor) used $D = 2.5, 5$ and 10 mm
- Maximum applicable load $= 30D^2$ (i.e. $30 \times 10^2 = 3000$ kg)
- Method of load application = push-button type (it can be computer controlled also).
- Capability of testing the lower hardness range = 1 BHN on application of $0.5D^2$ load

10.2.2 Test Procedure

This test employs a diamond or hardened steel ball as indentor. The ball is placed suitably in the upper housing of Brinell hardness testing machine shown in Figure 10.1. This machine is called a push–pull-button-type machine because the indenting load is applied by pushing a button.

There are several push buttons and each of them specifies a known load. Before conducting the test, the surface of the specimen is made free from oil, grease, dust and dirt. The indenting load P is applied on the specimen gradually for a minimum of 30 s. The load stage is different for various types of materials to be tested and is given by

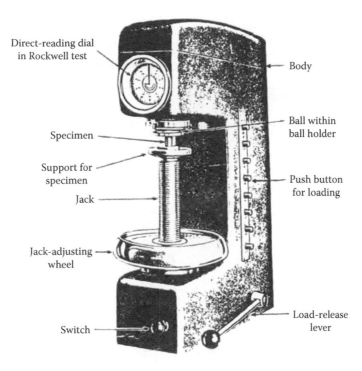

FIGURE 10.1
Details of push-button-type Brinell-cum-Rockwell hardness testing machine.

$$P = \lambda D^2 \tag{10.1}$$

where
  D is the diameter of the ball
  λ is a constant whose value varies between 0.5 and 30 as illustrated in Table 10.1

10.2.3 Observations and Calculations

The effect of this load is to make an indentation of depth t and diameter d as shown in Figure 10.2. The BHN is then calculated as follows after measuring d by an optical microscope:

$$
\begin{aligned}
\text{BHN} &= \frac{\text{Indenting loading (kgf)}}{\text{Spherical surface area of indentation (mm}^2\text{)}} \\
&= \frac{P}{(\pi d/2)\left[D - \sqrt{D^2 - d^2}\right]}
\end{aligned}
\tag{10.2}
$$

Result: BHN value of some materials and its comparison with other hardness scales is shown in Table 10.3.

TABLE 10.1

Hardness Range and Load Stages in Brinell Hardness Test

Material to Be Tested	Hardness Range (BHN) (kg/mm²)	Load Stage P (kgf)
• Soft iron, steel, steel casting, malleable iron, cast iron	67–500	$30D^2$
• Light alloys, die casting alloys, casting and forging alloys, brass and bronze	22–315	$5D^2$
• Al, Mg, Zn, cast brass	11–158	$5D^2$
• Bearing metals	6–78	$2.5D^2$
• Lead, tin, soft solder	3–39	$1.25D^2$
• Soft metal at elevated temperatures	1–15	$0.5D^2$

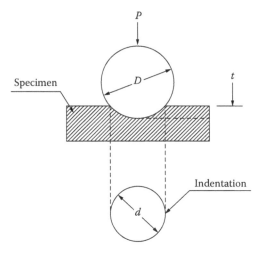

FIGURE 10.2
Brinell hardness test showing load, ball diameter, indented diameter and its thickness.

10.2.4 Test Requirements and Limitations

Test requirements, noticeable observations and limitations of Brinell hardness test can be enumerated as follows:

1. The steel ball indentor may be used to test the specimen of cast iron, unhardened steel and light alloys.
2. Standard diameters of the ball are 2.5, 5 and 10 mm.
3. Deformation of the ball during application of indenting load is neglected in calculating BHN.
4. Usually, the diameter of indentation $d = 0.2D$–$0.7D$.

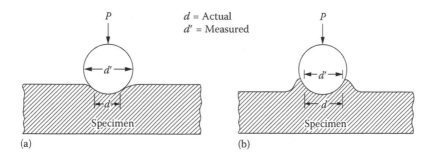

FIGURE 10.3
Defect in Brinell hardness test (a) sinking effect and (b) piling-up effect.

5. Thickness of the specimen should not be less than 10 times the expected depth t (Figure 10.2) of the impression.
6. If the impression of indentation is non-circular, the mean value should be taken from two diameters that are normal to each other.
7. Value of BHN is expressed in kgf/mm² or N/mm².
8. For most of the metals, BHN is proportional to their tensile strengths.
9. For steel, the tensile strength is 35 BHN (in N/mm²).
10. Brinell hardness test is not accurate for BHN > 500, as the ball itself deforms; hence, Equation 10.1 does not hold good for values above 500.

The drawbacks in Brinell hardness test are shown in Figure 10.3. These are

1. *Sinking effect* which occurs in manganese steel and austenitic stainless steel (Figure 10.3a.)
2. *Piling-up effect* which occurs in lead, tin, magnesium, etc. (Figure 10.3b)

Example 10.1

A hardened steel ball of 0.5 cm diameter is used to indent a steel specimen in Brinell hardness test. Diameter of indentation measured by an optical microscope of magnification 10× is observed to be 32.5 mm. Calculate the BHN of the steel specimen.

Solution: It is given that $D = 0.5$ cm = 5 mm.

Indentation diameter, magnified 10×, is 32.5 mm.
Therefore,

$$d = \frac{32.5}{10} = 3.25 \, \text{mm}$$

From Table 10.1, it can be seen that load P for steel specimen is

$$P = 30D^2$$

$$\therefore P = 30 \times (5)^2 = 750 \text{ kgf}$$

Using Equation 10.2, the hardness is obtained as

$$\text{BHN} = \frac{P}{(\pi d/2)\left[D - \sqrt{D^2 - d^2}\right]}$$

$$= \frac{750}{(\pi 5/2)\left[5 - \sqrt{5^2 - (3.25)^2}\right]}$$

$$= 79.5 \text{ kgf/mm}^2$$

It should be clear to our mind that the BHN is expressed in kgf/mm², but this unit is generally not written.

10.3 Rockwell Hardness Test

This test is more common due to its quick and simple method. There is no need of any calculation because the Rockwell hardness (HR) may be read directly on the dial (Figure 10.1). The test involves application of an initial load of 10 kgf on the specimen so that the effects of dust, dirt, oil, etc. are nullified. This makes the Rockwell test more accurate than the Brinell test.

10.3.1 Indentor and Test Procedure

This test employs a ball and a cone as indentors. The specimen is subjected to a major load for about 15 s, after the initial load. ASTM specifies 13 scales for testing of a wide range of materials ranging from very soft to very hard. These scales are named as *A, B, C, D, E, F, M, R,* etc. Of these, *B*-scale and *C*-scale are commonly employed. *B*-scale is preferred for soft steels and aluminium alloys, while *C*-scale is chosen for titanium and hard steel; *B*-scale employs a ball of 1/16 in. = 1.58 mm diameter. A cone indentor is used in *C*-scale with a cone angle of 120° and point of radius 0.2 mm. Hardness value determined from *B*-scale is referred to as HRB and from *C*-scale as HRC. Different scales, initial and major loads to be given on them, their suitability to the kind of materials and other related details are given in Table 10.2.

Result: If the depth of penetration t in mm is known, then the hardness may also be calculated from

TABLE 10.2

Various Scales in Rockwell Hardness Test

Scale	Type of Indentor (Dimension)	Initial Load (kgf)	Major Load (kgf) on Dial	Pointer Position	Kind of Material
A	Cone, 120°	10	50	0	Much harder such as carburized steel, cemented carbides
B	Ball, 1.58 mm	10	90	30	Soft steels, copper, aluminium, brass, grey cast iron
C	Cone, 120°	10	140	100	Hard steels, Ti, W, Va, etc.
D	Cone	10	90	—	Thin ferrous metals
E	Ball, 3.0 mm	10	90	—	Very soft such as metals, magnesium-bearing alloys
F	Ball, 1.58 mm	10	50	—	Soft such as babbits, bronze, brass

$$HRA = 100 - \frac{t}{0.002} \qquad (10.3a)$$

$$HRA = 130 - \frac{t}{0.002} \qquad (10.3b)$$

$$HRC = 100 - \frac{t}{0.002} \qquad (10.3c)$$

10.3.2 Suitable Applications

Rockwell hardness method may be used to determine the hardness of wires, blades and inside and outside cylindrical surfaces such as in IC engine cylinder and piston. Finished components can also be tested by this method as the indentation made is small. This method is suitable for hardness beyond the range of BHN. Rockwell hardness of some materials on scales B, C, M and R are given in Table 10.3.

10.4 Vickers Hardness Test

This test is similar to Brinell test but uses a different type of indentor. A square-based pyramid indentor of cone angle $\alpha = 136°$ between opposite faces of pyramid is used. The applied loads may be 5, 10, 30, 50, 100 or 120 kgf.

TABLE 10.3

Comparison of Different Hardness Scales and Hardness of Various Materials

Material	Mohs Scale	H_M	Brinell and Vickers Scales			Rockwell Scales		
			BHN and VHN	Material	B-Scale	C-Scale	M-Scale	R-Scale
→			3000					
	Diamond	10	2600	Diamond				
Covalent-bonded materials	Corundum	9	2000	Carbides				
			1000	Ceramics		80		
Ionic—↑ covalent-bonded material ↓	Topaz	8		Cutting tool Steel				
	Quartz	7	500	Alloy steels				
			300	Titanium ↑		40		
Ionic-bonded materials	Feldspar	6	200	Plain carbon steels	110	20		
	Apatite	5						
↑	Fluoride	4	100	→	80	0		
			80	Aluminium alloys				
	Calcite	3	40	Copper alloys	60			
Layered minerals	Gypsum	2		Magnesium alloys ↑	0		140	
	Talc	1	20	Plastics			70	130
→			10	Lead alloys			20	100
			5	Lead ↓				40
		0	0					

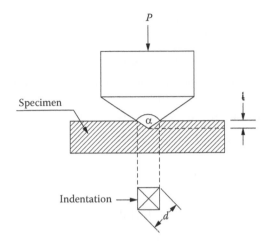

FIGURE 10.4
Square-based pyramid indentor in Vickers test and its indentation on the test piece.

The Vickers hardness H_V is calculated from

$$H_V = \frac{P}{\left[d^2 / 2 \sin \alpha / 2 \right]}$$

$$= \frac{1.8544 P}{d^2} \quad \text{for } \alpha = 136°$$

(10.4)

where
P is the applied load in kgf
d is the diagonal length in mm of indentation made by the pyramid

The indentor and the indentation are shown in Figure 10.4. This test is performed for smaller cross sections, very hard materials, polished and nitrided surfaces and very thin test pieces. Hardness of some materials on Vickers hardness number (VHN) scale is displayed in Table 10.3.

10.5 Introduction to Superhard Cutting Tool Materials

According to generally acceptable norms, a material is said to be superhard when its Vickers hardness number $H_V \geq 40$ GPa. The hardness value of 40 GPa implies that the material under consideration, when subjected to a force of 40×10^9 N, will create an indented area of 1 m² (i.e. 40×10^9 N/m²). This is equivalent to 40×10^2 kg/mm², (i.e. BHN = 4000). This number is beyond the range of Brinell scale which extends from 0 to 3000 only. The present trends in machining operations/manufacturing foresee towards following factors in order to increase the productivity:

- Ever-higher cutting speeds to increase the productivity
- Dry cutting (without using coolant/lubricants) in order to save the recycling costs of cutting fluids
- Green manufacturing which avoids the environmental problems
- Tool materials which can withstand the very severe operating conditions
- High wear resistant and high red hardness

10.5.1 Examples of Superhard Materials

To meet these requirements, the investigations/researches being carried out currently are inclined to search for new cutting tool materials which are not only hard or red-hard, but are superhard also. Some of the recent superhard materials in this regard are the following:

1. Ceramic composite tools, namely, SiC, WC and Si_3N_4 groups, alumina and zirconia
2. Cubic boron nitride c-BN having $H_V \approx 48$ GPa
3. Natural diamond having $H_V \approx 70–90$ GPa
4. Artificial diamond having $H_V \geq 90$ GPa
5. Superhard graphite
6. Osmium dinitride (OsN_2) with fluorite structure (CaF_2) having $H_V \approx 96$ GPa and OsO_2
7. Covalent metals, namely,
   a. Strong covalent compound fanned by light elements, for example C_3N_4, B_6O and C–BC_2N
   b. Partially covalent heavy transition metals, for example RuO_2 and OsB_2
8. WCoB–TiC-based hard materials
9. Double borides such as W–Co–B, Mo–Co–B and Mo–Fe–B
10. PtN_2 (fluorite)
11. Elemental osmium (Os)
12. Rhodium diboride (RhB_2)
13. Boron–osmium (B–Os)
14. Coating materials in machining of steel, for example TiN, TiC and ZrC

However, more investigation is required for some 'may be promising materials' like ZrB_2, TiB_2 and ZrN.

10.6 Present Trends in Machining

Present trends in machining operations foresee a quest for ever-higher cutting speeds to increase the productivity. Another relatively new trend is towards dry cutting, that is, machining without the use of coolants/lubricants in order to save the recycling cost of cutting fluids with their associated environmental problems. Hence, there has been a continued need for searching new cutting tool materials which can withstand the severe operating conditions at and near the cutting edge of the tool. Because of their high resistance to wear and high temperatures, the ceramics and ceramic composites remain attractive candidate materials for the cutting tool applications.

One might ask as to how is it possible to prepare materials with hardness approaching or even reaching that of diamond. The answer is quite simple: the presence of flaws such as dislocations and microcracks in engineering materials limits their strength to a small (10^{-4} to 10^{-2}) fraction of their ideal strength. The latter can be estimated at about $0.1G$ for materials which display crystal plasticity and about $(0.1–0.2)E$ for glasses where the dislocations are absent. Here, G and E are the shear and Young's modulus, respectively. Thus, the ideal strength of intrinsically strong materials lies in the range of several 10 GPa, but their practically achievable value seldom exceeds several 100 MPa. Furthermore, conventional hard materials are brittle and have a small elastic limit of ≤0.1% above of which fracture occurs. Therefore, the strengthening of conventional engineering materials is based on appropriate engineering of microstructure that hinders the formation and propagation of flaws. In spite of that, the strength of the strongest steels and metallic alloys in bulk reaches only 1–2 GPa and their elastic limit is below 0.5%. The upper limit of tensile strength of ≤2 GPa and elastic range of ≤2% was achieved in some metallic glasses.

10.6.1 Strengthening Methods to Enhance the Hardness of Materials and Coatings

Hardness of materials can be increased by adopting various methods such as the following:

1. By heat treatment such as quench hardening
2. By minimizing the dislocations, micro-cracks and other flaws from the material
3. By strengthening of microstructure that hinders the formation and propagation of flaws
4. By preparation of superhard coatings

10.7 Recent Developments in Superhard Materials

10.7.1 Advanced Ceramic and Ceramic Composite Cutting Tool Materials

Combustion synthesis is used to synthesize a large number of monolithic and composite organic materials. Different categories of these materials are as shown in Table 10.4.

10.7.1.1 Salient Applications

Some of these materials have been used or proposed to be used for the following applications:

- Abrasives, cutting tools, polishing powders (e.g. TiC, cemented carbides and carbonitrides)
- Elements for resistance heating furnaces (e.g. $MoSi_2$)
- High-temperature lubricants (e.g. chalcogenides of Mo)
- Neutron attenuators (e.g. refractory metal hydrides)
- Shape-memory alloys (e.g. TiNi)
- High-temperature structural alloys (e.g. Ni–Al intermetallic compounds)
- Steel-melting additive (e.g. nitride ferroalloys)
- Electrodes for electrolysis of corrosive media (e.g. TiN)
- Coating for containment of liquid metals and corrosive metals (e.g. products of aluminium and iron oxide thermite reactions)

TABLE 10.4

Various Categories of Superhard Materials

Category	Materials
Borides	CrB, HfB, NbB_2, TaB_2, TiB_2, LaB_6, MoB_2, WB, ZrB_2, VB_2
Carbides	TiC, ZrC, HfC, NbC, SiC, Cr_2C_2, SiC, B_4C, WC, TaC, VC, Mo_2C
Carbonitrides	TiC–TiN, NbC–NbN, TaC–TaN
Chalcogenides	MoS_2, $TaSe_2$, NbS_2, WSe_2, $MoSe_2$, MgS
Composites	TiC–TiB_2, TiB_2–Al_2O_3, B_4C–Al_2O_3, TiN–Al_2O_3, TiC–Al_2O_3, $MoSi_2$–Al_2O_3, Ni–ZrO
Hydrides	TiH_2, ZrH_2, NbH_2
Intermetallics	NiAl, $NiAl_3$, FeAl, NbGe, TiNi, CoTi, CuAl
Nitrides	TiN, ZrN, BN, AlN, Si_3N_4, TaN (cubic and hexagonal)
Silicides	$MoSi_2$, $TaSi_2$, Ti_5Si_3, $ZrSi_2$, $ZrSi_2$, WSi_2, $NbSi_2$
Oxides	ZrO, YSZ, $MgAl_2O_4$, $Bi_4V_4O_{11}$, $YBa_2Cu_3O_{7-x}$

- Powders for further ceramic processing (e.g. Si_3N_4, AlN)
- Thin films and coatings (e.g. silicides)
- Functionally graded materials (FGM) (e.g. TiC + Ni)
- Composite materials and cements (e.g. TiC + Al_2O_3, Ni + YSZ)
- Complex oxides with specific magnetic or electrical properties (e.g. $BaTiO_3$, $YBa_2Cu_3O_{7-x}$)

10.7.2 Superhard Materials and Superhard Coatings

The three groups (SiC, WC and Si_3N_4) are actually reflected in industrial practice since those phases with high chemical solubility are susceptible to excessive flank and crater wear, when used as cutting tool for the machining of steels. The chemically stable materials like alumina and zirconia are actually the constituent phases of successful cutting tool materials for high-speed turning of steel, whereas some of the phases of intermediate group (e.g. TiN, TiC and ZrC) are successfully used as coating materials in the machining of steel. Other groups (e.g. ZrB_2, TiB_2, ZrN) have yet not been used or properly investigated. Comparing the solubility of the Ti and Zr borides, nitrides and carbides, TiC is most soluble in pure iron, whereas TiN and ZrN are the least soluble. The solubility of TiB_2 is higher than that of ZrC and ZrB_2.

Superhard materials with Vickers hardness $H_V \geq 40$ GPa received recently increasing attention because of their already existing and potential future applications, for example as protective tribological coatings for machining and forming tools. The intrinsically superhard materials include diamond ($H_V \approx 70$–90 GPa; industrial diamonds may have a higher hardness due to substitutionally dissolved nitrogen) and cubic boron nitride c-BN ($H_V \approx 48$ GPa). Their application as protection coatings is still limited.

Three different approaches towards the preparation of superhard coatings are the following:

1. Deposition of intrinsically superhard materials, such as diamond and c-BN in a kinetically controlled regime
2. Hardening due to energetic ion bombardment during the deposition
3. Formation of an appropriate nanostructure which hinders the formation and propagation of flaws

10.7.3 Superhard Graphite

It is hard to imagine that graphite, the soft *lead* of pencils, can be transformed into a form that competes in strength with diamond. Researchers have speculated for years on the extreme conditions that might change the molecular

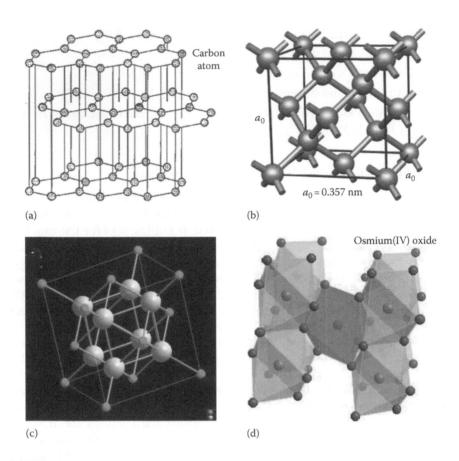

Carbon atom

a_0

a_0

$a_0 = 0.357$ nm

(a)

(b)

(c)

(d)

Osmium(IV) oxide

FIGURE 10.5
(a) Structure of graphite. (b) Structure of diamond. (c) Structure of CaF_2. (d) Structure of OsO_2. (From Fan, C.-Z. et al., _Phys. Rev. B_, 74(12), 125118, 2006.)

structure of graphite into a superhard form that rivals diamond. The graphite and diamond are both made of carbon. The geometric arrangement and spacing of the carbon atoms in them makes the difference in their appearance and strength. The atoms in graphite are arranged in layers (Figure 10.5a) that are widely spaced, whereas the atoms in diamond are tightly linked (Figure 10.5b) producing a strongly bonded structure. The scientists subjected the graphite to pressures of 17 GPa which is equivalent to 170,000 times the pressure at sea level, to see as to how the structure changes at atomic level when graphite is squeezed into superhard form. The graphite that resulted from this experiment was so hard that when the pressure was released, it had cracked the diamond anvil. The superhard form of graphite opens the door to a myriad of applications in industry, particularly as a structural component.

10.7.4 Superhard Osmium Dinitride with Fluorite Structure [1]

The search for novel hard materials compared to or even harder than diamond, which has the highest measured hardness of 96 GPa and bulk modulus of 443 GPa, has stimulated a variety of great achievements in high-pressure research. Consequently, many new superhard materials have been prepared by high-pressure technique, especially after the laser-heated diamond-anvil cells (DACs) were invented. In general, two groups of materials are powerful candidates for superhard materials, namely, the following:

- Strong covalent compounds fanned by light elements, such as polymorphy of C_3N_4, B_6O and c-BC_2N.
- Partially covalent heavy transition metal boride, carbide, nitride and oxide RuO_2 and OsB_2 are such examples.

Materials with high bulk or shear modulus are likely to be hard materials. The bulk modulus of osmium dinitride (OsN_2) with CaF_2 structure (Figure 10.5c) has a very high value (360.7 GPa), which is even higher than that of OsO_2 (Figure 10.5d) with the same structure (347.5 GPa) and is comparable with that of the orthorhombic OsB_2 (365–395 GPa).

10.7.5 WCoB–TiC-Based Hard Materials

Simple and ternary borides are very interesting materials widely used due to their high melting point, wear resistance, chemical inertness and high hardness values. The WCoB is a complex boride with extremely high hardness and excellent oxidation resistance. A series of double borides is obtained by mixing W and Mo with cobalt. These are mainly the ternary phase diagrams such as W–Co–B, Mo–Co–B and Mo–Fe–B. The properties of a coating with high percentages of WCoB generated at high temperatures from the reaction between TiH_2 and sintered compacts of WC–TiC–Co yield very good results. Different newer materials are emerging as recent superhard materials. They are in elemental forms, films and coatings. Various methods are adopted to enhance the hardness in materials, namely, these are mainly strengthening methods of materials, high-pressure technique and using superhard coatings. These details are described in the subsequent chapters.

10.7.6 Almost Incompressible Rhodium Diboride (RhB_2)

Though the diamond is the hardest known material, it is very expensive. Whereas the diamond powder is used in making drills for oil exploration and roads, for cutting holes in mountains, it gets ruined during cutting of steel. Therefore, there is a need to search for a substitute of diamond. There are two names in this regard, namely,

1. Cubic boron nitride (c-BN), but it is more costly than diamond, since it is made synthetically under very high-pressure and high-temperature conditions
2. Combination of boron with soft element osmium which is the most incompressible metal, but this transition metal has certain limitations of making scratches

However, the search of rhodium (Rh) is fruitful and it is used to make rhodium diboride (RhB_2) which is as incompressible as diamond in one direction, but slightly more incompressible in other direction. It is hard enough to scratch diamond and much harder than OsB_2. Since the RhB_2 can be made without applying high pressure, therefore it is further advantageous over diamond and c-BN, which are made by high-pressure techniques.

10.8 Comparison between the Properties of Diamond and Osmium

A comparison shown in Table 10.5 displays the comparative favourable features of diamond and osmium [2].

TABLE 10.5

Properties of Diamond versus Osmium

Characteristics	Diamond (Pure Crystal)	Osmium (Pure Crystal)	OsO_2	OsN_2
Density (kg/m³)	3,500	22,570	11,400	—
Young's modulus (GPa)	≈1140	645	579.3	283.1
Bulk modulus (GPa)	443	462	347.5	360.7
Poisson's ratio	0.07–0.08	0.24–0.25	0.22	0.37
Hardness	Hardest	Softer	—	—
Bonding	Covalent	Covalent	Ionic–covalent	Ionic–covalent
Structure	DC	FCC/HCP	—	Orthorhombic
Melting point (°C)	3550	2700	—	—
Shrinking under high compression pressure	More	Less	—	—
Equilibrium lattice parameter (Å)	3.546	3.824 (FCC) 2.731 (HCP)	4.816	4.819

10.9 Materials in Nonconventional Machining Processes

Nonconventional machining is preferred for harder work materials like Ti, stainless steels (SS), Nimonics, fibre reinforced concrete (FRC), ceramics and stellites (Co-based alloys). Various advanced machining processes adopted for this purpose are the following:

Abrasive jet machining (AJM): Abrasives such as Al_2O_3, SiC, glass beads and crushed glass and $Na_2(Co_3)_2$.

Ultrasonic machining (USM): Brass, SS, MS, etc. as tool material.

Abrasive flow machining (AFM): CBN, diamond, SiC, Al_2O_3, etc. and nylon and Teflon as inserts for restricting areas in finishing of the airfoil surfaces of impellers and nozzle of torch flame.

Magnetic abrasive finishing (MAF): Magnetic abrasive particles (MAPs) such as ferromagnetic particles with abrasives as cutting tool.

Water jet cutting (WJC): Nozzle material is a synthetic sapphire for cutting of fibreglass and polyethylene automotive parts, asbestos, etc.

Abrasive water jet machining (AWJM): WC, BC and sapphire as cutting materials.

Electro discharge machining (EDM): electrode material of Pt, W, Pt–W alloy, Pd, Rh, etc., dielectric fluids and tool materials such as graphite, Cu, Brass and W.

Laser beam machining (LBM): Solid-state lasers, ruby; Nd:YAG gaseous lasers, Co_2, H_2, N_2, etc.

10.10 Current Researches and Futuristic Trends

10.10.1 Hybrid TIN and CRC/C PVD Coatings Applied to Cutting Tools [3]

Wear-resistant composite coatings for cutting and forming tools consisting of multiple functional layers are currently used to improve tool performance in a variety of work materials. The machining of aluminium silicon alloys is a unique challenge due to high friction and adhesion of aluminium to the tool surface and the abrasiveness of silicon carbide particles in the aluminium matrix. It is desirable to have an outer surface layer that can reduce friction by preventing adhesion of the work material to the cutting tool and an inner refractory wear-resistant layer that can withstand abrasion.

A new development in cutting tools has been made for machining aluminium. Hybrid physical vapour deposition (PVD) coatings consisting of a superhard TiSiN base layer and a low-friction CrCx/a-C:H top layer are used on cutting tools where the combination of low friction and resistance to abrasion is required. When machining aluminium silicon, CrCx/a-C:H resists the adhesion and galling of aluminium to the cutting tool by reducing friction and TiSiN strongly resists abrasive wear.

10.10.2 Al₂O₃–Mo Cutting Tools for Machining Hardened Stainless Steel [4]

Alumina ceramic cutting tools are characterized with high hardness (~400 GPa) and wear resistance. The main drawback of using ceramic materials is their low fracture toughness. Low resistance to brittle fracture of ceramic is mainly due to lack of possibility of stress relaxation prior to the tip of the crack. Strengthening of composite material can be achieved through the influence of residual stresses with stress preceding the crack tip. Fracture toughness of ceramic materials can be enhanced, for example, through adding second phase which possesses much higher K_{IC} coefficient (i.e. metals). Such ceramic metal composites apart from having high mechanical properties can also have some other special properties.

10.10.3 Development of Micro Milling Tool Made of Single-Crystalline Diamond for Ceramic Cutting [5]

Micro aspheric glass lenses have been used in various devices, such as digital cameras, laser devices and medical devices. The micro glass lenses are generally mass produced in high temperature of 400°C–800°C by glass press moulding process with moulds or ceramic moulding dies made of tungsten carbide or silicon carbide. The aspheric ceramic moulds are mostly ground with micro diamond wheels and polished with loose abrasives. The workpiece form accuracy of 0.1–0.2 mm *P–V* and the surface roughness of 10–30 nm *Rz* are obtained by the grinding. However, the grinding wheel wears soon, and it is difficult to keep the original geometrical shape and surface of the wheel, and then, the diamond wheel must be trued carefully on the machine after some grinding. It is, therefore, expected that the ceramic dies and moulds could be finished with high accuracy if a proper diamond cutting tool is developed as the size of the moulds becomes smaller and the required accuracy becomes higher. The ultrasonic elliptical vibration cutting method is developed by Shamoto and is applied to mirror machining of hardened steels with single-crystalline diamond tool. The micro milling tools made of polycrystalline diamond (PCD) are developed and are applied to ultraprecision machining of tungsten carbide moulds. Butler-Smith developed microstructured tools made of chemical vapour deposition (CVD) diamond with a pulse laser and a focus ion beam, and fundamental grinding tests of hard materials such as Ti–Al–4V are carried out, and the tool performance are evaluated [6]. In these cutting methods, the tool wear is decreased by the interrupted cutting effects and the hard materials can be removed precisely.

10.11 Conclusions

Based on the literatures reviewed as discussed previously, the following conclusions can be drawn. Likely, futuristic superhard cutting tool materials are OsN_2 with CaF_2 structure, OsO_2, RhB_2, Os and B–Os.

1. TiN, TiC and ZrC are the likely coating materials for machining of steels.
2. ZrB_2, TiB_2 and ZrN may be the more promising materials and require more investigation.

References

1. C.-Z. Fan, S.-Y. Zeng, L.-X. Li, R.-P. Liu, W.-K. Wang, P. Zhang, and Y.-G. Yao, Potential superhard osmium di-nitride with fluorite structure: First-principles calculations, *Physical Review Series B*, 74, 2006, 125118 (arXiv:cond-mat/0610729 v1 26 Oct2006), China.
2. H. Cynn, J.E. Klepeis, C.-S. Yoo, and D.A. Young, Osmium has the lowest experimentally determined compressibility, *Physics Review Letter*, 88, 14 March 2002, 135701.
3. W.E. Henderer and F. Xu, Hybrid TiSiN, CrC/C PVD coatings applied to cutting tools, *Surface and Coatings Technology*, 215, 2013, 381–385.
4. K. Konopka, J.J. Bucki, S. Gierlotka,W. Zielinsk, and K.J. Kurzydłowski, Characterization of metal particles fraction in ceramic matrix composites fabricated under high pressure, *Materials Characterization*, 56, 2006, 394–398.
5. S. Suzuki, M. Okada, K. Fujji, S. Matsui, and Y. Yamagata, Development of micro milling tool made of single crystalline diamond for ceramic cutting, *Annals of the CIRP*, 62(1), 2013, 59–62.
6. P. Butler-Smith, D. Axinte, and M. Daine, Solid diamond micro-grinding tools from innovative design and fabrication to preliminary performance evaluation in Ti-6Al-4V, *International Journal of Machine Tools and Manufacture*, 59, 2012, 55–64.

11

Advances in Powder Metallurgy

11.1 Introduction

Powder metallurgy is a process by which the metallic objects are manufactured from individual, mixed or alloyed metallic powders with or without the addition of non-metallic constituents. For that, the metallic powders are first produced from bulk metal using different processes, then the elemental or alloy powders are mixed with additives and lubricants, and finally the powder is compacted or sintered in a punch-die set-up. For getting a good-quality product, it is essential to have suitable shape and size of powder particles and appropriate temperature and time for sintering (or compaction) pressure. The powder metallurgy process is a very suitable technique to work with refractory materials, sintered carbides and materials which are extremely difficult to machine. Powder metallurgy is also suitable for those materials which are not capable of making products by any other manufacturing method, such as diamond, cemented carbide and tungsten. In this process, the metal/alloy used in its raw form is solid which remains solid till the final product is obtained. Thus, there does not exist any stage when the metal/alloy has to be melted. The powder metallurgy technique is mainly helpful in producing the following specialities of parts:

- Highly porous components such as self-lubricated bearings and bushes
- Composite materials such as ferrous and non-ferrous metal matrix composites
- High-performance precise components made of superplastic superalloys
- Dispersion-strengthened alloys such as W powders with thoria (ThO_2) for making electric lamp filaments

11.1.1 Necessity of Powder Metallurgy Methods over Conventional Production Methods

The powder metallurgy methods are necessary to carry out various processes because of the limitations of conventional processes, as follows:

1. Considerable difference in melting points of the two elements which makes them not workable together. For example the melting point of W is 3410°C and that of Al is 657°C in which one element, the Al, will become gas even before the other element *W* has melted.
2. Some metals do not form liquid solution or have partial solubility. Hence, they cannot be alloyed to form an alloy having specific properties.
3. Identity of the constituents is generally lost during melting and solidification. For example tungsten carbide (WC) breaks down on melting.

11.2 Operations Involved in Powder Metallurgy

The basic operations involved in manufacturing of components/products from raw material stage to final stage are the following:

1. Production of powder from the metal. For that, we perform the following:
    a. Extraction of the metal from its ore or by converting it as an oxide
    b. Washing and grading (or screening) of powder
    c. Blending of powder particles
2. Compaction (i.e. pressing) of powder.
3. Sintering of compacted powder.
4. Secondary and other finishing operations. Most of these operations are performed in stages.

Compaction: Compaction is done to produce a *green compact* having sufficient strength so as to withstand the further handling operations. The green compact means pressed part, which is then taken for sintering.

Sintering: Sintering of green compacts is carried out to consolidate the bonded powders into a coherent body. It is carried out in a furnace under controlled atmosphere and temperature of about $1.7T_m$ where T_m is the absolute melting point. Sintering is done under protective atmosphere. The most commonly used protective atmospheres are the following:

- Endothermic and dissociated NH_3
- N_2

11.2.1 Raw Materials, Additives and Lubricants for Powder Production

The raw material used to produce powder metallurgy (P–M) products is the powder which may be of the following kinds:

- Pure element powder, namely,
  - Powder of ferrous metals, for example Fe
  - Powder of non-ferrous metals, for example Ti, Ni, Cu and Ag
  - Powder of refractory metals, for example W and Mo
- Elemental blended powder, namely, powders of insoluble metals, for example Cu–Fe, Ag–Cd, W–Cu and Ag–Cu
- Pre-alloyed powder, for example of stainless steel, superalloy and high-speed steels (HSS)

Powders of suitable shapes and size ranges and having other specific characteristics are chosen to obtain the desired property in the final product.

Additives: The additives used with raw materials are

- Composite materials powder such as
  - Metal–ceramic combinations, for example $Al–Al_2O_3$

Lubricants: Lubricants are added to powders along with additives to facilitate easy ejection of compact and to keep the wear of compaction tool to a minimum. Generally, the following lubricants are used for this purpose:

- Graphite
- Waxes
- Stearates

The quantity of lubricant to be added is normally 0.5%–2% of weight of the charge.

11.2.2 Secondary and Other Finishing Operations

Secondary operations are performed over sintered parts to ensure

1. Improved surface finish
2. Closer dimensional tolerance

3. Corrosion resistance
4. Increased density

Typical secondary operations in this regard are the following:

- Repressing
- Sizing
- Grinding
- Forging
- Re-sintering
- Machining
- Plating
- Tumbling
- Oil impregnation
- Metal infiltration
- Heat treatment

11.3 Methods of Powder Production

Metal powders are the main constituent of the products made by powder metallurgy technique. Their characteristics play important role in achieving the final properties of finished component. A variety of methods are used to produce powders so as to fulfill the diverse needs of the product. There are many ways in which the metals can be produced into powder form, the main among them are as given in Figure 11.1.

A brief detail of these methods are as follows.

Impact milling and grinding: In this method, the cast billet of a metal is first made into powder form by machining and then is reduced into fine powder form by impact milling or by grinding. This method is mainly suitable for brittle metals but can be used for ductile metals also. The examples are

- Brittle metals like Be
- Ductile metals like Ti alloys

Machining: In this method, the metals and alloys are first machined by turning or other machining operation, and the chips are then crushed or ground into powder form. The examples are

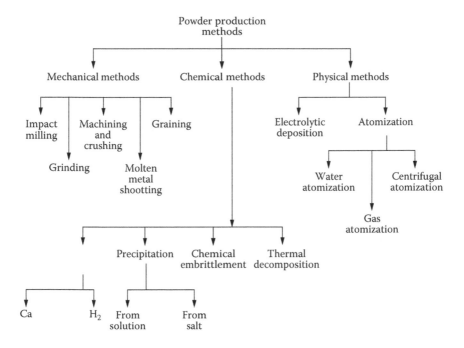

FIGURE 11.1
Various powder production methods.

- Mg
- Ag
- Solder
- Be
- Dental alloy

Shotting: In this case, the metal is first melted and then, a fine stream of molten metal is poured into air or other protective (inert) gas medium via a vibratory screen. On falling through the screen, the molten metal disintegrates and solidifies as spherical particles. The metal powders produced by this method, generally, are

- Al
- Ni
- Sn
- Cu
- Zn
- Pb
- Brass

Reduction of oxides: In this method, the reduction of oxides is done from those compounds whose powder is to be made. Reduction may be accomplished by different means such as by using reductants, electrolysis and thermal decomposition. The examples are

- Co
- Mo
- Ta
- Ti
- W
- Fe
- Ni
- Cu

Deposition from vapour phase: In this method, the powders are obtained as deposits from vapour phase. The examples include the low melting point metals such as

- Sn
- Zn

Atomization: In this method, the molten stream of metal is disintegrated using high-velocity fluid jets such as water, N_2 and/or argon gases. The use of water is very common. Water atomization is preferred for large-scale production of powders, whereas the gas atomization is desired when oxidation of powders is to be prevented. Gas atomization is also preferred when production of spherical powders are required. The examples are

- Stainless steel
- Superalloys
- Tool steel
- Iron
- Aluminium
- Brass

Among these methods, the atomization is the most common method of powder production. About 80% of total volume of powders is produced by this method.

11.4 Powder Production

This is a commercial method of powder production that employs mechanical forces to reduce the size from bulk metal. This size is further reduced to particle size. The mechanical forces of the following types do play a role to create particle size effect:

1. *Compressive force*, which is caused due to crushing or squeezing of particles
2. *Shear force*, which is caused due to cutting or cleaving of particles that results in fracture
3. *Impact force*, which is caused due to striking of one powder particle against the other
4. *Attrition force*, which refers to production of wear debris due to rubbing between two particles

This is the most widely used method of powder production. It is useful for both ductile and brittle metals as well as for soft and hard metals.

Salient applications: Mechanical milling method is being employed for producing the powder of the following metals and alloys:

- Bi
- Be
- Electrolytic Fe
- Metal hydrides
- Chemically embrittled alloy, for example sensitized stainless steel
- Niobium–Sn alloy for making superconducting magnet
- Composites reinforced by SiC whisker
- Alloys strengthened by oxide dispersion intermetallics, for example Ni, Fe and Ti aluminides

11.4.1 Equipment Used for Mechanical Powder Production

A variety of equipment are used to perform the job of producing metal powders. Main among them are the crushers and mills, as follows:

1. Crushers
   a. Jaw crusher
   b. Gyratory crusher
   c. Roll crusher

2. Mills
   a. Ball mill
   b. Vibratory ball mill
   c. Rod mill
   d. Hammer mill
   e. Planetary mill
   f. Attrition mill

These are shown in Figure 11.2. Their details are self-explanatory.

11.4.2 Ball Mill

Ball mills are mostly used for making powders and powder mixtures. A ball mill consists of a cylindrical vessel rotating horizontally along its axis. The length of the cylinder is almost equal to its diameter. The material to be milled is charged along with the grinding media such as tungsten carbide, hardened steel, zirconia, porcelain and alumina. When the drum rotates, the balls roll down or fall freely from certain height and thus grind the material by impact and attrition.

11.4.3 Dry and Wet Milling

The milling operation can be accomplished in two different ways, namely,

1. Dry milling
2. Wet milling

In *dry milling*, about 25% of powder by volume is added with about 1 wt.% lubricant such as oleic acid or stearic acid. In *wet milling*, about 30%–40% of powder by volume is added with about 1 wt.% of dispersing agent such as alcohol, water or hexane.

11.5 Chemical Methods of Producing Powder

Chemical methods of powder production are based on chemical reactions such as the following:

Reduction of compounds: In this method, the compounds such as oxides, hydrides, oxalates and formates are subjected to chemical reactions using reducing agents. The reducing agents may be solid or gaseous. Gases such as H_2, CO, coal gas, natural gas and associated NH_3 are used as reducing agent. General equations for reduction of metallic oxides (MeO) by H_2 and CO are given as follows:

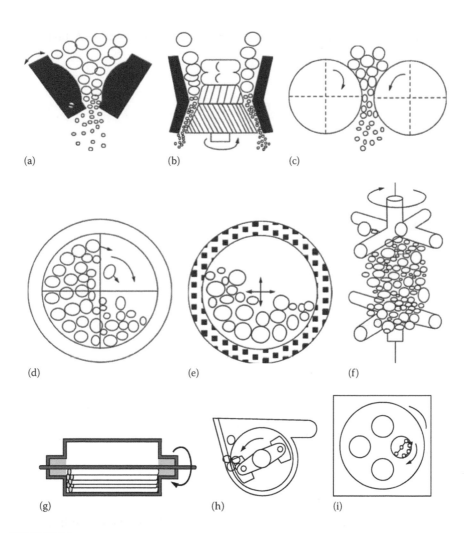

FIGURE 11.2
Different types of crushers and mills used for powder production: (a) jaw crusher, (b) gyratory crusher, (c) roll crusher, (d) ball mill, (e) vibratory ball mill, (f) attritor, (g) rod mill, (h) hammer mill and (i) planetary mill. (From Angelo, P.C. and Subramanian, R., *Powder Metallurgy: Science, Technology and Applications*, Prentice-Hall of India Pvt. Ltd., New Delhi, India, 2008.)

$$MeO + H_2 \uparrow \rightarrow Me + H_2O \uparrow$$

$$MeO + CO \uparrow \rightarrow Me + CO_2 \uparrow$$

The reduction is normally carried out in continuous mesh belt-type furnace for larger quantity production and in batch-type furnaces for smaller quantity production.

Applications: This method is widely used to produce powders of the following metals:

- Fe
- Ni
- CO
- Cu
- Ti
- Ta
- W
- Th
- Mo
- Sponge iron

11.6 Production

In this method, the metal carbonyl, for example iron carbonyl $Fe(CO)_5$ is first formed by passing CO over the heated metal and condensing the vapour as liquid, and then the carbonyl is decomposed to produce fine powders of high purity. The condensed vapour as liquid is volatile liquid, is stored under pressure, and is purified by fractional distillation before decomposing. The CO released during decomposition is recycled.

11.6.1 Carbonyl Reactions

Some typical carbonyl reactions for different metals are the following:

For nickel,

$$Ni + 4CO \uparrow \rightarrow Ni(CO)_4 \uparrow$$

Nickel tetracarbonyl

$$Ni(CO)_4 \rightarrow Ni + 4CO$$

For iron,

$$Fe + 5CO \rightarrow Fe(CO)_5 \uparrow$$

Iron pentacarbonyl

$$Fe(CO)_5 \rightarrow Fe + 5CO \uparrow$$

For tungsten,

$$W + 6CO \uparrow \rightarrow W(CO)_6$$

Tungsten hexacarbonyl

For cobalt,

$$2Co + 8CO \uparrow \rightarrow Co_2(CO)_8$$

Cobalt octacarbonyl

$$CoCo_2(CO)_8 \rightarrow 2CO + 8CO$$

11.6.2 Specific Applications

Powders produced by this method find the following specific applications:

1. Fe powder (a good soft magnetic material, perfectly spherical and 99.5% pure) is used to make magnetic cores.
2. Ni powder (irregularly shaped, porous and fine) is used to make electrodes of alkaline battery and fuel cells.
3. Fe–Ni alloy powder is used to make high-permeable core for communication applications.

11.7 Electrolytic Deposition Method

In this method, the metal powders are produced by electrodeposition from

1. Aqueous solution
2. Fused salts

High-purity powders of most metals such as Cu and Fe are precipitated from aqueous solutions, while the fused salts are used for producing Be and Ta powders. In electrodeposition, the final products are obtained from the cathode of an electrolytic cell. For producing Cu and Fe powders, generally, the following compounds are used:

- Sulphate solutions of Cu and Fe
- Chlorides of Cu and Fe
- Cyanides of Cu and Fe

Other related specific details are as follows:

Production of Cu powder: It is obtained from the solution of copper sulphate ($CuSO_4$), sulphuric acid (H_2SO_4) and crude copper (Cu) as anode. The involved reactions are

$$At\ anode: Cu \rightarrow Cu^+ + e^-$$

$$At\ cathode: Cu^+ + e^- \rightarrow Cu$$

Production of Fe powder: In this case, the low-carbon steel is used as anode, while the stainless steel as cathode. The final product is obtained in one of the following forms depending upon the processing conditions:

1. Hard brittle layer of pure metal such as Fe powder.
2. Soft spongy substance, loosely adherent and removable by scrubbing. It is used for obtaining a large quantity of powder.
3. Powder deposit from electrolyte that collects directly at the bottom of the electrolytic cell.

This method is widely used to produce large quantity of powder.

11.7.1 Suitable Metals for Producing Powder from Electrolytic Deposition Process

This method is mostly used to manufacture the powders of the following metals:

- Copper
- Silver
- Zinc
- Tin
- Nickel
- Tantalum
- Iron
- Tungsten
- Lead
- Cadmium
- Antimony
- Beryllium
- Magnesium
- Vanadium
- Thorium

However, the Cu is the most suitable metal for this process. The electrolytic powder thus produced is quite resistant to oxidation.

11.7.2 Advantages and Disadvantages of Electrolytic Deposition Process

The electrolytic deposition method has advantages and disadvantages both. These are listed as follows:

11.7.2.1 Advantages

Powders produced are of

- High purity
- Excellent compactibility
- Excellent sinterability
- A wide range of quality

11.7.2.2 Disadvantages

The following are the main disadvantages:

- Time-consuming process.
- Involvement of toxic chemicals.
- High operational cost due to washing, drying and annealing of powders.
- Waste disposal is problematic and results in pollution.

11.8 Atomization Method of Powder Production

By atomization, we mean breaking up of a molten metal stream into very fine droplets, by using the high-pressure fluid jets. Major steps involved in this process are the following:

1. Melting of metal/alloy
2. Atomization (i.e. disintegration) of the molten stream
3. Solidification of the droplets

This is a commercial method for producing a wide range of metallic, alloy and composite material powders. This process is used to produce bulk quantities of high-quality powders of the following metals and alloys:

- Stainless steel
- Tool steel
- Superalloys
- Iron
- Brass
- Aluminium

and reactive metals like

- Zirconium
- Titanium

11.8.1 Different Types of Atomization Processes

The atomization of molten metals can be accomplished in one of the following ways:

Water atomization	In which the high-pressure water jets are used for atomization of metals. The typical powders produced are of • Fe • Stainless steel
Gas atomization	In which the high-velocity He, Ar or N_2 gas jets are used for atomization of metals. The typical powders produced are of • Al alloys
Centrifugal atomization (or rotating electrode process)	In which the centrifugal force is used to break off the molten metal drops. The typical powders produced are of • Ni superalloys • Ti alloys • Co–Cr alloys
Rotating disc atomization	In which a stream of molten metal impinges upon the surface of a rapidly rotating disc for atomization of metal stream. The typical powders produced are of • Ni superalloys
Vacuum atomization	In which the molten metal supersaturated with a gas (usually [or Soluble gas atomization] H_2) under pressure is suddenly exposed to the vacuum, causing atomization of metal stream. The typical powders produced are of • Ni superalloys
Ultrasonic atomization	In which the ultrasonic waves are used to atomize the molten stream of metal.
Ultra-rapid solidification	In which a solidification rate of more than 1000°C/s is used to produce metallic glasses. The typical powders produced are of • Al–Li alloys

Among these processes, the water atomization and gas atomization are more popular methods. In them, the water atomization is cheaper than the gas atomization.

11.8.2 Inert Gas Atomization Process

This process is used to produce high-purity metal powders of spherical shape and high bulk density such as for Ni-based superalloys. Production of high-quality powders is accomplished under the control of parameters such as

- Viscosity, velocity and flow rate of atomizing media
- Viscosity, temperature and flow rate of molten metal
- Jet angle of atomizing system
- Nature of quenching media

Schematic diagram of an inert gas atomization system is shown in Figure 11.3. The arrangement depicts the horizontal gas atomization technique, the details of which are self-explanatory.

For effective production of powder, a careful design of nozzles and suitable control of pressure and temperature are required. However, the prevention of oxidation of metal powder is very essential. It can be done by either

1. Using an inert gas or
2. Using air, however, due to which the particles get little oxidized and form a thin protective coating over the surface (this prevents excessive occurrence of oxidation)

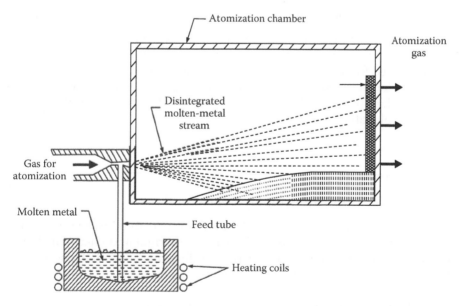

FIGURE 11.3
Schematic diagram of inert gas atomization.

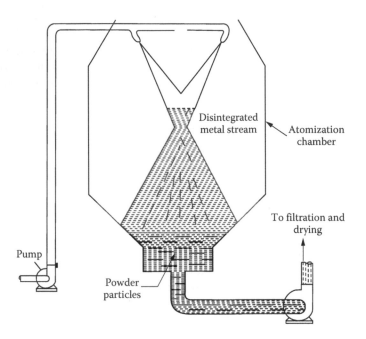

FIGURE 11.4
Schematic diagram of water atomization. (From Angelo, P.C. and Subramanian, R., *Powder Metallurgy: Science, Technology and Applications*, Prentice-Hall of India Pvt. Ltd., New Delhi, India, 2008.)

11.8.3 Water Atomization

The process of producing metal powders by disintegration of molten metal stream by a water jet is known as water atomization. This process is generally preferred for high-volume, low-cost production. Schematic diagram of a typical water atomization set-up is shown in Figure 11.4. The molten metal is poured either directly or by means of a runner into a tundish. Other details are self-explanatory.

11.9 Powder Conditioning (or Treatment)

The powders prepared by any of the methods explained earlier are conditioned to improve/modify their mechanical, thermal and chemical characteristics before fabrication (compaction or sintering). For that, the various treatments imparted to them are the following:

1. Drying of the wet powders to remove moisture.
2. Cleaning of powders to remove solid and gaseous contaminants.

3. Grinding of powders to obtain a finer size.

4. Distribution of particle size of powder to ensure the uniform properties in sintered part.

5. Annealing of the powders to improve softness of powders and hence their compaction.

6. Blending of powders of different particle sizes to minimize the porosity and powder mixes in order to achieve the homogeneity of mixture.

7. Mixing of powders to ensure uniform distribution of various constituents such as powders, lubricants and additives like graphite. It facilitates a good compaction and sintering.

8. Addition of lubricants to the powders to minimize inter-particle friction and for uniform density compaction. Commonly used lubricants are as follows:

   a. Stearic acid

   b. Synthetic waxes

   c. Zinc stearate

   d. Li stearate

Coating of powders to reduce porosity: Coating is done by electrodeposition, ball milling, hydrometallurgy, etc. Coating metals used are generally

1. Cu

2. Co

3. Ni

4. Sodium borohydride

11.9.1 Characteristics of Metal Powders

The properties of the final product depend on the powder characteristics. Important characteristics of powders are displayed in Figure 11.5.

Powder particles, produced by the aforementioned methods, must have the following characteristics for ease in fabrication and better-quality product:

1. Particle size of the powders generally between 4 and 200 μm

2. Shape of powder particles as

   a. Spherical

   b. Cuboid

   c. Parallelepiped

   d. Angular

   e. Irregular

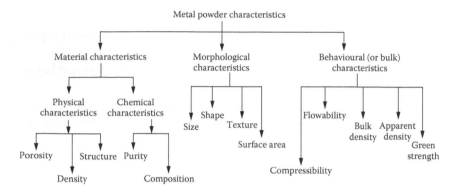

FIGURE 11.5
Important characteristics of metal powders.

    f. Dendritic

    g. Acicular

    h. Flake type

3. Flowability should be such that the free-flowing powders do not take much time in feeding and filling the die cavity.

4. Chemical condition of powder surface.

5. Microstructure of powder such as surface texture. A rough surface facilitates better interlocking of powders.

6. Specific surface area, that is the total surface area per unit weight, for example m^2/kg compactibility, should be high.

7. Toxicity should be the least to avoid undesirable health effects.

8. Pyrophoricity, that is susceptibility of metal powders to rapid oxidation, should be such as to provide safeguards against possible explosion.

11.9.2 Influence of Powder Characteristics

The compaction process depends on the powder characteristics to a great extent. For good cold compaction:

- Irregular-shaped particles are preferred as they provide better interlocking and hence higher green strength.

- Apparent density of the powders which dictates the die fill is another important factor. Powder characteristics such as particle size, shape and density in turn influence the apparent density.

- Flow rate is another factor which decides the time for filling the die. Flow rate, in turn, depends on the particle size, shape and apparent density of the powders.

The compaction pressure range for metals and ceramics generally ranges as the following:

- For metals   70–800 MPa
- For ceramics 100–400 MPa

11.10 Fabrication Methods of Products from Powder

Fabrication of products from powders involves the following processes:

Compaction involves application of pressure only. It can also be accomplished without pressure.

Sintering involves application of pressure as well as temperature.

Compaction is followed by sintering to make final product. After compaction, the density of powder increases due to consolidation which can reach up to 85%. During sintering, the particles are bonded, but density may or may not increase.

Sintering may be defined as the process of 'thermal treatment of powder or compact below melting point of main constituent of the powder, to increase the strength by binding the particles'. It is generally carried out at a temperature ranging between $0.7T_m$ and $0.9T_m$, where T_m is the melting point (in Kelvin) of the metal or alloy.

The *binders* hold the powder particles in desired shape. Mineral oil and polyethylene are used as binders.

11.11 Compaction of Metal Powders

The compaction of metal powders refers to *consolidation* of powder particles without application of heat. By compaction, the powders achieve a desired shape and adequate strength. There are a number of ways by which the powder compaction can be accomplished. Main among them are the following:

1. Methods without application of pressure (or pressureless compaction)
   a. Vibratory compaction
   b. Injection moulding

    c. Slip casting

    d. Slurry casting

    e. Loose powder compaction in mould

2. Methods with application of pressure (or pressure compaction)

    a. Unidirectional pressing (or die compaction)

       i. Single-acting pressing

      ii. Double-acting pressing

    b. Isostatic pressing

    c. Powder rolling

    d. Explosive compaction

    e. Powder extrusion

Among these, the unidirectional (die compaction) pressing is most widely used, and hence, we shall study it in detail.

11.11.1 Unidirectional Pressing (or Die Compaction) Method

This method of powder compaction involves external pressure to consolidate the powders. It is a widely employed method for mass production of powder metallurgy products. Schematic arrangement of die compaction is shown in Figure 11.6. It involves the following stages:

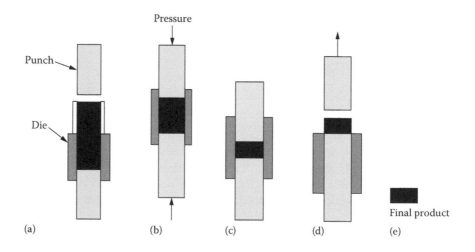

FIGURE 11.6

(a) Powder filling, (b) applying pressure, (c) releasing pressure, (d) ejection of compact from the die and (e) final product.

- Charging of powder mix, that is powder filling in die (Figure 11.6a).
- Application of load by means of a *punch* to compact the powders (Figure 11.6b).
- Removal of load by releasing the pressure and retracting the *punch* (Figure 11.6c).
- Ejection of compact from the die (Figure 11.6d)
- The final compact product is shown in Figure 11.6e.

Other details are self-explanatory.

11.12 Sintering

Sintering means the process of 'heating of compacted powder in a furnace below melting point of the metal powder' under controlled atmosphere. During this process, the metal powder is gradually heated up to sintering temperature (see Table 11.1) and is maintained there for certain duration, called sintering time (see Table 11.1). At this temperature and time, the powder particles bond themselves to form a coherent body/solid.

The International Standard Organization (ISO) defines sintering as 'the thermal treatment of a powder or compact at a temperature below the melting point of the main constituent, for the purpose of increasing its strength by bonding together of the particles'. Sintering operation is generally carried out at temperatures ranging from $0.7T_m$ to $0.9T_m$ (T_m is the melting point in Kelvin) of the metal or alloy. During sintering, the compact is heated in a protective atmosphere such as argon or hydrogen. The individual powder particles, which were either loose or physically bonded together, are metallurgically bonded to yield a solid structural part with the desired properties. Several changes take place during sintering like shrinkage, formation of solid solution and development of the final micro-structure. In many

TABLE 11.1

Sintering Temperature and Time of Some Metal Powders

Name of Metal Powder	Sintering Temperature (°C)	Sintering Time (min)
Alnico (Al alloy)	1200–1300	120–150
Cu and Cu alloys	750–900	10–50
Iron and steel	1000–1300	45–60
Ni	1000–1200	30–45
Stainless steels	1100–1300	30–60
W	2300–2400	≈480

cases, sintering results in a reduction/elimination of porosity, leading to densification. In other instances, sintering may not result in any densification, and in certain cases, sintering can result in the growth of the compact (i.e. swelling).

11.12.1 Sintering Temperature and Sintering Time

The sintering temperature and sintering time vary widely for different types of metal powders. They also depend on the magnitude of compressive load applied on the powder and the strength required in the finished product. Sintering temperature ($T_{sintering}$) and sintering time ($t_{sintering}$) for some commonly used metal powders are displayed in Table 11.1 for a ready reference.

11.12.2 Different Types of Sintering Process

Sintering may be performed in many ways; therefore, the sintering processes are of many kinds such as the following:

1. Solid-state sintering
2. Liquid phase sintering
3. Reaction sintering
4. Activated sintering
5. Microwave sintering
6. Rate-controlled sintering
7. Gas plasma sintering
8. Spark plasma sintering

Among these, the solid-state sintering is the most common. In this practice, the densification of metal and alloy powders is achieved by atomic diffusion. The sintering process is accomplished for the following systems:

- Single-component systems such as of pure metals and ceramics
- Multi-component systems involving more than one phase

11.12.3 Structure and Property Changes during Sintering

Since the purpose of sintering is densification of powder compact, the following changes in structure and properties generally occur during sintering:

- Densification is proportional to the amount of pores removed, that is shrinkage in single-component systems, for example metals.
- In multi-component systems, there is expansion rather than shrinkage.

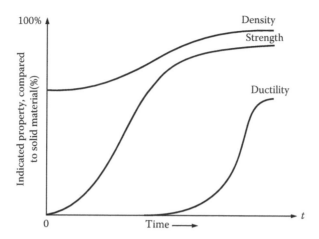

FIGURE 11.7
Property changes during sintering. (From Angelo, P.C. and Subramanian, R., *Powder Metallurgy: Science, Technology and Applications*, Prentice-Hall of India Pvt. Ltd., New Delhi, India, 2008.)

- Other changes taking place during sintering may include recrystallization and grain growth.
- Pore growth also occurs if sintering is carried out for longer times at high temperatures.
- The structural changes result in a change in properties such as density, strength and ductility.

These are shown in Figure 11.7.

11.12.4 Sintering of Blended Powders

Quite often a mixture of powders of different metals is to be sintered after compaction, and if their melting points are much different then consider the following:

1. Sintering is done at a temperature which is above the melting point of one of the components of metal powder. Consequently, the metal powder having low melting point will change to liquid phase.
2. The other component of metal powder, which is still in solid phase, is soluble in liquid phase metal; there occurs a marked diffusion of the solid phase metal through the liquid phase metal.

This results in an efficient bonding between the particles and hence produces a high-density blend.

11.12.5 Furnaces Used for Sintering

Different kinds of furnaces are used for performing sintering operation. These are

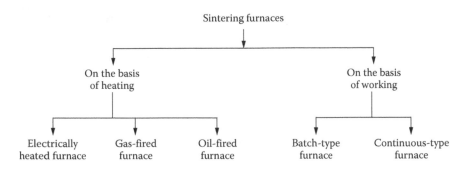

One such furnace is shown in Figure 11.8 as a ready reference. It is a continuous-type furnace. Continuous furnaces are used in the production of porous bearings, sintered carbides and structural parts made of steel.

11.12.6 Post-Sintering Operations

Although the sintered products are final products, yet sometimes, they are processed further to obtain better properties. The post-sintering operations and the induced betterment in the properties of the products are given as follows:

Infiltration: It is done to improve the porosity of the products by filling the pores. The pore filling takes place by means of capillary action. The metal

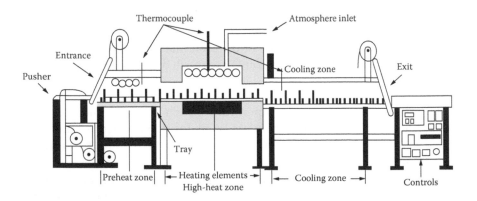

FIGURE 11.8
Schematic diagram of a continuous-type sintering furnace. (From Angelo, P.C. and Subramanian, R., *Powder Metallurgy: Science, Technology and Applications*, Prentice-Hall of India Pvt. Ltd., New Delhi, India, 2008.)

used for filling the pores should have a lower melting point than the melting point of metal powders for which the infiltrant is laid on the top of the metal powders. For example

- Cu (melting point 1083°C) is used as infiltrant to strengthen the steel

Sizing: Sizing is a cold working operation which is performed to improve dimensional accuracy and enhancement in density.

Coining: This operation is performed to improve the surface finish as well as mechanical properties. It also improves dimensional accuracy and increases the density.

Machining: The sintered items are sometimes machined to obtain high degree of accuracy in them.

Heat treatment: If desired to improve different specific properties in finally produced sintered items, they are heat-treated (annealed, normalized, hardened, tempered, etc.) as per the requirements of applications.

11.13 Applications of Powder Metallurgy

Powder metallurgy, nowadays, finds extensive use in production of small- to medium-sized products in automobile, electrical, tool, aircraft and other industries. Their main applications are enumerated as follows:

1. For making porous products such as
   a. Oil-impregnated (self-lubricating) bearings
   b. Filtering media for fuel, air and hydraulic fluids
   c. Electrical accumulator plates
2. For making *babbit alloys* in automobile engines such as
   a. Main bearings of automobile engines
   b. Bearings of connecting rod
3. For making electrical contact materials such as in
   a. Circuit breakers
   b. Relays
   c. Cu–graphite brushes
4. For making hard and soft magnetic materials such as
   a. Alnico (Al + Ni + Co + Fe alloy) which is difficult to produce by casting process and is not machinable also
   b. Permalloy (Ni + Fe alloy)

    c. Mn–bismuthide hard magnets

    d. Mg–Mn ferrite

    e. Permanent magnets using Fe, Ni, Co, Ba and ferrite

    f. Pole pieces using Fe, Si–Fe and Fe–Co

    g. Soft magnetic parts using Fe, Si–Fe and Ni–Fe

5. For making *cemented carbides* such as tungsten carbide (WC) which is used as

    a. Cutting tool/tool bits

    b. Dies for wire drawing, deep drawing operations, etc.

6. For making *aerospace products* such as

    a. Heat shields using Be and W powders

    b. Heat shield coating using Al powders

    c. Brake linings using Sn, Cu, graphite, Pb, Fe and high–Ni alloy powders

7. For making *electronic products* such as

    a. Printed circuits using Ag, Cu, Au, Pt and Pd powders

    b. Thin-film resistors using Ag and Pd powders

    c. Relays using Ni, Mo and Fe powders

8. For making *industrial products* such as

    a. Friction material using Cu, Zn, Pb, Sn, graphite and iron powders

    b. Sound deadening plastic using graphite and Pb powders

    c. Greases using Pb and graphite powders

    d. Bearing and bushings using Sn, Pb, Cu and bronze powders

9. For making *medical products* such as

    a. Surgical implants using Ni and Co alloys powders

    b. Dental amalgam using Au and Si powders

    c. Surgical pins and gauzes using Zr and Ta powders

    d. Prosthetics using superalloy powders

10. For making *other products* such as

    a. Recording tapes using Fe powders

    b. Pen points using W, Ru and Pt powders

    c. Paper coatings using Cu, Sn and Al powders

    d. Printed circuits using Ag and Cu powders

    e. Acoustical plastics for sound proofing using Pb and graphite powders

Some of the powder metallurgy–made parts are shown in Figure 11.9.

FIGURE 11.9
Typical powder metallurgy–made components. (From Angelo, P.C. and Subramanian, R., *Powder Metallurgy: Science, Technology and Applications*, Prentice-Hall of India Pvt. Ltd., New Delhi, India, 2008.)

11.14 Advantages and Disadvantages of Powder Metallurgy

The powder metallurgy methods and the articles produced by them possess several advantages but a few disadvantages also. These are listed as follows:

11.14.1 Advantages

1. Articles of any intricate shape and size limitation can be manufactured to a closely controlled tolerance of near-net shape.
2. Articles may be produced to any degree of porosity by controlling the particle size, powder distribution and compaction pressure.
3. Machining of articles is almost eliminated and material wastage is negligible.
4. Products have high quality of surface finish and dimensional accuracy.
5. Components having sharp corners and blind recesses can be produced, which otherwise are difficult to produce by conventional methods.
6. It is much convenient to mix metal powder with non-metal powder in any proportion to produce items of specialized nature, for

example 'copper particles + graphite powder' are used to produce dynamo brush.

7. Super-hard cutting tool bits of diamond, refractory materials, tungsten, etc. can be easily produced.
8. Materials with improved magnetic property can be produced.
9. Efficient utilization of materials (:: above 95% utilization).
10. Cost-effective, energy-efficient and environment-friendly process.
11. Powder metallurgy is a rapid process too.

11.14.2 Disadvantages and Limitations

This method has the following disadvantages also:

1. Heavy initial cost involving costly equipment for operation.
2. Highly accurate dies capable of sustaining high temperature and high pressures for small products are needed.
3. Ductility in the parts produced by powder metallurgy is generally poor.

11.15 Manufacturing of Cemented Carbide by Powder Metallurgy (Sintering) Method

Carbides are extremely hard compounds of tungsten, molybdenum, titanium, tantalum, etc. They are used to make

- Cutting tools
- Dies for wire drawing
- Dies for deep drawing operations

Among these, the tungsten carbide (WC) is widely used. Different kinds of carbides are also blended to suit the requirements of specific applications. While sintering these carbides, the Ni or Co is used as bonding agent.

Method of obtaining tungsten carbide (WC): To obtain WC, the following steps of manufacturing are executed:

1. The tungsten oxide (WO) and cobalt oxide (CoO) powders are thoroughly mixed with *lamp black* and heated in a stream of hydrogen.
2. WO and CoO are reduced to lots of metallic states, separately. The size of metallic powders is controlled by varying the rate of hydrogen flow and heating temperature and time.

The W powder is now grinded, mixed with lamp black and heated for several hours in a reducing atmosphere at about 1600°C. Thus, the WO is converted into WC.

11.15.1 Producing Cemented Tungsten Carbide–Tipped Tools

Further steps in manufacturing of WC are as follows:

1. The WC is mixed with Co powder in a suitable proportion.
2. For making uniform bond between WC and Co powders, the size of powders is not varied much. The two kinds of powders are also evenly distributed in each other.
3. The mixture of these powders is compacted in a suitable die made of alloy steel at about 5000 kg/cm² (500 MPa).
4. Now, sintering is done in two stages, namely, preliminary sintering which is done in a controlled atmosphere of hydrogen at about 900°C. By doing so, the Co fuses, whereas the WC remains intact. That is why the product is known as *cemented tungsten carbide*. The product thus produced is machined to obtain exact size.
5. Now, the second (final) stage sintering is done at about 1300°C for 2 h. The product is cooled gradually within the furnace so that the annealing occurs in it.

11.16 Comparison between Conventional and Powder Metallurgy Methods

Relative merits and demerits of conventional and powder metallurgy methods are compared as follows:

S. No.	Topic of Comparison	Powder Metallurgy Process	Conventional Method
1.	Requirement of machining	Nil or minimum	Yes
2.	Material utilization	Maximum	Not full
3.	Waste in scrap	Negligible	Much
4.	Quality of surface finish	Very good	Comparatively less
5.	Production of parts with controlled porosity	Yes, possible	No
6.	Dimensional accuracy	Too high	Less
7.	Microstructures	Uniform	Non-uniform
8.	Initial tooling cost	High	Low

11.16.1 Comparison between the Properties of Powder Metallurgy–Made Sintered Components and Conventionally Made Solid Components

Powder metallurgy–made sintered components have many favourable properties but some unfavourable properties also as compared to the components made by conventional processes. These are given as follows:

1. Sintered components are brittle than conventionally made components.
2. Since the tooling cost of producing sintered components is very high, therefore, they are suitable for large-scale production (generally 20,000 or more parts).
3. Sintered components are normally made up of particles of different sizes, a typical of which is given as follows:
   * Fifty percentage of particles of the size between 100 and 150 mesh size
   * Twenty-five percentage of particles of the size between 150 and 200 mesh size
   * Twenty-five percentage of particles of the size between 200 and 300 mesh size
4. Powder metallurgy–made parts can be oil-impregnated to produce self-lubricating bearings.
5. Powder metallurgy–made parts can be infiltrated with a metal having lower melting point so as to obtain greater strength and shock resistance. This is done to produce electrical contacts.
6. Powder metallurgy–made parts can be impregnated by resin to seal the inter-connected porosity and hence to improve the density.

11.17 Recent Advances in Powder Metallurgy

The technique of powder metallurgy is undergoing through tremendous advances. Some of these advances/trends are briefly summarized as follows.

Production of ultrafine powders: Ultrafine powders of 10–100 nm size are being produced. This size of powder enhances the properties.

Production of novel microstructures: Advanced processes like rapid solidification processing (RSP) and mechanical alloying (MA) have resulted in the production of novel microstructures.

Improvement in purity of powders: There is considerable improvement in production of ultra-pure powders, which has resulted in a better fatigue life of components.

Improved compaction techniques: Production of high-density components having much improved properties has become possible due to the use of cold and hot pressing and powder forging techniques.

Newer processing techniques: Owing to the advent of newer processing techniques, namely, laser sintering, metal injection moulding and spray forming, it has become possible to produce products with improved properties.

Reference

1. P.C. Angelo and R. Subramanian, *Powder Metallurgy: Science, Technology and Applications*, Prentice-Hall of India Pvt. Ltd., New Delhi, India, 2008.

12

Trends in the Development of Ferrous Metals and Alloys, and Effects of Alloying Elements on Them

12.1 Classification of Steels and Cast Irons

Plain-carbon steels. Carbon steels (or plain-carbon steels) are ferrous metals. Percentage of carbon in them varies from about 0.05% to 1.7%–2%. Accordingly, they are classified as mild steel (MS), medium-carbon steel (MCS) and high-carbon steel. MS is soft, MCS is tough and high-carbon steel is hard in comparison to each other. Classification of steels is given in Figure 12.1. It also shows different types of cast irons (CIs).

12.1.1 Low-Carbon Steels

Of all the different types of steels, those produced in greatest quantities are of this type. These generally contain less than about 0.25 wt.% C and are non-responding to heat treatments intended to form martensite. The strengthening is accomplished by cold work. Microstructures consist of ferrite and pearlite constituents. As a consequence, these steels are relatively soft and weak. They, however, have outstanding ductility and toughness. They are machinable, weldable and least expensive to produce. These can be easily rolled and forged to produce rod, bar, plate, etc.

Applications: Their typical applications include the following:

1. Automobile body components
2. Structural shapes such as I-beams, channels and angle irons
3. MS sheets that are used in pipelines, buildings, bridges
4. Tin cans

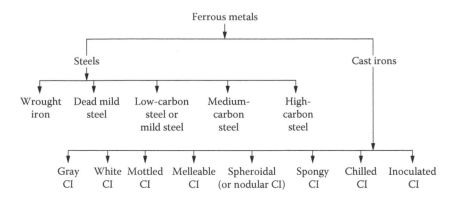

FIGURE 12.1
Classification of steels and CIs.

Tables 1.1, 1.2 and 10.1 given in earlier Chapters 1 and 10 present the compositions and mechanical properties of several plain low-carbon steels. They typically have

1. Yield strength of about 275 MPa
2. Tensile strengths between 415 and 550 MPa
3. Ductility of about 25%

12.1.2 Medium-Carbon Steels

The MCSs have carbon concentrations between about 0.25 and 0.65 wt.%. These can be heat treated by austenitizing and quenching. They can be tempered to improve their mechanical properties. It is shock-resisting steel. They are most often utilized in tempered condition, having microstructures of tempered martensite. The plain MCSs have low hardenabilities and can be heat-treated only in the form of very thin sections by very rapid quenching rates. Additions of chromium, nickel and molybdenum improve the capacity of these alloys to be heat-treated, giving rise to a variety of strength–ductility combinations. These heat-treated alloys are stronger than the low-carbon steels, but by sacrificing some ductility and toughness.

Applications: Main applications of MCSs include the following:

- Railway wheels and tracks
- Gears
- Crankshafts
- Machine parts
- High-strength structural components

TABLE 12.1

Plain-Carbon Steel–Based Commercial Steels (Alloys) and Their Uses

Commercial Name	Carbon Percentage (%)	Type	Use
Boiler steel	<0.25	MS	Boiler
Case hardening steel	<0.20	MS	Gear box parts
Deep drawing steel	<0.10	MS	Automobile bodies, stoves, refrigerators
Free machining steel	<0.15	MS	Bolts, nuts, screws
Machinery steel	0.30–0.55	MCS	Machine components
Structural steel	0.15–0.30	MS	Ships, bridges, buildings
Electrical steel	<0.05	Dead MS	Electrical items
Pressure vessel steel	0.10–0.15	MS	Pressure vessels, pipes
Spring steel	0.6–0.8	MCS	Helical spring, buffer spring, music wire
Rivet steel	0.10–0.15	MS	Boiler and structure joints
Forging steel	0.3–0.4	MCS	Axles, bolts, connecting rods

Such applications call for a combination of high strength, wear resistance and toughness. The compositions of several MCSs are given in Table 12.1.

12.1.3 High-Carbon Steels

The high-carbon steels normally have carbon contents between 0.65 and 1.5 wt.%. They are the hardest, strongest and least ductile among all plain-carbon steels. They are almost always used in a hardened and tempered condition. They are especially wear resistant and capable of holding a sharp cutting edge. The tool and die steels are high-carbon alloys, usually containing chromium, vanadium, tungsten and molybdenum. These alloying elements combine with carbon to form very hard and wear-resistant carbide compounds, for example $Cr_{23}C_6$, V_4C_3 and WC.

Applications: Some tool steel compositions and their applications are listed in Table 12.6 (later in the chapter). These steels are utilized as follows:

- Cutting tools and dies for forming and shaping materials
- Knives
- Razors
- Hacksaw blades
- Springs
- High-strength wire

12.1.4 Ultrahigh-Carbon Steel

Ultrahigh-carbon steels (UHCSs) are hypereutectoid steels having 1.0%–2.1% carbon. Despite their high carbon content, they have remarkable mechanical properties. They are superplastic at intermediate and high temperatures. At room temperatures, they have ultrahigh strengths with good ductility. They can be made exceptionally hard with high compression toughness by heat treatment.

12.1.5 Plain-Carbon Commercial Steels

Some common and important plain-carbon-based commercial steels indicating carbon percentage, properties and their uses are shown in Table 12.1. Examples of some other commercial steels are as follows:

- *Laminated steel* is an example of the bonding of low- and high-carbon cast strips.
- *Cladded steels* are made of stainless steel combined with low-carbon steels.

12.2 Cast Iron

CI is a type of iron that contains more than 1.7% carbon, usually in between 2% and 4.5% C. The pig iron is mixed with scrap of iron and steel to produce CI. CI is the cheaper metallurgical material available to the engineer. It is, in fact, a re-melted pig iron whose composition has undergone some changes during melting process. Apart from its low cost, other commendable properties of CI include the following:

1. Good rigidity
2. High compressive strength
3. Excellent fluidity so that it makes good casting impressions
4. Good machinability

While the ductility and tensile strength of ordinary grey CI are not very high, both of these properties can be considerably improved by treatments which modify the microstructures of suitable irons. CI is hard and brittle. Carbon content is always more than 1.7% and often around 3%. Carbon may be present in two forms, namely, (1) as free carbon or graphite and (2) as combined carbon or iron carbide (Fe_3C).

12.2.1 Types of Cast Irons

Ferrous metal containing about 2%–4.5% carbon is called CI. There are several kinds of CIs, each having some specific properties. These are listed as follows:

1. Grey CI
2. White CI
3. Mottled CI
4. Malleable CI
5. Spheroidal CI
6. Spongy CI
7. Chilled CI
8. Inoculated CI

12.3 Grey Cast Iron

Grey CI is obtained when CI is cooled slowly from the molten state. The combined carbon then has time to separate out and to remain present in free graphite form. Grey CI has a low tensile strength because of the presence of graphite flakes at the crystal boundaries. It is also self-lubricating and absorbs vibrations, has good resistance to compressive loads and is used for making frames that are not subjected to shock loads. It can be used in a cast form only and is not suitable for hot or cold working due to high carbon content. It can be welded, but with difficulty. It is commonly known as graphitic CI and finds maximum use among all other kinds of CIs. It contains flake form of graphite whose tips are sharp. Due to sharpness of tips, they become source of stress concentration. This makes strength of CI poor and is responsible for its brittle nature.

12.3.1 Mechanical Properties of Grey Cast Iron

The grey CI is very strong in compression, but poor in tension. Its stress–strain curves are shown in Figure 12.2. It shows $+\sigma$ versus $+\varepsilon$ plot for tension and ($-\sigma$ vs. $-\varepsilon$) plot in compression. Other salient properties of grey CI are the following:

- Tensile strength = 250 MPa
- Compressive strength = 1000 MPa
- Poisson's ratio = 0.17

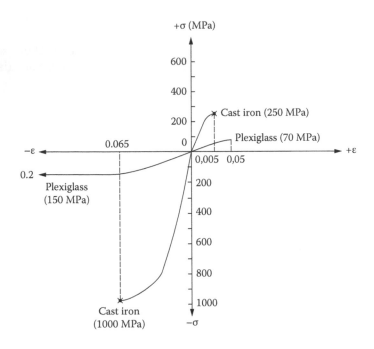

FIGURE 12.2
Stress–strain curves: both in tension and compression, compression shown as –σ versus –ε.

- Tensile modulus = 110 GPa
- Shear modulus = 51 GPa
- Plastic strain at fracture under tension = 0%–15%

12.3.2 Factors Affecting the Properties of Grey Cast Iron

It is a CI in which the carbon remains mainly in graphite form. As it is grey in colour, it is called grey CI. It is extensively used for machine parts because of being inexpensive. It can be given almost any desired shape. Since the graphite is an excellent lubricant, the grey CI can be easily machined, and the chips break off readily. The freedom with which the articles slide over a smooth surface of CI is largely due to the presence of graphite in the surface. However, brittleness, lack of ductility and poor shock resistance prohibit their use in parts subjected to high tensile stress or suddenly applied loads. Different factors that influence its properties are given as follows:

1. Silicon is used as a softener in CI. Increasing the silicon content of CI increases the free carbon and decreases the combined carbon.
2. Manganese tends to harden the CI as it promotes combined carbon. In castings, a balance has to be maintained between silicon and manganese contents so as to obtain a machinable but strong casting.

3. The rate of cooling has a considerable influence on the hardness of CI.
4. Castings of light section cool more rapidly than heavy castings.
   a. This result in formation of more combined carbon and less free carbon.
   b. The consequence is an increase in hardness.

For these reasons, for light castings, more silicon is required to encourage the formation of graphite.

12.3.3 Microstructure of Cast Iron

The microstructure shown in Figure 12.3 is for grey CI containing about 2%–4.3% C. The structure is marked by the presence of graphite flakes in ferrite/pearlite matrix. In the microstructure, the black image is of graphite flakes and the white image is of ferrite matrix. The graphite flakes are of about 0.05–0.1 mm length and occupy approximately 10% of the metal volume.

Uses: The grey CI is used for making the following components:

1. Body structures of machine tools
2. Underground gas and liquid-carrying pipelines
3. Cylinder head and block for IC engines
4. Manhole covers

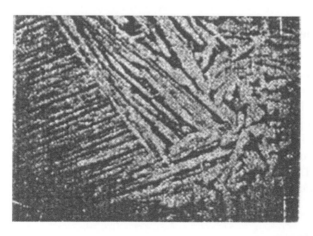

FIGURE 12.3
Microstructure of CI.

12.3.4 Effect of Si and Graphitization of Cast Iron

Slow rate of cooling and addition of Si promotes graphitization in CI. CI contains ferrite phase. Silicon imparts fluidity, lubricating nature and machinability in grey CI. It is used for structural parts such as columns, bed plates and pipes.

Silicon is always present in CI. It acts as a softening agent by promoting free carbon. It is also a reducing agent due to its affinity for oxygen. In grey CI, Si causes brittleness if present in proportions greater than 4% but has a little effect up to 2.5%. Phosphorus is also an impurity, but in small amounts, it is used to keep the iron as fluid in the mould and hence fill any small crevices.

12.4 Other Cast Irons

12.4.1 White Cast Iron

It comprises of cementite phase. White CI is obtained when CI is cooled rapidly and the carbon does not get time to separate out. It is hard and brittle. It can be machined by grinding only. White CI has a more even distribution of carbon within its structure and is heated for a prolonged time in the presence of iron oxide to reduce the carbon content. It is used to produce malleable CI. Due to the presence of cementite, it is extremely hard besides being strong and brittle.

12.4.2 Mottled Cast Iron

It contains a mixture of grey and white CIs.

12.4.3 Malleable Cast Iron

It is produced from white CI on prolonged heating at 900°C followed by slow cooling. Due to this process, cementite decomposes to ferrite. Content of Si is kept less than 1% so as to avoid graphitization. In malleable CI, the edges of graphite flakes act as stress raisers and the heating process changes the size and shape of flakes, making the material more ductile and malleable. Malleable CIs are a cheap substitute of steel forgings and are often used for pipe fittings. Pearlitic malleable CI has an increased manganese content (about 1%), which acts as carbide-stabilizing element and improves the shock-resistant properties. It is used as

- Parts of agriculture machinery and implements
- Pipe fittings
- Parts of railway rolling stock

12.4.4 Spheroidal Graphite Cast Iron (or Nodular Cast Iron)

It contains graphite in the form of spherical nodule. Since there are no sharp tips, condition of stress raiser does not exist, and brittleness is unnoticed. Spheroidal CI is produced by adding Mg or Ce to molten iron. Content of Si is about 2.5% in it. This has good casting properties and better mechanical properties than the grey CI. A graphitizing material such as magnesium mixed with nickel is added to the molten metal as it leaves the furnace and causes the graphite to assume spheroids shape rather than flakes. This is known as spheroidal graphite (SG) CI or SG iron. An SG iron can be partially hot- or cold-worked, machined or welded. It is a good substitute for steel forgings. Its uses include crankshafts, pump cases, gears, etc.

12.4.5 Inoculated Cast Iron

It is produced by inoculating molten pig iron by solution of silicon compounds such as calcium silicide. Addition of silicon produces better effects.

12.4.6 Alloy Cast Iron

Alloying elements are added to CI to produce special properties such as heat resistance, corrosion resistance, improved strength and hardness. For example,

- Nickel is added to promote graphitization, to promote grain uniformity and heat resistance.
- Typically, 5% Ni is added, although 15%–25% may be added to improve corrosion resistance.
- Chromium promotes the stability of combined carbon and is used with Ni to achieve balanced properties to produce hardness without brittleness.
- A small amount of copper on entering into solid solution of iron improves the corrosion resistance and promotes machining, strength and hardness.
- About 1% molybdenum with Ni produces a needlelike structure with greater strength and hardness, that too without loss of machinability.

12.5 Wrought Iron

Wrought iron is refined pig iron with a low carbon content, less than 0.1% and usually approximately 0.02%. It is produced in less quantities because of relatively high cost. Wrought iron is very ductile because of its low carbon content. It can be easily worked and welded although machining is more difficult. It is tough and resistant to atmospheric corrosion. It is used in those applications where a visual warning of failure is required, such as link chains where overloading causes a considerable increase in length of the links. Applications which earlier used wrought iron have now generally been substituted by MS which is easier to machine, cheaper and more reliable.

Wrought iron is a mechanical mixture of pig iron and uniformly distributed silicate slag. It possesses the important properties of ductility, malleability and toughness. It is suitable for machine parts to be shaped by forging. It also has excellent welding properties. With the introduction of steel, the use of wrought iron has decreased.

Applications. However, it is still used extensively for the following applications:

- Chains and crane hooks
- Bolts subjected to shock loads
- Pipe and pipe fittings
- Culvert plates

Its ultimate strength is about three quarters of that of structural steel, while the price is higher than that of MS.

12.6 Alloys, Alloying Elements and Their Effects

To meet the challenges of advancing technology, the development of newer materials, particularly the alloys, is very essential. In engineering applications, the materials are subjected to extreme conditions such as the following:

1. Vacuum (pressure) in steam condensers
2. Very high pressure in autoclave
3. Alkaline to acidic liquids of different pH values
4. Very low to very high temperatures in process industries, for example fertilizers, oil refineries and chemicals
5. Missiles, satellites, space-going vehicles and biomedical and nuclear applications

The metals and their alloys have to work satisfactorily against wear, abrasion, toughness, high-temperature hardness (red hardness) electrical conduction, etc. They have to be lightweight and economical too. Thus, the need arises to develop heat-resisting alloys, nuclear alloys, tool and die alloys, bearing alloys and others for specific uses.

12.6.1 Alloys and Alloying Elements

Pure metals generally suffer from one or the other handicap in their mechanical and other properties. They are, therefore, not suitable for most engineering applications. These metals, in their alloy forms, show remarkable improvement in properties. That is why alloys find wider use. As an illustration, consider aluminium. Its use in pure form is very rare. But its alloys such as alclad, hindalium and duralumin are vastly used.

An alloy is composed of two or more elements of which at least one element is a metal. Constitution of an alloy mainly consists of the following:

1. Base metal
2. Alloying elements

The metal constituent present in an alloy with highest proportion is known as the *base metal*. Other constituents, metallic or nonmetallic, present in alloy are called *alloying elements*. Nichrome, an alloy of nickel and chromium, contains 80% Ni and 20% Cr. Here, nickel is base metal while chromium is an alloying element. We may call this alloy as nickel based.

Alloying elements are deliberately added in the base metal to achieve specific properties which are not found in the base metal. If one or more elements inhibit into the base metal accidently, they are then referred to as *impurities*. Chances of such accidents are more during *smelting* and *refining processes*.

Marked changes in the properties of base metal are noticed on adding different percentage of alloying elements. The quantity of alloying element may be small or high. *Constantan* is chromium-based alloy. Percentage of nickel as an alloying element in it is as high as 45%, whereas proportion of silicon in *silicon steel* is only 4%. The nature and extent of changes in properties of base metals are influenced by the alloying elements. It depends on whether they dissolve or do not dissolve in the base metal or whether they form a new phase or not.

12.6.2 Purpose Served by Alloying Elements on Steel

In the preceding section, we described that an alloy is made by adding alloying elements in the base metal. The effects of various alloying elements are varying on different metals. Different purposes of adding alloying elements in steel are as follows:

- To form martensite without cracking of steel
- To enhance hardenability

- To increase red hardness (elevated-temperature hardness)
- To distribute carbides during tempering
- To provide wear and abrasion resistance
- To impart toughness
- To increase ductility

12.6.3 Austenite-Forming and Carbide-Forming Alloying Elements

Based on the purpose served, the alloying elements are divided into two categories, namely,

- Austenite formers
- Ferrite stabilizers or carbide formers

The elements such as Ni, Co, Mn, C, N and Cu belong to the first category, while the second type of elements include Mo, W, Va, Ti, Cr, Nb, Sn and Si. Silicon is a graphitizing element that forms ferrite but does not form carbide.

12.7 Effects of Alloying Elements on Steels

Chromium: It is used in steels as an alloying element to combine hardness with high strength and high elastic limit. It also imparts corrosion-resisting properties to steel. The most common chrome steels contain from 0.5% to 2% chromium and 0.1% to 1.5% carbon. The chrome steel is used for balls, rollers and races for bearings.

A *nickel–chrome steel* containing 3.25% nickel, 1.5% chromium and 0.25% carbon is much used for armour plates.

Chrome–nickel steel is extensively used for motor car crank shafts, axles and gears requiring great strength and hardness.

Manganese: It improves the strength of the steel in both the hot-rolled and heat-treated condition. The manganese alloy steels containing over 1.5% manganese with a carbon range of 0.40%–0.55% are used extensively in gears, axles, shafts and other parts where high strength combined with fair ductility is required. The principal uses of manganese steel are in machinery parts subjected to severe wear. These steels are all cast and ground to finish.

Cobalt: It gives red hardness by retention of hard carbides at high temperatures. It tends to decarburize steel during heat treatment. It increases hardness and strength and also residual magnetism and coercive magnetic force in steel for magnets.

Tungsten: It prohibits grain growth, increases the depth of hardening of quenched steel and confers the property of remaining hard even when heated to red colour. It is usually used in conjunction with other elements. Steel containing 3%–18% tungsten and 0.2%–1.5% carbon is used for cutting tools. The principal uses of tungsten steels are for cutting tools, dies, valves, taps and permanent magnets.

Molybdenum: A very small quantity (0.15%–0.30%) of molybdenum is generally used with chromium and manganese (0.5%–0.8%) to make molybdenum steel. These steels possess extra tensile strength and are used for airplane fuselage and automobile parts. It can replace tungsten in high-speed steels (HSSs).

Vanadium: It aids in obtaining a fine-grain structure in tool steel. The addition of a very small amount of vanadium (less than 0.2%) produces a marked increase in tensile strength and elastic limit in low- and MCSs without a loss of ductility. The chrome–vanadium steel containing about 0.5%–1.5% chromium, 0.15%–0.3% vanadium and 0.13%–1.1% carbon has extremely good tensile strength, elastic limit, endurance limit and ductility. These steels are frequently used for parts such as springs, shafts, gears, pins and many drop forged parts.

Silicon: The silicon steels behave like nickel steels. These steels have a high elastic limit as compared to ordinary carbon steel. Silicon steels containing from 1% to 2% silicon and 0.1% to 0.4% carbon and other alloying elements are used for electrical machinery, valves in IC engines, springs and corrosion-resisting materials.

12.7.1 Summary of the Effects of Various Alloying Elements on Steel

Table 12.2 shows the effects of some alloying elements on the properties of steel. The nose of the T–T–T curve shifts rightward due to the effect of these elements. Effects of various alloying elements on Brinell hardness of ferritic iron are shown in Figure 12.4.

12.8 Alloy Steels

Plain-carbon steels (i.e. pure metals) generally suffer from one or the other handicap in their mechanical properties for specific uses. But they show remarkable improvement when alloyed with one or more elements. That is why the alloys find wider use in industrial applications. Various elements are used to alloy with steel. Each of them has a particular effect on properties of steel. Alloy steels are more expensive to produce due to higher cost of

TABLE 12.2

Effects of Some Alloying Elements on the Properties of Steel

Property	Alloying Element												
	W	V	Co	Cr	Mo	Mn	Ti	Ni	Ta	P	S	Si	Pb
Ductility	3	3	—	—	3	3	—	2	2	—	5	3	—
Toughness	—	3	—	2	—	3	—	2	2	—	5	—	—
Hardenability	2	1	5	2	2	2	2	3	1	3	—	—	—
Red hardness	1	2	3	2	2	—	3	—	—	—	—	—	—
Abrasion resistance	2	3	—	3	3	3	3	—	—	—	—	—	—
Fine-grain distribution	3	1	—	4	3	—	2	2	—	—	—	—	—
Resistance to corrosion	—	—	—	1	3	—	—	2	2	—	—	2	—
Deoxidation	—	—	—	1	—	2	2	—	—	—	—	2	—
Desulphurization	—	—	—	—	—	2	—	—	—	—	—	—	—
Acid resistance	—	—	—	1	—	—	—	—	2	—	—	—	—
Wear resistance	—	—	—	2	—	1	2	—	—	—	—	—	—
Tensile strength	—	2	—	2	2	2	—	2	—	3	—	3	—
Weldability	—	—	—	—	—	4	—	—	—	—	—	—	—
Machinability	—	—	—	—	—	—	—	—	—	1	1	—	1

Note: 1, best; 2, better; 3, good; 4, bad; 5 undesirable.

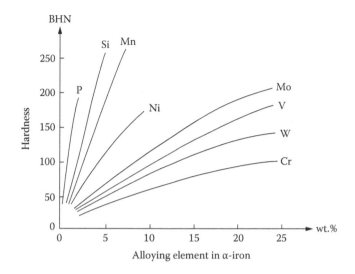

FIGURE 12.4
Effects of alloying elements on the hardness of ferritic iron.

alloying elements and processing and also, they are more difficult to fabricate and machine. Different alloying elements and their notable effects only are given as follows:

Tungsten	W	Imparts red hardness
Vanadium	V	Induces fine-grain distribution
Chromium	Cr	Improves resistance to corrosion and oxidation
Molybdenum	Mo	Increases hardenability
Manganese	Mn	Imparts wear resistance
Titanium	Ti	Increases abrasion resistance
Nickel	Ni	Improves tensile strength and toughness
Phosphorous	P	Enhances machinability

Based on these explanations, vivid varieties of alloy steels are described now in subsequent sections.

12.8.1 Classification of Alloy Steels

12.8.1.1 Low-Alloy Steels

These have a microstructure almost similar to plain-carbon steels, similar carbon contents and similar applications. Low-alloy steels usually contain up to 3%–4% of one or more alloying elements to improve the strength, toughness, hardenability, etc. Nickel improves the resistance to fatigue (Figure 12.5).

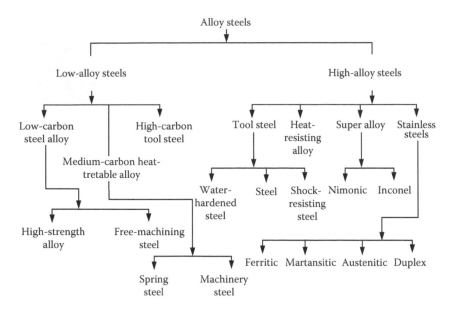

FIGURE 12.5
Classification of alloy steels.

12.8.1.2 High-Alloy Steels

These alloys contain the alloying elements more than about 4%. The structures of high-alloy steels and the heat treatment processes that are required are very different from plain-carbon steels. Examples of high-alloy steels are stainless steels and high-speed tool steels. These are special steels that have particular characteristics such as high corrosion resistance and low-temperature applications.

12.9 Stainless Steels (or Corrosion-Resistant Steels)

The stainless steels are highly resistant to corrosion (rusting) in a variety of environments, especially the ambient atmosphere. Their predominant alloying element is chromium in which a concentration of at least 11 wt.% Cr is required. Corrosion resistance may also be enhanced by adding nickel and molybdenum. Stainless steels are divided into three different classes on the basis of the phase constituent of the microstructure. These are

1. Ferritic stainless steel
2. Martensitic stainless steel
3. Austenitic stainless steel

The development of stainless steels has been mainly due to their corrosion-resistant properties. However, many of these steels can also be used at high temperatures where alloying additions to steel can reduce the problems of loss of strength, creep, oxidation and chemical attack.

12.9.1 Ferritic Stainless Steel

It contains about 16%–18% chromium. It has rustless iron with low carbon content. It has high resistance to corrosion but low impact strength and cannot be refined by heat treatment alone. If the chromium percentage lies between 25% and 30%, it is used for the purpose of furnace parts which are resistant to sulphur compounds.

12.9.2 Martensitic (Hardenable Alloys) Stainless Steel

It contains about 12%–14% chromium. It is characterized by high strength combined with considerable corrosion resistance. Further, stainless steel with 12% Cr and 0.1% C can be welded and easily fabricated. It is useful for making turbine blades. Stainless steel with 12% Cr and 0.2% C is difficult to weld and is corrosion resistant only when hardened and polished.

12.9.3 Salient Features of Different Types of Stainless Steels

Other main features of different classes of stainless steels are as follows:

1. Martensitic stainless steels are capable of being heat-treated in such a way that the martensite is the prime microconstituent.
2. Additions of alloying elements in significant concentration produce dramatic alterations in the iron–iron carbide phase diagram. For austenitic stainless steels, the austenite (or γ) phase is extended to room temperature.
3. Ferritic stainless steels are composed of a ferrite (BCC) phase.
4. Austenitic and ferritic stainless steels are hardened and strengthened by cold work because they are not heat treatable.
5. The austenitic stainless steels are the most corrosion resistant because of the high chromium contents and addition of the nickel. They are produced in the largest quantities.
6. Both martensitic and ferritic stainless steels are magnetic; the austenitic stainless steels are not.

12.9.4 Applications of Stainless Steels

Some stainless steels are frequently used at elevated temperatures and in severe environments because they resist oxidation and maintain their mechanical integrity under such conditions. Examples of such stainless steels are the following:

- The upper temperature limit in oxidizing atmospheres is about 1000°C. Equipment employing these steels includes gas turbines
- High-temperature steam boilers
- Heat-treating furnace
- Aircrafts, missiles and nuclear power generating units

There is an ultrahigh-strength stainless steel also which is unusually strong and corrosion resistant. Strengthening is accomplished by precipitation-hardening heat treatment.

Table 12.3 lists several stainless steels along with the composition and applications. A wide range of mechanical properties combined with excellent resistance to corrosion make stainless steels very versatile in their applicability.

12.10 Maraging Steel

Maraging steels are high-strength high-alloy steels containing high nickel content. The term maraging comes from age hardening (or ageing) of martensite, that is mar + ageing. Such steels are very costly; therefore, their use is limited to sophisticated technology such as rocketry. It is available in different grades that comprise of the following approximate compositions:

- Ni – 17%–18%
- Co – 8%–12%
- Mo – 3%–5%
- Ti – 0.2%–1.8%
- Al – 0.05%–0.15%
- Fe – rest percentage

Maraging steels possess the following mechanical properties:

- Tensile strength = 1500–2500 MPa
- Value at 0.2% proof stress = 1900 MPa
- Elongation = 5%–12%

TABLE 12.3

Compositions and Applications of Stainless Steels, Maraging Steel and Hastelloy and Some Ferrous Alloys

S.No.	Alloy	Composition (%)	Application
1.	Mn steel		
	• Silicomanganese	Mn 65%, Si 20%, Fe 15%	Railway crossings and switches, dredger buckets
	• Hadfield steel	Mn 13%, C 1.25%, rest Fe	Rock crushing machinery, excavators
2.	Mo steel		
	• Chlorimet	Mo 33%, Ni 65%, Fe 1%, Others 1%	Radio valves, mercury vapour lamps, high-speed cutting tools
3.	V steel		
	• Ferrovanadium	V 50%, Fe 50%	Springs, cutting tools
4.	Cr steel		
	• Ferritic stainless steel	Cr 16%, Mn 1%, Si 1%, C 0.12% max, S 0.04% max, P 0.04% max., rest Fe	Surgical instruments, cutleries, dairy industry, decorative pieces
	• Martensitic stainless steel	Cr 12.5%, Mn 1%, Si 1%, C 0.15% max, P 0.04% max, rest Fe	Utensils, springs, valves and valve seats
	• Austenitic stainless steel	Cr 17%, Ni 12%, Mn 2%, Mo 2.5%, Si 1%, C 0.08%	High-temperature components under severe max., rest Fe stresses
	• Duplex stainless steel	Cr 22.5%, Ni 4.8%, Mo 3.0%, Mn 1.3%, Si 0.7%,	Components of jet airplanes C 0.03%, N 0.1%, rest Fe
5.	Ni steel		
	• Maraging steel	Ni 18%, Co 7%, Mo 4.5%,	Aircraft undercarriage parts, portable bridges, booster motors in missiles
	• Hastelloy A	C 0.02%, Mn 0.1%, Si 0.1%,	Parts of chemical plants *silver* coin
	• Timken alloy	Ti 0.2%, B 0.003%, rest Fe	
6.	W steel	Ni 60%, Mo 30%, Fe 5%	Permanent magnets,
		Ni 25%, Cr 16%, Mo 6%,	cutting tools, spark plug
		C 0.1%, N2 0,15%, rest Fe	electrodes, gas turbine
		W 10%, Co 10%, Ni 59%,	blades
		Cr 9%, Hf 1.5%, Fe 0.25%,	
		others	

Maraging steel alloys are twice as hard as stainless steel, 85% harder than pure titanium and 35% stronger than the hardest titanium alloy. On the Rockwell scale of hardness (HRC), stainless steel is 23–26, titanium alloys 28–41 and maraging steel 52–55.

Applications. They are used for motor casing of booster rockets and face inserts for the thinnest supported face in golf.

12.11 Nickel Alloys

Nickel is a crystalline, nonferrous, ferromagnetic metal of silvery-white colour. Its hardness matches with the hardness of soft steel, but ductility is less than that. Other salient features of nickel are listed as follows:

1. It is capable of high-quality polishing, thereby providing lustre to the products on which it is polished.
2. It is reasonably malleable and can also be rolled provided the carbon content is in small amount (up to 0.05% or less).
3. It is resistant to acidic attacks, but dissolves readily in nitric acid.
4. Its melting point and specific gravity are 1453°C and 8.9, respectively.

Applications: Nickel is extensively used for nickel plating of metals to provide protective coating against corrosion. *Carbonized nickel* is used to make anodes of power tubes for rapid conduction of heat. It is widely used to make Ni–steel alloys along with many other elements. Important among them are described now.

12.11.1 Hastelloy: The Nickel–Molybdenum Steel Alloy

There are several nickel–molybdenum alloys which are very much resistant to corrosion. These alloys are called Hastelloys. They are subclassified as Hastelloy A, C and D according to their composition. Their details are discussed as follows.

Hastelloy A. It is nickel–molybdenum–iron alloy. It has high strength and ductility. Its strength does not decrease even at high temperature, and it can withstand a load of 110 kg/cm^2 at 900°C with a creep of 1% per year. Its other characterstics and properties are the following.

- It can be easily forged and rolled into sheet.
- It can be machined and can form good castings.
- It can be welded by oxyacetylene or electric arc method.
- Hastelloy A can resist the attack of even concentrated HCl at boiling point.
- It is resistant to concentrated sulphuric acid up to 70°C. It resist the attack of acetic and formic acids also, but not of nitric acid.
- Alkalies do not attack it. Salt solutions also have no effect on it.

Hastelloy C. It is nickel–molybdenum–chromium–iron alloy. Its malleability is less than that of Hastelloy A. It cannot be worked hot or cold.

- It is good for castings but cannot be machined at high speed.
- Hastelloy C is resistant to strong oxidizing agents such as nitric acid, free chlorine and acid solutions of salts such as cupric and ferric. In this respect, it is an improvement over Hastelloy A.

Hastelloy D. It is cast alloy consisting of nickel, silicon, copper and aluminium. This alloy is strong and tough and has high transverse strength.

- It is good for castings, but cannot be hot- or cold-worked.
- Its machinability is poor.
- It can be welded with oxyacetylene or electric arc method, but addition of a flux makes the operation easy.
- It is resistant to all acids except nitric acid.

12.12 Heat-Resisting Alloys

Need: Quite often, we desire temperature measurement of furnaces and ovens in different industries for a proper process control. We invariably desire use of heat-resisting alloys in applications such as in gas turbines. Besides using common alloys, nickel-based superalloys are widely used for this purpose. These are depicted in Table 12.4.

Characteristics: Heat-resisting steels resist oxidation at high temperatures and, in some cases, attack by sulphurous gases. For this reason, they contain as much as 30% chromium together with small amounts of Si. The general tendency of chromium to induce grain growth along with its brittleness is offset by adding nickel.

These steels are supposed to be reasonably strong at high temperatures; therefore, they are stiffened by adding small amounts of carbon, tungsten, molybdenum, titanium and aluminium. This causes formation of carbides or intermetallic compounds. Metallurgical principles involved in the heat-resisting steels also apply to the *Nimonic* series of alloys. These are not steels but contain basically 75% Ni + 20% Cr, stiffened with small amounts of carbon, titanium, aluminium, cobalt, molybdenum or niobium.

12.13 Superalloys

The superalloys have superlative combinations of properties. Most of these are used in aircraft turbine components, which withstand exposure to severely oxidizing environments and high temperatures for reasonable

TABLE 12.4

Some Heat-Resisting Alloys

Alloy of	Name of Alloy	Composition	Application
Ni	Nichrome	Ni 80%, Cr 20%	Thermocouple, heating element of furnace, strain gauges
	Inconel	Ni 80%, Cr 14%, Fe 6%	Food-processing machine components
Fe	Martensitic Stainless steel	Cr 11.5%–18% C 0.15%–1.2% Ni 25%–2.25%, rest Fe	Cutting tools
	Ferritic Stainless steel	Cr 11.5%–27% C 0.08%–0.2%, rest Fe	Kitchen spoons, forks, cutlery
	Austenitic	Cr 16%–26%, Ni 6%–22%,C 0.05%–0.25%, Mo 2%–4%, rest Fe	Roller and ball races
High-temperature composites (HTC)			
• Co matrix HTC, Cr_7C_3		TaC reinforcing phase	Fibre as reinforcement
• Nb matrix HTC		Nb_2C reinforcing phase	Fibre as reinforcement
• Ni matrix HTC		TaC reinforcing phase	Fibre as reinforcement
• Ni_3Al matrix HTC		Ni_3Nb reinforcing matrix HTC	Plates as reinforcement
Glass–ceramic Si + Al + N	Pyroceram Sialon	$SiO_2 + TiO_2$ $Si_3N_4 + Al_2O_3$	Cookware Gas turbine blades

period of time. Mechanical integrity under these conditions is critical. In this regard, the density is an important consideration because the centrifugal stresses are diminished in rotating members when density is reduced. These materials are classified on the basis of dominant metal in the alloy. It may be cobalt, nickel or iron. Other alloying elements include the refractory metals such as Nb, Mo, W, Ta, chromium and titanium. In addition to turbine applications, these alloys are also utilized in nuclear reactors and petrochemical equipment.

12.13.1 Types of Superalloys

Superalloys are a group of high-performance alloys designed to meet very demanding requirements for strength and resistance to surface degradation at high service temperatures. These may be iron based, nickel based and cobalt based as given in the following:

Iron-based superalloys: The composition of such superalloys are

- Iron 30%–50% + nickel 20%–30% + cobalt 20% + chromium 15%–21% + other alloying elements such as titanium and molybdenum

- Incoloy having 46% Fe, 32% Ni, 21% Cr, <1% Ti and <1% other elements

Nickel-based superalloys. The composition of such superalloys are

- Nickel 40%–76% + chromium 16%–21% + cobalt 8%–11% + other alloying elements such as aluminium, titanium, molybdenum, niobium and iron
- *Nimonic* having 3% Fe, 76% Ni, 20% Cr, <1% Ti and <1% other elements

Cobalt-based superalloys. The composition of such superalloys are

- Cobalt 39%–53% + chromium 20%–30% + nickel 3%–22% + other alloying elements such as molybdenum, tungsten and iron

Application of superalloys may be found in following main areas:

- Gas turbines
- Jet and rocket engines
- Steam turbines
- Aircraft turbine components
- Nuclear power plant
- Petrochemical equipment

More illustrations are given in Table 12.5.

TABLE 12.5

Compositions of Some Nickel-Based Superalloys

Wt.% of	Alloy				
	Waspaloy	Astroloy	Nimonic	Inconel	Incoloy
Cr	19	15	14.5	15	—
Co	13	17	13.5	—	15
Ti	3	3.5	3.8	2.4	1.4
Al	1.4	4.0	5	1.0	0.7
Mo	—	5	3.3	—	4.0
Nb	—	—	—	1.0	3.0
Ta	—	—	—	1.0	—
Fe	—	—	—	7.0	—
Zr	0.06	0.04	0.04	—	—
C	0.08	0.02	0.15	0.04	0.05
B	—	0.02	0.01	—	—
Ni	Balance	Balance	Balance	Balance	Balance
Service temperature (°C)	910	—	770–870	820	—

12.14 Cryogenic Steels (or Extremely Low Temperature Purpose Alloys)

The use of steels at low temperatures usually requires that the metal maintains its properties especially its toughness at low temperatures and also when reheated to normal temperatures. Steels of the 18–8 type are found to be most suited to these types of applications. Nickel-based cryogenic steels have been developed for use in cryogenic industries. Cryogenic processes involved in these industries are carried out at temperatures below −157.5°C. The storage and transport of liquified gases for industrial and domestic purposes are done at much lower temperatures. To site some examples, these are as follows:

1. Oxygen liquifies at −183°C.
2. Nitrogen liquifies at −196°C.
3. Ammonia is liquified at about −190°C in fertilizer industry.
4. Skin, semen, kidney and other vital body parts are stored at such low temperatures.

12.14.1 Ni-Based Cryogenic Steels

To serve these purposes, various nickel-based cryogenic steels are used. Their nickel composition and service temperature are as follows:

Percentage of nickel	3.5	5.0	9.0
Service temperature (°C)	−100	−120	−190

12.15 Tool and Die Steels

Manufacturing techniques make use of tools as an essential need. Depending on the requirements of production, a tool may be just an ordinary one or shock-resisting or a cutting tool. Among wide categories of tools, cutting tools are most important. They are used in machining operations such as turning, drilling, milling and broaching. Different types of commonly employed tool steels are the following:

Water-hardening tool steels: These contain 0.7%–1.3% carbon. They are used for drills, files, chisels, hammers, forging dies, etc.

TABLE 12.6

Important Cutting Tool Materials

Material	Composition	Application
High-carbon tool steel	C 0.8%–1.3%, Si 0.1%–0.4%, Mn 0.15%–0.4%, rest Fe	Milling cutters, twist drills and turning tools for soft metals
Low-alloy carbon tool	C 1.2%, Mn 0.3%–0.7%, W 1.5%–4%, Mo 0.3%–0.5%, S 0.3%, rest Fe	High-speed milling drilling and turning tools All kinds of cutting tools
HSSs		
• 18–4–1 steel	W 18%, Cr 4%, V 1%, C 0.7%, Co, rest Fe	Useful at high-temperature operations
• High-Mo steel	Mo 8.5%, Cr 4%, W 1.5%, V 1%, C 0.8%, rest Fe	Useful for high-temperature cutting
• W–Mo steel	W 6%, Mo 5%, Cr 4%, V 2% C 0.8%, rest Fe	For machining highly abrasive materials, rapid machining of hard metals
Stellites	Co 40%–50%, Cr 25%–30%, W 15%–30%. C 2%–3%	
Cemented carbide		
• UCON	TaC 10%, TiC 15%, Co	As throwaway tool inserts
• Borazon (CBN) (cubic boron nitride)	Nb 50%, Ti 30%, W 20%	As grinding wheel on HSS and stellites
Diamond	—	All kinds of cutting tools

Hot-worked tool steels: These are suitable for use in dies for casting, forming, blanking, extrusion, etc.

Shock-resisting tool steels: These are Cr–W, Si–Mo or Si–Mn alloys. Due to their outstanding toughness, they are used in making dies.

Cutting tool steels: These are either W based or Mo based, each having addition of Co in them. Important types of cutting tool materials are illustrated in Table 12.6.

12.15.1 High-Speed Tool Steels

The addition of tungsten and chromium to a high-carbon steel causes the formation of very hard and stable carbides. It results in subsequent increase in critical temperature and softening temperature. These steels can then be used to produce hard-wearing metal-cutting tools which can be operated up to about 600°C. A typical composition is 18% W, 4% Cr, 1% Va and 0.8% C. It is known as 18–4–1 steel.

12.16 Special Purpose Alloy Steels

The list of varieties of alloy steels is too lengthy. But keeping in view the quanta of syllabus, we shall restrict our studies to some more alloy steels, given as follows.

12.16.1 High-Strength Low-Alloy Steel

High-strength low-alloy (HSLA) steel is a product of recent technology aimed at producing strong, lightweight steel at a competitive price with that of carbon steels. It has a higher strength of 415 versus 275 kPa for carbon steel. Consequently, the overall cost may be cheaper for the HSLA, and significant weight savings are realized. The transportation industry, especially the automotive section, employs HSLA steel in numerous structural applications.

12.16.2 Microalloyed Steels

Some HSLA steels, called *microalloyed steel*, contain micro amounts of elements such as nitrogen, titanium, aluminium or niobium that form carbides or nitrides in the austenite. It, thus, prevents the growth of the austenite grains. These steels gain their strength without the need for heat treating through the ability to control their grain growth. Weighing less than regular alloy steels but with increased yield strengths (290–480 MPa), these economical steels are ideal for the following applications:

- Bridges
- Off-highway vehicles
- Ships
- Machinery

Possessing good weldability, they are produced for sheet, plate, pipe and forging applications.

12.16.3 Free-Cutting Steel

A free-cutting steel is one in which an element like lead or sulphur is deliberately added to promote rapid machining. When sulphur is added to steel, it forms a brittle constituent with manganese, which is known as manganese sulphide (MnS). The MnS, being brittle, allows the chip cracks to propagate and to break the chips into easily handleable lengths. Contrary to this, the lead (Pb) does not chemically combine with other elements. It gets distributed throughout the mass as minute droplets. Thus, the chip cracks are propagated easily, and the chips come out in easily handled lengths.

12.16.4 Wear-Resisting Steels

It is an austenitic manganese steel containing 12%–14% Mn, 0.75% Si and 1% C. It may also contain some Cr or V which form carbides and improve the strength of steel. The high Mn contents make the steel austenitic and magnetic at all temperatures. It can be obtained as castings, forgings or rolled sections, but will work-harden if machined. It has excellent resistance to abrasion and is used to make crushing machinery and pneumatic drill bits.

12.16.4.1 Hadfields Steel

These contain carbon from 1% to 1.2%. Addition of 12.5% manganese makes austenite to be retained after heating and quenching. This alloy is known as *Hadfields* austenitic steel, which responds to work hardening and becomes harder as a result of rough treatment. Accordingly, it is used for track work, dredging equipment and crushing machinery. Steels with 1.4% chromium and 0.45% manganese can be oil quenched and are very hard. They are used for making roller and ball races.

12.17 Some Common Steel Alloys

Details of many common steel alloys are given in Table 12.7. It depicts their base element, composition, special characteristics and applications.

12.18 Overheated and Burnt Steel

Overheated steel is the outcome of defect in steel developed during heating. If the overheating is excessive, die grain boundaries of steel may get burnt. Such steel is called burnt steel. The overheating and burning of steel have certain effects on its characteristics, which are given as follows. In *overheated steel*, the overheating causes

- Coarse-grained microstructure
- Fracture
- Reduced ductility
- Substantial reduction in impact strength

TABLE 12.7

Some Common Ferrous Alloys

Alloy	Major Constituent (Base of Element)	Salient Composition	Special Characteristics	Application
Invar	Ni-based steel	Ni 36% Fe 64%	Coefficient of linear expansion α is almost zero	Precision measuring instruments, survey measuring tapes and chains
Elnivar	Ni-based steel	Ni 32%, Fe 68%	Low α	Springs of watches and instruments
Delta metal	Cu based	Cu 55%, Zn 40% Pb 3%, Fe 2%	—	Parts of mining industry and chemical plants, motor bushes
Mischmetal	Ce based	Ce 50%, La 40% Fe 7%, others 3%	—	As high strength alloy steel
Nodular iron (spheroidal graphite iron)	Cl based	Molten iron Si 2.5%, PO.1%, Mg 10%, Ce 0.05%	—	High strength and easily machinable castings
Constantan	Cu based	Cu 60%, Ni 40%	—	Strain gauges, electrically heated appliances

In *burnt steel*, the burning causes

- Burning of those grain boundary regions which are enriched in carbon (this burning is called first stage burning)
- Burning of nonoxidized cavities and blow holes (this burning is called second stage burning)
- Burning of iron-oxide inclusions (this burning is called third stage burning)
- Brittle fracture, similar to the fracture of stone
- Low ductility

12.18.1 Causes and Remedies of Overheating and Burning of Steel

The causes and possible remedies to avoid overheated and burnt steels during heat treatment are suggested as follows:

S.No.	Cause of	Possible Remedies
1.	*Overheating* • Heating for longer duration at considerably higher than normal temperature	• If only slight overheating is required, the heating should be done at normal temperature specified for annealing, normalizing, etc. • If considerable overheating is required, the heating should be performed in two steps, namely, the following: 　i. First, heat to about 100°C–150°C above normal temperature (see Figure 16.8 for normal temperature of different processes). 　ii. Then, heating should be done at normal temperature.
2.	*Burning* • Heating for longer duration at higher temperature in an oxidizing atmosphere	• To avoid first stage burning, a homogeneous heating followed by two-step heating should be accomplished. • To avoid second stage burning, die forging process should be carried out on burnt steel, followed by annealing. • Burnt steel of third stage is a waste, and this defect cannot be remedied.

12.19 Temper Brittleness: Its Causes and Remedies

Tempering is done in steel to reduce the brittleness, cracks, etc. that has cropped in after quench hardening. But if still the brittleness persists in steel even after its tempering, it indicates a defect. This defect is called temper brittleness. The effects, causes and possible remedies of temper brittleness are as follows:

S.No.	Effects	Causes	Possible Remedies
1.	Excessive hardness and brittleness after tempering	i. Tempering temperature is lower than its normal value. ii. The product is held for insufficient time during tempering.	i. Redo tempering at appropriate temperature. ii. Redo tempering by holding it for sufficient duration.
2.	Insufficient hardness but greater brittleness after tempering	i. Tempering temperature is too high.	i. Redo the annealing, hardening and tempering at appropriate temperatures.

13

Recent Non-Ferrous Metals and Alloys

13.1 Non-Ferrous Metals and Alloys

13.1.1 Copper

Copper has many applications, both as pure metal and an alloy base metal. It is malleable and ductile metal and can be easily rolled, drawn or forged. The tensile strength and hardness of copper can be improved by cold working although the ductility is reduced. Conversely, the annealing can improve the ductility at the expense of the tensile strength and hardness. The importance of copper is due mainly to its very high coefficient of electrical conductivity. The metal also has very good thermal conductivity and corrosion resistance. The impurities present in copper can have serious effects upon the properties of the metal. The electrical conductivity is reduced by 25% due to the presence of 0.04% phosphorus. Although the addition of 1% Cd improves the strength when used in telephone wires, it has minimal effect on the electrical conductivity. Copper is rapidly attacked by sulphuric acid, hydrochloric acid, nitric acid, ammonia, sodium hydroxide, potassium hydroxide and amines. Copper can be safely used with sulphurous acid (in paper industry); neutral salts, for example sodium chloride; hydrocarbons; alcohols; acetic acid; aldehydes; ketones; ethers and lactic and tannic acid.

Uses: Copper is available in the form of wire, bar, billet, rod, plate, sheet, strip and foil.

1. Various grades of pure copper are used for electrical windings and wiring, for cladding and castings (from sheet), and for heat exchangers and domestic installations (as tubing).
2. With copper pipes, there are no threaded joints, but for low-temperature duty, soldering/brazing is used.

3. For high-temperature applications and for medium-sized pipes, thin flanged connections are used.

4. Copper is used extensively in chemical plants for evaporators, kettles, stills and heaters, mainly due to its corrosion resistance, ease of cleaning and high thermal conductivity.

13.1.1.1 Copper Alloys

Several alloys of copper fall in the category of low-resistivity conducting materials. Main among them are brass and bronze. Brass is basically an alloy of copper and zinc, while bronze is an alloy of copper and tin. But they are prepared with the addition of other elements also. Brass and bronze are available in different compositions and so possess varying properties. Consequently, they are of different types, namely, admiralty brass, Muntz metal, delta metal, phosphor bronze, bell metal and gun metal. But among these, each type is not suitable for electrical applications. Various kinds of brass and bronze and other copper alloys that find use in electrical industry are summarized in the following.

13.1.1.1.1 Brass

Brass is a general name used for alloys of copper containing up to 40%–45% zinc. Brasses are easily tinned. They may be joined by soft soldering using tin-based solders with an antimony content below 0.5%. As the zinc content decreases, the risk of cracking during tinning or soldering is reduced. Due to the tendency of zinc vaporization, welding operation requires an oxidizing oxyacetylene flame and a borax–boric acid flux. A brass filler rod containing silicon is used, forming a protective layer over the weld.

Its conductivity is lower than that of copper. It has high tensile strength and is fairly resistant to corrosion. It can be easily pressed into desired shape and size, can be drawn into wires and can be easily brazed. Brasses are widely used in the following applications:

- Plug points
- Socket outlets
- Lamp holders
- Fuse holders
- Switches
- Knife switches
- Sliding contacts for rheostats and starters

13.1.1.1.2 Bronze

Bronze is the general name given to alloys of copper and tin. It has a composition of 10% Sn in 90% Cu. Its conductivity is lower than that of pure copper.

Bronze components are generally made by forging process. It is corrosion resistant and possesses high strength. Different types of bronze are generally used in the following applications:

- Beryllium bronze for making current-carrying springs, sliding contacts, knife, switchblades, etc.
- Phosphor bronze for making springs, bushings, etc.
- Cadmium bronze for making commutator segments

Bronzes have superior mechanical properties and corrosion resistance properties to brasses. They are high-strength alloys and are often used as bearing alloys materials in the cast iron form. Some important bronzes are described in the following:

1. *Phosphor bronzes* contain residual phosphorus to deoxidize the alloy. Phosphor bronzes have good cold-working properties and are resistant to coefficient of friction, but also cause loss of ductility. Up to 5% lead is added to bronzes to improve machinability of the alloy. Gunmetal is a bronze usually containing 10% tin and 2% zinc. Zinc acts as a deoxidizer.
2. *Silicon bronzes* are alloys of copper, silicon and zinc. Si is added to other copper-based alloys to increase the toughness, hardness and tensile strength of the alloy. Si improves the fluidity of the casting bronzes and allows intricate shapes to cast. They are rolled, forged and extruded to produce hydraulic pressure pipeline equipment, pressure vessels, filter screen and marine components.
3. *Nickel bronzes* are tin bronzes containing small amounts of nickel and zinc. They are more accurately termed as nickel gunmetals. Ni improves the mechanical properties and corrosion and wear resistance. These are used for valve and pump parts for boiler feed water.
4. *Aluminium* bronzes are similar to tin bronzes. These alloys are used as sheet or tube for condensers and heat exchangers. Heat treatment is not often required for aluminium bronzes. They exhibit excellent corrosion resistance due to surface oxide film. They retain their mechanical properties at elevated temperatures and have good wearing properties. The cast alloys are used for pump castings, valve parts and gears.

13.1.1.1.3 Other Copper Alloys

Besides brass and bronze, other important copper alloys are the following:

1. *Monel* is an important alloy containing 68% nickel. It has strength of steel and the corrosion resistance of copper. It is stronger than mild steel even when annealed. But thermal conductivity is only

15% of the value of copper, although it is higher than that of mild steel. Monel is tough and shock resistant, retains its strength at higher temperatures and is difficult to cast due to blow holes. Monel is used for valves and heat exchangers, sometimes in food industry, for high-temperature applications except in the SO_2 or oxidizing environments.

2. The cupro-nickels are fabricated by rolling, forging, pressing, drawing and spinning and are used for heat exchangers and condenser tubing.

3. *Cadmium copper* contains about 1% cadmium which improves the strength, toughness and fatigue resistance as well as the raising of softening temperature. This quantity of cadmium slightly reduces the electrical conductivity of the copper and is useful in telephone and overhead wires.

4. *Arsenic copper* contains 0.4% arsenic, which raises the softening temperature of cold-worked copper from 200°C to 550°C. Even though the arsenic greatly reduces the electrical conductivity of copper, it is still used for high-temperature and steam plant applications.

5. *Silver copper* contains approximately 0.08% silver which also increases the softening temperature, consequently the strength, hardness and creep resistance also.

6. *Tellurium copper* contains approximately 0.5% tellurium which is insoluble in copper and therefore does not affect the conductivity. The presence of small tellurium globules makes the machining easier by breaking up the chips, having a similar effect of that of the lead in steels.

7. *Beryllium copper* (or beryllium bronze) contains up to 2% beryllium and small amount of cobalt. Only a small amount of Be (less than 0.5%) dissolves in copper.

13.1.2 Aluminium

Aluminium is obtained from the ore *bauxite* by an electrolytic process. The pure metal is relatively expensive, is soft and ductile and has a low tensile strength. This strength can be doubled by cold working or by addition of alloying elements. Aluminium-based alloys are generally used for applications requiring improved strength. It has high thermal conductivity and low density. It also possesses good electrical conductivity. It is a lighter material. It has a very high coefficient of linear expansion, approximately four times that of common metals.

Aluminium reacts readily with oxygen forming a thin, dense oxide film on the metal surface that is impermeable to oxygen, and therefore, it exhibits excellent corrosion resistance against atmosphere and dilute acids. The metal is non-magnetic and non-sparking and can be polished to reflect both heat and light. Aluminium and its oxide are non-toxic and are used extensively in food-processing industries. Aluminium is malleable and ductile and can be easily fabricated, making it a useful material for domestic cooking utensils. The metal can be riveted, brazed and welded, although molten aluminium absorbs carbon dioxide and nitrogen and forms blow holes when cooled. Aluminium is widely used in the following:

- Heat exchangers.
- Aircraft.
- Chemical plant and the electrical industry.
- High purity aluminium (more than 99.5%) is used as a corrosion-resistant lining on other base metals.

13.1.2.1 Aluminium Alloys

Aluminium is alloyed with a number of elements to produce a range of materials with specific properties. Main alloying elements are copper, silicon, zinc, magnesium, iron, manganese and nickel. The alloys are usually classified as either wrought alloy or cast alloy. Wrought alloys are available in the form of rolled sheets, strips, plates, wires, rods and tubes. Cast alloys are shaped by sand casting or gravity die casting and in a few cases by pressure die casting.

1. Alloys containing up to 10% magnesium are particularly useful in marine environments.

   They have good corrosion resistance that increases with magnesium content.
2. Alloys containing approximately 1.25% manganese and a total of 2.5% alloying elements also possess improved properties.
3. Heat-treatable wrought aluminium alloys contain up to 4% copper and up to 2% silicon and magnesium. The increase in strength is due to the formation of an intermetallic compound such as $CuAl_2$. Other elements such as iron and zinc may be added to these alloys to increase their strength.
4. Nickel is added for use at high temperatures.
5. Commercially pure aluminium containing small amounts of silicon and magnesium can be cast, being both ductile and corrosion resistant.

6. Addition of more than 5% silicon improves the fluidity of the alloy. The main alloys that are used *as cast* and derive negligible benefit from heat treatment are those containing either 10%–14% silicon, or 10% silicon and 1.5% copper or 4.5% magnesium and 0.5% manganese. These alloys have medium strength and good corrosion resistance. The addition of 0.01% of metallic sodium to the molten alloy just before casting produces a finer-grain structure.

7. Cast aluminium alloys that respond to heat treatment usually contain either 4% copper or 2% nickel and 1.5% Mg or 3% Ni (forming $NiAl_3$). Other elements are often present that act as hardening agents.

8. Aluminium and its alloys are widely used in chemical plant as distillation columns for organic solvents, acetic acid plant, dairy industry and pharmaceuticals.

13.1.3 Magnesium

Magnesium and aluminium are both light metals, and they have similar properties, for example melting points of pure metal and an affinity for oxygen. Magnesium would probably be more widely employed because of its low density, except that the metal is less ductile than aluminium and is more difficult to cold-work. This is due to hexagonal crystal structure that prohibits slip, although the metal can be easily hot-worked.

13.1.3.1 Magnesium Alloys

Due to its low tensile strength, poor oxidation resistance and difficulties in cold working, magnesium is not used in pure state. It is alloyed with other elements to improve the strength and other properties. Alloying elements include Mn and Zn in small quantities and Al and Ag in larger quantities:

1. The solid solubility of Al or Zn in magnesium increases with increasing temperature. Rare earth metals are often added to Mg for similar reasons.

2. Silver is added to speed up the ageing process.

3. Manganese improves the corrosion resistance and zirconium acts as grain refiner.

4. The addition of thorium produces a creep-resistant alloy.

Magnesium alloys can be cut by sharp high-speed tools, but care must be taken to remove the inflammable powder. Magnesium alloys can be bolted, riveted and welded. Alloy surfaces are often treated to prevent corrosion. Cast alloys are used for petrol tanks and engine castings and in the aircraft industry due to low density. Wrought alloys are used for railings, ladders and brackets. Magnesium alloys are now used as a canning material in nuclear reactors because of the negligible neutron reaction.

13.1.4 Various Types of Brass, Bronze, Aluminium and Magnesium Alloys: Their Composition and Applications

Brass and bronze are alloys of copper. Brass is made of Cu + Zn and bronze is made of Cu + Sn. Details of different kinds of brass and bronze are summarized later. Few important and common non-ferrous alloys of aluminium and magnesium are also given in Table 13.1. It shows their composition and applications.

TABLE 13.1

Compositions and Applications of Some Non-Ferrous Alloys

Alloy	Composition	Application
Mg alloy		
Elektron	Al 3%–12%, Mn 0.03%, Zn 2%, rest Mg	Parts of aircrafts and automobiles, collapsible tubes, coating on steel utensils, solder, coating of petroleum containers and mild steel sheets
Al alloy		
Duralumin	Cu 4%, Mg 0.5%, Si 0.7%, rest Al	Electrical cables, components of airplanes and automobiles
Y alloy	Cu 4%, Mg 1.5%, Ni 2%, Fe 0.6%, rest Al	Casting of engine parts
Cu alloy: Brass		
Muntz metal	Cu 60%, Zn 40%	Brass castings, stampings and extruded parts
Yellow brass	Cu 90%, Zn 10%	Screw, wires, hardwares
Admiralty brass	Cu 70%, Zn 29%, Sn 1%	Marine uses, pump parts, ship parts
Cartridge brass	Cu 70%, Zn 30%	Bullet shots, cistern, storage batteries, military ammunition, foils
Arsenical copper	As 0.3%, rest Cu	Heat exchangers, condensers
Cu alloy: Bronze		
Ounce metal	Cu 85%, Sn 5%, Zn 5%, Pb 5%	Ornamental fixtures, pipe fittings, low pressure
Beryllium bronze	Be 1.5%–2%, Sn 8%, rest Cu	Bellows, diaphragms, springs
Aluminium bronze	Al 10%; small quantities of Ni, Mn, Fe; rest bronze	Parts containing corrosive liquids
Alpha bronze	Cu 95%, Sn 5%	Coin making, springs
Admiralty gun metal	Cu 88%, Sn 10%, Zn 2%	Bearings, steam pipe fittings
Phosphor bronze	Cu 84%–89%, Sn 10%–15%, P 0.1%–0.3% + Ni + Pb	Parts exposed to seawater, boiler fittings
Bell metal	Cu 70%, Sn 30%	Casting of bells

13.2 Other Non-Ferrous Metals

13.2.1 Nickel

Nickel is a crystalline, non-ferrous, ferromagnetic metal of silvery-white colour. Its hardness matches with the hardness of soft steel, but ductility is less than that. Other salient features of nickel are listed as follows:

1. It is capable of high-quality polishing and thereby provides lustre to the products on which it is polished.
2. It is reasonably malleable and can also be rolled provided the carbon content is in small amount (up to 0.05% or less).
3. It is resistant to acidic attacks, but dissolves readily in nitric acid.
4. Its melting point and specific gravity are 1453°C and 8.9, respectively.

Nickel is extensively used for nickel plating of metals to provide protective coating against corrosion.

Carbonized nickel is used to make anodes of power tubes for rapid conduction of heat.

13.2.1.1 Nickel Alloys

The applications of nickel alloys include items requiring corrosion resistance, high-temperature and low-temperature applications and low expansion coefficients. Important alloying elements with nickel are iron, molybdenum, copper and chromium and their alloys are as follows:

Nickel–iron alloys are used for thermostats, glass-to-metal seals, precision equipment, etc.

Nickel–molybdenum alloys are used mainly at room temperature for extremely good corrosion resistance. Alloy such as coronel has high strength and hardness.

Nickel–copper alloys are used mainly in chemical plants and steam turbines. These alloys are easily fabricated, machined and joined. Monel is an alloy containing 68% Ni, 30% Cu and up to 2% Fe or Mn. Steel can be cladded with Monel for a combination of strength and corrosion resistance.

Nickel–chromium alloys generally possess a high electrical resistance and melting point and are resistant to high-temperature oxidation. These alloys are suitable for use as resistance wires and heater elements at temperatures up to bright red heat.

13.2.2 Zinc

Pure zinc is relatively soft and weak and is brittle at room temperature. It is therefore difficult to cold work, but can be easily rolled into sheets at temperatures between 100°C and 150°C. In this form, it is used for battery cases and roofing and provides corrosion resistance due to the formation of a dense protective surface layer. Zinc is also used for sacrificial anodes to protect ship's hulls and buried pipes. A major use of zinc is in providing a coating, known as galvanizing, for corrosion protection on ferrous materials.

13.2.2.1 Zinc Alloys

The low melting point of zinc-based alloys makes them an excellent material for die casting purposes, and so they can be cast in relatively inexpensive dies. A small amount of aluminium on diffusing into solid solution with Zn lowers the melting temperature. Alloys used for die casting usually contain 4% Al and 1%–2% Cu, and the balance is zinc, producing a reasonably strong material.

13.2.3 Titanium

Titanium is the fourth most abundant metal to be found in the Earth's crust and is 50 times more plenty than copper. It has an excellent strength-to-weight ratio, placing in between aluminium and steel. Its corrosion resistance is similar to that of 18–8 stainless steel, with added advantage of chloride corrosion resistance. Oxygen and nitrogen can enter into solid solution with titanium. As these quantities increase, the tensile strength of titanium also increases, although the metal becomes less ductile.

13.2.3.1 Titanium Alloys

Titanium is alloyed with various amounts of (less than 6% of each element) other metals to produce specific properties. Copper–manganese or aluminium–manganese is added to improve tensile strength. Tin is added to make the alloy easier to shape and weld. Vanadium or zirconium–molybdenum–silicon is added to produce an alloy having high tensile strength and creep resistance. Titanium and its alloys have been mainly employed in jet engines, aircraft, spacecraft and nuclear rockets and also in chemical reactors. These applications reflect the high strength, corrosion resistance and high-temperature advantages of these alloys.

13.2.4 Lead and Lead-Based Alloys

Lead is a soft and malleable metal with good corrosion resistance. It is mainly used as an alloy for solders, printer's type and bearing materials, as a radiation shielding material, for electrical cable sheathing and for storage batteries. The strength and hardness of lead are increased by alloying with antimony which is a hard, brittle and crystalline metal. Tin improves the toughness, lowers the melting point and increases the fluidity of the lead–antimony alloys. Lead–tin alloys are used as soft solders, because of the low melting points. It is used in making babbits (see Section 13.2.6).

13.2.4.1 Lead and Lead Alloys as Sheathing Materials

Sheathing materials are used to protect cables against moisture, corrosion and chemical actions. Sheathing of cables is necessary because the insulation of cables will break on moisture penetration through the cracks. The following materials are generally used for this purpose:

- Lead
- Lead alloys with Sb, Sn, Cu and Cd
- PVC

Among these, the latter two materials are more common as they do not suffer from the deficiencies that a pure lead sheathing suffers from. The pure lead suffers from several deficiencies such as the following:

- Due to higher specific gravity (11.36), it adds to the weight of the cable.
- It is mechanically weak (strength is about 50 MPa only).
- It cannot withstand vibrations at high temperatures.
- Cracks are developed in them when lead-sheathed cables are laid near the railway tracks and on the bridges.
- It is corroded by various acids, lime and many chemical substances.
- It is also corroded by fresh concrete, tanners and chalk in the presence of air and water.

As an alternative to pure lead, the lead alloys and PVC coverings are preferred for use now. They possess much improved properties. PVC is lighter in weight also (specific gravity = 1.35–1.45 only).

13.2.5 Tin and Tin-Based Alloys

Tin is soft and weak corrosion-resistant metal having a low melting point. It is used as a coating on sheet steel known as tin plate and is used in the food industry because it is not attacked by fruit juices. Tin is used in bearing metals and low melting point alloys such as solders. It is added to copper alloys such as bronzes, gun metals and brasses, mainly to improve corrosion resistance. It is also used in making babbits (see Section 13.2.6).

13.2.6 Bearing Materials (or Babbits): Their Composition, Properties and Uses

These alloys are tin based, lead based and silver based. These are used as bush bearings to support rotating shafts. Bearing metals should possess toughness, hardness and high compressive strength. Tin-based and lead-based bearing metals are popularly known as babbit metals or babbits. Tin reduces brittleness and increases compressive strength. Tin-based bearing metals have better corrosion resistance against acidic oils. Babbits possess excellent embeddability and conformability. They are very suitable for high-speed and fluctuating load applications. Some bearing metals along with other related details are shown in Table 13.2.

13.2.7 Common Non-Ferrous Alloys

Some non-ferrous alloys commonly used are given in Table 13.3 along with all related details, properties and applications.

13.2.8 Properties of Silver, Platinum, Palladium and Rhodium

Contact materials are used as *contact points* and *contact surfaces* in electrical equipments, appliances, devices and instruments. They operate under severe conditions, aggressive constituents of surrounding medium and frequent reversal of mechanical rubbing while *making and breaking* the electrical circuits. They are also subjected to *arcing* and *sparking* due to ionization of surrounding medium between the contacts, when the contacts are separated.

13.2.8.1 Salient Applications

Some important applications of contact materials are the following:

- Push-button contacts in telephones
- Commutator segments in small DC motors
- Circuit breaker contacts

TABLE 13.2

Some Bearing Metals and Their Related Details

Name	Composition	Application
Tin babbit	Sn 80%, Sb 12%, Cu 8%	High-speed bearing bushes
Lead babbit	Pb 85%, Sb 10%, Sn 5%	Railway wagon bearings
Silver bearing metal	Ag 95%, Pb 4%, Sn 1%	Antifriction instrument bearings
Copper bearing alloy	Cu 80%,	Pb 10%, Si 10%, heavy-duty bearings
Nickel–cadmium bearing metal	Ni 97%, Cd 3%	High-temperature bearings

TABLE 13.3

Common Non-Ferrous Alloys

Alloy	Major Constituent (Base of Element)	Salient Composition	Special Characteristics	Application
Invar	Ni-base steel	Ni 36%, Fe 64%	Coefficient of linear expansion α is almost zero	Precision measuring instruments, survey measuring tapes and chains
Elinvar	Ni-base steel	Ni 32%, Fe 68%	Low α	Springs of watches and instruments
Muntz metal	Cu base	Cu 60%, Zn 40%	—	Fittings of pumps and marines
Red brass	Cu base	Cu 90%, Zn 10%	—	Rivets, hardwares, plumbing
Delta metal	Cu base	Cu 55%, Zn 40%, Pb 3%, Fe 2%	—	Parts of mining industry and chemical plants, motor bushes
Gun metal	Cu base	Cu 83%, Sn 10%, Zn 2%	—	Valves, fittings in corrosive environments
Mischmetal	Ce base	Ce 50%, La 40%, Fe 7%, others 3%	—	As high-strength alloy steel
Nichrome	Ni base	Ni 80%, Cr 20%	Tensile strength $\sigma = 1000$ MPa	Thermocouple, strain gauge and automobiles
Manganin	Cu base	Cu 87%, Mn 13%	High-resistivity alloy	Resistors
Constantan	Cu base	Cu 60%, Ni 40%	Melting point = 1300°C	Strain gauge, electrically heated appliances

- Contacts at brushes and collector rings
- Magnet contacts in magneto system of automobiles and airplanes
- Contacts for oil-immersed motor starters
- Contacts in computer and its peripherals
- Auxiliary contacts of switches
- Contacts in electronic devices
- Contacts in domestic appliances
- Contacts in domestic wiring systems

Contact materials in pure metal form are less wear resistant than the alloy form, but possess superior conductivity and corrosion resistance.

Hence, they are preferred for those applications where wear resistance is not of greater importance. Properties of some common contact metals are briefly explained as follows.

13.2.8.2 Fine Silver of 96.99c Purity

It is used where a high resistance against wear is not required.

- It possesses a superior electrical conductivity ($6.6 \times 10^7 / \Omega$ m).
- It possesses a superior corrosion resistance (anodic to Pt and Au only).
- It has a tendency to arc.
- It oxidizes in presence of arc.
- Its oxide (AgO) breaks at about 20°C and therefore affects the stability of contact resistance.
- It combines with sulphur which is undesired.
- It is used to form alloys with copper, tungsten, etc.

Silver contacts are suitable for low contact pressure applications such as in light and precise devices.

13.2.8.3 Platinum

It is a milky-white-coloured metal. It is next to gold at cathodic end of electromotive-force series. Therefore, it corrodes the least. Its other features are the following:

- It does not oxidize in air.
- It has no tendency to arc.
- It may form *needles* and *bridges* with low currents.
- It is malleable and hence can be transformed to sheet form, which is desirable for use as contact material.
- It is corrosion and chemical resistant.

It is a noble metal and is anodic to gold only. Its other features are the following:

- Specific gravity = 21.45
- Melting point = 1769°C
- Coefficient of thermal expansion = $9 \times 10^{-9} /°C$

Platinum in the wire form is the most commonly used metal in electrical resistance thermometers because of the following favourable features:

1. Its resistance increases uniformly with rise in temperature from 200° to 1200°C. Therefore, it offers a high accuracy in temperature measurement in this range.
2. It does not react chemically with other substances.
3. Being anodic to gold only, it is highly corrosion and oxidation resistant.

It is used to measure temperatures in the range of –200°C to 1200°C satisfactorily. However, temperatures above 1200°C cannot be measured accurately as platinum begins to evaporate above this temperature.

13.2.8.4 Palladium

It is moderately hard metal having a specific gravity of 12.02. Its melting point is 1552°C. Its other features are as follows:

- It is cheap and is used to make alloy with silver.
- It may oxidize in air.
- It has no tendency to arc.
- It has less resistance to erosion.

13.2.8.5 Tungsten

It is a very hard metal having a high melting point of 3410°C. Its specific gravity is 19.3 and hardness above 150 Brinell hardness number (BHN). It possesses the following salient properties:

- It has no tendency to arc.
- It has a least tendency of erosion.
- It does not weld during service.
- It reacts with air and forms the oxide film.
- It enhances red hardness, that is hardness at high temperatures.

13.2.8.6 Molybdenum

It is a hard metal having a melting point of 2610°C. Its specific gravity is 10.22 and hardness above 200 BHN. Its other features are as follows:

- It suffers from erosion problem.
- It forms loosely packed oxide in the air which obstructs normal working by closing the contacts.

13.2.8.7 Rhodium

It is a hard metal and possesses the following features:

- Melting point = 1966°C.
- Young's modulus = 372 GPa.
- Specific gravity = 12.44.
- Structure is face centered cubic (FCC).

13.2.9 Other Non-Ferrous Metals

Tantalum: It is a crystalline, non-ferrous metal of specific gravity 16.6 and melting point 2996°C. Its hardness is about 500 BHN. It is used to make *electrolytic capacitors*. The electrolytic capacitor is made by placing tantalum and aluminium in a suitable solution and making them positive electrode of a galvanic cell. Consequently, a thin insulating surface film is formed that can withstand a considerable voltage and is also capable of a high electrostatic capacity. It is resistant to most chemicals. It is classified as a refractory metal and is suitable for high-temperature applications. It is used for pumps, stills and agitators and in human surgery as replacements.

Beryllium: This word comes from the mineral *beryl*. Beryllium resembles aluminium in many properties. It is light but strong (stronger than steel) metal. It has excellent thermal conductivity and high heat capacity. Its uses are in making the following:

- Crew cabins and cones of spacecraft due to excellent thermal conductivity and high heat capacity.
- As rocket propellant for space flights to distant planets as beryllium has the capability of releasing vast amounts of heat while burning.
- Be is least affected by fatigue. When added to steel, beryllium reduces the metal fatigue of steel to zero. Therefore, Fe + Be springs can withstand more than 10 million impacts without failure. So these are useful for making springs of rapid-firing aircraft machine guns.
- Windows of x-ray tubes as beryllium allows x-rays to pass through better than any other metal.

Hafnium: It is similar to zirconium. It has good corrosion resistance and strength. It is heavier and harder and less malleable than zirconium. It can be forged, rolled and drawn. It can be machined and welded. Main uses are as control rods in nuclear reactors because of neutron absorption ability.

Niobium: It is scarce and expensive metal. It is malleable and ductile. It can be rolled and welded. It has resistance to most chemicals and molten metals. It is used as nuclear fuel canning material, for gas turbine blades and sometimes as an alloying addition to austenitic stainless steels.

13.2.10 Thermocouple Materials

These are made of a pair of metals or alloys to form a hot and a cold junction for flow of thermoelectric current in accordance with Seebeck effect. Although various pairs of metals and alloys can be used, the following pairs are more common:

- Fe + Cu
- Fe + constantan (an alloy of 60% Cu and 40% Ni)
- Fe + Ni
- Ni + nichrome (an alloy of Ni and Cr)
- W + Mo
- Pt + Pt–Ir alloy
- Sb + Bi
- Pt + Pt–rh alloy (rh is rhodium)
- Chromel + alumel
- Ir + Ir–rb alloy (rb is rubidium)

Each of these pairs finds favour in different thermometers, but Sb–Bi pair is the most common. It is because it produces a higher thermo-emf as compared to other pairs for the same temperature difference between hot and cold junctions.

14

Emerging and Futuristic Materials

14.1 Introduction to FGMs

A functionally graded material (FGM) is a two-component composite characterized by a compositional gradient from one component to the other. In contrast to traditional composites that are homogeneous mixtures and involve a compromise between the desirable properties of component materials, the significant proportions of FGMs contain the pure form of each component; hence, the need for compromise is eliminated. Thus, the properties of both components can be fully utilized. For example, the toughness of a metal can be mated with the refractoriness of a ceramic, without any compromise in the toughness of the metal or the refractoriness of the ceramic. The materials can be designed for specific functions and applications. Various approaches based on the particulate processing, preform processing, layer processing and melt processing are used to fabricate the FGMs.

14.1.1 Potential Applications of FGMs

There are many areas of application for FGM. The concept is to make a composite material by varying the microstructure from one material to another material with a specific gradient. This enables the material to have the best of both materials. If it is for thermal or corrosive resistance, or malleability and toughness, both strengths of material may be used to avoid corrosion, fatigue, fracture and stress corrosion cracking. The aerospace industry and computer circuit industries need the materials that can withstand very high thermal gradients. This is normally achieved by using a ceramic layer connected with a metallic layer. This type of functionally graded (or layered) fibre-reinforced concrete (FG FRC) can potentially be utilized in designing more economical structural systems, for example, in airport and highway pavements, commercial and precast slabs and structural elements.

14.2 FGMs in Construction Applications

FGMs are a novel class of materials having unique characteristics. They are graded property materials and are used as medical implants, for thermal protection of space vehicles, as thermoelectric converter for energy conservation, etc. Due to the versatility of their behaviour, they are now used as construction and building materials also. The development of FG FRC, fibre cement, FGM cement composites and functionally graded particulate composite can be quoted in this regard. These FGMs can be effectively used as precast slabs, airport and highway pavements, various structural elements, etc. An FG FRC beam and epoxy–TiO_2 particulate-filled functionally graded composites are other examples.

The functionally graded (or layered) FRC can be used to design more economical structural systems. Varying concrete constituents (e.g., fibres, aggregate type or air voids) can provide extra flexibility in designing the efficient structural systems. Functionally graded fibre cement is a new concept in fibre cement technology. Among other potential benefits, grading polyvinyl alcohol (PVA) fibre content is an effective way to produce functionally graded fibre cement, which allows for a reduction of total fibre volume without significant reduction on modulus of rupture of composite. The experimental study for TiO_2 particulate-filled epoxy-graded composite has shown that the TiO_2 particle additions on epoxy-graded composites have a dramatic effect on the flexural strength, tensile modulus and impact strength as compared to homogeneous composites.

14.3 Functionally Graded Fibre-Reinforced Concrete [1]

Fibre-reinforced concrete (FRC) composites are utilized to improve the performance of plain concrete materials, to repair and retrofit the structures. Fibres can be spatially (or functionally) distributed to improve the structural performance while minimizing the amount of fibres. These spatially varied microstructures created by nonuniform distributions of reinforcement phase are called FGMs. Recently, functionally graded fibre-reinforced cement composites have been developed using the extrusion technique and the Hatschek process in order to improve the structural performance of a component while reducing its material cost. The functionally layered FRC can subsequently be applied to concrete slabs, pavements or structural elements.

14.4 Functionally Graded Fibre Cement [2]

Most man-made materials such as concrete and ceramics are manufactured with uniform properties in their entirety. With structural materials, the mechanical properties are usually high enough to resist maximum stress, which most often occurs in a small area of the structural element. This approach is effective in guaranteeing the structural safety, but it is inefficient in the use of natural resources because it results in some of the structural elements receiving a higher volume of such resources than is necessary. In turn, it produces higher mechanical properties than are actually needed.

FGMs are a class of materials in which the properties change over the volume with an aim to obtain a desired performance. In FGM components, the material properties are engineered to change locally in a controlled way to meet the actual need. Therefore, they are intrinsically more efficient. Some natural materials present the property gradients as the result of an evolutionary process when adapting to environmental conditions. Bamboo, for instance, has stronger phases located at the more stressed regions of its cross sections. In these regions, there is higher fibre content and lower porosity.

Asbestos and PVA are useful materials for the development of fibre cement. However, one of the main drawbacks in replacing asbestos in fibre cement components is higher production cost of the substitute material. In a typical formulation, PVA fibre, a common substitute for asbestos fibre in air-cured asbestos-free fibre cement, can account for approximately 40% of the total cost of the raw material. Consequently, a reduction in PVA fibre content will reduce production costs.

14.4.1 Mixture Design for Choosing Fibre Cement Formulations

The production of functionally graded fibre cement involves controlling the distribution of properties to the extent that defined areas are better suited to fulfill specific tasks. To obtain such a sophisticated product, mixture rules are established for the correlation of the relevant (including mechanical) properties of the composite with the raw material proportion. This is a difficult task due to the high number of raw materials being utilized in the fibre cement industry. For example, the cement, mineral admixtures, cellulose fibres, synthetic fibres and organic admixtures compose a typical fibre cement formulation.

14.5 Processing of the FGM Cement Composites [3]

The processing of FGMs is generally performed using two basic steps: gradation and consolidation. Gradation consists of building a non-homogeneous structure into the material using a suitable method such as pancaking or continuous stacking. The consolidation step involves assuring the non-homogeneous structure of the final product. The gradation step in fibre cement components is probably the most difficult step to perform. Consolidation, on the other hand, appears to be easy since it occurs by cement hydration, which is a very well-known process.

The Hatschek process is the most widely employed one in producing fibre cement components. It consists of producing fibre cement sheets by stacking thin laminas made from a suspension (slurry) of cement, fibres, mineral admixtures and water, in a process that resembles the production of paper. The first step is to prepare the slurry, which consists of mixing an adequate proportion of solid materials, with water in a low solid concentration (10%–15% of total mass). Cement, asbestos, cellulose fibres, limestone filler and water are the most commonly employed materials by the asbestos cement technology. The slurry is then transported to the vats with sieve cylinders where wet solid material is deposited. Sequentially, the running felt removes the material from the sieve cylinders, thus forming a green lamina. Vacuum is applied to remove water from the lamina before it is transferred to the formation cylinder where the stacking is performed. Finally, the green sheet is cut, shaped (corrugated sheets and accessories) and submitted to curing. By using different mixtures in different vats and activating and deactivating the sieve cylinders of the Hatschek machine, we can control the composition layer by layer, producing a thickness gradation.

14.6 Functionally Graded Particulate Composite for Use as Structural Material [4]

FGMs are attracting considerable attention in modern engineering applications due to their increasing performance demands. Such materials are also used in biomedical, sensor and energy applications. FGMs contain either a gradual or a stepwise change in material properties along a given direction. In particulate composites, a graded structure can be obtained by changing either (1) the particle volume fraction or (2) the particle size along the thickness of the composite. The properties of the FGMs have shown improved mechanical properties compared to their homogeneous counterparts. However, the improvement in mechanical properties is achieved at the expense of dimensional stability at varying temperature and moisture conditions.

Gradients in the matrix volume fraction and matrix-particle interfacial area can lead to warping, localized swelling and interfacial fracture when these composites are subjected to high-temperature or moisture conditions.

14.7 Epoxy–TiO$_2$ Particulate-Filled Functionally Graded Composites [5]

Polymer composites containing different fillers and/or reinforcements are frequently used for applications like automotive parts and gear assemblies used inside home appliances in which the friction and wear are critical issues. In particular, they are now being used as sliding elements, which were formerly composed of metallic materials only. In this work, an epoxy–titania particulate-filled FG composite has been developed. TiO$_2$ (titania) is a conventional filler material and is used extensively for developing new materials. It has a strong potential as coating material with an elastic modulus of 283 GPa. The TiO$_2$ has also been used in combination with hydroxyapatite for developing biomaterials for implants because of its favourable biological effects and improved corrosion resistance. TiO$_2$ is also frequently used to reinforce Al$_2$O$_3$ wear-resistant coating on metal substrate. In this study, TiO$_2$-filled epoxy resin–based FGM has been synthesized using vertical centrifugal casting, and their mechanical and wear characterization has been performed.

14.7.1 Hardness Properties

The experimental results show that the density of the composites increases with increasing volume fraction of particulates. Increase in density indicates that the particle breakage may not have any significant influence on the composites. The variations of porosity level in these composites indicate that an increasing amount of porosity is observed by increasing the volume fraction, especially for low particle sizes of composites, because of decrease in inner-particle spacing. The trend of variation in micro-hardness with filler concentration is represented in Figure 14.1. The hardness profile of different composites filled with 0, 10 and 20 wt.% of TiO$_2$ is shown in Figure 14.2. As the percentage of TiO$_2$ enhances, the hardness of the composites also enhances. About 20 wt.% of TiO$_2$-filled epoxy composite has exhibited the maximum hardness.

14.7.2 Strength Properties

The tensile and flexural strengths of homogeneous and graded TiO$_2$–epoxy composites are shown in Figures 14.3a and b, respectively. It can be seen for graded composites that the tensile strength decreases considerably as the

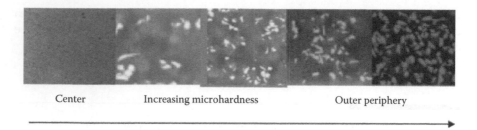

Center   Increasing microhardness   Outer periphery

FIGURE 14.1
Trend of micro-hardness variation of the composites. (From Siddhartha, W.M., *Mater. Design*, 32, 615, 2011.)

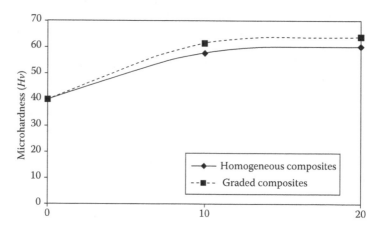

FIGURE 14.2
Variation of micro-hardness of the composites with TiO_2 content. (From Siddhartha, W.M., *Mater. Design*, 32, 615, 2011.)

weight fraction of TiO_2 is increased. There is an approximate 75% reduction in tensile strength for the composite having 0 wt.% particulate filled (neat epoxy) in comparison to that of epoxy resin composite.

14.8 Conclusions

From the studies carried out in the previous sections, the following conclusions are drawn:

1. The results illustrate that the functionally graded (or layered) FRC can potentially be used to design more economical structural systems. Functionally graded (or layered) FRC is one application of

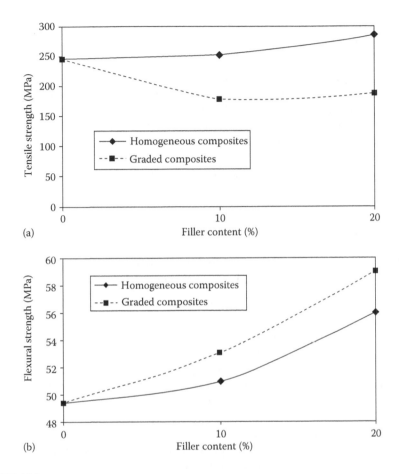

FIGURE 14.3
(a) Variation of tensile strength with TiO_2 content. (b) Variation of flexural strength with TiO_2 content. (From Siddhartha, W.M., *Mater. Design*, 32, 615, 2011.)

multifunctional and functionally graded concrete materials for the civil infrastructure. Varying concrete constituents (e.g., fibres, aggregate type or air voids) can provide extra flexibility in designing the efficient structural systems.

2. Functionally graded fibre cement is a new concept in fibre cement technology. Among other potential benefits, grading PVA fibre content is an effective way to produce functionally graded fibre cement, which allows for a reduction of total fibre volume without significant reduction on modulus of rupture of composite.

3. Processing of FGMs in the laboratory depends not only on the materials involved but also on the type and extension of the gradient and the geometry of the required component. New challenges in

the future such as cost-effectiveness of production processes, qual-, ity control and modelling of production processes may contribute to solve some problems.

4. The experimental study for TiO_2 particulate-filled epoxy-graded composites has shown that the TiO_2 particle additions on epoxy-graded composites have a dramatic effect on the flexural strength, tensile modulus and impact strength as compared to homogeneous composites. And also, the SEM observation shows that in the centrifugal casting, particulate plays a major role as compared to particulate-filled homogeneous composite.

14.9 Functionally Graded Nanoelectronic, Optoelectronic and Thermoelectric Materials

Due to their versatility of behaviour, they are now used as nanoelectronic, optoelectronic and thermoelectric materials also. Future applications demand for the materials having extraordinary mechanical, electronic and thermal properties which can sustain different environment conditions and are easily available at reasonable prices. The carbon nanotube (CNT)-reinforced functionally graded-composite materials (FGCM) are expected to be the new-generation material having a wide range of unexplored potential applications in various technological areas such as aerospace, defence, energy, automobile, medicine and structural and chemical industry. They can be used as gas adsorbents, templates, actuators, catalyst supports, probes, chemical sensors, nanopipes, nanoreactors, etc.

In this regard, the research developments of newer materials by different investigators, namely, the development of CNT-reinforced FGCM, FGM in optoelectronic applications and FG thermoelectric materials, have been presented here. In these elaborations, it is shown by the respective investigators that these FGMs can be effectively used as MRI scanner cryogenic tubes, tools and dies, solar panels, photodetectors, high-temperature joining materials, various structural elements, etc.

14.10 Applications of CNT in FGM [6]

CNT-reinforced metal matrix functionally graded composites due to their unique combination of hardness, toughness and strength are universally used in cutting tools, drills, machining of wear-resistant materials, mining and geothermal drilling. CNT-reinforced FGCMs have the ability to generate new features and perform new functions that are more efficient than larger structures and machines. Due to functional variation of FGM materials, their physical/chemical properties (e.g., stability, hardness, conductivity, reactivity, optical sensitivity, melting point) can be manipulated to improve the overall properties of conventional materials. Some of the current and futuristic applications of FGM are listed as follows:

- *Commercial and industrial*: Pressure vessels, fuel tanks, cutting tool inserts, laptop cases, wind turbine blades, firefighting air bottles, MRI scanner cryogenic tubes, eyeglass frames, musical instruments, drilling motor shaft, x-ray tables and helmets
- *Automobiles*: Combustion chambers (SiC–SiC), engine cylinder liners (AI–SiC), diesel engine pistons (SiCw/AI alloy), leaf springs (E-glass–epoxy), drive shafts (AI–C), flywheels (AI–SiC), racing car brakes and shock absorbers (SiCp/AI alloy)
- *Aerospace equipment and structures*: (TiAl–SiC fibres) rocket nozzle, heat exchange panels, spacecraft truss structure, reflectors, solar panels, camera housing, turbine wheels (operating above 40,000 rpm) and nose caps and leading edge of missiles and space shuttle
- *Submarine*: Propulsion shaft (carbon and glass fibres), cylindrical pressure hull (graphite–epoxy), sonar domes (glass–epoxy), composite piping system and diving cylinders (AI–SiC)
- Other applications of CNT-reinforced FGM

Potential applications of FGM are both diverse and numerous. Some more applications of CNT in FGM, having recent applications, are the following:

1. CNT-reinforced functionally graded piezoelectric actuators
2. As furnace liners and thermal shielding elements in microelectronics
3. CNT-reinforced functionally graded tools and dies for better thermal management, better wear resistance, reduce scrap and improved process productivity
4. CNT-reinforced functionally graded polyester–calcium phosphate materials for bone replacement.

14.11 FGM in Optoelectronic Devices [7]

FGMs are materials in which some particular physical properties are changed with dimensions. Properties of such materials can be described by material function $f(x)$. In homogenous materials, this function is constant, like in Figure 14.4a. In case of a junction of two different materials, function

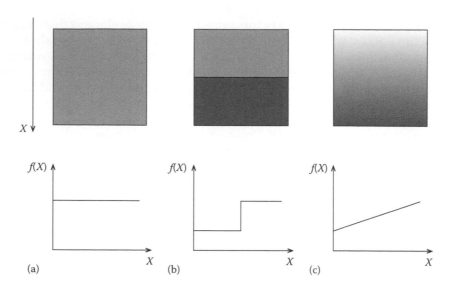

FIGURE 14.4
Schematic representation of material function in different structures: (a) homogenous material, (b) junction and (c) FGM. (From Gupta, N. et al., Novel applications of functionally graded nano, optoelectronic and thermoelectric materials, in *Proceedings of the International Conference on Nano and Material Engineering (ICNME* 2013), Bangkok, Thailand, 08–09 April 2013.)

$f(x)$ has a stair shape (Figure 14.4b). In FGM, this material function should be continuous or quasi-continuous. It means that the particular properties change continuously or quasi-continuously along one direction, like it is shown in Figure 14.4c. In many cases, FGM can be presented as a composition of several connected thin layers.

Depending on the number of directions the properties changed, it is discriminated as 1D, 2D or 3D FGM. It can be mathematically described for 3D FGM as

$$\frac{dF}{dx} \neq 0, \quad \frac{dF}{dy} \neq 0, \quad \frac{dF}{dz} \neq 0$$

where $F(x, y, z)$ is the material function.

14.11.1 Possible Applications of FGM in Optoelectronics

Nowadays, the graded materials are widely used for anti-reflective layers, fibres, GRIN lenses and other passive elements made from dielectrics and also for sensors and energy applications. For example, the modulation of refractive index can be obtained in such components through the change in material composition. Another possibility is to apply concept of gradation in semiconductor active devices. In semiconductors, the material function can

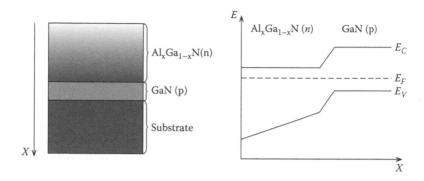

FIGURE 14.5
Schematic structure of *p-n* photodiode with graded layer electric field; therefore, the diffusion length of carriers should be longer than junction depth. (From Gupta, N. et al., Novel applications of functionally graded nano, optoelectronic and thermoelectric materials, in *Proceedings of the International Conference on Nano and Material Engineering* (*ICNME* 2013), Bangkok, Thailand, 08–09 April 2013.)

describe energetic bandgap, refractive index, carrier concentration, carrier mobility, diffusion length, built-in electric field and other properties which influence the parameters of optoelectronic devices.

14.11.2 High-Efficient Photodetectors and Solar Cells [7]

The fundamental limitation of the efficiency of homogenous silicon solar cells is the constant energetic bandgap width in bulk material. Because the high-energetic radiation is absorbed in a shallow layer under surface, it is necessary to form electric field in close vicinity to the surface. Generated carriers can effectively be separated in electric field; therefore, the diffusion length of carriers should be longer than junction depth.

Another factor which decreases carrier generation efficiency is the difference of energetic bandgap and absorbed photon energy. By using materials with gradation of energetic bandgap, it is possible to match the absorption edge with bandgap, which improves generation efficiency. The appliance of cascade of junctions with different energetic bandgap width can be one of the solutions. Another way to overcome this limitation is the use of graded material. The idea of such device is shown in Figure 14.5.

14.12 FGM Thermoelectric Materials [7]

A good thermoelectric material possesses a large Seebeck coefficient, high electrical conductivity and low thermal conductivity. A high electrical conductivity is necessary to minimize the Joule heating, while a low thermal conductivity helps to retain heat at the junctions and maintain a large temperature gradient. These materials, however, have a constant ratio of electrical to thermal conductivity (Wiedemann–Franz–Lorenz law), so it is not possible to increase one without increasing the other. Metals best suited to thermoelectric applications therefore possess a high Seebeck coefficient. Unfortunately, most of these metals possess Seebeck coefficients in the order of 10 µV/K, resulting in generating efficiencies of only fractions of a percent. Therefore, the development of formulated semiconductors with Seebeck coefficients in excess of 100 µV/K increased the interest in thermoelectricity. Earlier, it was not known that the semiconductors were superior thermoelectric materials due to their higher ratio of electrical conductivity to thermal conductivity, when compared to metals.

A higher figure-of-merit Z of a thermoelectric material shows a higher performance at a specific narrow temperature range. On the other hand, the specific temperature can be shifted to higher temperature by increasing the carrier concentration. Bismuth telluride (Bi_2Te_3), lead telluride (PbTe) and Si–Ge alloy (SiGe) are used for low-, medium- and high-temperature range, respectively. Usually, a monolithic and uniform thermoelectric material is used, though a temperature gradient exists in the thermoelectric material. Therefore, each part has not proper carrier concentration for each temperature. Two times of higher performance than a traditional thermoelectric material can be expected, if the proper carrier concentration gradient is performed to fit with the temperature gradient. Performing stepwise change of carrier concentration is also a performing method for practical application. That is a fundamental concept of energy-converting FGM.

It is essential to choose a proper material for each part to fit the temperature gradient. The proper material is a material with proper carrier concentration and a proper compound to match the temperature of each part along the temperature gradient. FGM joining of these materials and fitting electrodes with FGM interface are also core technique, because thermal stress relaxation caused by the difference of thermal expansion coefficient is important at a high temperature.

14.12.1 PbTe-Based FGM Thermoelectric Material [7]

Performing higher Z over a wide temperature range is essential to obtain higher thermoelectric efficiency η. Figure 14.6 shows the temperature dependence of Z for five kinds of n-type PbTe with different carrier concentration, which was induced from a report of ZT-data by Goff and Lowney as the parameter of carrier concentration. Every Z-curve has a maximum value, and the corresponding temperature shifts with the carrier concentration. If these materials are joined to fit with the temperature gradient, a higher performance than traditional thermoelectric materials can be expected. When the

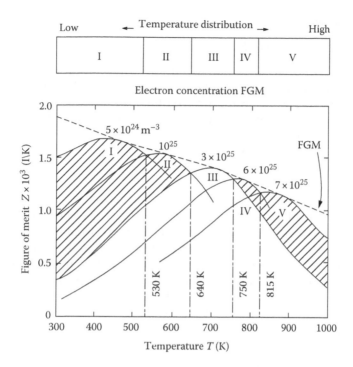

FIGURE 14.6
Temperature dependence of Z for *n*-type PbTe as parameters of carrier concentration. (From Gupta, N. et al., Novel applications of functionally graded nano, optoelectronic and thermoelectric materials, in *Proceedings of the International Conference on Nano and Material Engineering (ICNME 2013)*, Bangkok, Thailand, 08–09 April 2013.)

FGM structure is applied to PbTe, the conversion efficiency can be improved remarkably. Five kinds of PbTe in Figure 14.6 are joined in sequence at the intercept temperatures of 530, 640, 750 and 815 K. This segmented PbTe has five kinds of different carrier concentration and shows five maximum values of Z at different temperatures.

Comparing with the ordinal monolithic PbTe with a carrier concentration of $3 \times 10^{25}/m^3$, which shows the highest efficiency in monolithic PbTe, the FGM exhibits a larger Z value as shown in the hatched area. The broken line in Figure 14.6 shows an estimation on the assumption of continuous change of the carrier concentration for PbTe. This gradient PbTe has an ideal FGM structure and exhibits a higher Z than the segmented PbTe over a wide temperature range. It is estimated that the average Z of the ideal FGM should improve by 50% in comparison with a monolithic PbTe and maximum thermoelectric efficiency η at $T_h = 950$ K should extend to 19%.

This concept can be applied not only to the carrier concentration gradient but also to joining different materials, such as Bi_2Te_3–PbTe–SiGe, corresponding to the temperature gradient.

14.13 Thermoelectric Materials

The good thermoelectric materials possess large Seebeck coefficient, high electrical conductivity and low thermal conductivity. A high electrical conductivity is necessary to minimize the Joule heating, while a low thermal conductivity helps to retain heat at the junctions and maintain a large temperature gradient.

14.13.1 Metals

Although the properties favoured for good thermoelectric materials were known, the advantages of semiconductors as thermoelectric materials were neglected, and research continued to focus on metals and metal alloys. These materials however have a constant ratio of electrical to thermal conductivity (Wiedemann–Franz–Lorenz law), so it is not possible to increase one without increasing the other. Metals best suited to thermoelectric applications therefore posses a high Seebeck coefficient. Unfortunately, most possess Seebeck coefficients in the order of 10 μV/K, resulting in generating efficiencies of only fractions of a percent.

14.13.2 Semiconductors

The development of synthetic semiconductors with Seebeck coefficients in excess of 100 μV/K increased the interest in thermoelectricity. Earlier, it was not known that the semiconductors were superior thermoelectric materials due to their higher ratio of electrical conductivity to thermal conductivity, when compared to metals.

14.14 Applications [8]

14.14.1 Thermoelectric Generation

The simplest thermoelectric generator (TEG) consists of a thermocouple, comprising a p-type and n-type thermoelement connected electrically in series and thermally in parallel. Heat is pumped into one side of the couple and rejected from the opposite side. An electrical current is produced, proportional to the temperature gradient between the hot and cold junctions (Figure 14.7).

Uses: Some of the uses of thermoelectric generation are as follows:

- Generally, the auto-vehicle makers such as Volkswagen and BMW have developed TEGs that recover waste heat from a combustion engine.

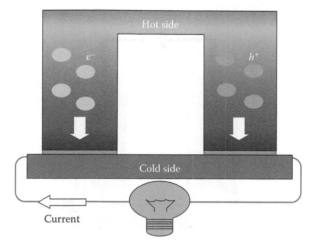

FIGURE 14.7
Electrical power output illustration of thermoelectric generation (Seebeck effect). (From Tiwari, P., *Int. J. Adv. Mater. Res.*, 685, 161, 2013.)

- According to a report by Prof. Rowe of the University of Wales in the International Thermoelectric Society, Volkswagen claims 600 W output from the TEG under highway driving condition. The TEG-produced electricity meets around 30% of the car's electrical requirements, resulting in a reduced mechanical load (alternator) and reduction in fuel consumption of more than 5%.
- BMW and DLR (German Aerospace) have also developed an exhaust-powered TEG that achieves 200 W maximum and has been used successfully for more than 12,000-km road use.
- Space probes to the outer solar system make use of the effect in radioisotope TEGs for electrical power.

14.14.2 Thermoelectric Cooling

If an electrical current is applied to the thermocouple as shown, heat is pumped from the cold junction to the hot junction. The cold junction will rapidly drop below ambient temperature provided heat is removed from the hot side. The temperature gradient will vary according to the magnitude of current applied (Figure 14.8).

Uses: Some of the uses of thermoelectric cooling are as follows:

- Peltier devices are commonly used in camping and portable coolers and for cooling electronic components and small instruments. Some electronic equipment intended for military use in the field is thermoelectrically cooled. The cooling effect of Peltier heat pumps can also be used to extract water from the air in dehumidification.

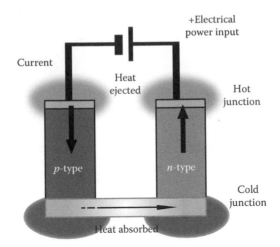

FIGURE 14.8
Illustration of thermoelectric cooling (Peltier effect). (From Tiwari, P., *Int. J. Adv. Mater. Res.*, 685, 161, 2013.)

- The effect is used in satellites and spacecraft to counter the effect of direct sunlight on one side of a craft by dissipating the heat over the cold-shaded side, whereupon the heat is dissipated by thermal radiation into space.
- Photon detectors such as CCDs in astronomical telescopes or very high-end digital cameras are often cooled down with Peltier elements. This reduces the dark counts due to thermal noise. (A dark count is the event that a pixel gives a signal although it has not received a photon but rather mistook a thermal fluctuation for one. On digital photos taken at low light, these occur as speckles [or *pixel noise*].)
- Thermoelectric coolers can be used to cool computer components to keep temperatures within design limits without the noise of a fan or to maintain stable functioning when over clocking. A Peltier cooler with a heat sink or water block can cool a chip to well below ambient temperature.

14.14.3 Thermoelectric Module

A typical thermoelectric module is shown in Figure 14.9. The module consists of pairs of *p*-type and *n*-type semiconductor thermoelements forming thermocouples which are connected electrically in series and thermally in parallel. In cooling mode, an electrical current is supplied to the module. Heat is pumped from one side to the other (Peltier effect), and the result is that one side of the module becomes cold. In generating mode, a temperature gradient is maintained across the module. The heat flux passing through the module is converted into electrical power (Seebeck effect).

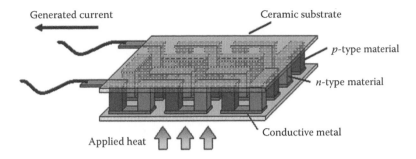

Generated current

Ceramic substrate

p-type material

n-type material

Applied heat

Conductive metal

FIGURE 14.9
Thermoelectric module. (From Tiwari, P., *Int. J. Adv. Mater. Res.*, 685, 161, 2013.)

References

1. K. Park, G.H. Paulino, and J. Roesler, Cohesive fracture model for functionally graded fibre reinforced concrete, *Cement and Concrete Research*, 40, 2010, 956–965.
2. C.M.R. Dias, H. Savastano, and V.M. John, Exploring the potential of functionally graded materials concept for the development of fibre cement, *Construction and Building Materials*, 24, 2010, 140–146.
3. B. Kieback, A. Neubrand, and H. Riedel, Processing techniques for functionally graded materials, *Materials Science and Engineering A*, 362(1–2), 2003, 81–106.
4. N. Gupta, S.K. Gupta, and B.J. Mueller, Analysis of a functionally graded particulate composite under flexural loading conditions, *Materials Science and Engineering*, 485(1–2), 2008, 439–447.
5. W.M. Siddhartha, S.A. Patnaik, and A.D. Bhatt, Mechanical and dry sliding wear characterization of epoxy–TiO_2 particulate filled functionally graded composites materials using Taguchi design of experiment, *Materials and Design*, 32, 2011, 615–627.
6. G. Udupal, S. Shrikantha Rao, and K.V. Gangadharan, Future applications of carbon nanotube reinforced functionally graded composite materials, in *IEEE International Conferences on Advances in Engineering, Science and Management* (ICAESM-2012), Nagapattinam District, Tamil Nadu, India, 2012.
7. N. Gupta, I. Bharti, and K.M. Gupta, Novel applications of functionally graded nano, optoelectronic and thermoelectric materials, in *Proceedings of the International Conference on Nano and Material Engineering* (ICNME 2013), Bangkok, Thailand, 08–09 April 2013.
8. P. Tiwari, N. Gupta, and K.M. Gupta, Advanced thermoelectric materials in electrical and electronic applications, *International Journal of Advanced Materials Research*, 685, 2013, 161–165, Trans Tech Publications, Switzerland.

15

Special Materials in Specialized Applications

15.1 Materials for Pumps and Valves in Various Industries

Nature of fluids has an impact on choice of materials. A fluid can be either a liquid or a gas, acidic or alkaline, cold or hot, chemically active or inactive and less or more viscous. Comparative selection of materials for all these situations is given in Table 15.1 for different working fluids in various industries.

15.2 Materials in Robots

Robotics is an interdisciplinary subject. The technologies in its construction and working involve the following engineering:

1. Mechanical engineering for design of manipulators, robot arms, joints, slides, bearings, etc.
2. Electrical engineering in design of drives and controls
3. Electronics engineering in design of circuits, controls, feedback, etc.
4. Computer engineering in design, hardware and programming
5. Civil engineering in design of structural components
6. Specialized fields, namely, production, marketing and finance

The design considerations involve knowledge of statics, dynamics, computations, cybernetics, biological and sensing. Nowadays, a robot functions in almost each field of life. A robot may be used to serve the purpose of an industry, civic sanitation, music and dangerous and hazardous activities. The function of an industrial robot may be in the field of material handling, welding, machine loading, painting, assembly, inspection and humanoid. Its reach may be of few metres, payload in many 10 kg, speed in few m/s and

TABLE 15.1

Comparative Suitability of Pumps and Valves Materials for Various Working Fluids Used in Different Industries

Working Fluid	Excellent	Good	Fair	Unsatisfactory
Acetic acid	Bronze, Cu, Zn, Pb, Monel metal Ni, SS, PTFE	Cast aluminium	Nickel	CI, brass
NH$_3$ gas	Cast Al, CS, CI, carbon steel	—	—	Bronze, brass
NH$_3$ anhydrous liquid	Cast Al, CS, SS, PTFE	Neoprene	—	Bronze, gun metal
Antibiotic	Glass-lined SS	—	Ni, naval bronze (65% Cu)	Bronze, cast Al, CI, cast steel
Apple juice	Ni, naval bronze, SS, glass-lined PTFE	—	—	—
Aviation fuel	Bronze, CS, Ni, SS, PTEF	CI	—	—
Freon 12	Ni, SS, glass lined	Cast Al, CS, CI, PTFE	—	—
Epoxy resin	Glass lined	SS, CS, CI	—	—
Refined gasoline	Cast Al, CS, CI, SS, glass-lined PTFE	Aluminium bronze	Naval bronze	—
Lime and slurries	Al bronze, Buna-N, PTFE, neoprene	—	—	—
Paper pulp	Bronze, Ni, SS, PTFE, rubber lined	Cast Al, Al bronze	Gun metal	CS, Pb lined
Seawater	Bronze, Al bronze, Ni, naval bronze, PTFE	Gunmetal, SS	—	Cast Al, CI, CS
Sewage	CI, PTFE, rubber lined	—	Cast Al, CS, SS	—
Superheated steam	Ni, Monel metal	SS	CS	Bronze, cast Al, CI, rubber
Urea	Glass lined, Pb lined	SS, CI, bronze	—	—

Notes: SS, stainless steel; CI, cast iron; CS, cast steel; PTFE, polytetrafluoroethylene (Teflon).

accuracy in ±0.01 mm. Humanoid functions approximate the performance of human in a task.

The study of robotics involves various subjects. As the purpose of this section is to acquaint the reader with robotic materials, hence other details are omitted here and can be referred to in specific texts. The choice of materials in robot construction is unlimited. A schematic robot is shown in Figure 15.1. Important and commonly used materials are illustrated in Table 15.2.

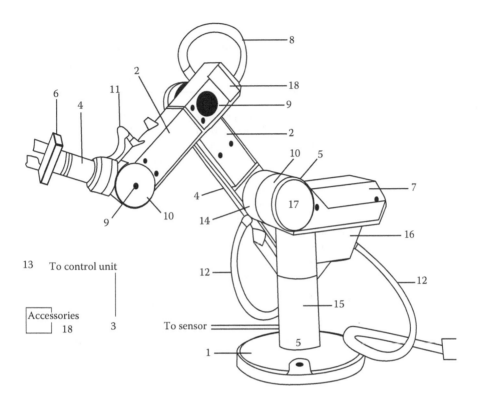

FIGURE 15.1
Materials used in different system of a typical industrial robot.

15.3 Materials for Rocket and Missile

A missile employs several materials. Important among them are given as follows:

- Ti alloy (Ti + 6% Al + 4% V) for high-pressure air bottles and combustion chamber.
- Maraging steel ($\sigma_{Tensile} = 170$ kgf/mm^2) for booster rocket motor casing.
- Hg–Cd–Te crystals for guiding the missile by imaging infrared technology and detecting a temperature difference of 0.1°C in anti-tank missile. The strategical technology uses focal plane array (FPA) sensor-based seeker. A mercury–cadmium–telluride crystal can house 100,000 sensor elements per square centimetre:

TABLE 15.2

List of Some Commonly Used Materials Employed in Robot Construction

System	Material	Specialty	Code Number in Figure 15.1
Manipulator			
Structure and base	Aircraft grade Al alloys, graphite fibre-reinforced plastics (GrRPs)	$\sigma_y = 500$ MPa $\sigma_u = 580$ MPa	1
Robot arm	Li–Al alloys, boron carbide (BC), Lockalloy (Be 62% + Al 38%)	$E = 1.9 \times 10^5$ MPa $\gamma = 2100$ kg/m^3	2
Sensor	Nichrome (Ni + Cr) thermocouple, Hall effect sensor (of semiconducting materials)		3
Links	Ceramics		4
Bearings	Elastomer, rubber		5
End of arm device	Mg, Al, Ti		6
Guideways	Ball, roller and needle bearings of stainless steel		7
Damping devices	TiC, alloyed W		8
Joints	Brass bushings and nylon sleeves		9
Gears	Aluminium alloys	$\sigma_y = 150$ MPa $\sigma_u = 165$ MPa	10
Combination links	Incramute (Cu + Mn + Al)	$E = 0.95 \times 10^5$ MPa $\gamma = 7630$ kg/m^3	11
Cables and hoses	Nitinol (Ni + Ti + Al)	$E = 0.8 \times 10^5$ MPa $\gamma = 6600$ kg/m^3	12
Control unit			
Input/output	*npn* and *pnp* switches		13
Safety	Collision protection (software and electrical drive)		13
CPU	Microprocessor and microcomputer (semiconducting materials)		13
Coordinate system	XYZ and robot joints		14
Programming language	Real-time robotic language		Not shown
Drive system			
Actuator	DC servomotor, hydraulic, pneumatic, AC servomotor		15
Transmission	Harmonic device		16

(Continued)

TABLE 15.2 (*Continued*)

List of Some Commonly Used Materials Employed in Robot Construction

System	Material	Specialty	Code Number in Figure 15.1
Feedback	Optical encoders		17
Grippers	Vacuum (pneumatic), DC (electrical)		—
Gripper connections	Optical fibres, metal conductors		—
Accessories			
Pendant	Handheld (direct control) Computer-using materials as in computers		18
Rotary table	Aluminium alloy		—
Conveyor	Asbestos, rubber, flax		—
Linear slide base	Antifriction materials		—

Notes: σ_y, yield strength; σ_u, ultimate strength; E, Young's modulus; γ, density.

- (Al + Mn + Si) alloy for liquid propellant tanks
- Mg alloy plates for missile wings
- Carbon–carbon composite for control surfaces and nose tip
- Ceramic–Kevlar fibre composite for radomes

A typical missile shown in Figure 15.2 presents glimpse of materials used for various parts. Figure 15.3 shows different parts of a typical rocket made of composites. It also depicts the products and the processes used for the aforementioned rocket parts.

15.4 Materials in Safety System against Explosion and Fire (or Fusible Alloys)

Fusible alloys are the alloy of bismuth and have low melting points. They have the eutectic composition of Pb, Sn, Sb and Cd. Due to their low melting points, they find applications in the plugs of pressure cookers, boilers, safety systems of explosion and fire, electrical fuses and sprinkler system in LPG refuelling plant, etc. Some of the fusible alloys with their approximate compositions are given in Table 15.3.

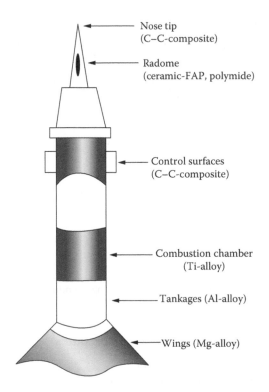

FIGURE 15.2
Materials used in various components of a missile.

15.5 Metals and Alloys for Nuclear Industry

The advancement in nuclear science and technology and their widening area of applications have necessitated advent of new materials. A nuclear material should be high-temperature resisting and least affected by radiations. Various materials used in nuclear power reactors may be summarized as follows:

1. *Fuel materials*: Uranium, uranium oxide, thorium, thorium oxide, etc. Uranium rod is inserted in airtight Al cylinder.
2. *Reaction control materials*: Hafnium oxide, cadmium, boron steel, gadolinium oxide and samarium oxide.
3. *Moderator–reflector materials*: Water, heavy water (H_3O or D_2O), graphite, beryllium, beryllium oxide and metal hydrides.
4. *Materials for cooling*: Liquid metal of sodium–potassium alloy, mercury, water, lead–bismuth, nitrogen, carbon dioxide and helium.

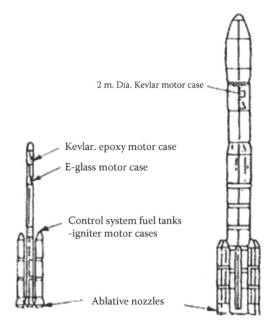

2 m. Dia. Kevlar motor case

Kevlar. epoxy motor case

E-glass motor case

Control system fuel tanks
-igniter motor cases

Ablative nozzles

Product		Process
Motor cases pressure vessels		Filament winding
Nozzles	Convergent	Rosette moulding six pressure curing
	Divergent	Tape winding six pressure curing

FIGURE 15.3
Components of rockets made up of composite materials.

TABLE 15.3

Bismuth-Based Fusible Alloy

Element Wt.%	Wood's Metal	Lipowitz Metal	Rose Metal	Newton Metal
Bi	50	50	50	50
Pb	25	27	2S	31
Sn	12	13	22	19
Cd	13	10	—	—
Melting point (°C)	70	60	93	95

5. *Shielding materials*: Reinforced cement concrete (RCC), lead, boron, bismuth, iron and tantalum.

6. *Materials for structures*: Cermets, silicon carbide, stainless steel, nickel alloys, molybdenum, aluminium, zirconium, magnesium, titanium, plastics and elastomers.

TABLE 15.4

Nuclear Materials

Metal/Alloy	Chemical Symbol	Characteristics and Properties
Zirconium	Zr	Silver white in colour
Hafnium	Hf	Good in neutron absorption, free from radiation attack, suitable for control rod in nuclear reactors
Beryllium	Be	Nonreactant with neutrons, light in weight (specific gravity 1.84)
Lead	Pb	As liquid metal cooling
Magnox	Mg–Al	Casing in Magnox reactor
Plutonium oxide	PuO_2	As fuel in fast-breeder reactors that do not use moderators
Zircaloy-2 and zircaloy-4	$Zr + 2.5\%$ Nb, Si, SO_3, $Na + U + Pu + Zr$	For fuel cladding and as fuel in likely future
Intermetallic compounds	UAl_4, $PuAl_4$	As fuel in likely future
Multifilamentary superconductor	47% Nb + Ti filament in Cu matrix	Cable-in-conduit conductors
Synroc	(TiO_2 + perovskite + zirconolite + hollandite mineral)	For high-level waste form in future

Common metals and alloys suitable for use in nuclear industries are given in Table 15.4.

15.6 Criteria for Selection of Acidic and Alkaline-Resistant Materials

Materials in contact with the liquids are also selected on the basis of pH value which is a measure of relative acidity or alkalinity of a liquid (Figure 15.4).

- A liquid having pH value = 0 is highly acidic while it is highly alkaline at pH = 14.
- Liquid having pH = 7 is neutral and is very suitable for most metals.
- Bronze is not suitable for alkaline liquids as they can react together while stainless steel and cast iron cannot.
- As stainless steel is costly and difficult to machine, cast iron may be selected as a cheaper metal. Suitable selection of materials on the basis of pH value can be made as follows.

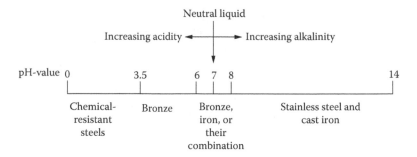

FIGURE 15.4
Selection of materials on the basis of pH value.

Such considerations in materials selection become essential in pumps and valves used in services like firefighting, slurry disposal, tube well and dredge.

15.7 Superheavy Elements

Mendeleev's periodic table incorporates 103 elements in which the lawrencium of actinide series is the last. Elements of atomic number 104–109 were reported thereafter. Recently, elements of atomic number 110–118 have also been reported; hence, further extension of the periodic table is desired. Elements 104–109 belong to d-block. These elements and some related details are given as follows. The following matter has been taken from http://www.periodni.com (Ref. *Pure Appl. Chem.* 81, No. 11. 2131–2156 (2009)):

Atomic Number	Element (Electron Configuration)	Symbol	Relative Atomic Mass
104	Rutherfordium ($4f^{14}6d^27s^2$)	Rf	267
105	Dubnium	Db	268
106	Seaborgium	Sg	271
107	Bohrium	Bh	272
108	Hassium	Hs	277
109	Meitnerium	Mt	276
110	Darmstadtium	Ds	281
111	Roentgenium	Rg	280
112	Copernicium	Cn	285
113	Ununtrium	Uut	—
114	Flerovium	Fl	287

(Continued)

Atomic Number	Element (Electron Configuration)	Symbol	Relative Atomic Mass
115	Ununpentium	Uup	—
116	Livermorium	Lv	291
117	Ununseptium	Uus	—
118	Ununoctium	Uuo	—

Source: http://www.periodni.com. Ref. *Pure Appl. Chem.* 81(11), 2131–2156, 2009.

15.8 Material Identification of Common Industrial Components

Items made of different materials may be industrial belt, gear, tableware, lens, bearing bush, wooden furniture, floppy disc, rope, welding electrode, table tennis ball, lathe tool, bulb filament, x-ray plate, etc. Materials are classified into metals and nonmetals, alloys and composites, polymeric and ceramic, etc. They possess different mechanical, physical, chemical, electrical and many other properties. They are used in construction and manufacturing of machines, structures, devices, instruments, consumable products, etc. It is expected from an engineer to be able to identify them for their suitability and appropriate selection in specific applications. Some of the engineering components, the materials of which they are made, and other related details are given as follows:

S. No.	Components	Material	Other Details
1.	Lathe tool	High-speed steel HSS (18–4–1 steel)	Alloy containing 18% W + 4% Cr + 1% V + 0.7% C + rest Fe
2.	Industrial belt	Neoprene	Rubber, non-metal
3.	Floppy disc	Flexible mylar plastic	Aluminized polyester
4.	Strain gauge	Nichrome, constantan	Ni + Cr, Ni + Cu alloys
5.	Plywood	Wooden veneers	Has directional property
6.	Container of a battery	Polypropylene, ebonite	Thermoplast, highly vulcanized rubber
7.	Bearing bush	Phosphor bronze, babbit	About 85% Cu + 15% Sn alloy, Sn–Sb or Pb–Sb alloy
8.	Lens	Borosilicate glass	Ceramic, noncrystalline
9.	Water-carrying pipe	Galvanized iron (GI)	Iron with Zn coating
10.	Lamp filament	Tungsten	m.p. = 3410°C
11.	Urea bag	High-density polyethylene (HDPE)	Semicrystalline, specific gravity = 0.97
12.	Structural steel bar	Mild steel	Ductile, E = 200 GPa
13.	Bolts, nuts, screws	Plain-carbon steel	0.10%–0.15% C

(Continued)

S. No.	Components	Material	Other Details
14.	Cutleries	Ferrite stainless steel	16% Cr + 1% Mn + 1% Si + 0.12% C+ rest Fe
15.	Contact material for magneto of a scooter	W, Mo, Pt	Arc and spark resistant
16.	Solder wire	60% Pb + 40% Sn	Highly creeping alloy
17.	Motor winding wire	Annealed Cu	$\rho = 17 \times 10^{-9}\ \Omega\,m$
18.	Car windshield	Plexiglass (PMMA)	A kind of thermoset
19.	Copper hardwares	Yellow brass	90% Cu + 10% Zn
20.	X-ray plate	$AgNO_3$-coated plastic	For radiographic images
21.	Grinding wheel	Diamond, corundum, or Al_2O_3 particles in shellac or vitrified matrix	Particulate composite
22.	Domestic plastic crockery	Melamine	A kind of thermoset
23.	Photocopier paper	Fibrous cellulose,	s.g. = 0.8–0.9
24.	Piston ring of a diesel engine	Malleable and grey CI	Hardened by nitriding
25.	Telephone receiver	Phenolics	A kind of thermoset
26.	Coiled spring	Medium-carbon steel	0.6%–0.8% C, resilient
27.	Artificial teeth	Special ceramic	A biomedical material
28.	Tin cans for food packing	Sn-coated iron	Corrosion-preventive coating
29.	Temple bell	Bell metal	70% Cu + 30% Sn, least sound damping ability
30.	Rope	Nylon 66	Tensile strength = 77–90 MPa
31.	Coin	Alpha bronze	95% Cu + 5% Sn
32.	Table	Teak wood	Amorphous, non-isotropic
33.	Granite	Ceramic	A kind of igneous rock
34.	Welding electrode	Similar to parent metal which is to be welded	May be bared or flux coated
35.	Computer screen	Fluorescent plastic	Phosphor coating of oxides and sulphides of Zn, Cd, Be, BaO, etc.
36.	Table tennis ball	Nitrocellulose	A kind of thermoplast
37.	Baked tiles	Terra-cotta	A type of ceramic
38.	Helmet	Fibreglass–epoxy	Polymeric composite
39.	Transistor	GaAs, InSb, etc.	Compound semiconductor
40.	Electrical fuse holder	Porcelain (i.e. white ware)	Made from china clay + feldspar
41.	Bottle	Soda-lime glass, polypropylene	Ceramic, thermoplast
42.	Beam of a building	MI5 grade RCC	Inorganic composite
43.	Cloth fabric	Cotton, terylene, silk	Woven fibre

(Continued)

S. No.	Components	Material	Other Details
44.	Blackboard duster	Felt	Amorphous fibres
45.	Needle bearing	Steel, bronze	Manufactured by powder metallurgy technique
46.	Tube of a moped	Elastomer with sulphur	Butyl rubber cross-linked tyre
47.	Talcum powder	Talc	Noncrystalline white sea product
48.	Permanent magnet	Alnico	Al + Ni + Co alloy, a hard magnetic material
49.	Agricultural implements	Malleable CI on heating at 900°C	Produced from white CI on heating at 900°C followed by slow cooling

Notes: m.p., melting point; s.g., specific gravity; *E*, Young's modulus; ρ, electrical resistivity; PMMA, polymethyl methacrylate.

15.9 Important Properties and Main Uses of All Known Elements

Some of the important properties and main uses of all known 109 elements (atomic number 1–109) are shown in Table 15.5. Their specific gravity vary from a lowest 0.53 for lithium to a highest 22.5 for iridium. Melting point varies from −270°C for helium to 3550°C for carbon in graphite form. Range of Young's modulus is from 1.7 GPa in francium to 540 GPa in osmium.

15.10 Electronic Systems/Materials Used to Telecast a Cricket Match from Cricket Field to Worldwide Televisions

Whether it is a test match or a 1-day cricket match, every interested viewer sits in front of his or her television and enjoys the thrills of cricket. The thrill reaches to a climax when very close judgement for a runout appeal, leg before wicket (LBW), stumping or to catch for a bleak click sound is to be announced. The decision mainly comes through very minute vision by electronic systems. Here, we shall know as to how the complete telecast of the game including these thriller moments is transmitted.

The information regarding the game of cricket is transmitted live through an electronic system shown in Figure 15.5. It includes a miniature camera fitted inside the middle stump and its connection with various computers (installed outside the field) through the cables and control box.

TABLE 15.5

Properties and Main Uses of All Known Elements

Element	Symbol	Atomic Number	Specific Gravity	Melting Point (°C)	Young's Modulus (GPa)	Major Use
Actinium	Ac	89	—	1050	35	Nuclear fuel
Aluminium	Al	13	2.70	660	71	Aircrafts, electrical cables, utensils
Americium	Am	95	11.7	1170	—	Source of α-particles
Antimony	Sb	51	6.62	630	55	Alloying with Sn, Pb and doping with Si, Ge
Argon	Ar	18	—	−189	—	Provides inert atmosphere for welding stainless steel, Ti, Mg, Al, etc.
Arsenic	As	33	5.72	817	39	Dopant to make light-emitting diodes
Astatine	At	85	—	302	—	Nuclear reaction
Barium	Ba	56	3.5	714	12.7	In x-ray of the intestine and kidney
Berkelium	Bk	97	14.8	986	—	Available in milligrams only
Beryllium	Be	4	1.85	1277	289	Moderator in nuclear reactors
Bismuth	Bi	83	9.80	271	34	Automatic fire sprinkler systems
Boron	B	5	2.34	2030	440	To make impact-resistant steel, e.g. carbide abrasive
Bromine	Br	35	3.12	−7	—	Flame retardants, pesticides, fireproof fabrics
Cadmium	Cd	48	8.65	321	62	Storage batteries, paints, steel plating
Calcium	Ca	20	1.55	838	19.5	To control carbon in cast iron, reducing agent in metal production
Californium	Cf	98	—	900	—	Available in milligrams only
Carbon (graphite form)	C	6	2.25	3550	8.3	Tyres, newspaper ink, carbon black
Cerium	Ce	58	6.77	804	30	Glass polishing, as coating in ovens
Cesium	Cs	55	1.90	28	1.75	Forms halides such as CsCl, CsBr, CsI
Chlorine	Cl	17	—	−101	—	Bleaching textiles, papers, purifying drinking water

(Continued)

TABLE 15.5 (*Continued*)

Properties and Main Uses of All Known Elements

Element	Symbol	Atomic Number	Specific Gravity	Melting Point (°C)	Young's Modulus (GPa)	Major Use
Chromium	Cr	24	7.19	1875	243	Making oxidation-resistant ferrous alloys
Cobalt	Co	27	8.85	1495	206	Making high-temperature-resisting steel alloys
Copper	Cu	29	8.96	1083	124	Electrical industry
Curium	Cm	96	13.5	1340	—	Available in milligram only
Dysprosium	Dy	66	8.55	1407	63	In making compounds
Einsteinium	Es	99	—	860	—	Available in milligram only
Erbium	Er	68	9.15	1497	73	In making complexes
Europium	Eu	63	5.25	826	15	In making complexes
Fermium	Fm	100	—	—	—	Available in micrograms only
Fluorine	F	9	—	−220	—	Coating for nonstick pans, toothpaste
Francium	Fr	87	—	27	1.7	As radioactive element
Gadolinium	Gd	64	7.86	1312	56	In making garnets
Gallium	Ga	31	5.91	30	92.5	Semiconductor devices
Germanium	Ge	32	5.32	937	99	Semiconductor devices
Gold	Au	79	19.32	1063	78	Ornaments, electrical contacts on computer boards
Hafnium	Hf	72	13.09	2222	137	Nuclear reactors used in submarines
Hahnium	Ha	105	—	—	—	Radioactivity
Helium	He	2	—	−270	—	Cryogenics, chromatography
Holmium	Ho	67	6.79	1461	67	As paramagnetic material
Hydrogen	H	1	—	−259	—	Industrial fuel, to make ammonia, in oil refineries
Indium	In	49	7.31	156	10.5	Doping crystals to make transistors
Iodine	I	53	4.94	114	—	Antiseptic, tincture, animal feeds, seeding artificial clouds
Iridium	Jr	77	22.5	2454	528	Pivots for instruments
Iron	Fe	26	7.87	1535	210	Ship building, girders, machine tools, rails

Element	Symbol	Atomic number	Density	Melting point		Uses
Krypton	Kr	36	—	-157.2	—	In discharge tube to give colours
Lanthanum	La	57	6.19	920	38	Crooke's lens to protect against ultraviolet light
Lawrencium	Lr	103	—	—	—	Radioactivity
Lead	Pb	82	11.36	327	15.7	Paints, pigments for road signs, crown glass
Lithium	Li	3	0.53	181	11.5	Making automobile grease, armour plate, in medicine, to toughen glass, aircraft parts
Lutetium	Lu	71	9.85	1652	84	As precipitator and separator, forms protective film
Magnesium	Mg	12	1.74	650	44	Lightweight structures, aircraft parts
Manganese	Mn	25	7.43	1245	198	Rock-crushing machinery
Mendelevium	Md	101	—	—	—	Radioactivity, low-scale power source
Mercury	Hg	80	13.55	-38	—	Extraction of precious metals, street light, thermometers
Molybdenum	Mo	42	10.22	2610	328	Making cutting steel
Neodymium	Nd	60	7.00	1019	38	As a liquid laser
Neon	Ne	10	—	-249	—	Neon signs
Neptunium	Np	93	—	637	100	As power source in satellites
Nickel	Ni	28	8.90	1453	198	Alloys, magnets, coins, radiators
Niobium[a]	Nb	41	8.57	2468	105	In stainless steels
Nitrogen	N	7	—	-210	—	Fertilizers, explosives
Nobelium	No	102	—	—	—	As oxide coatings, radioactivity
Osmium	Os	76	22.57	2700	540	Making hard alloys
Oxygen	O	8	—	-229	—	As rocket fuel, oxyacetylene welding
Palladium	Pd	46	12.02	1552	124	Chemical catalyst
Phosphorus	P	15	1.83	44	4.6	Fertilizers, phosphor bronze, phosphoric acid
Platinum	Pt	78	21.45	1769	170	Jewellery, catalytic convertors on cars
Plutonium	Pu	94	19.5	640	96.5	Nuclear fuel, bomb, producing electricity for *pacemaker*
Polonium	Po	84	9.14	254	25.5	Metallic conduction, radioactivity, biodegradable detergent
Potassium	K	19	0.86	64	3.5	Breathing apparatus, submarines, photography

(*Continued*)

TABLE 15.5 (*Continued*)

Properties and Main Uses of All Known Elements

Element	Symbol	Atomic Number	Specific Gravity	Melting Point (°C)	Young's Modulus (GPa)	Major Use
Praseodymium	Pr	59	6.77	919	33	As fluorides
Promethium	Pm	61	—	1027	42	Complex formation
Protactinium	Pa	91	15.4	1230	100	Shiny malleable metal tarnishes in air
Radium	Ra	88	5.0	700	16	Radiotherapy for cancer
Radon	Rn	86	—	−71	—	Radioactivity
Rhenium	Re	75	21.04	3180	460	Electrical furnace winding, catalyst for making lead-free petrol
Rhodium	Rh	45	12.44	1966	372	Catalyst in control of automobile emissions
Rubidium	Rb	37	1.53	39	2.7	Formation of superoxides
Ruthenium	Ru	44	12.2	2500	10	Powder metallurgy
Rutherfordium	Rf	104	—	—	—	Radioactivity
Samarium	Sm	62	7.49	1072	34	High temperature superconductors
Scandium	Sc	21	2.99	1539	79	Metallurgy
Selenium	Se	34	4.79	217	58	Xerox-type photocopiers, decolourizing glass
Silicon	Si	14	2.33	1410	103	Semiconductor devices
Silver	Ag	47	10.49	961	80.5	Photographic emulsion, ornaments, microcircuits
Sodium	Na	11	0.97	98	8.9	Making soap, glass, liquid Na as nuclear coolant
Strontium	Sr	38	2.60	768	13.5	Formation of compounds
Sulphur	S	16	2.07	119	19.5	Viscose rayon, vulcanizing rubber, gunpowder

Tantalum	Ta	73	16.6	2996	181	Capacitors for electronics industries
Technetium	Tc	43	11.5	2130	370	Radiographic scanning of the human liver and other organs
Tellurium	Te	52	6.24	450	41	Making ferrous and nonferrous alloys
Terbium	Tb	65	8.25	1356	57.5	Biological tracer for drug in humans and animals
Thallium	Tl	81	11.85	303	8	Toxic and poisonous, ceramic high temperature (HT) superconductor
Thorium	Th	90	11.66	1750	74	Incandescent gas mantles
Thulium	Tm	69	9.31	1545	75	As paramagnetics, formation of complexes
Tin (grey)	Sn	50	7.30	232	52	Electroplating, food cans
Titanium	Ti	22	4.51	1668	106	Jet and gas turbine engines, airframes
Tungsten	W	74	19.3	3410	396	Cutting tools, bulb filaments
Uranium	U	92	19.07	1132	186	Nuclear fission—bomb, power reactors
Vanadium	V	23	6.1	1900	132	Catalyst in oxidation reaction
Xenon	Xe	54	—	-112	—	Formation of halides, oxides and oxyanions
Ytterbium	Yb	70	6.96	824	18	High-temperature superconductors
Yttrium	Y	39	4.47	1509	65	Red phosphor for TV tubes, synthetic garnet in radar, gemstones
Zinc	Zn	30	7.13	420	92	Dry batteries, coating iron articles and pipes
Zirconium	Zr	40	6.49	1852	92	Cladding of nuclear fuel, chemical plants

[a] Niobium was previously known *as columbium* (Cb).

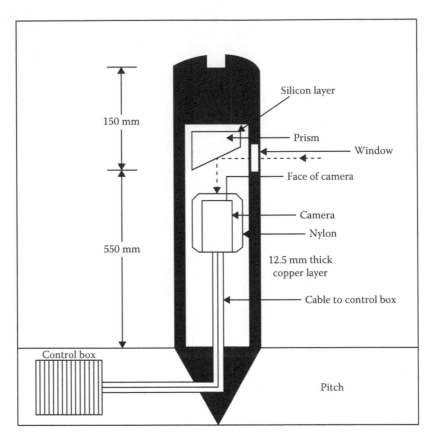

FIGURE 15.5
Schematic layout showing transmission of cricket game from a cricket field to worldwide TVs.

The stump camera (or stump vision) is about 12.5 mm wide and 50 mm long and is fitted 550 mm above the pitch surface. Weighing about 455 gm, this camera faces a prism at 60°. To safeguard against any damage, the internal parts of the stump are covered with a nylon and copper layer. A control box is located at about 1 m distance from the camera. It remains hidden under the wickets and covered below the Astroturf. Its one end is connected to the camera, while the other end is attached with broadcasting system, placed outside the boundary lines, through underground wires/cables.

When the bowler delivers the ball, its image is formed on the prism through a window-type gate on it. This image is then reflected to the camera from where it reaches the control box. Signal of the image is then transferred to the broadcasting system (vehicle) for live telecast. The stump vision is highly accurate and can see the minutest actions occurring between *bat, pad* and *ball*. Its construction is so robust that neither the wicket nor the camera breaks due to the impact of the ball.

16

Vivid Fields of Ongoing Researches

16.1 Palmyra Fibre Extraction, Processing and Characterization [1]

16.1.1 Introduction

There are several sources to obtain natural fibres. Plant fibres are one among them. They are obtained from different positions (leaves, fruits, seeds, etc.) of a plant. Accordingly, they are known as surface fibres, soft (or bast) fibres and hard fibres. Surface fibres (i.e. plant trichomes) are obtained from the surface of leaves, fruits and seeds; soft fibres are obtained from tissues of primary stem; and the hard fibres are the strands of entire vascular bundles that have moderate mechanical properties, but possess low specific weight. This results in a higher specific strength and stiffness. The use of these fibres cuts down the cost of raw materials and causes no irritation to the skin, and their thermal recycling is also possible. Natural fibres are good substitute also for synthetic fibres due to their lower cost of raw material, ecological recycling, weight reduction, etc.

Although a large variety of natural fibres have been investigated in the past, many more are yet to be utilized. Therefore, in this work, the palm (i.e. palmyra) fibres have been investigated. Palm plants are abundantly available in India, Sri Lanka, Southeast Asia and New Guinea. Hence, to utilize the locally available natural resource, the following properties have been determined by conducting appropriate tests on palmyra fibres:

1. Density by taking five different single fibres
2. Tensile testing on 10 single fibres in natural condition without any treatment

16.1.2 Obtaining Palmyra Fibres and Their Processing

Palmyra fibres are produced from the leaf sheath of palmyra tree shown in Figure 16.1. These are obtained from the junction of leaf sheath bottom and trunk of the tree. For that, the basic processing consists of beating the sheath

FIGURE 16.1
Palmyra tree and its leaf sheath. (From Self Photograph from MNNIT Campus, Allahabad, India.)

with a wooden mallet to separate the fibres. The leaf sheaths are first soaked in water for 1–2 days to facilitate the fibre extraction; otherwise, extraction will be very difficult. After separating the fibres from the sheath, the lignin part is removed with the help of a dry cloth. The fibres thus obtained are shorter and finer in nature, wiry and strong. They are nonuniform in shape and size. They have different cross sections. Even for a single fibre, the cross section varies along its length. These cross sections are of circular, rectangular, elliptical and other shapes. However, in testing, these fibres have been assumed circular as most of them were of this cross section, when viewed through a profilometer.

16.1.3 Determining Fibre Density

To determine the density of palmyra fibres, the weight and volume have been obtained for five individual single fibre. Its weight has been found by weighing on an electronic balance of 0.001 g accuracy. The fibre cross-sectional area has been measured by a profilometer with the help of a magnifying lens of magnifying power 50×. The diameter is measured at five different sections along the length of the fibre and then the mean diameter of fibre is calculated. The observations recorded in this regard are shown in Table 16.1.

$$\text{Hence, the average density} = \frac{\begin{matrix}0.685 + 0.730 + 0.606 \\ + 0.789 + 0.684\end{matrix}}{5} = 0.699\,\text{g}/\text{cm}^3$$

TABLE 16.1

Determination of Density of Individual Single Fibres

Fibre No.	Weight W (g)	Length l (mm)	Mean Diameter D (mm)	Cross-Sectional Area A (cm²)	Density $= W/Al$ (g/cm³)
1	0.019	200	0.21	0.0014	0.685
2	0.019	200	0.20	0.0013	0.730
3	0.018	200	0.21	0.0015	0.606
4	0.022	200	0.21	0.0014	0.789
5	0.012	200	0.10	0.0004	0.684

16.1.4 Tensile Testing of Fibres

For determining the tensile strength of palmyra fibres, the tensile test has been conducted on 10 individual single fibres. Since different fibres of the same leaf sheath have different cross-sectional area and also the cross-sectional area does not remain the same along the length of the fibre, the fibres taken are of varying shapes and sizes and from different locations. For more authentic prediction, 10 different single fibres in natural condition and straight configuration have been tested. These fibres are of different lengths and cross-sectional areas. Tensile tests have been done using a Hounsfield tensometer. The scale used is 0–30 kg (0–300 N) having a least count of 0.2 kg (2 N). Elongation is measured with the help of a dial gauge indicator of least count 0.1 mm. The observed values of load elongation for four different individual fibres are depicted in Tables 16.2 through 16.10 along with the calculated stress–strain (σ–ε) values. However, the stress–strain curves for all 10 fibres are shown in Figure 16.2.

TABLE 16.2

Tensile Test of Single Fibre No. 1

$l = 84$ mm, $d = 0.59$ mm, $A = 0.28$ mm²					
Load P		**Elongation**	**Stress**		
(kg)	(N)	(mm)	(kg/cm²)	(MPa)	Strain
0.0	0	0.0	0.00	0.00	0.000
0.4	4	5.9	480.77	48.07	0.039
0.8	8	9.0	961.54	96.15	0.060
1.2	12	12.0	1442.31	144.23	0.079
1.6	16	16.0	1923.08	192.30	0.106

TABLE 16.3

Tensile Test of Single Fibre No. 2

$l = 106$ mm, $d = 0.42$ mm, $A = 0.14$ mm^2					
Load P		Elongation	Stress		Strain
(kg)	(N)	(mm)	(kg/cm^2)	(MPa)	
0.0	0	0.0	0.00	0.00	0.000
0.4	4	5.0	285.71	28.57	0.047
0.8	8	8.0	571.42	57.14	0.075
1.2	12	12.0	857.14	85.71	0.113
1.6	16	16.0	1142.85	114.28	0.151
2.0	20	18.0	1428.57	142.85	0.170
2.2	22	20.0	1571.42	157.14	0.189

TABLE 16.4

Tensile Test of Single Fibre No. 3

$l = 97$ mm, $d = 0.55$ mm, $A = 0.24$ mm^2					
Load P		Elongation	Stress		Strain
(kg)	(N)	(mm)	(kg/cm^2)	(MPa)	
0.0	0	0.0	0.00	0.00	0.000
0.2	2	5.1	83.33	8.33	0.053
0.4	4	7.1	166.67	16.66	0.073
0.6	6	7.3	250.00	25.00	0.075
0.8	8	7.4	333.33	33.33	0.076
1.0	10	7.5	416.67	41.66	0.077
1.2	12	8.0	500.00	50.00	0.082

TABLE 16.5

Tensile Test of Single Fibre No. 4

$l = 84$ mm, $d = 0.60$ mm, $A = 0.288$ mm^2					
Load P		Elongation	Stress		Strain
(kg)	(N)	(mm)	(kg/cm^2)	(MPa)	
0.0	0	0.0	0.00	0.00	0.000
0.4	4	4.0	268.45	26.84	0.035
0.8	8	5.6	536.91	53.69	0.049
1.2	12	6.8	805.36	80.53	0.060
1.6	16	13.8	1073.82	107.38	0.122
2.0	20	15.0	1342.28	134.22	0.132
2.2	22	20.0	1474.53	147.45	0.176

TABLE 16.6

Tensile Test of Single Fibre No. 5

$l = 91$ mm, $d = 0.42$ mm, $A = 0.14$ mm^2					
Load *P*		Elongation	Stress		
(kg)	(N)	(mm)	(kg/cm^2)	(MPa)	Strain
0.0	0	0.0	0.00	0.00	0.000
0.4	4	4.3	285.71	28.57	0.047
0.8	8	6.8	571.43	57.14	0.075
1.2	12	9.3	857.14	85.71	0.102
1.6	16	10.0	1000.00	100.00	0.109

TABLE 16.7

Tensile Test of Single Fibre No. 6

$l = 84$ mm, $d = 0.59$ mm, $A = 0.28$ mm^2					
Load *P*		Elongation	Stress		
(kg)	(N)	(mm)	(kg/cm^2)	(MPa)	Strain
0.0	0	0.0	0.00	0.00	0.000
0.2	2	4.5	69.44	6.94	0.054
0.4	4	6.5	138.89	13.88	0.077
0.6	6	7.5	208.33	20.83	0.089
0.8	8	8.5	277.78	27.77	0.101
1.0	10	8.8	347.22	34.72	0.105
1.2	12	9.0	500.00	50.00	0.107

TABLE 16.8

Tensile Test of Single Fibre No. 7

$l = 152$ mm, $d = 0.42$ mm, $A = 0.144$ mm^2					
Load *P*		Elongation	Stress		
(kg)	(N)	(mm)	(kg/cm^2)	(MPa)	Strain
0.0	0	0.0	0.00	0.00	0.000
0.4	4	5.7	277.78	27.77	0.038
0.8	8	9.3	555.56	55.55	0.061
1.0	10	12.0	694.44	69.44	0.079
1.2	12	18.0	833.33	83.33	0.118

TABLE 16.9

Tensile Test of Single Fibre No. 8

$l = 153$ mm, $d = 0.04$ mm, $A = 0.0015$ mm²					
Load P		**Elongation**	**Stress**		
(kg)	**(N)**	**(mm)**	**(kg/cm²)**	**(MPa)**	**Strain**
0.0	0	0.0	0.00	0.00	0.000
0.2	2	4.5	133.33	13.33	0.029
0.6	6	6.0	400.00	40.00	0.039
0.8	8	7.8	533.33	53.33	0.052
1.0	10	9.8	666.67	66.66	0.064
1.2	12	18.0	800.00	80.00	0.118

TABLE 16.10

Tensile Test of Single Fibre No. 9

$l = 150$ mm, $d = 0.035$ mm, $A = 0.00098$ mm²					
Load P		**Elongation**	**Stress**		
(kg)	**(N)**	**(mm)**	**(kg/cm²)**	**(MPa)**	**Strain**
0.0	0	0.0	0.00	0.00	0.000
0.2	2	3.2	204.08	20.40	0.021
0.4	4	4.8	408.16	40.81	0.032
0.6	6	5.2	612.24	61.22	0.035
0.8	8	6.8	816.32	81.63	0.045
1.2	12	17.5	1224.49	122.44	0.117

16.1.5 Results and Discussion

From the present investigation, we observe the following salient features:

1. For individual fibres of 200 mm length having mean diameter 0.10–0.21 mm and weights varying from 0.018 to 0.022 g, the density (ρ) is found to vary between 0.6067 and 0.7893 g/cm³ (i.e. 606.7 and 789.3 kg/m³). This gives an average density of 0.699 g/cm³ (699 kg/m³).
2. Maximum sustained loads P vary between 1.2 and 2.2 kg (12 and 22 N).
3. Tensile strength σ_{ult} varies from 500 to 1923 kg/cm² (i.e. 50 to 192.3 MPa).
4. Maximum elongation δl varies between 8 and 20 mm.
5. Longitudinal strains ε vary from 0.082 to 0.189.
6. It is readily seen that the nature of stress–strain curves for different fibres is widely varying. Some of them are linear, while most of them are nonlinear.

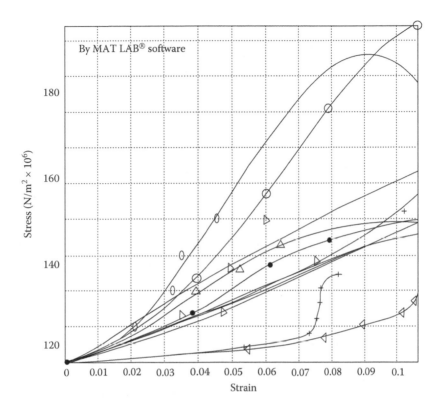

FIGURE 16.2
Stress–strain curves for 10 different individual single fibres.

7. Other important results are the following:
   a. Tensile modulus $E = 6{,}078.9$–$17{,}492.5$ kg/cm² (607.8–1,749.2 MPa).
   b. Secant modulus $E_{sec} = 4{,}672.9$–$18{,}142.2$ kg/cm² (467.2–1,814 MPa) at the stresses of 500–1,923 kg/cm²).
   c. Specific tensile modulus $E/\rho = 869.5$–2502.4 MPa/(g/cm³).
   d. Specific tensile strength $\sigma_{ult}/\rho = 71.5$–275.1 MPa/(g/cm³).
   e. Elongation at failure $\delta l = 8.24\%$–23.8%.

16.1.6 Comparison of Properties of Present Work (Palmyra Fibre) with Other Natural Fibres

Important properties of different fibres are compared with palmyra fibre as displayed in Table 16.11.

It can be seen that whereas the specific strength and specific modulus of a palmyra fibres are comparable to coir fibre (as shown in the table), its density

TABLE 16.11

Important Properties of Different Fibres Are Compared with Palmyra Fibre

Property	Glass	Flax	Hemp	Jute	Coir	Sisal	Cotton	Palmyra (Present Work)
Density (g/cm³)	2.55	1.4	1.48	1.46	1.25	1.33	1.51	0.7
Tensile strength (N/mm²), i.e. MPa	2400	800–1500	550–900	400–800	220	600–700	400	49–189.5
Tensile modulus (GPa)	73	60–80	70	10–30	6	38	12	7.78
Specific strength (N m/kg × 10³)	941	571–1075	370–620	270–550	176	50–530	265	70–270
Specific modulus (GPa/[g/cm³])	29	26–46	47	7–21	5	29	8	11.8
Elongation at break (%)	3	1.2–1.6	1.6	1.8	15–25	2–3	3–10	8–24

is least among all known fibres. It is 0.7 only as compared to 1.25 for coir, 1.33 for sisal, 1.40 for flax, 1.46 for jute, 1.48 for hemp and 1.51 for cotton fibres. This is a highly favourable property from the viewpoint of lightweightness. This makes the palmyra fibres the most suitable natural fibre for making reinforcement in composite materials.

16.1.7 Conclusions

Based on the aforementioned findings, we conclude that the palmyra–glass hybrid composite (PGHC) is suitable for low-pressure applications such as pipes carrying sewage and industrial waste; low-strength applications such as tables, boards and supporting pads; and other salient applications such as in

- Car bodies
- Footwear
- Windmill blades
- Tubes and ducts
- Ceiling tiles
- Office partitions
- Core materials for doors
- Low-pressure containers
- Greenhouse framework

- Low-density insulation board
- Lightweight furnitures
- Louvre shutter assembly for railway coaches

16.2 Development and Characterization of Natural (Palmyra) Fibre–Reinforced Composite [2]

16.2.1 Preface

Composites made of conventional fibres (glass, carbon, graphite, boron, Kevlar, etc.) are of high cost. If made of cheaper fibres, they will cut down the cost of components for which they are used. Natural fibres are such materials. They are not only cheap but are abundantly available also. However, they are light in weight and hence they possess high specific strength and low specific weight and are ecologically favourable too. Their strength is not as high as those of synthetic fibres. Hence, natural fibre composites are likely to be a blend of lightweight and strong material. The present work has been undertaken with this aim.

The adoption of natural fibre for composite applications is mainly due to low price, less weight and easy processability of removable resources. It prevents depletion of forest resources as well as ensures good economic returns for the cultivation of natural fibres. The use of natural fibres in automotive industries has grown rapidly in recent years. Different types of natural fibres in common use are jute, coir, sisal, pineapple leaf fibre, bamboo, banana, ramie cardanol as an additive from cashew nut industry, hemp (Felna 34), flax, cotton, cornstalks, lignin from paper pulp as filler in polyolefins, agrowaste such as jute sticks, rice husk and acacia mangium tree fibre.

16.2.2 Natural (Cellulose Fibres) Plant Fibres

Natural fibres can be classified as seed fibres (such as cotton, capok), bast fibres (such as flax, hemp, jute, kenaf, ramie), hard fibres (such as sisal, pineapple, manila, hemp, yucca), fruit fibres (such as coconut), wood fibres (such as pinewood) and leaf fibres (such as sisal, banana). These fibres consist of amorphous matrix of hemicellulose (approximately 60%–80%), lignin (approximately 5%–20%) and moisture (approximately 20%). Hemicellulose is responsible for biodegradability and moisture absorption, while the lignin is responsible for thermal stability and ultraviolet degradability. Bast plants are characterized by long, strong fibre bundles that comprise the outer portion of the stalk. The word *bast* refers to the outer portion of the stem of these plants. This vascular portion comprises 10%–40% of the stem mass. The remaining portion is known as *core*. The bast fibres generally have a relatively low specific gravity and high specific strength. They are typically the best for improvement in tensile and bending strength and modulus.

16.2.2.1 Advantages and Disadvantages of Natural Fibres

Various advantages of natural fibres are low specific weight, higher specific strength and stiffness than glass, much better bending stiffness, producible with low investment at low cost, easy processing with no wear of tooling, no skin irritation or biological problems, renewable resource with possible thermal recycling and better thermal and acoustic insulating properties. However, they possess various disadvantages also. These are lower strength particularly the impact strength, variable quality due to unpredictable influence of weather, swelling of fibre due to moisture absorption, limited maximum processing temperature, lower durability, poor fire resistance, irregular fibre lengths (therefore, spinning is required to obtain continuous yarns for weaving or winding) and price is affected by harvest result and agricultural policies.

16.2.3 Palmyra Fibre

These are palm fibres obtained from palm (toddy) plant. Palmyra is also known as toddy, wine siwalan, lontar palm. It is a bast fibre. Its botanical name is _Borassus flabellifer_. Palmyra fibres are produced from the leaf sheath (petioles) of palmyra tree. The fibre is strong and wiry, shorter and finer in nature. It lies in the category of hard fibres. Palmyra grows wild from the Persian Gulf to the Southeast Asia, predominantly in India, Sri Lanka, New Guinea and West Africa. It is easily available at Allahabad too. The palmyra tree is a large tree up to 30 m in height. There may be 30–40 leaf sheaths in a tree.

16.2.3.1 Fibre Processing Details

The basic processing consists of beating the sheath with a wooden mallet to separate the fibres. In this work, fibre has been extracted from the leaf sheath of young palmyra tree. The leaf sheaths are first soaked in water for 1–2 days to facilitate the extraction. After separating the fibres from sheath, the lignin part is removed with the help of a dry cloth. The fibres are found from the junction of leaf sheath bottom and trunk of adult palmyra tree. Palmyra fibres are shown in Figure 16.3.

These fibres are not available in uniform shape and size. They have different cross sections. Even for a single fibre, the cross section varies along its length. These cross sections may be of circular, rectangular and elliptical shapes. In testing, these fibres have been assumed circular as most of them were of this section when viewed through a profilometer. Basic properties of palmyra fibre are given in Table 16.12.

FIGURE 16.3
Individual palmyra fibre.

TABLE 16.12

Properties of Palmyra Fibre

Properties	Palmyra Fibre
Mean fibre diameter (mm)	0.1–0.3
Aspect ratio (L/d)	2500
Specific density	0.699
Tensile strength (kgf/cm²) (MPa)	500–1930
	50–190
Elongation at failure (%)	8–24

16.2.3.2 Testing of Palmyra Fibre–Reinforced Composite: Tensile Test, Compression Test, Bending Test and Impact Test

A unidirectional (U/D) palmyra fibre–reinforced composite (PFRC) sheet has been prepared by laying successive layers of Araldite and palmyra fibres. Wax has been used on inner surfaces of the mould for better surface finish and easy withdrawal of the composite. A metal roller has been rolled over each layer to remove entrapped air, if any. The palmyra fibres used are 300 mm in length. The density of the composite has been determined by measuring its weight and volume and is found to be 1110 kg/m³.

Tensile test: The tensile test has been performed on the specimens having fibre volume fraction of $V_f \approx 58\%$ in accordance with Indian Standards. These are shown in Figure 16.4. The rectangular specimen is of 16 mm × 3 mm cross section, overall length = 150 mm and gauge length = 40 mm. The specimen sustained a maximum load of 80 kg (800 N) and underwent an elongation of 0.9 mm.

FIGURE 16.4
Tensile test specimens of natural fibre composite.

TABLE 16.13

Stress–Strain Value for PFRC and GRP in Tension

Stress		Linear Strain × 10⁻⁶	
(kg/cm²)	(MPa)	For PFRC	For GRP
0	0	0	0
27	2.7	250	330
44	4.4	500	540
63	6.3	770	840
82	8.2	940	1140
100	10.0	1140	1380
120	12.0	1410	1600
145	14.5	1700	1970
166	16.6	1940	2220
184	18.4	—	2420
208	20.8	—	2670
236	23.6	—	2930
256	25.6	—	3190
270	27.0	—	3360
289	28.9	—	3600
312	31.2	—	3870
333	33.3	—	—

The observations are taken at a load interval of 10 kg. Thus, the ultimate strength is calculated as 166.6 kg/cm² (16.6 MPa), linear strain $= 6220 \times 10^{-6}$ and lateral strain $= 4700 \times 10^{-6}$. The observations for PFRC are shown in Table 16.13, and stress–strain curve is shown in Figure 16.5, along with that of glass fibre–reinforced composite (GRP) for comparison.

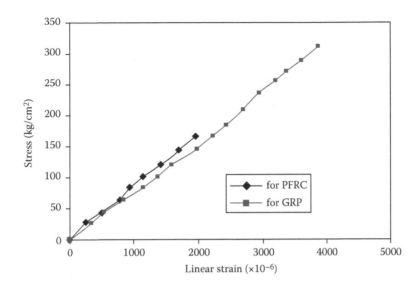

FIGURE 16.5
Tensile stress–strain curves of PFRC and GRP.

Compression test: This test has been performed on the specimen having palmyra volume fraction of $V_f \approx 58\%$. The test has been conducted on universal testing machine (UTM). The specimen used is of 25 mm × 25 mm cross section and 30 mm height. The specimen sustained a maximum load of 1.15 t before fracture, where the contraction recorded is 0.2 mm. Thus, the ultimate strength is found to be 200 kg/cm² (20 MPa) and strain $= 6450 \times 10^{-6}$. The observed data are shown in Table 16.14, and stress–strain diagram is shown in Figure 16.6 along with that of GRP for comparison.

Bending test: This test has been performed on a U/D specimen of $V_p \approx 58\%$. The specimen is of 70 mm span and 20 mm × 3 mm cross section, simply supported and subjected to a concentrated midspan load. Maximum deflection is recorded as 4.4 mm under a load of 40 kg (400 N). The load–deflection observations are given in Table 16.15, and the curve is shown in Figure 16.7 along with that of GRP for comparison.

Impact test: This test has been performed on a U/D specimen having $V_p \approx 58\%$. The Charpy and Izod tests have been performed on specimens of width = 10 mm and thickness = 10 mm, having depth of V-notch = 2 mm. The length of the specimen in the Charpy test is 50 mm, while in the Izod test, it is 75 mm. The impact strength in the Charpy test is found to be 0.8 kg (8 N m), while in the Izod test, it is 0.2 kg (2 N m).

TABLE 16.14

Stress–Strain Values for PFRC and GRP in Compression

For PFRC		For GRP	
Stress (kg/cm²)	Linear Strain × 10⁻⁶	Stress (kg/cm²)	Linear Strain × 10⁻⁶
0	0	0	0
40	320	16	200
80	640	32	350
120	1010	48	590
160	1160	64	810
200	1370	80	970
—	—	96	1210
—	—	112	1340
—	—	128	1490
—	—	144	1770
—	—	160	2000
—	—	176	2140
—	—	192	2270
—	—	208	2630
—	—	224	2820
—	—	240	2990
—	—	256	3220
—	—	272	3370
—	—	288	3610
—	—	312	3780
—	—	330	3860

16.2.3.3 Results and Discussion

PFRC is a bimodulus material, since the modulus in tension and compression is different. These are as follows:

- In tension, $E = 8.332 \times 10^4$ kg/cm² = 8.173 GPa.
- In compression, E_{sec} at $\sigma = 80$ kg/cm² = 1.25×10^5 kg/cm² = 12.26 GPa.
- In compression, E_{sec} at $\sigma = 200$ kg/cm² = 1.45×10^5 kg/cm² = 14.22 GPa.

It shows linear as well as nonlinear (σ–ε curve) behaviour in tension, but nonlinear only in compression. Various physical and mechanical properties are as given as follows:

- Specific gravity = 1.11.
- Tensile strength = 166.6 kg/cm² (16.6 Mpa).

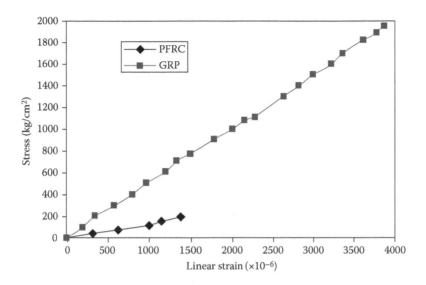

FIGURE 16.6
Compressive stress–strain curves of PFRC and GRP.

TABLE 16.15

Load–Deflection Curves for PFRC and GRP in Bending

For PFRC		For GRP	
Load (kg)	Deflection (mm)	Load (kg)	Deflection (mm)
0	0	0	0
5	1.4	10	0.3
10	1.9	20	0.5
15	2.3	30	0.8
25	2.7	40	1.1
30	3.2	50	1.4
35	3.6	60	1.7
40	4.1	70	2.1
—	4.4	80	2.5
—	—	90	2.9
—	—	100	3.2
—	—	110	3.6

- Compressive strength $= 200.0$ kg/cm^2 (20 Mpa).
- Bending strength $= 28.7$ MPa.
- Shear strength $= 4.3$ MPa.
- Impact strength in the Charpy test $= 800$ J/m^2 and in the Izod test $= 200$ J/m^2.

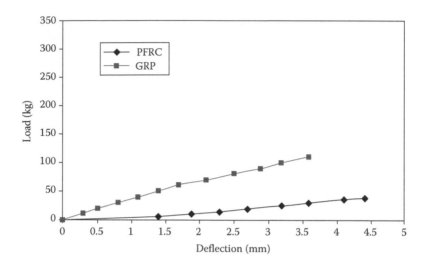

FIGURE 16.7
Load–deflection curves of PFRC and GRP in bending.

TABLE 16.16

Comparison of Strengths and Density of PFRC with GRP

Types of Composites → Properties ↓	PFRC	GRP
Density (kg/m³)	1110	1850
Tensile strength (MPa)	16.6	33.3
Compressive strength (MPa)	20.0	31.2
Flexural strength (MPa)	28.7	48.2
Impact strength (kJ/m²)	0.8	12.0

From these observations, we notice the following salient features. The PFRC shows lightweightness and higher tensile and compressive moduli, as compared to GRP. These are given in Tables 16.16 and 16.17.

16.2.3.4 Conclusion

Based on the aforementioned findings, we conclude that PFRC is suitable for low-pressure applications such as pipes carrying sewage and industrial waste and low-strength applications such as tables, boards and supporting pads. Other salient applications may be in car bodies, louvre shutter assembly for railway coaches, footwear, greenhouse framework, low-density insulation board, ceiling tiles, lightweight furniture, office

TABLE 16.17

Comparison of Specific Strengths and Specific
Moduli of PFRC as Compared to GRP

Properties	PFRC	GRP
Tensile modulus (GPa)	8.17	7.56
Compressive modulus (GPa)	14.22	7.84
Specific tensile strength (MPa)	14.95	18.00
Specific compressive strength (MPa)	18.01	16.86
Specific tensile modulus (GPa)	7.36	4.08
Specific compressive modulus (GPa)	12.81	4.36

partitions, core materials for doors, windmill blades, low-pressure containers, tubes and ducts.

They can also be used as wood substitutes in housing and construction sector, as substitutes for plywood and medium-density fireboards, as panel and flush doors for low-cost housing, as panel and roofing sheets, as door frames, and as geotextiles for prevention of soil erosion, leaching, etc., and in automobile coach interior, in packaging tray for automobile parts, and in storage devices, water tanks, and latrines.

16.3 Development and Characterization of Human Hair–Reinforced Composite [3]

Composites made of conventional fibres (glass, carbon, graphite, boron, Kevlar, etc.) are of high cost. If made of cheaper fibres, they will cut down the cost of components for which they are used. Human hair is one such material. It is not only cheap but is abundantly available also. It is a waste material. If used as fibre to produce composites, an alternative material can be developed from biological wastes. Conversion of waste material into meaningful material has ever been the cherish desire of materials scientists. Therefore, if the properties of human hair may be utilized to an optimum extent, the cost of machine and structural components can be reduced to a lower value as compared to present-day composites. The present work has been undertaken with this aim.

In this work [3], a human hair–reinforced composite (HHRC) has been developed using hand lay-up process. Epoxy (Araldite AY-103) has been used as matrix constituent. The human hair in bundle form has been tested to determine its physical and elastic properties in tension. Epoxy specimens have been prepared and tested for their properties under tension, compression, impact and bending. Similarly, U/D reinforced and chopped HHRC specimens have also been fabricated and tested under tension, compression, impact and bending to characterize their properties. It is concluded that HHRC will prove to be a cheaper alternative material for applications such as supports of industrial piping and lightweight machines and equipments and for making economical hybrid (composite) components for vivid structural applications.

16.3.1 Tensile Testing of Human Hair

Long human hair (i.e. continuous fibre) has been procured from beauty parlours. These are first washed in water to remove dust, dirt, etc., and then dried in air. Tensile testing of human hair has been performed on a Hounsfield tensometer. As the testing of a single hair was not possible, a bundle of 100 hairs has been tested. The bundle of 100 hairs was tested under 3 different conditions. In the first case, the human hairs in the bundle were straight; in the second case, the bundle was twisted and in the third case, the bundle of hairs dipped in mustard oil was tested. The observation recorded is shown in Tables 16.18 through 16.20, respectively. The initial length, average diameter and cross-sectional area of this bundle are 50 mm, 0.91 mm and 0.0065 cm^2, respectively. When subjected to a gradually increasing load up to 2 kg (20 N), an elongation of 39.2 mm is recorded in it. Finally, the applied load falls to 1.6 kg (16 N) and elongation becomes 48.4 mm. Thus, the ultimate stress σ and strain ε are found as $\sigma_{ult}=414$ kg/cm^2 (41.4 MPa) and $\varepsilon_{ult}=0.96$. The stress–strain curve is shown in Figure 16.8. It is of nonlinear nature.

16.3.2 Testings of Epoxy: Tensile Test, Compression Test, Bending Test and Impact Test

Epoxy resin (Araldite AY-103) is casted in glass mould, and specimens for different tests have been prepared in accordance with Indian Standards.

TABLE 16.18

Tensile Strength of 100 Dry Hairs

S. No.	Load (kg)	Elongation (mm)	Stress (kg/cm²)	Strain
1	0.2	1.0	28	0.02
2	0.4	1.5	83	0.03
3	0.8	4.2	138	0.084
4	1.0	19.0	221	0.38
5	1.4	25.8	276	0.516
6	1.8	36.0	359	0.72
7	2.0	39.2	414	0.784
8	1.8	47.8	469	0.956
9	1.6	48.4	524	0.96

TABLE 16.19

Tensile Strength of 100 Twisted Hairs

S. No.	Load (kg)	Elongation (mm)	Stress (kg/cm²)	Strain
1	0.2	0.6	28	0.012
2	0.6	1.2	83	0.027
3	1.0	2.3	138	0.046
4	1.6	4.1	221	0.082
5	2.0	6.0	276	0.120
6	2.6	10.4	359	0.204
7	3.0	15.6	414	0.312
8	3.4	22.0	469	0.440
9	3.8	30.0	524	0.600

TABLE 16.20

Tensile Strength of 100 Hairs Dipped in Mustard Oil

S. No.	Load (kg)	Elongation (mm)	Stress (kg/cm²)	Strain
1	0.2	0.2	35	0.004
2	0.6	1.2	106	0.024
3	1.0	2.4	176	0.048
4	1.6	4.0	282	0.080
5	2.0	6.7	352	0.134
6	2.6	18.0	458	0.360
7	3.0	24.1	529	0.482
8	3.2	28.0	564	0.560
9	3.0	35.5	529	0.700

Tensile test: This test has been performed on a rectangular specimen using a Hounsfield tensometer. The total length of the specimen $= 150$ mm, width $= 16$ mm, thickness $= 5$ mm, cross-sectional area $A = 0.8$ cm² and gauge length $= 4\sqrt{A} = 35.8$ mm. When subjected to a load of 125 kg, an elongation of 45 mm is recorded by a dial gauge of 0.01 mm least count. The lateral strain is measured by a digital strain indicator. The results obtained are stress $= 156.26$ kg/cm², linear strain $= 0.125$ and lateral strain $= 4710 \times 10^{-6}$. The stress–strain curve is shown in Figure 16.8. It is a linear curve.

Compression test: This test has been performed on a universal testing machine using a load scale of 0–10 t. The specimen length and width are 20 mm $\times$ 20 mm and the depth (or thickness) $= 25$ mm. The specimen sustained a maximum load of 9 t (90 kN) and the compression recorded is 3.8 mm. Thus, the ultimate compressive strength $\sigma_{ult} = 2250$ kg/cm² and the strain $= 0.152$. The stress–strain curve is shown in Figure 16.9. It is nonlinear in nature.

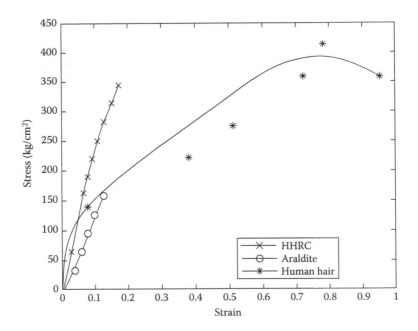

FIGURE 16.8
Tensile stress–strain curve of dry human hair, epoxy and U/D HHRC.

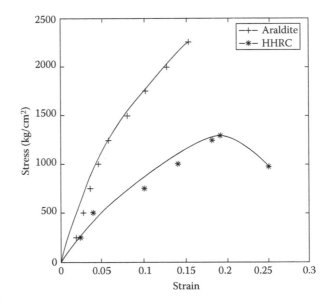

FIGURE 16.9
Compressive stress–strain curve of epoxy and chopped hair–reinforced composite.

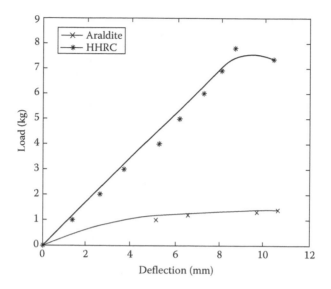

FIGURE 16.10
Load–deflection curve of epoxy and HHRC in bending.

Bending test: It has been performed on a specimen of 70 mm span, which is simply supported and subjected to a concentrated load at its midspan. The cross section of the beam is 20 mm × 4 mm, and the maximum deflection is recorded as 9.7 mm under a load of 1.3 kg (13 N). The load–deflection characteristic is shown in Figure 16.10.

Impact test: The Charpy and Izod tests have been performed on specimens of width = 10 mm and thickness = 10 mm, having depth of V-notch = 2 mm. The length of the specimen in the Charpy test is 50 mm, while in the Izod test, it is 75 mm. The impact strength in both cases is found to be 1 kg (10 N m = 10 J).

16.3.3 Testing of Human Hair–Reinforced Composite

Two kinds of HHRC sheets (U/D and chopped) have been prepared by laying successive layers of Araldite and human hair. Wax has been used on inner surfaces of the mould for a better surface finish and easy withdrawal of the composite. A metal roller has been rolled over each layer to remove entrapped air, if any. The human hairs used are 175 mm in length. The density of the composite has been determined by measuring its weight and volume and is found to be 1800 kg/m³.

Tensile test: The tensile test has been performed on the specimens made of U/D HHRC having volume fraction of hair $V_h = 36\%$ in accordance with Indian Standards. The rectangular specimen is

TABLE 16.21

Observations of Tensile Test of HHRC

S. No.	Load (kg)	Stress (kg/cm²)	Elongation (mm)	Linear Strain	Lateral Strain (10^{-6})
1	0	0.0	0.0	0.0	0
2	25	31.25	0.5	0.014	280
3	50	62.50	1.0	0.028	560
4	75	93.75	1.5	0.042	850
5	100	125.00	2.0	0.056	1140
6	125	156.25	2.5	0.069	1390
7	150	187.50	2.8	0.078	1560
8	175	218.75	3.0	0.093	1650
9	200	281.25	3.8	0.106	2530
10	225	281.25	4.5	0.126	3060
11	250	312.50	5.5	1.153	3455
12	275	343.75	6.1	0.170	3855

of 16 mm × 5 mm cross section, overall length = 150 mm and gauge length = 35.8 mm. The observations recorded are shown in Table 16.21. The specimen sustained a maximum load of 275 kg (2750 N) and underwent an elongation of 6.1 mm. The observations are taken at load interval of 25 kg. Thus, the ultimate strength is calculated as 343.75 kg/cm², linear strain = 0.17 and lateral strain = 3855×10^{-6}. The stress–strain curve is shown in Figure 16.8. The curve is almost linear.

Compression test: This test has been performed on the specimen made of chopped hair. The volume fraction of hair is $V_h \approx 27\%$. The test has been conducted on UTM. The specimen used is of 20 mm × 20 mm cross section and 25 mm height. The specimen sustained a maximum load of 5.2 t before fracture, when the contraction recorded is 6.2 mm. Thus, the ultimate strength is found to be 1300 kg/cm² and strain = 0.18. The stress–strain diagram is shown in Figure 16.9.

Bending test: This test has been performed on a U/D specimen of $V_h = 29\%$. The specimen is of 70 mm span and 20 mm × 4 mm cross section, simply supported and subjected to a concentrated midspan load. Maximum deflection is recorded as 8.6 mm under a load of 7.8 kg (78 N). The load–deflection curve is shown in Figure 16.10.

Impact test: This test has been performed on a U/D specimen having $V_h \approx 36\%$. The Charpy and Izod tests have been performed on specimens of width = 10 mm and thickness = 10 mm, having depth of V-notch = 2 mm. The length of the specimen in the Charpy test is 50 mm, while in the Izod test, it is 75 mm. The impact strength in both cases is found to be 1 kg.

16.3.4 Discussions

From these observations, we notice the following features. *For hair*: since the hairs show a nonlinear stress–strain relation, its (1) yield strength is determined by 0.2% offset method. It is $\sigma_y = 350$ kg/cm^2 (34.33 MPa), (2) secant modulus at 200 kg/cm^2 is found as $E_{sec} = 2173.9$ kg/cm^2 (213.26 MPa) and (3) tangent modulus at 200 kg/cm^2 is $E_{tan} = 892.75$ kg/cm^2 (87.5 MPa). *For epoxy*: (1) its strength in compression (220.72 MPa) is about 14.4 times more than its strength in tension (15.32 MPa). *For HHRC*: (1) as the compressive strength is about 1000 kg/cm^2 (98.1 MPa), this is adequate for applications capable of sustaining light to medium compressive loads.

16.3.5 Conclusions

Based on these findings, we conclude that the HHRC is suitable for low-pressure applications such as pipes carrying sewage and industrial waste and low-strength applications such as tables, boards and supporting pads. With a higher volume fraction of hair, better will be to achieve, higher strength in composites.

16.4 Development and Characterization of a Novel Biohybrid Composite Based on Potato Starch, Jackfruit Latex and Jute Fibre [4]

Shifting from synthetic fibres to natural fibres provides only half-green composites. A green composite will be achieved if the matrix component is also eco-friendly. Keeping this in view, a detailed literature surveyed has been carried out through various issues of journals related to this field. In these literatures, attempts have been made to develop biocomposites and biohybrid composites reinforced with natural fibres. Hence, the starch- and latex-based biohybrid composites have been developed in this investigation. The material systems used are potato starch, jackfruit latex and jute fibres. Some epoxy has also been added for stabilization of and drying of the biocomposite system. The fabrications of these biohybrid composites have been done with $W_{fibres} \approx 25\%$ and $V_m \approx 75\%$. The dual matrix is mixed in different weight fraction compositions such as 75% epoxy and 25% jute fibre composite (i.e. $E_{75}J_{25}$); 15% jackfruit latex, 60% epoxy and 25% jute fibre biohybrid composite (i.e. $Jl_{15}E_{60}J_{25}$); 30% jackfruit latex, 45% epoxy and 25% jute fibre biohybrid composite(i.e. $Jl_{30}E_{45}J_{25}$); 15% potato starch, 15% jackfruit latex, 45% epoxy and 25% jute fibre biohybrid composite (i.e. $Jl_{15}S_{15}E_{45}J_{25}$) and 30% potato starch, 45% epoxy and 25% jute fibre biohybrid composite (i.e. $S_{30}E_{45}J_{25}$). Their static behaviour under tension, compression, bending and impact loadings has been investigated by conducting experiments using strain gauge rosette technology. The stress–strain curves and load–deflection characteristics are obtained. The tensile, compressive, flexure and impact strengths and specific strength have been calculated. Water absorption behaviour has also been done for every composition. Various strengths and specific strengths of hybrid composites have been found for different compositions. Based on the findings of the experimental results, it is concluded that the selection of the best proportion of composites in biohybrid composite plays a vital role in achieving favourable mechanical properties.

16.4.1 Introduction

The present-day technology is hard pressed with pollution problems and demands environmental-friendly developments. Whereas the importance of the applications of composites and hybrid composites is well known, thrust on the use of natural fibres in them for reinforcement has been given priority for some times. But shifting from synthetic fibres to natural fibres provides only half-green composites. A green composite will be achieved if the matrix component is also eco-friendly. The present instigation has been undertaken with this aim. The natural fibres are of jute (Figure 16.11a), latex is of jackfruit (Figure 16.12a) and starch is of potato (Figure 16.13a). Jute is obtained from jute fibre plant (Figure 16.11b). Its botanical name is *Corchorus*. Fibres come from the stem and ribbon (outer skin) of the jute plant. The jute fibre is strong and wiry, longer and finer in nature. It lies in the category of bast (or soft) fibres.

The botanical name of jackfruit is *Artocarpus heterophyllus*. Jackfruit latex serves as birdlime. The heated latex is employed as household cement for mending chinaware and earthenware and to caulk boats and holes in buckets. It contains 82.6%–86.4% resins, which may have value in varnishes. Its bacteriolytic activity is equal to that of papaya latex. The gum of the jackfruit is obtained during cutting of the fruits from the plants (Figure 16.12b) and also at the time when fruits are cut for use.

The potato is a starchy, tuberous crop of the Solanaceae family. Potatoes are the world's fourth-largest food crop, following rice, wheat and maize. In terms of nutrition, the potato is best known for its carbohydrate content (approximately 26 g in a medium potato). Potato starch has been used in powder form (Figure 16.13b).

All the aforementioned compositions of composites/hybrids have been prepared using hand lay-up technique. For this purpose, an open type mould made of mild steel plate has been used. A Mylar sheet is placed on the lower part of the mould to obtain a good surface finish and easy withdrawal of the composite from the mould. In addition to it, the wax is also used to cover the surface of the mould for easy withdrawal of the composite. The jackfruit latex and epoxy, potato starch and epoxy, and jackfruit latex, potato starch and epoxy are mixed uniformly for different types of biohybrid composite

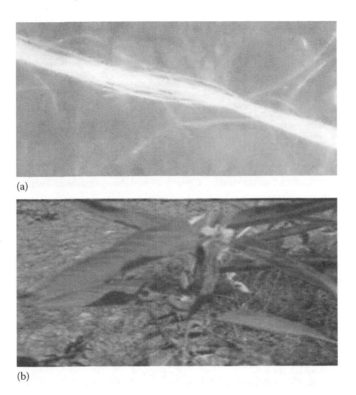

(a)

(b)

FIGURE 16.11
(a) Natural jute fibres. (b) Jute fibre plant. (From Gupta, K.M. and Mishra, N.K., Development and characterisation of natural fibre (palmyra fibre) reinforced polymeric composite, MTech thesis, Department of Applied Mechanics, Motilal Nehru National Institute of Technology, Allahabad, India, 2003.)

matrix. Then the matrix has been layered on the mould in 1.0–1.5 mm thickness, and the jute fibres are placed U/D on it. The weights are hung on both sides to maintain tension in the fibres. The entrapped air is removed with the help of metal roller rolled on the layer.

16.4.2 Testing of Specimens: Tensile Test, Compression Test, Bending Test, Impact Test, Water Absorption Test and Shore Hardness

Specimens of different compositions have been tested under tension, compression, bending and impact to characterize their properties. The testings have been done in accordance with Indian Standards. The tensile test has been performed on a Hounsfield tensometer (Figure 16.14), the compression test on a universal testing machine, the bending test on a Hounsfield tensometer using the bending attachment and the impact test on the Izod impact testing machine. Their details are given in the subsequent sections.

(a)

(b)

FIGURE 16.12

(a) Gummy latex flows from the jackfruit stalk. (b) Jackfruit tree when the slightly underripe fruit is harvested. (From Gupta, K.M. and Srivastava, A.K., Development and characterisation of palm and glass fibre hybrid composite, MTech thesis, Department of Applied Mechanics, Motilal Nehru National Institute of Technology, Allahabad, India, 2004.)

Tensile test: The tensile test is performed on a Hounsfield tensometer using the scale range of 0–1000 kgf (0–9810 N). The specimen of size 150 mm length × 30 mm width × 3 mm thickness has been used. The stress–strain behaviour of different material systems is shown in Figure 16.15.

Compression test: Compression tests have been performed on the specimens of size 25 mm ×25 mm × 30 mm. The load scale chosen is of 0–2000 kg (0–19.62 kN). To record strains, a three-element rectangular strain gauge rosette is mounted on the specimen. Strains are measured with the help of a strain indicator. The compression test results are shown in Figure 16.16 on a stress–strain plot.

Bending test: Bending test has been performed on the specimen of size 75 mm × 25 mm × 3 mm. The test piece is placed in a simply supported position and load is applied at the centre. Strain is measured

(a)

(b)

FIGURE 16.13
(a) Potato. (b) Potato starch.

FIGURE 16.14
The experimental setup showing the Hounsfield tensometer, tensile test specimen and strain gauge indicator.

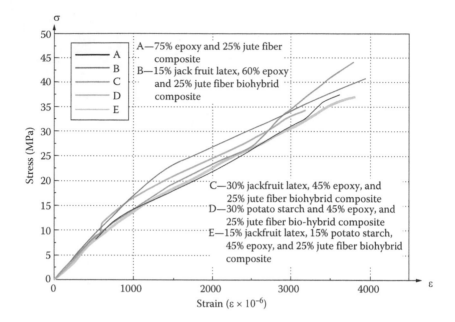

FIGURE 16.15
Tensile stress–strain curves for composite and biohybrids during tension test.

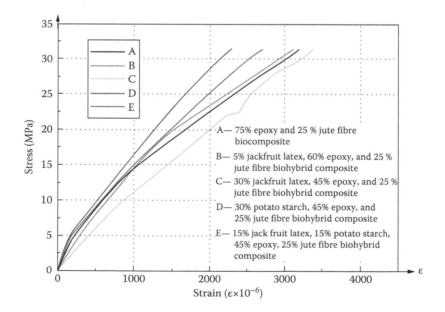

FIGURE 16.16
Comparative stress–strain curves for composites and biohybrid composites during compression test.

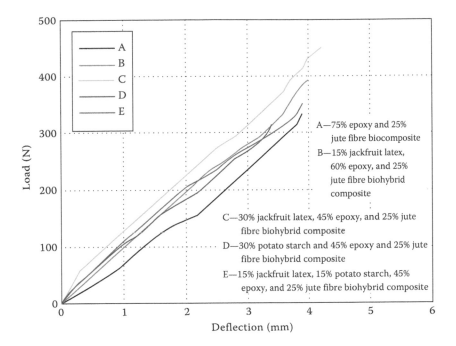

FIGURE 16.17
Comparative load–deflection curves for composites and biohybrid composites during bending test.

by digital strain indicator and then deflection is measured by dial gauge. Least count of the dial gauge is 0.01 mm. To record strains, a three-element rectangular strain gauge rosette is mounted on the specimen. The results are shown in Figure 16.17 on a load–deflection plot.

Impact test: The impact test has been performed to assess the shock-absorbing capacity of different material systems. Two types of impact tests, namely, the Charpy test and Izod test have been conducted. In the Charpy test, the specimen is placed as a simply supported beam, while in the Izod test as a cantilever beam. Results obtained by testings are presented in Figure 16.18 with the help of a bar chart.

Shore hardness: The hardness of the composites has been measured by using a durometer. We apply the durometer on composite, noting down the hardness from display of the durometer. Comparative bar charts of shore *D* hardness of various composite and biohybrid composites are shown in Figure 16.19.

Water absorption test: Water absorption test is done for different composites. First, each specimen is weighed and dipped in water and then,

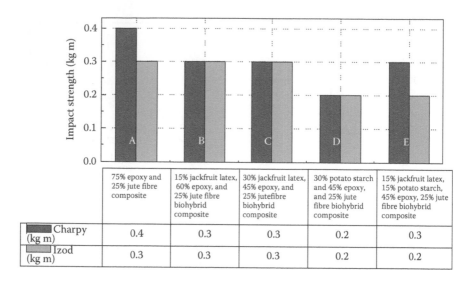

	75% epoxy and 25% jute fibre composite	15% jackfruit latex, 60% epoxy, and 25% jute fibre biohybrid composite	30% jackfruit latex, 45% epoxy, and 25% jutefibre biohybrid composite	30% potato starch and 45% epoxy, and 25% jute fibre biohybrid composite	15% jackfruit latex, 15% potato starch, 45% epoxy, 25% jute fibre biohybrid composite
Charpy (kg m)	0.4	0.3	0.3	0.2	0.3
Izod (kg m)	0.3	0.3	0.3	0.2	0.2

FIGURE 16.18
Comparative impact strength (Izod and Charpy) for composites and biohybrid composites during impact test.

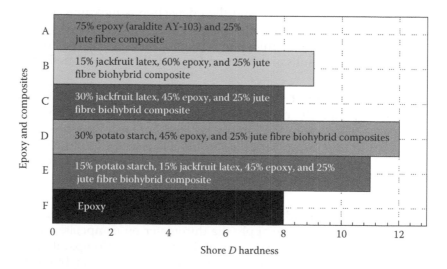

FIGURE 16.19
Comparison of shore *D* hardness of various composites.

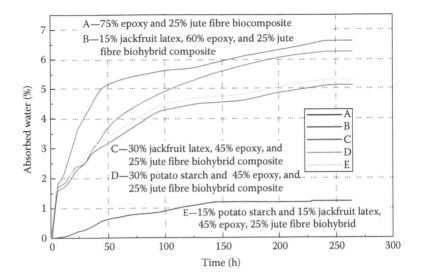

FIGURE 16.20
Comparative water absorption curves for composites and biohybrid composites.

after fixed interval, the weight is taken. The weight is taken until the weight of the specimen stops increasing. The weight of the specimen is measured using an electronic balance. Comparative water absorption curves for composites and biohybrid composites are given in Figure 16.20.

16.4.3 Summary of Results

From these experimental observations, the following results given in Table 16.22 are obtained for different material systems.

16.4.4 Conclusions

It is concluded that out of the five types of composites and biohybrid composites developed in this investigation, the following outcome are observed:

1. $Jl_{30}E_{45}J_{25}$ (30% jackfruit latex, 45% epoxy and 25% jute fibre) biohybrid composite is best for tensile application.

2. $S_{30}E_{45}J_{25}$ (potato starch, 45% epoxy and 25% jute fibre) biohybrid composite is best for compression application.

3. $Jl_{15}E_{60}J_{25}$ (30% jackfruit latex, 45% epoxy and 25% jute fibre) biohybrid composite is best for bending application.

TABLE 16.22

Mechanical Properties of Different Composites

Types of Composites → Properties ↓	75% Epoxy and 25% Jute Fibre Composite ($E_{75}J_{25}$)	15% Jackfruit Latex, 60% Epoxy and 25% Jute Fibre Biohybrid Composite ($Jl_{15}E_{60}J_{25}$)	30% Jackfruit Latex, 45% Epoxy and 25% Jute Fibre Biohybrid Composite ($Jl_{30}E_{45}J_{25}$)	15% Jackfruit Latex, 15% Potato Starch, 45% Epoxy and 25% Jute Fibre Biohybrid Composite ($Jl_{15}S_{15}E_{45}J_{25}$)	30% Potato Starch, 45% Epoxy and 25% Jute Fibre Biohybrid Composite ($S_{30}E_{45}J_{25}$)
Specific density	1.1520	1.1140	1.0758	1.0359	0.999
Tensile strength (MPa)	37.57	40.83	44.10	36.75	34.30
Specific tensile strength (MPa/kg m^{-3})	32.61	36.65	40.99	35.48	34.33
Tensile modulus (GPa)	17.84	16.83	18.11	17.36	15.81
Specific tensile modulus (MPa/kg m^{-3})	15.49	15.10	16.83	16.75	15.83
Tensile strain at fracture (%)	0.362	0.395	0.378	0.319	0.378
Poisson's ratio	0.316	0.358	0.333	0.362	0.322
Compressive strength (MPa)	47	45	45	54	58
Specific compressive strength (MPa/kg m^{-3})	40.80	36.65	41.83	52.13	58.06
Bending strength (MPa)	88.85	104.53	120.21	94.08	83.37
Specific bending strength (MPa)	77.13	93.83	111.74	90.82	83.45
Shore D hardness	7	9	8	11	12
Impact (kg m) Izod	0.3	0.3	0.2	0.2	0.2
Charpy	0.4	0.3	0.2	0.3	0.3
Water absorption (wt.%)	1.210	5.090	6.238	5.291	6.615

4. $S_{30}E_{45}J_{25}$ (30% potato starch, 45% epoxy and 25% jute fibre) biohybrid composite is best for hardness application.

5. $Jl_{30}E_{45}J_{25}$ (30% jackfruit latex, 45% epoxy and 25% jute fibre) biohybrid composite is best for impact application.

6. $E_{75}J_{25}$ (75% epoxy and 25% jute fibre) composite is best for low-absorption application; however, keeping in view the higher specific strength and higher specific modulus of $Jl_{30}E_{45}J_{25}$ (30% jackfruit latex, 45% epoxy and 25% jute fibre) biohybrid composite, the latter may also be opted for such applications.

7. Hence, it is concluded that the biocomposites developed in this work are likely to get applications in the packaging industry such as packaging of food products, packaging of fruits and grains, packaging of agricultural feed stock and packaging of electronics items etc.

16.5 Experimental Investigation of the Behaviour of Dual Fibre Hybrid Composites under Different Stacking Sequences [5]

Summary: Composites made of natural fibres are cheaper, eco-friendly and lighter in weight, but their strength is lower than the composites made of synthetic fibres. Hence, for optimum utilization of lightweightness and strength, a polymeric hybrid composite having two different kinds of fibres has been developed in this investigation. The material system consists of natural flax fibre, synthetic glass fibre and epoxy polymer. Consequently, the flax fibre–reinforced composite (FFRC) lamina and GRP lamina have been fabricated. The hybridization of these laminae has been done in two different stacking sequences, namely, GRP lamina sandwiched between FFRC laminae and FFRC lamina sandwiched between GRP laminae. The two kinds of hybrids thus produced are named as glass–flax–glass hybrid (GFGH) and flax–glass–flax hybrid (FGFH).Their static behaviour under tension, compression, bending and impact loadings has been investigated by conducting experiments using strain gauge rosette technology. The stress–strain curves and load–deflection characteristics are obtained. The tensile, compressive, flexure and impact strengths and specific strength have been calculated. Various strengths and specific strength of hybrid composites have considerably increased due to hybridization and stacking sequences. Based on the findings of the experimental results, it is concluded that the selection of stacking sequence of laminae in a hybrid composite plays a vital role in achieving favourable mechanical properties.

16.5.1 Introduction

Composites made of conventional fibres is of high cost. If made of cheaper fibres, they will cut down the cost of components for which they are used. Natural fibres are such materials. They are not only cheap but are abundantly available also. They possess high specific strength and low specific weight and are ecologically favourable too. However, their strength is not as high as those of synthetic fibres. Hence, if natural fibre composites are made hybrid with synthetic fibre composites, the resulting hybrid composite is likely to be a blend of lightweight and strong material. The present work has been undertaken with this aim. One fibre is natural and the other is synthetic. The natural fibre is of flax (Figure 16.21), while the synthetic fibre is of glass (Figure 16.22).

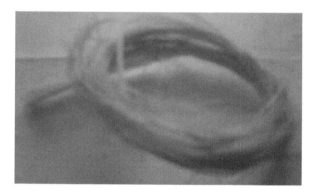

FIGURE 16.21
Natural flax fibre.

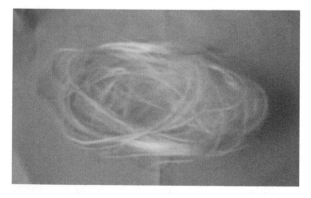

FIGURE 16.22
Synthetic glass fibre.

16.5.2　Fabrication of Composites and Hybrid

In this work, a U/D FFRC and a U/D GRP have been fabricated using hand lay-up process. Epoxy (Araldite AY-103) has been used as matrix constituent. The hybrid is made up of two different composite laminae, namely, flax polymer composite and glass fibre composite. They have been stacked in two different sequences as shown in Figures 16.23 and 16.24. Laminae of both of these composites are glued alternately to produce GFGH and FGFH. A U/D FFRC sheet has been prepared by laying successive layers of Araldite and flax fibre. Mylar sheet has been used on inner surfaces of the mould for a better surface finish and easy withdrawal of the composite. Heavy weights are hung on both sides of the fibres to maintain unidirectionality in them. A metal roller has been rolled over each layer to remove entrapped air, if any.

16.5.3　Testing of Composites and Hybrids

Specimens of FFRC, GRP, GFGH and FGFH composites have been prepared and tested under tension, compression, bending and impact to characterize their properties. The testings have been done in accordance with Indian Standards. The tensile test has been performed on a Hounsfield tensometer similar to as shown in Figure 16.14. The compression test was done on a universal testing machine, the bending test on a Hounsfield tensometer using the bending attachment and the impact test on the Izod impact testing machine. Their details follow in the subsequent sections.

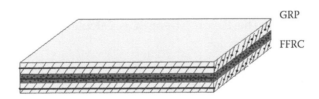

FIGURE 16.23
GFGH epoxy plastic.

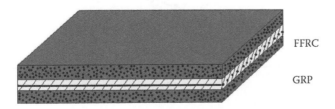

FIGURE 16.24
FGFH epoxy plastic.

16.5.4 Testing of Flax Fibre–Reinforced Composite: Tensile Test, Compression Test, Bending Test and Impact Test

The flax fibres used are 300 mm in length. The density of the composite has been determined by measuring its weight and volume and is found to be 1648 kg/m³.

Tensile test: The tensile test has been performed on the specimens having volume fraction of fibres $V_{fl} \approx 54\%$ in accordance with Indian Standards. The rectangular specimen is of 16 mm × 3 mm cross section, overall length = 150 mm and gauge length = 28 mm. The specimen sustained a maximum load of 150 kg (1500 N) and underwent an elongation of 0.14 mm. The observations are taken at a load interval of 10 kg. Thus, the ultimate strength is calculated as 312.5 kg/cm² (31.25 MPa), linear strain = 5130 × 10⁻⁶ and lateral strain = 4190 × 10⁻⁶. The stress–strain curve is shown in Figure 16.25.

Compression test: This test has been performed on the specimen having flax volume fraction of $V_{fl} \approx 54\%$. The test has been conducted on UTM. The specimen used is of 25 mm × 25 mm cross section and 30 mm height. The specimen sustained a maximum load of 1.95 t before fracture, where the contraction recorded is 0.18 mm. Thus, the ultimate strength is found to be 320 kg/cm² (32 MPa) and linear strain = 6050 × 10⁻⁶. The stress–strain diagram is shown in Figure 16.26.

Bending test: This test has been performed on a U/D specimen of $V_{fl} \approx 54\%$. The specimen is of 70 mm span and 20 mm × 3 mm cross section, simply supported and subjected to a concentrated midspan load. The flexure strength is calculated to be 40.8 MPa.

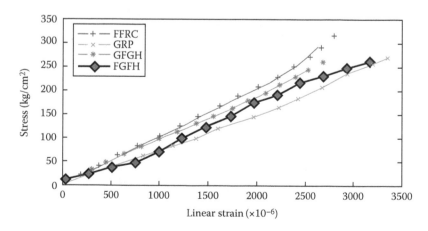

FIGURE 16.25
Tensile stress–strain curves of FFRC, GRP, GFGH and FGFH.

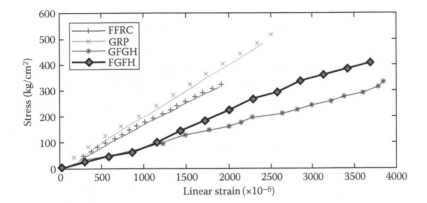

FIGURE 16.26
Compressive stress–strain curves of FFRC, GRP, GFGH and FGFH.

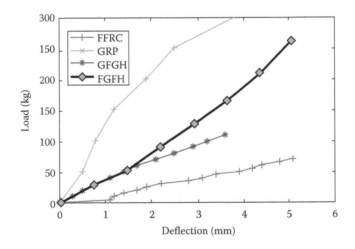

FIGURE 16.27
Load–deflection curves of FFRC, GRP, GFGH and FGFH in bending.

Maximum deflection is recorded as 5.1 mm under a load of 70 kg (700 N). The load–deflection curve is shown in Figure 16.27.

Impact test: This test has been performed on a U/D specimen having $V_{fl} \approx 54\%$. The Charpy and Izod tests have been performed on specimens of width = 10 mm and thickness = 10 mm, having depth of V-notch = 2 mm. The length of the specimen in the Charpy test is 50 mm, while in the Izod test, it is 75 mm. The impact strength in the Charpy test is found to be 0.9 kg (9 N m), while in the Izod test, it is 0.4 kg (4 N m).

16.5.5 Testing of Glass Fibre–Reinforced Composite: Tensile Test, Compression Test, Bending Test and Impact Test

A U/D GRP sheet has been prepared by laying successive layers of Araldite and glass fibres. Glass fibres used are 300 mm in length. The density of the composite has been determined by measuring its weight and volume and is found to be 1898 kg/m^3.

Tensile test: The tensile test has been performed on the specimens having volume fraction of glass fibres as $V_g \approx 59\%$ in accordance with Indian Standards. The rectangular specimen is of 16 mm×3 mm cross section, overall length=150 mm and gauge length=40 mm. The specimen sustained a maximum load of 150 kg (1500 N) and underwent an elongation of 1.0 mm. Thus, the ultimate strength is calculated as 333.3 kg/cm^2 (33.3 MPa), linear strain=6750×10^{-6} and lateral strain=4780×10^{-6}. The stress–strain curve is shown in Figure 16.25.

Compression test: This test has been performed on the specimen having glass fibre volume fraction of $V_g \approx 59\%$. The specimen used is of 25 mm×25 mm cross section and 30 mm height. The specimen sustained a maximum load of 1.95 t before fracture, where the contraction recorded is 0.17 mm. Thus, the ultimate strength is found to be 312.3 kg/cm^2 (31.2 MPa) and linear strain=5710×10^{-6}. The stress–strain diagram is shown in Figure 16.26.

Bending test: This test has been performed on a U/D specimen of $V_g \approx 59\%$. The specimen is of 70 mm span and 20 mm×3 mm cross section, simply supported and subjected to a concentrated midspan load. The flexure strength is found to be 48.2 MPa. Maximum deflection is recorded as 3.6 mm under a load of 110 kg (1100 N). The load–deflection curve is shown in Figure 16.27.

Impact test: This test has been performed on a U/D specimen having $V_g \approx 59\%$. The Charpy and Izod tests have been performed on specimens of width=10 mm and thickness=10 mm, having depth of V-notch=2 mm. The length of the specimen in the Charpy test is 50 mm, while in the Izod test, it is 75 mm. The impact strength in the Charpy test is found to be 12 kg (120 N m), while in the Izod test, it is 5.8 kg (58 N m).

16.5.6 Testing of Hybrid having Epoxy Reinforced with Flax–Glass–Flax Fibre: Tensile Test, Compression Test, Bending Test and Impact Test

A U/D FGFH sheet has been prepared by laying successive layers of FFRC and GRP sheets glued with Araldite. Mylar sheet has been used on the inner surfaces of the mould for easy withdrawal of hybrid. Epoxy resin has been sprayed on the bottom side of the mould and then glass fibres are placed unidirectionally. A metal roller has been rolled over each layer to remove entrapped air, if any, between the layers. The hybrid produced has three plies of 14 mm thick, consisting of a ply of GRP sandwiched between two plies of FFRP. The density of the hybrid composite has been determined by measuring its weight and volume and is found to be 1880 kg/m^3.

Tensile test: The tensile test has been performed on the specimens in accordance with Indian Standards. The rectangular specimen is of 14 mm × 24.5 mm cross section, overall length = 270 mm and gauge length = 168 mm. The specimen sustained a maximum load of 640 kg (6400 N) and underwent an elongation of 0.68 mm. The observations are taken at load interval of 40 kg. Thus, the ultimate strength is calculated as 186.5 kg/cm^2 (18.6 MPa), linear strain = 3780 × 10^{-6} and lateral strain = 3580 × 10^{-6}. The stress–strain curve is shown in Figure 16.25.

Compression test: This test has been performed on the U/D hybrid specimen on UTM. The specimen used is of 25 mm × 25 mm cross section and 30 mm height. The specimen sustained a maximum load of 3.2 t before fracture, where the contraction recorded is 0.14 mm. Thus, the ultimate strength is found to be 512 kg/cm^2 (51.2 MPa) and strain = 4700 × 10^{-6}. The stress–strain diagram is shown in Figure 16.26.

Bending test: This test has been performed on a U/D specimen of 270 mm span and 20 mm × 14 mm cross section, simply supported and subjected to a concentrated midspan load. The flexure strength is found to be 309.9 MPa. Maximum deflection is recorded as 3.8 mm under a load of 300.0 kg (3000 N). The load–deflection curve is shown in Figure 16.27.

Impact test: The Charpy and Izod tests have been performed on U/D specimens of width = 10 mm and thickness = 30 mm, having depth of V-notch = 2 mm. The length of the specimen in the Charpy test is 50 mm, while in the Izod test, it is 75 mm. The impact strength in the Charpy test is found to be 48 kg (480 N m), while in the Izod test, it is 16 kg (160 N m).

16.5.7 Testing of Hybrid having Epoxy Reinforced with Glass–Flax–Glass Fibre: Tensile Test, Compression Test, Bending Test and Impact Test

A U/D GFGH sheet has been prepared by laying successive layers of FFRC and GRP sheets glued with Araldite. Mylar sheet has been used on the inner surfaces of the mould for easy withdrawal of hybrid. Epoxy resin has been sprayed on the bottom side of the mould and then glass fibres are placed unidirectionally. A metal roller has been rolled over each layer to remove entrapped air, if any, between the layers. The hybrid produced has three plies of 14 mm thick, consisting of a ply of FFRC sandwiched between two plies of GRP. The density of the hybrid composite has been determined by measuring its weight and volume and is found to be 1940 kg/m^3.

> *Tensile test*: The tensile test has been performed on the specimens in accordance with Indian Standards. The rectangular specimen is of 14 mm $\times$ 24.5 mm cross section, overall length = 270 mm and gauge length = 168 mm. The specimen sustained a maximum load of 560 kg (5600 N) and underwent an elongation of 0.75 mm. The observations are taken at load interval of 40 kg. Thus, the ultimate strength is calculated as 228.5 kg/cm^2 (22.8 MPa), linear strain = 3550 $\times$ 10^{-6} and lateral strain = 3710 $\times$ 10^{-6}. The stress–strain curve is shown in Figure 16.25.

> *Compression test*: This test has been performed on the U/D hybrid specimen on UTM. The specimen used is of 25 mm $\times$ 25 mm cross section and 30 mm height. The specimen sustained a maximum load of 3.5 t before fracture, where the contraction recorded is 0.16 mm. Thus, the ultimate strength is found to be 560 kg/cm^2 (56.0 MPa) and strain = 3100 $\times$ 10^{-6}. The stress–strain diagram is shown in Figure 16.26.

> *Bending test*: This test has been performed on a U/D specimen of 270 mm span and 20 mm $\times$ 14 mm cross section, simply supported and subjected to a concentrated midspan load. The flexure strength is found to be 361.6 MPa. Maximum deflection is recorded as 3.5 mm under a load of 350 kg (3500 N). The load–deflection curve is shown in Figure 16.27.

> *Impact test*: The Charpy and Izod tests have been performed on U/D specimens of width = 10 mm and thickness = 30 mm, having depth of V-notch = 2 mm. The length of the specimen in the Charpy test is 50 mm, while in the Izod test, it is 75 mm. The impact strength in the Charpy test is found to be 57 kg (570 N m), while in the Izod test, it is 21 kg (210 N m).

16.5.8 Summary of Results and Conclusions

From the aforementioned experimental observations, the following results given in Table 16.23 are obtained for different material systems.

It is clearly evident that the effect of hybridization is to enhance the compressive strength, flexural strength and impact strength as compared to non-hybrid composites. However, the tensile strength lowers down. Further, the effect of different stacking sequence of dual fibre composite is to improve all the properties as compared to nonhybrid composites. It is also concluded that the stacking configuration of GFGH is better than the stacking configuration of FGFH in all respects. It results in the following improvements:

TABLE 16.23

Mechanical Properties of FFRC, GRP, FGFH and GFGH

Types of Composites/Hybrids → Property ↓		FFRC	GRP	FGFH	GFGH
Density (kg/m³)		1648.0	1898.0	1880.0	1940.0
Tensile strength (MPa)		31.2	33.3	18.6	22.8
Compressive strength (MPa)		32.0	31.2	51.2	56.0
Flexure strength (MPa)		40.8 (with one ply)	48.2 (with one ply)	309.9 (with three ply)	361.6 (with three ply)
Impact strength (N m)	Charpy	8.8	120.0	480.0	570.0
	Izod	3.9	—	160.0	210.0
Specific tensile strength (MPa/kg/m³)		0.0189	0.0175	0.0098	0.011

- Increase in tensile strength = 18.4%.
- Increase in compressive strength = 8.5%.
- Increase in flexural strength = 14.2%.
- Increase in impact strength = 15.7%.
- Increase in specific tensile strength = 10.9%.

Hence, it is concluded that *GFGH* will prove to be a better material for applications such as supports, lightweight machines and equipment and vivid structural elements, for low-pressure applications such as pipes carrying sewage and industrial waste and for low-strength applications such as tables, boards and supporting pads.

16.6 Fabrication and Experimental Investigation of the Behaviour of a Novel Dual Green Fibre Hybrid Composite with Different Compositions of the Material System [6]

16.6.1 Introduction

Composites made of conventional fibres are of high cost. If made of cheaper fibres, they will cut down the cost of components for which they are used. Natural fibres are such materials. They are not only cheap but are abundantly available also. They possess high specific strength and low specific weight and are ecologically favourable too. However, their strength is not as high as those of synthetic fibres. Hence, if natural fibre composites are made hybrid with two different kinds of natural fibres, one that has a low density such as flax and other with greater strength, the resulting dual fibre green hybrid composite is likely to be a blend of lightweight and strong material. The present work has been undertaken with this aim.

The two natural fibres are of flax (Figure 16.28a) and palmyra (Figure 16.29b). Flax is obtained from flax fibre plant (Figure 16.28b). Its botanical name is *Linum usitatissimum*. It is commonly known as *patsan*, which is a substitute of *jute*. The flax fibre is strong and wiry, longer and finer in nature. It lies in the category of bast (or soft) fibres. Palmyra fibre is also known as *B. flabellifer*. It is obtained from a tall and swaying tree (Figure 16.29a). The word 'borassus' is derived from a Greek word that describes the leathery covering of the fruit and *flabellifer* means *fan bearer*.

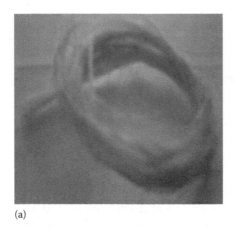
(a)

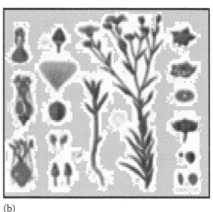
(b)

FIGURE 16.28
(a) Natural flax fibres. (b) Flax fibre plant.

(a)

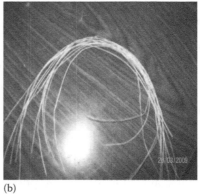
(b)

FIGURE 16.29
(a) Palmyra tree. (b) Natural palmyra fibre.

16.6.2 Fabrication of Specimen

Various specimens having the material systems of natural fibre composites and dual natural fibre hybrids have been fabricated using the volume fraction of dual fibres as $V_{fibres} \approx 54\%$ and volume fraction of polymeric matrix as $V_m \approx 46\%$. The dual fibres are mixed in the following volume fraction compositions. All specimens are U/D, that is the fibre lay-up is U/D:

1. 100% flax fibre polymeric composite (FPC)
2. 70% flax + 30% palmyra fibre polymeric fibre hybrid $(FPPH)_{70-30}$
3. 50% flax + 50% $FPPH_{50-50}$
4. 30% flax + 70% $FPPH_{30-70}$
5. 100% palmyra polymeric fibre composite (PPC)

All these compositions of composites/hybrids have been prepared using hand lay-up technique. For this purpose, an open type mould made of mild steel plate has been used. A Mylar sheet is placed on the lower part of the mould to obtain a good surface finish and easy withdrawal of the composite from the mould. In addition to it, the wax is also used to cover the surface of the mould for easy withdrawal of the composite. The flax and palmyra fibres are mixed uniformly. Then the epoxy resin (Araldite AY-103) has been layered on the mould in 1.0–1.5 mm thickness, and the mixture of flax fibres and palmyra fibres is placed U/D on it. The weights are hung on both sides to maintain tension in the fibres. The entrapped air is removed with the help of a metal roller rolled on the layer.

16.6.3 Testing of Specimens: Tensile Test, Compression Test, Bending Test, Impact Test and Water Absorption Test

Specimens of different compositions have been tested under tension, compression, bending and impact to characterize their properties. The testings have been done in accordance with Indian Standards. The tensile test has been performed on a Hounsfield tensometer similar to what is shown in Figure 16.14. The compression test is done on a universal testing machine, the bending test on a Hounsfield tensometer using the bending attachment, and the impact test on the Izod impact testing machine. Their details are given in the subsequent sections.

Tensile test: The tensile test is performed on a Hounsfield tensometer using the scale range of 0–1000 kgf (0–9810 N). The specimen of size 150 mm length × 30 mm width × 3 mm thickness has been used. The stress–strain behaviours of different material systems are shown in Figure 16.30.

Compression test: Compression tests have been performed on the specimens of size 25 mm × 25 mm × 30 mm. The load scale chosen is of

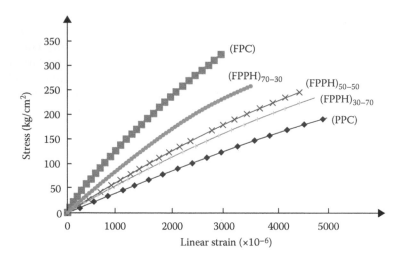

FIGURE 16.30
Tensile stress–strain curves for flax and palmyra fibre composites and hybrids during tension test.

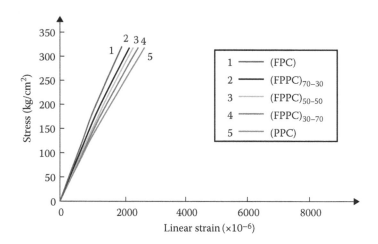

FIGURE 16.31
Comparative stress–strain curves for epoxy, flax and palmyra fibre composites and hybrids during compression test.

0–2000 kg (0–19.62 kN). To record strains, a three-element rectangular strain gauge rosette is mounted on the specimen. Strains are measured with the help of a strain indicator. The compression test results are shown in Figure 16.31 on a stress–strain plot.

Bending test: The bending test has been performed on the specimen of size 75 mm ×25 mm×3 mm. The test piece is placed in a simply

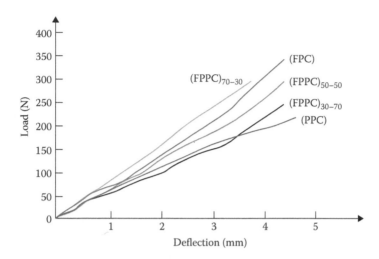

FIGURE 16.32
Comparative load–deflection curves for flax and palmyra fibre composites and hybrids during bending test.

supported position and the load is applied at the centre. Strain is measured by a digital strain indicator and then deflection is measured by dial gauge. Least count of the dial gauge is 0.01 mm. To record strains, a three-element rectangular strain gauge rosette is mounted on the specimen. The results are shown in Figure 16.32 on a load–deflection plot.

Impact test: The impact test has been performed to assess the shock-absorbing capacity of different material systems. Two types of impact tests, namely, the Charpy test and the Izod test, have been conducted. In the Charpy test, the specimen is placed as a simply supported beam, while in the Izod test as a cantilever beam. Results obtained by testings are presented in Figure 16.33 with the help of a bar chart.

Water absorption test: The water absorption test is done for different composites. First, each specimen is weighed and dipped in water and then, after fixed interval, the weight is taken. The weight is taken until the weight of the specimen stops increasing or change in weight is minimum. The weight of the specimen is measured using an electronic balance. Comparative water absorption curves for epoxy, flax, and palmyra fibre composites are given in Figure 16.34.

16.6.4 Summary of Results

From these experimental observations, the following results given in Table 16.24 are obtained for different material systems.

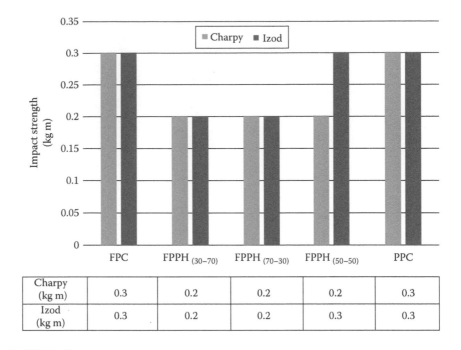

| Charpy (kg m) | 0.3 | 0.2 | 0.2 | 0.2 | 0.3 |
| Izod (kg m) | 0.3 | 0.2 | 0.2 | 0.3 | 0.3 |

FIGURE 16.33
Comparative impact strength (Izod and Charpy) for flax and palmyra fibre composites.

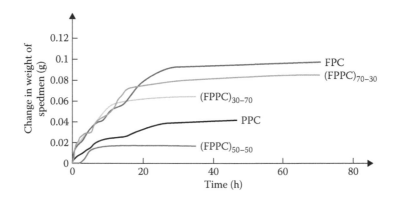

FIGURE 16.34
Comparative water absorption curves for epoxy, flax and palmyra fibre composites.

TABLE 16.24

Experimental Observations for Different Material Systems

Material Systems → Properties ↓	FPC	PPC	30% Flax and 70% Palmyra Composite (FPPC)$_{30-70}$	70% Flax and 30% Palmyra Composite (FPPC)$_{70-30}$	50% Flax and 50% Palmyra Composite (FPPC)$_{50-50}$
Specific density	1.230	0.930	1.064	1.011	1.132
Tensile strength (kg/cm^2)	328.88	205.00	233.30	263.30	251.11
Tensile modulus (kg/cm^2) × 10^4	13.51	4.21	5.68	8.13	6.41
Compressive strength (kg/cm^2)	975.50	528.16	591.02	833.46	713.46
Bending strength (kg/cm^2)	101.92	67.72	75.78	86.24	81.05
Impact energy (kg m) Izod	0.30	0.20	0.20	0.20	0.30
Charpy	0.30	0.20	0.30	0.20	0.30
Water absorption (% by weight)	2.70	1.10	1.76	2.10	2.20
Specific tensile strength (kg/cm^2)	267.38	220.43	219.26	260.43	221.82
Specific tensile modulus (kg/cm^2) × 10^4	10.98	4.53	5.34	8.04	5.66

16.6.5 Conclusion

1. It is clearly evident that the effect of hybridization is to enhance the tensile strength, compressive strength, flexural strength and impact strength as compared to nonhybrid polymeric composites.

2. Tensile strength increases with increase in percentage of flax fibre and decreases with increase in percentage of palmyra fibre. Flax fibre composite has a higher tensile strength. Tensile strength varies from 205 kg/cm^2, which is of palmyra fibre composite, to 328.88 kg/cm^2, which is of flax fibre composite.

3. Compressive strength increases with increase in percentage of flax fibre and decreases with increase in percentage of palmyra fibre.

4. Bending strength increases with increase in percentage of flax fibre and decreases with increase in percentage of palmyra fibre. Flax fibre composite has a higher flexural strength. Flexural strength varies from 67.72 kg/cm^2, which is of palmyra fibre composite, to 101.92 kg/cm^2, which is of flax fibre composite.

5. Impact strength has no noticeable change after addition of fibre. Impact strength varies from 0.1 to 0.3 kg. The specimen with higher flax content has more impact strength.

6. Water absorption is lowest in palmyra fibre composite and is highest for flax fibre composite. It increases with increase in percentage of flax fibre. Water absorption varies from 1.1% for palmyra fibre composite to 2.7% by weight for flax fibre composite.

7. Hence, it is concluded that the dual green fibre hybrid composites will prove to be a better material for applications such as small fishing boats, biogas containers, grain storage silos, lightweight machines and equipment, and vivid structural elements, for low-pressure applications such as pipes carrying sewage and industrial waste, and for low-strength applications such as tables, boards, panels, partitions and false ceilings.

16.7 Development and Characterization of Hybrid Composites Reinforced with Natural (Palmyra) Fibre and Glass Fibre [2]

16.7.1 Introduction

Composites made of conventional fibres (glass, carbon, graphite, boron, Kevlar, etc.) are of high cost. If made of cheaper fibres, they will cut down the cost of components for which they are used. Natural fibres are such materials.

They are not only cheap but are abundantly available also. They possess high specific strength and low specific weight and are ecologically favourable too. However, their strength is not as high as those of synthetic fibres. Hence, if natural fibre composites are made hybrid with synthetic fibre composites, the resulting hybrid composite is likely to be a blend of lightweight and strong material. The present work has been undertaken with this aim.

In this work [2], a U/D palmyra (natural palm) fibre–reinforced composite (PFRC) and a GRP have been fabricated using hand lay-up process. Epoxy (Araldite AY-103) has been used as a matrix constituent. Laminae of both of these composites are glued alternately to produce PGHC. Specimens of PFRC, GRP and PGHC composites have been tested under tension, compression, bending and impact to characterize their properties. The observations are compared, and it is concluded that *PGHC* will prove to be a cheaper alternative material for applications such as supports of industrial piping, for lightweight machines and equipment, and for making vivid structural elements.

16.7.2 Testing of Palmyra Fibre–Reinforced Composite: Tensile Test, Compression Test, Bending Test and Impact Test

A U/D PFRC sheet has been prepared by laying successive layers of Araldite and palmyra fibre. Wax has been used on the inner surfaces of the mould for a better surface finish and easy withdrawal of the composite. A metal roller has been rolled over each layer to remove entrapped air, if any. The palmyra fibres used are 300 mm in length. The density of the composite has been determined by measuring its weight and volume and is found to be 1110 kg/m^3.

> *Tensile test*: The tensile test has been performed on the specimens having volume fraction of fibres $V_p \approx 58\%$ in accordance with Indian Standards. The rectangular specimen is of 16 mm × 3 mm cross section, overall length = 150 mm and gauge length = 40 mm. The specimen sustained a maximum load of 80 kg (800 N) and underwent an elongation of 0.9 mm. The observations are taken at a load interval of 10 kg. Thus, the ultimate strength is calculated as 166.6 kg/cm^2 (16.6 MPa), linear strain = 6220×10^{-6} and lateral strain = 4700×10^{-6}. The stress–strain curve is shown in Figure 16.35.
>
> *Compression test*: This test has been performed on the specimen having palmyra volume fraction of $V_p \approx 58\%$. The test has been conducted on UTM. The specimen used is of 25 mm × 25 mm cross section and 30 mm height. The specimen sustained a maximum load of 1.15 t before fracture, where the contraction recorded is 0.2 mm. Thus, the ultimate strength is found to be 200 kg/cm^2 (MPa) and strain = 6450×10^{-6}. The stress–strain diagram is shown in Figure 16.36.

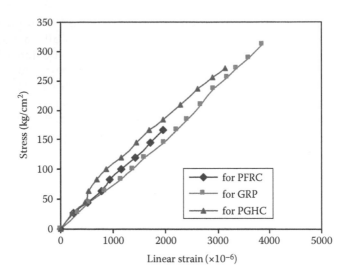

FIGURE 16.35
Tensile stress–strain curves of PFRC, GRP and PGHC.

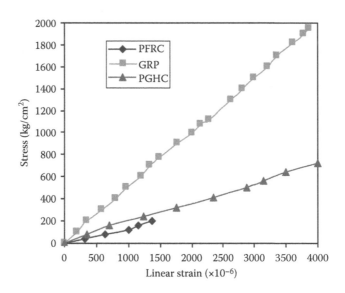

FIGURE 16.36
Compressive stress–strain curves of PFRC, GRP and PGHC.

Bending test: This test has been performed on a U/D specimen of $V_p \approx 58\%$. The specimen is of 70 mm span and 20 mm × 3 mm cross section, simply supported and subjected to a concentrated midspan load. Maximum deflection is recorded as 4.4 mm under a load of 40 kg (400 N). The load–deflection curve is shown in Figure 16.37.

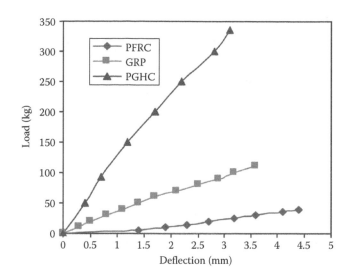

FIGURE 16.37
Load–deflection curves of PFRC, GRP and PGHC in bending.

Impact test: This test has been performed on a U/D specimen having $V_P \approx 58\%$. The Charpy and Izod tests have been performed on
specimens of width = 10 mm and thickness = 10 mm, having depth
of V-notch = 2 mm. The length of the specimen in the Charpy test is
50 mm, while in the Izod test, it is 75 mm. The impact strength in the
Charpy test is found to be 0.8 kg (8 N m), while in the Izod test, it is
0.2 kg (2 N m).

16.7.3 Testing of Glass Fibre–Reinforced Composite: Tensile Test, Compression Test, Bending Test and Impact Test

A U/D GRP sheet has been prepared by laying successive layers of Araldite
and glass fibres. Wax has been used on the inner surfaces of the mould
for better surface finish and easy withdrawal of the composite. A metal
roller has been rolled over each layer to remove entrapped air, if any. The
glass fibres used are 300 mm in length. The density of the composite has
been determined by measuring its weight and volume and is found to be
1898 kg/m³.

Tensile test: The tensile test has been performed on the specimens having volume fraction of glass fibres $V_g \approx 59\%$ in accordance with
Indian Standards. The rectangular specimen is of 16 mm × 3 mm
cross section, overall length = 150 mm and gauge length = 40 mm.
The specimen sustained a maximum load of 150 kg (1500 N)
and underwent an elongation of 1.0 mm. The observations are

TABLE 16.25

Stress–Strain Values for PFRC, GRP and PGHC in Tension

Stress		Linear Strain × 10⁻⁶		
(kg/cm²)	(MPa)	For PFRC	For GRP	For PGHC
0	0	0	0	0
27	2.7	250	330	230
44	4.4	500	540	470
63	6.3	770	840	530
82	8.2	940	1140	680
100	10.0	1140	1380	870
120	12.0	1410	1600	1150
145	14.5	1700	1970	1440
166	16.6	1940	2220	1680
184	18.4	—	2420	1950
208	20.8	—	2670	2280
236	23.6	—	2930	2620
256	25.6	—	3190	2890
270	27.0	—	3360	3150
289	28.9	—	3600	—
312	31.2	—	3870	—
333	33.3	—	—	—

taken at a load interval of 10 kg. Thus, the ultimate strength is calculated as 333.3 kg/cm², linear strain = 6750×10^{-6} and lateral strain = 4780×10^{-6}. The stress–strain curve is shown in Figure 16.35 and the calculated values are shown in Table 16.25 along with those of PFRC and PGHC.

Compression test: This test has been performed on the specimen having glass fibre volume fraction $V_g \approx 59\%$. The test has been conducted on UTM. The specimen used is of 25 mm × 25 mm cross section and 30 mm height. The specimen sustained a maximum load of 1.95 t before fracture, where the contraction recorded is 0.17 mm. Thus, the ultimate strength is found to be 312.3 kg/cm² and strain = 5710×10^{-6}. The calculated stress–strain values are given in Table 16.26; the diagram is shown in Figure 16.36 along with those of PFRC and PGHC.

Bending test: This test has been performed on a U/D specimen of $V_g \approx 59\%$. The specimen is of 70 mm span and 20 mm × 3 mm cross section, simply supported and subjected to a concentrated midspan load. Maximum deflection is recorded as 3.6 mm under a load of 110 kg (1100 N). The observed load–deflection data are given in Table 16.27, and the curve is shown in Figure 16.37 along with those of PGHC and PFRC.

TABLE 16.26

Stress–Strain Values for PFRC, GRP and PGHC in Compression

For PFRC		For GRP		For PGHC	
Stress (kg/cm²)	Linear Strain × 10⁻⁶	Stress (kg/cm²)	Linear Strain × 10⁻⁶	Stress (kg/cm²)	Linear Strain × 10⁻⁶
0	0	0	0	0	0
40	320	16	200	80	350
80	640	32	350	160	710
120	1010	48	590	240	1240
160	1160	64	810	320	1760
200	1370	80	970	415	2350
—	—	96	1210	502	2880
—	—	112	1340	560	3140
—	—	128	1490	640	3500
—	—	144	1770	720	4000
—	—	160	2000	—	—
—	—	176	2140	—	—
—	—	192	2270	—	—
—	—	208	2630	—	—
—	—	224	2820	—	—
—	—	240	2990	—	—
—	—	256	3220	—	—
—	—	272	3370	—	—
—	—	288	3610	—	—
—	—	312	3780	—	—
—	—	330	3860	—	—

Impact test: This test has been performed on a U/D specimen having $V_g \approx 59\%$. The Charpy and Izod tests have been performed on specimens of width = 10 mm and thickness = 10 mm, having depth of V-notch = 2 mm. The length of the specimen in the Charpy test is 50 mm, while in the Izod test, it is 75 mm. The impact strength in the Charpy test is found to be 12 kg, while in the Izod test, it is 5.8 kg.

16.7.4 Testing of Hybrid Having Epoxy Reinforced with Palmyra–Glass Fibre: Tensile Test, Compression Test, Bending Test and Impact Test

A U/D PGHC sheet has been prepared by laying successive layers of PFRC and GRP sheets glued with Araldite. Wax has been used on the inner surfaces of the mould for easy withdrawal of the hybrid. A metal roller has been rolled over each layer to remove entrapped air, if any, between the layers. The density of the hybrid composite has been determined by measuring its weight and volume and is found to be 1230 kg/m³.

TABLE 16.27

Load–Deflection Observations for PFRC, GRP and PGHC in Bending

For PFRC		For GRP		For PGHC	
Load (kg)	Deflection (mm)	Load (kg)	Deflection (mm)	Load (kg)	Deflection (mm)
0	0	0	0	0	0
5	1.4	10	0.3	50	0.4
10	1.9	20	0.5	93	0.7
15	2.3	30	0.8	150	1.2
25	2.7	40	1.1	200	1.7
30	3.2	50	1.4	250	2.2
35	3.6	60	1.7	300	2.8
40	4.1	70	2.1	335	3.1
—	4.4	80	2.5	—	—
—	—	90	2.9	—	—
—	—	100	3.2	—	—
—	—	110	3.6	—	—

Tensile test: The tensile test has been performed on the specimens in accordance with Indian Standards. The rectangular specimen is of 13.5 mm × 30 mm cross section, overall length = 150 mm and gauge length = 40 mm. The specimen sustained a maximum load of 995 kg (9950 N) and underwent an elongation of 0.6 mm. The observations are taken at load interval of 80 kg. Thus, the ultimate strength is calculated as 245.6 kg/cm², linear strain = 3950×10^{-6} and lateral strain = 3020×10^{-6}. The stress–strain values are shown in Table 16.25 and the curve in Figure 16.35.

Compression test: This test has been performed on the U/D hybrid specimen on UTM. The specimen used is of 25 mm × 25 mm cross section and 30 mm height. The specimen sustained a maximum load of 4.2 t before fracture, where the contraction recorded is 0.2 mm. Thus, the ultimate strength is found to be 672 kg/cm² and strain = 6400×10^{-6}. The observed stress–strain values are shown in Table 16.26, and the diagram is shown in Figure 16.36.

Bending test: This test has been performed on a U/D specimen of 70 mm span and 20 mm × 30 mm cross section, simply supported and subjected to a concentrated midspan load. Maximum deflection is recorded as 3.1 mm under a load of 350 kg (3510N). The load–deflection data are given in Table 16.27 and the curve is shown in Figure 16.37.

Impact test: The Charpy and Izod tests have been performed on U/D specimens of width = 10 mm and thickness = 30 mm, having depth of V-notch = 2 mm. The length of the specimen in the Charpy test is 50 mm, while in the Izod test, it is 75 mm. The impact strength in the Charpy test is found to be 62 kg, while in the Izod test, it is 18 kg.

16.7.5 Results and Discussion

PGHC is a bimodulus material, since the modulus in tension and compression is different. These are as follows:

- In tension, $E = 8.33 \times 10^4$ kg/cm^2 = 8.17 GPa.
- In compression, $E = 1.25 \times 10^5$ kg/cm^2 = 12.26 GPa.

It shows linear as well as nonlinear (σ–ε curve) behaviour in tension, but linear only in compression. Various physical and mechanical properties are given in the following:

- Specific gravity = 1.23
- Tensile strength = 270 kg/cm^2 (27 MPa)
- Compressive strength = 720 kg/cm^2 (72 MPa)
- Bending strength = 77.2 MPa
- Impact strength in the Charpy test = 12 kg, and in the Izod test = 5.8 kg

16.7.6 Comparison with Other Works

From these observations, we notice the following salient features. The PGHC shows improved strengths, higher moduli and increased specific strengths and moduli as compared to PFRC and GRP. These are given in Tables 16.28 and 16.29 and Figure 16.38.

16.7.7 Conclusion

Based on the aforementioned findings, we conclude that the PGHC is suitable for low-pressure applications such as pipes carrying sewage and industrial waste and low-strength applications such as tables, boards and supporting pads. Other salient applications may be in car bodies, louvre shutter assembly for railway coaches, footwear, greenhouse framework, low-density insulation board, ceiling tiles, lightweight furniture, office partitions, core materials for doors, windmill blades, low-pressure containers, tubes and ducts.

They can also be used as wood substitutes in housing and construction sector, as substitutes for plywood and medium-density fireboards, as panel and flush doors for low-cost housing, as panel and roofing sheets, as door

TABLE 16.28

Comparison of Strengths and Density of PGHC with PFRC and GRP

Types of Composites → Properties ↓	PFRC	GRP	PGHC
Density (kg/m^3)	1110	1850	1230
Tensile strength (MPa)	16.6	33.3	24.5
Compressive strength (MPa)	20.0	31.2	67.2
Flexural strength (MPa)	28.7	48.2	77.2
Impact strength (kJ/m^2)	0.8	12.0	62.0

TABLE 16.29

Comparison of Specific Strengths and Specific Moduli of PGHC with PFRC and GRP

Properties	PFRC	GRP	PGHC
Tensile modulus (GPa)	8.17	7.56	10.36
Compressive modulus (GPa)	14.22	7.84	22.42
Specific tensile strength (MPa)	14.95	18.0	22.07
Specific compressive strength (MPa)	18.01	16.86	60.54
Specific tensile modulus (GPa)	7.36	4.08	8.42
Specific compressive modulus (GPa)	12.81	4.36	18.22

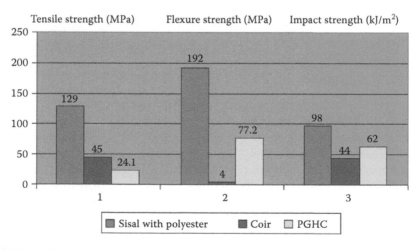

FIGURE 16.38
Comparison with other works.

frames, and as geotextiles for prevention of soil erosion, leaching, etc., and in automobile coach interior, in packaging tray for automobile parts, and in storage devices, water tanks, and latrines.

16.8 Development and Characterization of Banana Fibre–Reinforced Rice–Potato Biocomposites [7]

16.8.1 Introduction

The present-day technology is hard pressed with pollution problem and demands environment-friendly developments. Whereas the importance of the applications of composites and biocomposites is well known, the thrust on the use of natural fibres in them for reinforcement has been given priority for quite sometimes. But shifting from synthetic fibres to natural fibres provides only half-green composites. A green composite will be achieved if the matrix component is also eco-friendly. The present investigation has been undertaken with this aim.

Keeping this in view, a detailed literature surveyed has been carried out through various issues of journals related to this field. In these literatures, attempts have been made to develop biocomposites reinforced with natural fibres. The natural fibre taken in this work is of banana (Figure 16.39a). Banana fibre is obtained from banana plant (Figure 16.39b). Its botanical name is *Musa acuminata*. Fibres come from the stem of the banana plant. The banana fibre is strong and wiry, long and fine in nature. It lies in the category of bast (or soft) fibres.

The botanical name of natural rubber is *Hevea brasiliensis*. The heated latex is employed as household cement for mending chinaware and

(a)

(b)

FIGURE 16.39
(a) Natural banana fibres. (b) Banana fibre plant.

(a)   (b)

FIGURE 16.40
(a) Rubber tree. (b) Gummy latex flows from the rubber plant.

FIGURE 16.41
Boiled potato.

earthenware and to caulk boats and holes in buckets. It contains 82.6%–86.4% resins, which may have a useful value in varnishes. Its bacteriolytic activity is equal to that of papaya latex. Latex, that is gum of natural rubber (Figure 16.40b), is obtained from rubber trees during cutting of the plants (Figure 16.40a).

The potato is a starchy, tuberous crop of the Solanaceae family. Potatoes are the world's fourth-largest food crop following rice, wheat and maize. In terms of nutrition, the potato is best known for its carbohydrate content (approximately 26 g in a medium potato). Boiled potato is being used in the form of paste (Figure 16.41). Rice is the seed of the monocot plants *Oryza sativa* (Asian rice) or *Oryza glaberrima* (African rice). As a cereal grain, it is the most important staple food for a large part of the world's human population, especially in Asia and the West Indies (Figure 16.42); the rice plant can grow from 1 to 1.8 m tall.

FIGURE 16.42
Boiled rice.

16.8.2 Fabrication of Specimen

Various specimens have been fabricated using the weight fraction of fibres as $W_{fibres} \approx 30\%$ and weight fraction of polymeric matrix as $W_e \approx 20\%$–40%. All specimens are U/D, that is the fibre lay-up is U/D. The following types of composites have been prepared:

1. 15% boiled rice, 15% boiled potato, 40% epoxy and 30% banana fibre biocomposite (called biocomposite 1, now onward)

2. 15% boiled rice, 15% boiled potato, 10% natural rubber latex, 30% epoxy and 30% banana fibre biocomposite (called biocomposite 2, now onward)

3. 15% boiled rice, 15% boiled potato, 20% natural rubber latex, 20% epoxy and 30% banana fibre biocomposite (called biocomposite 3, now onward)

All these compositions of composites have been prepared using hand lay-up technique. For this purpose, an open type mould made of mild steel plate has been used. A Mylar sheet is placed on the lower part of the mould to obtain a good surface finish and easy withdrawal of the composite from the mould. In addition to it, the wax has also been used to cover the surface of the mould for easy withdrawal of the composite. The natural rubber latex and epoxy, boiled rice–potato and epoxy, and natural rubber latex, boiled rice–potato, and epoxy are mixed uniformly for different types of biocomposite matrix. Then the matrix has been layered on the mould in 1.0–1.5 mm thickness, and the banana fibres are placed unidirectionally on it. The weights are hung on both sides to maintain tension in the fibres. The entrapped air is removed with the help of a metal roller rolled on the layer.

16.8.3 Testing of Specimens

Specimens of different compositions have been tested under tension, compression, bending, and impact to characterize their properties. The testings have been done in accordance with Indian Standards. The tensile test has been performed on a Hounsfield tensometer similar to as shown in Figure 16.14. The compression test is done on a universal testing machine, the bending test on a Hounsfield tensometer using the bending attachment, and the impact test on the Izod impact testing machine. Their details are given in the subsequent sections.

Tensile test: The tensile test is performed on a Hounsfield tensometer using the scale range of 0–1000 kgf (0–9810 N). The specimen of size 150 mm length × 30 mm width × 3 mm thickness has been used. The stress–strain behaviours of different material systems are shown in Figure 16.43.

Compression test: Compression tests have been performed on the specimens of size 25 mm × 25 mm × 30 mm. The load scale chosen is of 0–2000 kg (0–19.62 kN). To record strains, a three-element rectangular strain gauge rosette is mounted on the specimen. Strains are measured with the help of a strain indicator. The compression test results are shown in Figure 16.44 on a stress–strain plot.

Bending test: The bending test has been performed on the specimen of size 75 mm × 25 mm × 3 mm. The test piece is placed in a simply supported position and load is applied at the centre. Strain is

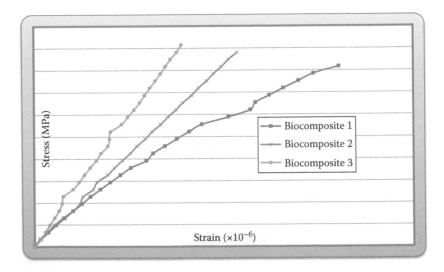

FIGURE 16.43
Tensile stress–strain curves for biocomposites.

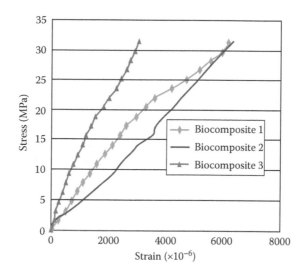

FIGURE 16.44
Comparative stress–strain curves for biocomposites under compression test.

measured by a digital strain indicator and then deflection is measured by a dial gauge. Least count of the dial gauge is 0.01 mm. To record strains, a three-element rectangular strain gauge rosette is mounted on the specimen. The results are shown in Figure 16.45 on a load–deflection plot.

Impact test: The impact test has been performed to assess the shock-absorbing capacity of different material systems. Two types of impact tests, namely, the Charpy test and the Izod test, have been conducted. In the Charpy test, the specimen is placed as a simply supported beam, while in the Izod test as a cantilever beam. Results obtained by testings are presented in Figure 16.46 with the help of a pie chart.

Shore hardness test: The hardness of the composites has been measured by using a durometer. For that, the durometer is kept on the specimens and the hardness is noted from the pointer of the display dial of the durometer. Comparative bar charts of shore D hardness of various composites and biohybrid composites are shown in Figure 16.47.

Water absorption test: Water absorption test is done for different biocomposites and biohybrids as shown in Table 16.30. First of all, each specimen is weighed in a dry condition. It is then dipped in water and after fixed intervals, the weights are taken again. The weights are taken until the weights of specimens stop increasing. The weights of the specimens have been measured using an electronic balance. Comparative water absorption curves for composites and biohybrid composites are given in Figure 16.48.

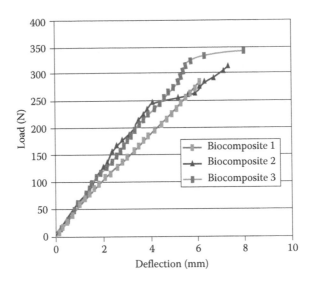

FIGURE 16.45
Comparative load–deflection curves for biocomposites under bending test.

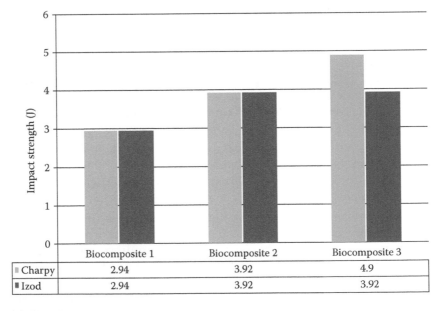

	Biocomposite 1	Biocomposite 2	Biocomposite 3
Charpy	2.94	3.92	4.9
Izod	2.94	3.92	3.92

FIGURE 16.46
Comparative impact strength (Izod and Charpy) for biocomposite during impact test.

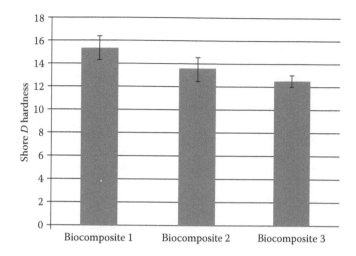

FIGURE 16.47

Comparison of shore *D* hardness of various biocomposites.

TABLE 16.30

Water Absorption Values of Biocomposites at Different Time Intervals

| Time Interval (h) | Water Absorption Values of Biocomposites at Different Time Intervals (% Weight Gain) | | |
	Biocomposite 1	Biocomposite 2	Biocomposite 3
0	0	0	0
3	0.936	0.901	1.112
12	1.812	1.824	2.132
21	2.256	2.325	3.201
36	2.846	2.937	4.483
60	3.513	4.066	5.286
72	3.833	4.357	5.425
120	4.587	5.236	5.675
192	4.812	5.976	6.275

16.8.4 Summary of Results

From these experimental observations, the following results given in Table 16.31 are obtained for different material systems.

16.8.5 Conclusions

It is concluded that out of the three types of biocomposites developed in this investigation, the following outcome are observed:

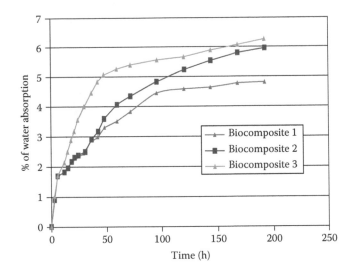

FIGURE 16.48

Comparative water absorption curves for biocomposites.

TABLE 16.31

Mechanical Properties of Different Biocomposites

Composition → Properties ↓	Biocomposite 1	Biocomposite 2	Biocomposite 3
Specific density	1.08	1.01	0.99
Tensile strength (MPa)	42.46	44.91	46.55
Specific tensile strength (MPa/kg m^{-3})	39.06	44.15	46.64
Tensile modulus (GPa)	3.54	3.92	4.46
Specific tensile modulus (GPa/kg m^{-3})	3.25	3.86	4.47
Tensile strain at fracture (%)	1.91	1.33	0.97
Poisson's ratio	0.32	0.21	0.20
Compressive strength (MPa)	44.00	46	47
Specific compressive strength (MPa/kg m^{-3})	40.47	45.23	47.09
Bending strength (MPa)	73.17	104.53	120.21
Specific bending strength (MPa)	67.37	102.78	120.45
Shore *D* hardness	15.30	13.5	12.5
Impact strength (J) Izod	2.94	3.92	3.92
Impact strength (J) Charpy	2.94	3.92	4.9
Water absorption (wt.%)	4.81	5.97	6.27

1. 15% boiled rice, 15% boiled potato, 20% natural rubber latex, 20% epoxy and 30% banana fibre biocomposite (i.e. $R_{15}P_{15}E_{20}NR_{20}BF_{30}$ biocomposite) is best for tensile application.

2. 15% boiled rice, 15% boiled potato, 20% natural rubber latex, 20% epoxy and 30% banana fibre biocomposite (i.e. $R_{15}P_{15}E_{20}NR_{20}BF_{30}$ biocomposite) is best for compression application.

3. 15% boiled rice, 15% boiled potato, 20% natural rubber latex, 20% epoxy and 30% banana fibre biocomposite (i.e. $R_{15}P_{15}E_{20}NR_{20}BF_{30}$ biocomposite) is best for bending application.

4. 15% boiled rice, 15% boiled potato, 40% epoxy and 30% banana fibre biocomposite (i.e. $R_{15}P_{15}E_{40}BF_{30}$ biocomposite) is best for hardness application.

5. 15% boiled rice, 15% boiled potato, 20% natural rubber latex, 20% epoxy and 30% banana fibre biocomposite (i.e. $R_{15}P_{15}E_{20}NR_{20}BF_{30}$ biocomposite) is best for impact application.

6. 15% boiled rice, 15% boiled potato, 40% epoxy and 30% banana fibre biocomposite is best for low-absorption application. However, keeping in view the higher specific strength and higher specific modulus of 15% boiled rice, 15% boiled potato, 20% natural rubber latex, 20% epoxy and 30% banana fibre biocomposite, the latter may also be opted for such applications.

7. Hence, it is concluded that the biocomposites developed in this work are suitable for applications in the packaging industry such as packaging of food products, packaging of fruits and grains, packaging of agricultural feed stock and packaging of electronics items.

16.9 Development and Characterization of a Jute–Cane Dual Green Fibre Hybrid Composite [8]

Composites made of natural fibres are cheaper, eco-friendly and lighter in weight, although their strength is lower than the composites made of synthetic fibres. Hence, for optimum utilization of lightweightness and strength, dual green fibre hybrid composite having two different kinds of natural fibres has been developed in this investigation. The material system consists of natural jute fibre, natural cane fibre and epoxy polymer. Consequently, two different kinds of fibres in different compositions have been used to fabricate hybrid fibre–reinforced composite laminae. The fabrications of these laminae have been done with $V_{fibres} \approx 55\%$ and $V_m \approx 45\%$. The dual fibres are mixed in different volume fraction compositions such as 100% jute and 0% cane, 0% jute and 100% cane, 70% jute and 30% cane, 30% jute and 70% cane and 50% jute and 50% cane. Their static behaviour under tension, compression, bending and impact loadings has been investigated by conducting experiments using strain gauge rosette technology. The stress–strain curves and load–deflection characteristics are obtained. The tensile, compressive, flexure and impact strengths and specific strength have been calculated. Water absorption behaviour has also been done for every composition. Various strengths and specific strengths of hybrid composites have been found for different compositions. Experimental results are shown in Figures 16.49 and 16.50.

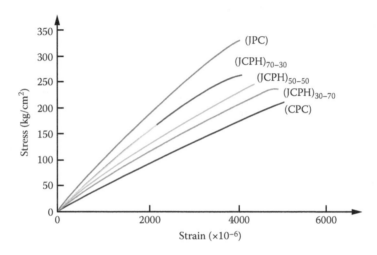

FIGURE 16.49
Tensile stress–strain curves for jute and cane fibre composites.

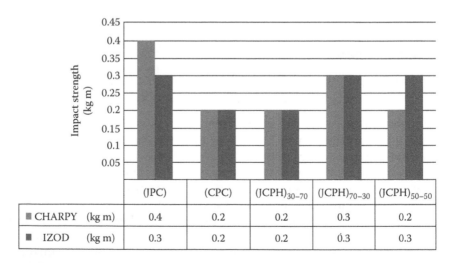

		(JPC)	(CPC)	(JCPH)$_{30-70}$	(JCPH)$_{70-30}$	(JCPH)$_{50-50}$
■ CHARPY	(kg m)	0.4	0.2	0.2	0.3	0.2
■ IZOD	(kg m)	0.3	0.2	0.2	0.3	0.3

FIGURE 16.50
Comparative impact strength (Izod and Charpy) for jute and cane fibre composites.

References

1. K.M. Gupta and N.K. Mishra, Development and characterisation of natural fibre (palmyra fibre) reinforced polymeric composite, MTech thesis, Department of Applied Mechanics, Motilal Nehru National Institute of Technology, Allahabad, India, 2003.
2. K.M. Gupta and A.K. Srivastava, Development and characterisation of palm and glass fibre hybrid composite, MTech thesis, Department of Applied Mechanics, Motilal Nehru National Institute of Technology, Allahabad, India, 2004.
3. K.M. Gupta, R. Kumar, and M. Lal, Material characterization of human-hair reinforced composite, MTech thesis, Department of Applied Mechanics, Motilal Nehru National Institute of Technology, Allahabad, India, 2002.
4. K.M. Gupta and B. Yadav, Development and characterization of starch based bio-composite, MTech thesis, Department of Applied Mechanics, Motilal Nehru National Institute of Technology, Allahabad, India, 2011.
5. K.M. Gupta, Nagarjun, and Rakesh Kumar, Development and characterisation of flax—Glass fibre epoxy hybrid composite, MTech thesis, Department of Applied Mechanics, Motilal Nehru National Institute of Technology, Allahabad, India, 2006.
6. K.M. Gupta and A. Kumar, Dual matrix—Single fibre hybrid composite, MTech thesis, Department of Applied Mechanics, Motilal Nehru National Institute of Technology, Allahabad, India, 2009.
7. K.M. Gupta, S.J. Pawar, and J.P. Yadav, Development and characterization of banana fiber reinforced rice-potato bio-composite, MTech thesis, Department of Applied Mechanics, Motilal Nehru National Institute of Technology, Allahabad, India, 2013.
8. K.M. Gupta and A. Srivastava, Mechanical characterization of jute-epoxy hybrid composite, MTech thesis, Department of Applied Mechanics, Motilal Nehru National Institute of Technology, Allahabad, India, 2009.

17

Trends in the Research of Natural Fibre–Reinforced Composites and Hybrid Composites

17.1 Bamboo Fibre–Reinforced Composites

Wong et al. [1] have studied 'Fracture characterisation of short bamboo fibre reinforced polyester composites' in 2010. In the study, fracture behaviour of short bamboo fibre–reinforced polyester composites is investigated. The matrix is reinforced with fibres ranging from 10 to 50, 30 to 50 and 30 to 60 vol.% at increments of 10 vol.% for bamboo fibres at 4, 7 and 10 mm lengths, respectively. The results reveal that at 4 mm fibre length, the increment in fibre content deteriorates the fracture toughness. As for 7 and 10 mm fibre lengths, positive effect of fibre reinforcement is observed. The optimum fibre content is found to be at 40 vol.% for 7 mm fibre and 50 vol.% for 10 mm fibre length. The highest fracture toughness is achieved at 10 mm/50 vol.% fibre-reinforced composite, with 340% of improvement compared to neat polyester. Fractured surfaces investigated through the scanning electron microscopy (SEM) describing different failure mechanisms are also reported. Tensile strength (TS) is improved only at 10 mm fibre length. Highest TS with improvement of 25% compared to neat polyester is achieved for 10 mm/40 vol.% composite. Young's modulus is deteriorated by fibre reinforcement except at 10 mm/40 vol.% of reinforcement. Fracture toughness of all types of composites is higher compared to neat polyester. Maximum increment of 340% is achieved at 10 mm/50 vol.% of fibre reinforcement. The toughening mechanisms involved are crack-tip blunting, crack deflection and crack pinning, which lead to energy dissipation through matrix plastic deformation, fibre debonding, fibre pullout and fibre damage. LEFM approach is applied with notable success.

Liu et al. [2] have studied 'Starch composites reinforced by bamboo cellulosic crystals' in 2010. Using a method of combined HNO_3–$KClO_3$ treatment and sulphuric acid hydrolysis, bamboo cellulose crystals (BCCs) were prepared and used to reinforce glycerol plasticized starch. The structure and morphology of BCCs were investigated using x-ray diffraction (XRD), electron microscopy and solid-state 13C NMR. Results showed that BCCs were of typical cellulose I structure and the morphology was dependent on its concentration in the suspension. BCCs of 50–100 nm were assembled into leaf nervations at low concentration (i.e. 0.1 wt.% of solids) but congregated into a micro-sized *flower* geometry at high concentration (i.e. 10.0 wt.% of solids). TS and Young's modulus of the starch/BCC composite films (SBC) were enhanced by the incorporation of the crystals due to reinforcement of BCCs and reduction of water uptake. BCCs at the optimal 8% loading level exhibited a higher reinforcing efficiency for plasticized starch plastic than any other loading level. BCCs were prepared using a combined HNO_3–$KClO_3$ treatment and sulphuric acid hydrolysis. The thus obtained nanoscaled crystals showed typical cellulose I structure, and the morphology was dependent on concentration in the suspension. At low concentration (i.e. at 0.1 wt.% solids of BCC), crystals of 50–100 nm assembled into leaf nervations; at high concentration (i.e. at 10.0 wt.% solids of BCC), crystals congregated into a micro-sized *flower* geometry. The different geometries resulting from aggregation intensity of BCCs were due to high surface electrostatic energy and large surface area. TS and Young's modulus of SBC8 were 12.8 and 210.3 MPa, respectively, which were much higher than their counterparts for glycerol plasticized starch without bamboo crystals. Results from DMTA proved that the reinforcing effect of BCC on the starch composites was indeed due to size effect and to the reduction of water uptake. The dispersion and polymorph of cellulose crystals were severely influenced by different treatments and the surrounding matrix, which ultimately affected the reinforcing effect on the plasticized starch-based biocomposites.

Abdul Khalil et al. [3] have studied 'Bamboo fibre reinforced biocomposites: A review' in 2012. Due to the reduction in harmful destruction of ecosystem and to produce low-cost polymeric-reinforced composites, the researchers are emerging with policies of manufacturing the composites using natural fibres (NFs) which are entirely biodegradable. These policies had generated safe strategies to protect our environment. The utilization of bamboo fibres as reinforcement in composite materials has increased tremendously and has undergone high-tech revolution in recent years as a response to the increasing demand for developing biodegradable, sustainable and recyclable materials. The amalgamation of matrix and NFs yields composite possessing best properties of each component. Various matrices used currently are soft and flexible in comparison to NFs; their combination leads to composite formation with high strength-to-weight ratios. The rapid advancement of the technology for making industry products contributes consumer the ease of making a suitable choice and own desirable tastes. Researchers have expanded their expertise in the product design by applying the usage of raw materials like bamboo fibre which is stronger as well as can be utilized in generating high-end quality sustainable industrial products. Thereby, this article gives a critical review of the most recent developments of bamboo fibre–reinforced composites and the summary of main results presented in literature, focusing on the processing methodology and ultimate properties of bamboo fibres with polymeric matrices and applications in well-designed economical products. The exploitation of bamboo fibres in various applications has opened up new avenues for both academicians and industries to design a sustainable module for future use of bamboo fibres. Bamboo fibres have been extensively used in composite industries for socioeconomic empowerment of peoples. The fabrication of bamboo fibre–based composites using different matrices has developed cost-effective and eco-friendly biocomposites, which directly affect the market values of bamboo. To design such composites, thorough investigation of fundamental, mechanical and physical properties of bamboo fibres is necessary. Thus, this review has made an attempt to gather information for both basic properties of bamboo fibre–based composites and their economic utilization. The scientists all over the globe have conducted a wide range of studies with novel ideas to provide basic support to working as well as employing communities. Current researches on bamboo fibre–based composites includes both basic and applied science, either in terms of modification or mechanophysical, thermal and other properties. But the ultimate goal of utilizing the bamboo fibre to its full extent is far behind than its projected milestone, particularly in Malaysia, although other countries such as India and China have moved far ahead in utilizing bamboo fibre in socioeconomic way. The sustainable future of bamboo-based composite industry would help in utilizing the bamboo in a way other than usual traditional mode. The effective characterization of bamboo fibre as well as bamboo fibre–based composites should be more advanced in terms of analysis and testing. In this review article, we have tried to gather the information about the analysis and testing methods used. However, the scientists have already done a lot of works on bamboo-based composites, but they are still required to do more research and innovation in this area to overcome potential challenges ahead. These things will make life easy for both urban and rural people who are more dependent on synthetic-based composites. The sustainable tomorrow of future generation lies in the present industrial development towards eco-efficiency of industrial products and their process of manufacturing. High-performance, biodegradable materials and renewable plant materials can form new platform for sustainable and eco-efficient advance technology products and compete with synthetic/petroleum-based products presently dominated in market which are diminishing natural petroleum feedstock. NFs and biocomposites made from natural sources integrate the sustainable, eco-friendly and well-designed industrial products which can replace dominance of petroleum-based products in the future. Bamboo fibre is obtained from a source which is known for its renewability in terms of fast growth and better mechanical properties. The utilization of bamboo fibre for fabrication of biocomposites by using advance technology transforms the future of the coming generation. The well-designed and engineered products from the bamboo fibre can help in making new revolution to sustain our natural resources. Thereby, based on this brief review, the bamboo fibres can be utilized for advance and engineered product development for different applications. It will be an alternative way to develop the biocomposites which can be particularly used for the daily needs of common people whether it is household furniture, house, fencing, decking, flooring and lightweight car components or sports equipments. Their low cost, easy availability and aesthetic designs will be the main driving force to transform the dependent present to sustainable future.

17.2 Banana Fibre–Reinforced Composites

Deepa et al. [4] have studied 'Structure, morphology and thermal characteristics of banana nano fibers obtained by steam explosion' in 2011. In this work, cellulose nanofibres were extracted from banana fibres (BFs) via a steam explosion technique. The chemical composition, morphology and thermal properties of the nanofibres were characterized to investigate their suitability for use in bio-based composite material applications. Chemical characterization of the BFs confirmed that the cellulose content was increased from 64% to 95% due to the application of alkali and acid treatments. Assessment of fibre chemical composition before and after chemical treatment showed evidence for the removal of noncellulosic constituents such as hemicelluloses and lignin that occurred during steam explosion, bleaching and acid treatments. Surface morphological studies using SEM and atomic force microscopy (AFM) revealed that there was a reduction in fibre diameter during steam explosion followed by acid treatments. Percentage yield and aspect ratio of the nanofibre obtained by this technique are found to be very high in comparison with other conventional methods. Thermogravimetric analysis (TGA) and differential scanning calorimetry (DSC) results showed that the developed nanofibres exhibit enhanced thermal properties over the untreated fibres. BFs in various forms were analyzed to assess their thermal behaviour, chemical composition and morphology and therefore to understand the variations introduced by the different treatments on the chemical structure of NFs. The chemical composition of raw, steam exploded and bleached BFs was determined. Chemical analysis of the fibres revealed the partial removal of hemicellulose and lignin due to the success of the chemical treatment applied. The major constituent of these fibres is found to be cellulose. Chemical analysis showed that the cellulose content of the BFs was increased from 64% to 95%, while hemicellulose and lignin content was significantly decreased to 0.4% and 1.9%, respectively. Morphological studies revealed that there is a reduction in the fibre diameter during the steam explosion followed by acid treatments. The fine structural changes of the fibres can be seen from the respective SEM and AFM micrographs. The SEM and AFM analysis also substantiates the dissolution of the noncellulosic components present in the fibre cell wall by the acid correlated steam treatment process, which enhances the extraction of crystalline cellulose components from the fibre. The aspect ratio and percentage yield of nanocellulose fibres obtained by this technique have been found to be very high as compared to conventional methods. TGA and DSC analyses revealed that thermal degradation of BFs depends mainly on the cellulose structure and the content of noncellulosic constituents that were present in the fibre. The component of the fibres and the nature of the cellulose contribute significantly to their thermal stability. Derivative thermogravimetry (DTG) and DSC results showed that the developed nanofibres exhibit enhanced thermal properties. The higher thermal stability of the developed nanofibres has been related to the higher crystallinity of the cellulose obtained after the removal of hemicellulose and lignin components from the fibre. Therefore, it can be concluded from these results that the produced nanofibres exhibit enhanced thermal properties than the untreated fibres, making them suitable as reinforcing elements in biorenewable composite material preparation.

Wickramarachchi et al. [5] have studied 'Preservation of Fiber-Rich Banana Blossom as a Dehydrated Vegetable' in 2005. Banana blossom is an excellent source of crude fibre in the human diet. Hot water blanching adopted at cottage level is found ineffective for preserving the banana blossom due to enzymatic browning which reduces market demand of the processed product. Therefore, attempts were made to develop a ready to-cook dehydrated product from the banana blossoms, while maintaining the quality and minimizing enzymatic browning and use of controversial sulphating agents. Cutting the banana blossoms into slices of 3 mm directly into a 0.2% citric acid solution and keeping the slices immersed for 30 min followed by drying at 50°C for 6 h gave an acceptable product with respect to appearance, flavour and overall quality. The quality of the product remained almost unchanged when stored in aluminium foil laminated with high-density polyethylene (Al/HDPE) for more than a month. Oriented polypropylene laminated with cast polypropylene (OPP/CPP) was by far inferior for storage of the dehydrated banana blossom, of which moisture content increased by 2.9% and L' value decreased from 41.23 to 37.42. Results of this study showed that the banana blossom could be processed to reduce browning, which is a major defect observed. Increased temperatures aiming at inhibiting polyphenol oxidase enzyme and its reactions were ineffective as it increased browning and reduced rehydration ratio. As a pretreatment, emersion of the banana blossom slices in 0.2% citric acid solution for 30 min followed by drying at 50°C for 6 h gave the dehydrated end product with reduced browning, which was ready to cook and acceptable with respect to appearance, flavour and overall quality. Al/HDPE was highly superior to OPP/CPP as a promising packaging material for the dehydrated banana blossom. The product could also be stored for an extended period beyond 1 month since the current storage was conducted for 1 month and the product remained unchanged during the 1-month period.

Belewu and Belewu [6] have studied 'Cultivation of mushroom (Volvariella volvacea) on banana leaves' in 2005. Following the solid-state fermentation of banana leaves (*Musa sapientum* Linn) by lignin degrading mushroom (*Volvariella volvacea*), yield of fruiting bodies and compositional changes of the substrate were evaluated using a student parametric T test model. The biological efficiency was 5.21, while the total weight of fruit yield was 2.5 kg. The percentage biomass loss was 18.20%. The banana leaves treated with *V. volvacea* exhibited losses primarily in the polysaccharide components and with a greater percentage of the fibre components being degraded. The crude protein content was enhanced by the incubation of the mushroom due probably to the addition of microbial protein. The acid detergent lignin (ADL) was significantly reduced in the fungus-treated sample. The acid detergent fibre (ADF) and neutral detergent fibre (NDF) followed similar trend, but the cellulose and hemicellulose increased. The development of this simple technology is expected to improve the yield of mushroom as well as provide sustainable feed (spent substrate) for ruminant animals. The mycelia covered the banana leaves in about 12 days, while full colonization of the substrate was observed in 15 days. Total weight of the fruits was 2.5 kg. Biological value was put at 15.21%. The yield and the pileus diameter observed in this study agreed with the observation of Oei (2003) for similar mushroom. There was a reduction in the weight of the banana leaves (16.30%) used as substrate, and this shows that the mushroom has the ability to degrade lignocellulosic materials during the idiophase stage following severe nitrogen and carbon depletion (Mason et al., 1989). Banana biomass loss was 18.20%. Significantly different degradation and solubilization was more intensive in the banana leaves–based substrate. The crude protein, crude fibre, ether extract, ADF, NDF and cellulose and hemicellulose contents of the untreated banana leaves agreed with the report of Sompson et al. (2004). However, the dry matter decomposition after inoculation with fungus spawn (% of original sample) was 18.75%. The protein content of the fungus-treated banana was significantly higher than the untreated sample due probably to the addition of fungal proteins during solubilization and degradation. This agrees with the report of Farkas (1979) and Jacqueline and Visser (1996) who reported that the extracellular enzymes secreted by the fungus contain amorphous homo- and heteropolysaccharides, which are often in association with protein (fungal protein). The protein content of the fungus-treated sample increased from 7.08% to 10.26%. The increasing crude protein content could be compared to the protein content of most cereal crops. It was, however, higher than common straw and grasses. The higher crude protein content will likely increase the importance of the leaves as ruminant diet. The fibre fraction decreased significantly in the fungus-treated leaves compared to the untreated leaves. The ADL decreased from 6.01% to 2.15%, and this was followed closely by ADF and NDF in that order. The decrease in the fibre fractions could be

due to the production of various enzymes during the vegetative and reproductive phases with lignocellulose degrading properties. The solubilization
of the lignin occurs during the vegetative phase, and enzymes like laccase,
manganese peroxidases and lignin peroxidases are secreted, while cellulose
degrading enzymes are secreted during the reproductive phase (Tamara
et al., 1995). The higher hemicellulose content recorded for the fungus-
treated banana leaves indicates that it is a valuable product for the lignin
degrading fungus (for it provides the organism with energy source for better functioning). Also, the higher cellulose content recorded for the fungus-
treated sample will provide more glucose for ruminant animals since the gut
of the animal is well equipped with microbes that can convert the cellulose
to glucose. Conclusively, the study revealed the potential of banana leaves as
a good substrate for the cultivation of *V. volvacea* and the spent substrate as a
viable ingredient in ruminant feed.

Kumar et al. [7] have studied 'Banana fiber-reinforced biodegradable soy
protein composites' in 2008. BF, a waste product of banana cultivation, has
been used to prepare BF–reinforced soy protein composites. Alkali-modified
BFs were characterized in terms of density, denier and crystallinity index.
Fourier transform infrared (FTIR) spectroscopy, SEM and TGA were also
performed on the fibres. Soy protein composites were prepared by incorporating different volume fractions of alkali-treated and untreated fibres
into soy protein isolate (SPI) with different amounts of glycerol (25%–50%)
as plasticizer. Composites thus prepared were characterized in terms of
mechanical properties, SEM and water resistance. The results indicate that
at 0.3 volume fraction, TS and modulus of alkali-treated fibre-reinforced soy
protein composites increased to 82% and 963%, respectively, compared to
soy protein film without fibres. Water resistance of the composites increased
significantly with the addition of glutaraldehyde, which acts as cross-linking
agent. Biodegradability of the composites has also been tested in the con-
taminated environment, and the composites were found to be 100% biodegradable. Alkali treatment of the BFs decreased the lignin component and
crystallinity and increased the roughness of the surface and mechanical
properties of the fibres. Mechanical properties of the fibre-reinforced composites were strongly dependent on the volume fraction of the BFs and the
amount of plasticizer used. It can also be concluded that alkali treatment of
the fibres is necessary to get composites with moderate mechanical properties as well as better adhesion between fibres and matrix.

Memon et al. [8] have studied 'Banana Peel: A Green and Economical
Sorbent for Cr(III) Removal' in 2008. Banana peel, a common fruit waste,
has been investigated to remove and preconcentrate Cr(III) from industrial
wastewater. It was characterized by FTIR spectroscopy. The parameters pH,
contact time, initial metal ion concentration and temperature were investigated, and the maximum sorption was found to be 95%. The binding of
metal ions was found to be pH dependent with the optimal sorption occurring at pH 4. The retained species were eluted using 5 mL of 2 M HNO_3.

The mechanism for the binding of Cr(III) on the banana peel surface was also studied in detail. The Langmuir and Dubinin–Radushkevich (DR) isotherms were used to describe the partitioning behaviour of a system at different temperatures. Kinetic and thermodynamic measurements of the banana peel for chromium ions were also studied. The method was applied for the removal and preconcentration of Cr(III) from industrial wastewater. The present work explores a new cheaper, economical and selective adsorbent as an alternative to costly adsorbents for the removal of Cr(III) ions. The main advantages of procedure are

- Cost of process
- Ease and simplicity of preparation of the sorbent
- Sensitivity

Rapid attainment of phase equilibration and good enrichment. FTIR analysis of banana peel showed the presence of various functional groups, indicating the complex nature of the banana peel. The kinetics of sorption for chromium follows a pseudo-first-order rate equation. The positive value of ΔH and negative values of ΔG indicate the endothermic and spontaneous nature of the sorption process. For the sorption of Cr(III) ions, most of the cations and anions can be tolerated up to 1:10 and 1:50 concentration ratios except with Al(III), Cd(II), Bi(III), Ni(II), Mn(II), PO_4^{3-}, and $C_6H_5O_7^{3-}$ ions, which can be tolerated up to ratios of 1:1 and 1:10. Study shows that the banana peel has the ability to extract Cr(III) from industrial wastewater.

Kiruthika et al. [9] have studied 'Experimental Studies on the Physicochemical Properties of Banana Fibre from Various Varieties' in 2009. Natural cellulosic fibres from various varieties of banana plants such as red banana, Nendran, Rasthaly, Morris, and Poovan have been extracted manually, and the physicochemical properties of these fibres are investigated. The TS of these fibres varies from 176 to 525 MPa. The untreated fibres have more TS than the treated one. The thermal properties of these fibres are studied by DSC. Two DSC thermal peaks, one is around 25°C–180°C and the other is around 155°C–240°C, are noticed. The TSs have a direct correlation with the area of the lower thermal peak (enthalpy) and activation energy of the DSC, and also with the moisture absorption characteristics. The FTIR shows characteristic bands corresponds to cellulose. The reflections of the x-ray fibre diffraction pattern recorded for the BF have been correlated with the mechanical strength. The TS of the cellulosic fibres from different varieties of banana plant varies from 176 to 525 MPa. The untreated BFs showed the highest TS when compared with the treated fibres. The variation in TS is reflected on the DSC thermal peaks, moisture absorption characteristics and x-ray fibre diffraction. The TS along with the thermal properties can be taken as a parameter while considering these materials for industrial applications.

Wachirasiri et al. [10] have studied 'The effects of banana peel preparations on the properties of banana peel dietary fibre concentrate' in 2009. Four different preparation methods of banana peel—dry milling, wet milling, wet milling and tap water washing and wet milling and hot water washing—were investigated on their effects on the chemical composition and properties of the banana peel dietary fibre concentrate (BDFC). The dry milling process gave the BDFC a significant higher fat, protein and starch content than the wet milling process, resulting in a lower water holding capacity (WHC) and oil holding capacity (OHC). Washing after wet milling could enhance the concentration of the total dietary fibre by improving the removal of protein and fat. Washing with hot water after wet milling process caused a higher loss of soluble fibre fraction, resulting in a lower WHC and OHC of the obtained BDFC when compared to washing with tap water. Wet milling and tap water washing gave the BDFC the highest concentration of the total and soluble dietary fibre, WHC and OHC. Banana peel is a good source of dietary fibre exhibiting 50 g/100 g dry matter. The different preparation method had an influence on the properties of BDFC obtained subsequently after removal of fat, protein and starch fraction. Dry milling process yielded the banana dietary fibre concentrates with significantly higher fat, protein and starch residue than wet milling processes ($P < 0.05$), resulting in the lower dietary fibre concentration. Among the wet milling processes, washing with either tap or hot water contributed to the removal of protein fraction, enhancing the total dietary fibre concentration. Although the wet milling and hot water washing process could reduce higher starch fraction in the dietary fibre, the higher loss of soluble fibre fraction was observed. Wet milling and tap water washing gave the BDFC having the highest WHC and OHC. This result indicated that this treatment was the most effective method to provide an opportunity to enhance the functionality of dietary fibre concentrate and hence to use the BDFC as a low-caloric functional ingredient for fibre enrichment, although the incorporation of them within the food system may slightly affect the colour of the final product.

Nayak et al. [11] have studied 'Degradation and flammability behaviour of PP/banana and glass fiber-based hybrid composites' in 2009. Hybrid composites of polypropylene (PP) reinforced with short banana and glass fibre were fabricated using melt-blending technique followed by injection moulding. Maleic anhydride grafted polypropylene (MAPP) has been used as a coupling agent to promote the interfacial adhesion between the fibres and the PP matrix. The degradation behaviour of the composites and hybrid composites were studied employing FTIR spectroscopy. Test results indicated an increase in the carbonyl and hydroxyl regions with the incorporation of fibres within the PP matrix. Further, the banana fibre polypropylene (BFPP) composites exhibited higher degradation tendency as compared with the virgin polymer as well as the hybrid composites. Extent of biodegradation in the irradiated samples showed increased weight loss in the BFPP samples, thus revealing effective interfacial adhesion upon hybridization with glass fibres.

Horizontal rate of burning test indicated 50% reduction in rate of burning in BFPP and the banana-glass polypropylene (BGPP) composites with the incorporation of $Mg(OH)_2$ flame retardant. Limiting oxygen index tests also showed more consumption of oxygen for combustion in case of BGPP hybrid composites, indicating improved flame retardant characteristics. BFPP and BGPP composites and hybrid composites were successfully prepared employing melt blending technique. The photodegradation behaviour of virgin PP, BFPP and BGPP composites and hybrid composites were investigated using FTIR spectroscopy. The biodegradability in the samples was measured as % weight loss in samples after composting. The flammability characteristics of BFPP composites and BGPP hybrid composites were also investigated, and the following conclusion has been drawn. The carbonyl and hydroxyl regions increase with the incorporation of fibres within the PP matrix. BFPP composites displayed maximum degradation in these regions during irradiation, which indicates increased photodegradation tendency as compared with BGPP hybrid composites. Biodegradation tendency in the BFPP and BGPP composites and hybrid composites was higher than that of the virgin matrix. The samples irradiated for 100 h were unrecoverable after 5 months and showed brittleness characteristics during UV treatment. BFPP composite samples could not be recovered after 4 months, whereas the degradation tendency in hybrid composites was less, thereby confirming the improved adhesion with the incorporation of glass fibre. SEM micrographs indicated that in case of BFPP, the presence of BF results in rapid fragmentation and specimens could not be recovered. Horizontal rate of burning reduced substantially with the incorporation of $Mg(OH)_2$ in BFPP composites. BGPP hybrid composites with and without $Mg(OH)_2$ showed lesser burning rate (BR) than BFPP. Limiting oxygen index tests also showed more consumption of oxygen for combustion in case of BGPP hybrid composites, indicating improved flame retardant characteristics. Thus, BFRP and BGPP composites at optimum content of banana and glass in the presence of MAPP are cost-effective and can be used for commercial applications.

Barreto et al. [12] have studied 'Chemically Modified Banana Fiber: Structure, Dielectrical Properties and Biodegradability' in 2010. BFs, as well as other lignocellulosic fibres, are constituted of cellulose, hemicellulose, lignin, pectin, wax and water-soluble components. The abundance of this fibre combined with the ease of its processing is an attractive feature, which makes it a valuable substitute for synthetic fibres that are potentially toxic. In this work, the structure characterization of the BF modified by alkaline treatment was studied. Some important properties of this fibre changed due to some chemical treatments, such as the crystalline fraction, dielectric behaviour, metal removal (governed by solution pH) and biodegradation. Our results showed that treated BF is a low-cost alternative for metal removal in aqueous industry effluents. Thus, for regions with low resources, the biosorbents are an alternative to diminish the impact of pollution caused by local industries, besides being a biodegradable product. It was observed by XRD that chemical treatment with sodium hydroxide (NaOH) increased the crystalline fraction of the BF, due to partial removal of the lignin (amorphous phase). The vibrational modes obtained by IR spectroscopy did not suffer significant changes after this alkaline process and the main bands appeared approximately in the same range wave number. The dielectric permittivity and the loss factor are dependent on the alkaline solution concentration. It was obtained 12.59 for dielectric permittivity at 10 Hz with the major concentration. The values for the dielectric loss were approximately between 10^{-1} and 10^{-2} depending on the sample. These values are reasonable and could also be utilized as an electronic device in conjunction with other materials to do a composite phase. The dielectric relaxation processes observed by the electrical modulus were dependent on the temperature measurements and related to water molecules present in hydrophilic groups of the anhydroglucose. The chemical treatment employed in this work contributed to increase the metal removal, and the values were governed by solution pH. The results showed that treated BF is a low-cost alternative for metal removal in aqueous industry effluents. Thus, for regions with low resources, the biosorbents are an alternative to diminish the impact of the pollution caused by local industries, besides being a biodegradable product.

Majhi et al. [13] have studied 'Mechanical and fracture behaviour of banana fiber reinforced Polylactic acid biocomposites' in 2010. Polylactic acid (PLA)/ BF biocomposites were fabricated using melt compounding technique. The effect of variable fibre composition, mercerization of fibre, as well as incorporation of maleic anhydride (MA) as compatibilizer and glycerol triacetate ester (GTA) on the properties of PLA/BF composites was studied. MA compatibilized biocomposites exhibited improved tensile modulus to the tune of 62%, whereas GTA plasticized composites showed improvement in impact strength by 143%. TGA revealed an increase in the stability of MA incorporated systems, and DSC showed increase in Tc with the incorporation of BF and compatibilizer. Increase in storage modulus of the composites as determined from dynamic mechanical analysis (DMA) also confirmed effective stress transfer from the matrix to BF. The morphological investigations using SEM indicated improved interfacial adhesion in compatibilized and plasticized systems. From the aforementioned results, it can be summarized that the addition of NFs enhances the mechanical performance in biopolymer. It was found that TS could be increased proportionally to the fibre content up to a critical fibre loading of 30 wt.%. A further enhancement in the TS and modulus coupled with the impact strength can be noticed when the NFs were mercerized. Additionally significant improvement in tensile modulus and strength to the tune of 62% and 29% was observed in PLA/MBF/MA biocomposites. Elongation and impact strength were also found to increase in PLA/MBF/GTA biocomposite. The morphology of the interface region indicated uniform dispersion of fibres within the PLA matrix as noticed in MA and GTA incorporated systems. FTIR analysis confirmed the interactions between constituent polymers and modifiers. The DSC results showed that 1 wt.% MA compatibilized composite showed the optimum Tg at 58.56°C,

Tc at 124.09°C and Tm at 149.94°C with a crystallinity of 25.95% among all the biocomposite. The TGA thermograms also confirmed improved thermal stability of PLA/MBF/MA composite. DMA studies showed that the storage modulus increased in the case of MA compatibilized PLA/MBF composites. PLA/MBF/MA biocomposites also exhibited an increase in glass transition temperature to 65°C as studied from loss modulus peaks. The tan δ peaks revealed minimal attenuation in energy for PLA/MBF/MA composites.

Sathasivam and Haris [14] have studied 'Adsorption Kinetics and Capacity of Fatty Acid-Modified Banana Trunk Fibers for Oil in Water' in 2010. Oil spill leaves detrimental effects to environment, living organisms and economy. As such, it is of considerable interest to find an effective, simple and inexpensive method to treat this calamity. This work reports the use of banana trunk fibres (BTFs) modified with oleic acid, stearic acid, castor oil and palm oil for oil spill recovery. The maximum sorption capacity, effect of oil-to-water ratio, effect of light oil fractions and effect of dissolved organic compounds in weathered oil-contaminated seawater were studied. It is found that BTF treated with oleic acid exhibited the best sorption capacity for engine oil, dissolved organic compounds in weathered oil, and light oil fractions. The equilibrium process was described well by the Freundlich isotherm model, and the kinetic studies show good correlation coefficients for a pseudo-second-order kinetic model. This present work proves that the oil sorption capacity of modified BTF with oleic acid, stearic acid, castor oil and palm oil increased markedly as compared with that of the unmodified fibres. Among all the modified BTF, oleic acid–grafted BTF is found to have the best sorption capacity for engine oil, dissolved organic compounds in weathered oil and light oil fractions. Furthermore, the reusability of the modified BTF was impressive, namely, the material was able to retain its original efficiency to sorb oil for at least three repeated cycles. The sorption of engine oil and dissolved organic compounds in weathered oil fits best in the Freundlich isotherm and pseudo-second-order kinetic models.

Shih and Huang [15] have studied 'Polylactic acid (PLA)/banana fiber (BF) biodegradable green composites' in 2011. In this study, PLA/BF composites were prepared by melt blending method. The BF was conjugated onto PLA chains through the use of a coupling agent and chemical modification. Consequently, the thermal stability and mechanical properties of the PLA were dramatically elevated through the incorporation of BF. Mechanical tests showed that the tensile and flexural strengths of the composites markedly increased with the fibre content, reaching 78.6 and 65.4 MPa when reinforced with 40 phr fibre, approximately 2 and 1.66 times higher, produced by pristine PLA. However, the impact strengths of composites are somewhat decreased with the increased content of fibres. The addition of 40 phr BF into the composite increased the heat deflection temperature (HDT) of pure PLA from 62°C to 139°C, an improvement of about 122%. Apart from enhancing the mechanical properties and thermal stability, the incorporation of BF can reduce the production cost of materials while meeting the demands of environmental protection agencies. As a result, when the MBFs are conjugated to the main chain of the PLA, the compatibility and stability of PLA can be substantially raised. Furthermore, the thermal stability and mechanical properties of PLA can be effectively elevated. The rates of increase in both the thermal deformation temperature and the TS can exceed 100%. This MBF-reinforced PLA composite can not only reduce domestic dependence on synthetic fibres but can also relieve cost pressure in industrial circles. In addition, because reinforced PLA can endure higher stress changes and be applied to products under high temperature, such as containers for hot food or cases for electronic and photoelectronic products, both applicability and an extra premium are added to the materials. Moreover, because of the innate biodegradability of BF, this type of reinforced PLA is more environmental friendly than manmade reinforced fibres.

Akmal Hadi Ma' Radzi and Noor Akmal Mohamad Saleh [16] have studied 'Banana Fiber Reinforced Polymer Composites' in 2011. Biological fibre–reinforced composite has the potential to improve the physical or mechanical properties for a wide range of potential application. The properties of some NFs have been investigated, and the reported results showed promising utilization of some of them as an alternative to glass fibre in many application. The BTF is a local waste material that can be utilized as a composite material in order to reduce this local waste in Malaysia. The banana plant has long been a source of fibre for high-quality textiles because it produces fibres with varying degrees of softness and differing qualities of specific use. This research studies the use of PLA reinforced with banana leaf fibre (BAF) in producing bioplastic composite with better mechanical properties compared to PLA-based plastic. The parameters that have been considered in this study are fibre loading and types of reinforcement. The fibre loading involved are 10%, 20% and 30% of fibre ratio over polymer. The mechanical properties were studied on both untreated and treated BaF-reinforced composites for fibre loading that have been mentioned earlier. Addition of BF to PLA reduces the raw material cost of PLA, as BF is an abundance local waste in Malaysia. The increasing utilization of renewable resources will be one of the strong drivers for sustainable products. From a waste, it becomes a useful thing to work on. Thus, BaF/PLA composite can be commercialized into a broad scope in the future. Treated BaF-PLA composite represents the best reinforced composite with good flexural and impact properties. This brittle treated BaF-PLA composite verified that alkaline treatment on BaF was competent to improve the bonding between filler and the matrix to produce bioplastic with good mechanical properties. Treated BaF-PLA composite also possessed low thermal stability compared to pure PLA plate.

Venkateshwaran et al. [17] have studied 'Mechanical and Dynamic Mechanical Analysis of Woven Banana/Epoxy Composite' in 2012. In this work, three types of woven (plain, twill and basket) composites are made of BFs/epoxy resin using hand lay-up method. Mechanical properties (tensile, flexural and impact) of three types of woven composites are compared to evaluate the effect of weaving architecture on the properties. DMA is carried out to analyze the viscoelastic behaviour of the composite. DMA is carried out in the temperature range of 40°C–140°C at 0.1, 1 and 10 Hz frequencies. SEM analysis is carried out to investigate the fracture behaviour of the composite. From the investigation, it is found that plain type of woven composite has better mechanical and dynamic mechanical behaviour. From the aforementioned study, it is found that the weave architecture plays a significant role in determining the mechanical properties of the composite. It also revealed that the composite made of plain weave pattern has better mechanical properties than twill and basket types because of better interlacing of fibres in warp and weft direction in plain weave. DMA shows that for all the patterns, storage modulus decreases with increase in the temperature. Further, it also indicates that the incorporation of fibres improves the storage modulus and decreases the damping property. The Cole–Cole plot and comparative study of experimental and theoretical models also shows good fibre–matrix adhesion for plain weave type of composite, which is also evident from SEM image analysis. Hence, plain weave pattern provides better mechanical and viscoelastic properties of the composite. The storage modulus of all the patterns is well above the untreated jute/HDPE and PALF/polyester composite.

Oliveira et al. [18] have studied 'Surface Modification of Banana Fibers by DBD Plasma Treatment' in 2012. BFs, environmentally friendly raw material freely available, were physically modified by atmospheric dielectric barrier discharge (DBD) plasma treatment of different dosages. The influence of the plasma treatment applied on the BFs was performed, considering the mechanical properties, wettability, chemical composition and surface morphology. These properties were evaluated by tensile tests, static and dynamic contact angle, FTIR spectroscopy, energy-dispersive spectroscopy (EDS), XRD, conductivity and pH of aqueous extract, DSC, and SEM images. We compare untreated and treated fibres with three different DBD plasma dosages. The results of this study showed considerable modifications in BFs when these are submitted to plasma treatment. Atmospheric plasma treatment can effectively modify physically and chemically the surface of BFs. According to obtained results, DBD at atmospheric pressure may ensure the production of fibres with improved TS, elastic modulus, chemisorptions and adhesive properties. The plasma treatment increased the content of hydrophilic functional groups on the fibre surface with increasing dosage applied. The treated BFs showed significant improvement in hydrophilicity and spreading of distilled water after treatment in the fibre's surface. The static and dynamic contact angle value was found to depend on the dosage applied—higher dosage lowers the contact angle and the time of water absorption. According to the FTIR and EDS measurements, plasma reactions changed the surface chemistry of banana samples. EDS analysis of the fibre surface revealed an increase of oxygen and nitrogen and a decrease in carbon contents, related with the intensity of the plasma treatment, among the treated samples and control sample. The increase of oxygen and nitrogen atoms can promote surface modifications in accessible polar groups at the BFs, which can explain the decrease of static and dynamic contact angle. The plasma treatment also removes some impurities from the cellulose fibre, as was observed in the SEM images. The use of DBD plasma to create changes in the surface of BFs can be crucial to expand the field of application of this natural cellulosic fibre.

Jandas et al. have studied [19] 'Renewable Resource-Based Biocomposites of Various Surface Treated Banana Fiber and Poly Lactic Acid: Characterization and Biodegradability' in 2012. Eco-friendly and completely biodegradable biocomposites have been fabricated using PLA and BF employing melt blending technique followed by compression moulding. BFs were surface treated by NaOH and various silanes, namely, 3-aminopropyltriethoxysilane and *bis*[3-(triethoxysilyl)propyl] tetrasulphane (Si69), to improve the compatibility of the fibres within the matrix polymer. Characterization studies have been suggested for a better fibre–matrix interaction because of the newly added functionalities on the BF surface as a result of chemical treatments. In comparison with the untreated BF biocomposite, an increase of 136% in TS and 57% in impact strength has been observed for Si69-treated BF biocomposite. DSC thermograms of surface-treated BF biocomposites revealed an increase in glass transition and melting transition due to the more restricted macromolecular movement as a result of better matrix–fibre interaction. The thermal stability in the biocomposites also increased in case of biocomposite made up of BF treated with Si69. Viscoelastic measurements using DMA confirmed an increase of storage modulus and low damping values for the same biocomposite. Biodegradation studies of the biocomposites have been investigated in *Burkholderia cepacia* medium through morphological and weight loss studies. Short randomly oriented PLA/BF composites with different fibre loading were prepared. Biocomposites having 30 wt.% of fibre showed better mechanical properties. The stiffness of PLA increased to the tune of 31% with the incorporation of 30 wt.% BF. Both TS and modulus increased for surface-treated BF biocomposites, which is an indication of improved interfacial addition between the fibres and the matrix. Si69 surface-treated BF composites possessed superior mechanical properties. Si69 treatment of BF tends to increase the impact strength of PLA to 29.95 J/m. DSC analysis of the biocomposites showed double melting peak wherein Si69-treated biocomposites exhibited higher Tg, Tc and Tm values. Among all of surface-treated fibre biocomposites, Si69-treated fibre composite showed better thermal stability shown by TGA analysis. DMA results of biocomposites also revealed that surface-treated fibre-reinforced biocomposites have higher storage moduli than UBF-reinforced composites, which showed better adhesion between treated BF and polymer matrix. Si69-treated fibre-reinforced biocomposites have exhibited optimum storage modulus. All of the biocomposites showed 80%–100% degradation on inoculation in bacterial medium. Test results showed that surface modification of fibres shows effect on microbial activity.

Hence, PLA/BF biocomposites can be effectively used for various end-use applications at optimal concentration of fibre and coupling agents.

Kiruthika et al. [20] have studied 'Preparation, Properties and Application of Tamarind Seed Gum Reinforced Banana Fibre Composite Materials' in 2012. Technology has been developed to prepare a biodegradable and environmental friendly composite material from tamarind seed gum and banana fibre. Tamarind seed gum is prepared from the endosperm of roasted seeds of the tamarind tree. The different temperature conditions maintained for roasting the seeds are 130°C, 160°C and 180°C. BFs are extracted from different varieties of banana trees, which are used for the preparation of composite material. The TS of the composite material is measured and shows dependency on the variety of BF used in the preparation and also the roasting temperature condition of the tamarind seed. Tamarind seed gum (the seed roasted at the temperature condition of 130°C) and red BF composite shows the highest TS of 3.97 MPa, and Poovan fibre composites show the lowest TS of 1.90 MPa. The composite material of other varieties shows TS in between these two values. The percentage of moisture absorption of the composite material has a direct correlation to the TS. In addition, investigation on fire-retardant (FR) test of tamarind seed gum and banana fibre composite material (TSGBFCM) revealed that untreated and varnish-coated BF composite material has good FR characteristics. This is an important feature to promote the use of this composite material as a false roofing material instead of thermocole. A technology has been developed to prepare the environmental friendly and biodegradable TSGBFCMs using the BF from various varieties and by varying the roasting temperature condition of the tamarind seed. The TS measurement is carried out for the prepared composite materials and the varnish, paint-coated and latex-dipped composite materials. In the untreated one, red banana has the highest TS. However, varnish coating enhances the strength of the composite material. Moisture absorption measurement shows a direct correlation with the TS. FR test performed on the composite material shows that untreated BF composite material and varnish-coated BF composite material have considerable FR characteristics. It is established that the prepared TSGBFCM has the practicability of using as a false roofing material. Its potential applications as a false roofing can replace the thermocole material, giving better environmental protection.

Sathasivam and Haris [21] have studied 'Thermal properties of modified banana trunk fibers' in 2012. Thermal decomposition of an agrowaste, namely, BTFs, was investigated by thermogravimetry (TG) and DTG up to 900°C at different heating rates (from 5°C to 100°C/min). The BTF was subjected to modification by means of various known chemical methods (mercerization, acetylation, peroxide treatment, esterification and sulphuric acid treatment). Various degradation models, such as the Kissinger, Friedman and Flynn–Wall–Ozawa, were used to determine the apparent activation energy. The obtained apparent activation energy values (149–210 kJ/mol) allow in developing a simplified approach to understand the thermal decomposition behaviour of NFs as a function of polymer composite processing. Dynamic TGA was employed to investigate the thermal decomposition of five treated BTF. The treatments are mercerization, acetylation, peroxide treatment, esterification and sulphonation. The common isoconventional methods (Kissinger, Friedman and Flynn–Wall–Ozawa) were employed to determine the apparent activation energy of these fibres. A similar degradation pattern was observed for all the treated BTF (except for BTF-5), as the base material is cellulosic fibres. The obtained curves had distinct peak and a high temperature tail, which are attributed to cellulose and lignin, respectively. Decomposition temperature of the treated fibres increased in the order of esterification > untreated > mercerization > peroxide treated > acetylation. The apparent activation energy of the treated fibres is within the polymer processing temperature range, indicating that the modification is suitable for polymer composite processing.

Venkateshwaran and ElayaPerumal [22] have studied 'Mechanical and Water Absorption Properties of Woven Jute/Banana Hybrid Composites' in 2012. This work aims to predict the mechanical properties of woven jute/banana hybrid composite. Woven fabrics are arranged in three layers of different sequence. Resin used in this work is epoxy LY556 with hardener HY951. Composite specimen is prepared by hand lay-up techniques. The effect of layering sequence on the mechanical properties, namely, tensile, flexural and impact, was analyzed. It is found that the tensile and flexural strengths of hybrid composite (banana/jute/banana) are higher than that of individual composites. Similarly, the impact strength of jute/banana/jute hybrid composite is better than other types of composite. It is found that the moisture absorption of woven BF composite is lesser than the hybrid composite. Fractography study of the fractured specimen is carried out using SEM to analyze the fracture behaviour of the hybrid composite. The effect of layering sequence and hybridization of jute fabrics with banana fabric on the mechanical and water absorption properties was studied. It is observed that the layering pattern has significant effect on the tensile, flexural and impact properties of the composite. Even though woven jute fibres were hybridized with woven BF, the effect of hybridization is lesser when compared with the effect of layering sequence. Further, it is observed that composite of banana/jute/banana is suitable for applications encountering tensile and bending loads. For application that encounters impact loading, composite of jute/banana/jute trilayer is suitable. From the SEM image analysis, it is found that the failure of composite occurs at the interface between the layers of the fabrics than fibre pullout. ANOVA also shows the layering pattern has significant effect on the mechanical strength of the composite. Further, Duncan's multiple range test was conducted to predict the effect of means with the different layer arrangement.

Indira et al. [23] have studied 'Thermal Stability and Degradation of Banana Fibre/PF Composites Fabricated by RTM' in 2012. The influence of chemical modifications of BFs and fibre/PF composites fabricated by RTM technique was investigated by TGA. The kinetic studies of thermal degradation of untreated and treated fibres have been performed using Broido method. The treatment causes variation in the surface topography of the fibres. Therefore, more energy is needed for the degradation of fibres and hence higher activation energy for decomposition. Fibre-reinforced composites with 40 wt.% fibre loading was found to be more thermally stable. Furthermore, the treated fibre-reinforced composites with 40 wt.% fibre loading posses superior thermal stability with respect to untreated fibre-reinforced composites, especially with the alkali-treated fibre-reinforced composites. Thermal stability of untreated and treated BFs and their composites has been investigated by TGA. As the fibre loading increased, the weight loss at a particular temperature was geared up. This happened at all temperatures due to the low thermal stability of BF compared to PF resin. A gradual weight loss was observed starting at around 300°C for all the composites. This initial weight loss is due to the degradation of moisture, lignin, hemicellulose, etc., present in the fibres. The thermal stability of the fibres was increased by chemical treatment. The composite samples reinforced with alkali, $KMnO_4$ and formic acid–treated fibre provide a higher degradation temperature in comparison with the untreated composites. On the other hand, in benzoylation, acetylation and silane treatment, the hydrophilicity of the fibre surface is decreased, and the void content increased. This leads to weak chemical bonding at the fibre–matrix interface. Hence, the degradation takes place only at temperature comparable to untreated composite.

Khurana et al. [24] have studied 'Statistical Modelling on Grafting and Hydrolysis for Prediction of Absorbency of Banana Fibres' in 2013. The plant fibres or cellulosic fibres such as banana, flax, hemp, jute, and sisal posses tremendous resource potential but have remained unexploited due to easy availability and advantageous properties of cotton, which is the purest form of cellulose naturally available. Many efforts have been made to explore potential of these nonconventional fibres for various applications. In this study, the use of statistical modelling has been done to understand the effect of various parameters of grafting for increasing absorbency of BF. The parameters that have been taken into consideration are monomer concentration, initiator concentration and time of grafting. Grafting of the fibres was followed by hydrolysis, and the effect of parameters such as alkali concentration, temperature and time of hydrolysis was also evaluated. The experiments were carried out according to Box–Behnken response surface design. Relationship between the parameters of grafting such as time of grafting, initiator concentration and monomer concentration followed by hydrolysis parameters such as time of treatment, temperature and alkali concentration was established with absorbency as output response. The relationship was tested for predictability, and the observed as well as calculated values showed good agreement. The optimized conditions for grafting are treatment time 6 h with 7.5% of initiator concentration and monomer concentration three times of weight of fibre, while that of hydrolysis are 3 h hydrolysis time with 5% of NaOH concentration at 90°C.

Zaman et al. [25] have studied 'Banana Fiber-Reinforced Polypropylene Composites: A Study of the Physico-Mechanical Properties' in 2013. BF-reinforced PP matrix composites were prepared by compression moulding, and their mechanical properties were evaluated. BFs and matrices were irradiated with UV radiation at different intensities. Mechanical properties of irradiated BFs and matrix-based composites were found to increase significantly compared to untreated counterparts. Optimized BFs were treated with 2-hydroxyethyl methacrylate (HEMA) solution and were cured in an oven at different temperatures for different curing times, and then composites were fabricated. Monomer concentration, curing temperature and curing time were optimized with the extent of polymer loading and mechanical properties and showed better mechanical properties over untreated composites. Water uptake and simulating weathering test of the composites were also investigated. In this study, it was found that irradiated BF/irradiated PP (75 UV passes)-based composites showed better results than untreated ones. Irradiated BFs were treated with monomer (HEMA) in MeOH solution along with 2% benzyl peroxide and cured under thermal curing method. Significant improvement of the mechanical properties was observed after monomer treatment. The TS, TM and IS of the monomer-treated BF-reinforced composites were higher than those of the untreated ones. It was also observed that water uptake values by the 15% HEMA-treated composite are lower than that of the control specimens. Simulated weathering effect revealed that treated sample showed better weathering durability than untreated sample.

Jandas et al. [26] have studied 'Thermal properties and cold crystallization kinetics of surface-treated banana fiber (BF)-reinforced poly(lactic acid) (PLA) nanocomposites' in 2013. Nucleation capacity of organically modified natural montmorillonite within the surface-treated BF-reinforced PLA biocomposites has been studied using DSC analysis in the present investigation. Both the surface treatments and nanoclays play vital roles in the variation in nucleation process of PLA during cold crystallization process. Biocomposite made up of silane-treated BF and its bionanocomposite prepared using Cloisite 30B (C30B) showed superior nucleation parameters, n and K values, in the Avrami plots. Enhanced equilibrium melting point and lower Ea suggest the reinforcing effect imparted by the BF surface treatments and C30B within the PLA matrix. Even though Louritzen–Hoffmann theory revealed that there is no change in crystallization regimes of PLA even after the biocomposite and bionanocomposite preparation, TGA revealed better heat barrier capacity for all the biocomposites and bionanocomposites in comparison with virgin PLA (V-PLA). Increased storage modulus values for biocomposites and bionanocomposites also confirm the reinforcing effects of the fillers. HDT and the flammability studies concluded better application window for newly developed materials than that of V-PLA. BF has been successfully surface functionalized using various surfactants and reinforced within the PLA matrix using melt mixing method. PLA/Si69-BF showed superior thermal transition temperatures like Tg and Tm in comparison with other nanocomposites and even better than V-PLA. Intercalation/exfoliation of C30B further enhances the thermal transition temperatures of the biocomposites. Tg of PLA has been enhanced from 57.3°C to 62.6°C in the Si69-reinforced nanocomposite. Nucleation capacity of Si69-BF and C30B has been evaluated using isothermal crystallization kinetics using DSC analysis. Both the fillers Si69-BF and C30B have been proved to be very good nucleating agents where they have showed much superior n and K value (5%–15%) from Avrami plots than that for V-PLA. Similarly, Ea of biocomposite and bionanocomposite observed as much lower than that of V-PLA indicates easy nucleation process in the respective biocomposites and nanocomposites. L–H fitting reveals a double-regime crystallization process for both PLA and its composites, from regime II to regime III. Thermal stability on the other hand enhanced significantly for all the biocomposites and bionanocomposites due to the modified interface by surface modification and well discussed heat barrier capacity of nanosilicates, and PLA/Si69-BF/C30B showed the highest thermal stability among all. Increased storage modulus during DMA analysis reveals the reinforcing effect of Si69-BF and C30B imparts within the PLA matrix. Also the flammability study using LOI and UL94 concludes better FR character for PLA/Si69-BF and PLA/Si69-BF/C30B.

Ibrahim et al. [27] have studied 'Cellulose and microcrystalline cellulose from rice straw (RS) and banana plant waste: preparation and characterization' in 2013. As part of continuing efforts to prepare cellulose and microcrystalline cellulose (MCC) from renewable biomass resources, RS and banana plant waste were used as the available agricultural biomass wastes in Egypt. The cellulose materials were obtained in the first step from RS and banana plant waste after chemical treatment, mainly applying alkaline–acid or acid–alkaline pulping, which was followed by hypochlorite bleaching method. The results indicate a higher cellulose content, 66.2%, in case of acid–alkaline treatment for RS compared to 64.7% in case of alkaline–acid treatment. A low degree of polymerization (DP), 17, was obtained for the cellulose resulting from acid–alkaline treatment for banana plant waste indicating an oligomer and not a polymer, while it reached 178 in case of the cellulose resulting from alkaline–acid treatment for the RS. MCC was then obtained by enzymatic treatment of the resulting cellulose. The resulting MCC shows an average diameter ranging from 7.6 to 3.6 lm compared to 25.8 lm for the Avicel PH101. On the other hand, the morphological structure was investigated by SEM, indicating a smooth surface for the resulting cellulose, while it indicates that the length and the diameter appeared to be affected by the duration of enzyme treatment for the preparation of MCC. Moreover, the morphological shape of the enzyme-treated fibres starts to be the same as the Avicel PH101, which means different shapes of MCC can be reached by the enzyme treatment. Furthermore, FTIR spectroscopy was used to indicate characteristic absorption bands of the constituents, and the crystallinity was evaluated by XRD measurements and by iodine absorption technique. The reported crystallinity values were between 34.8% and 82.4%

for the resulting cellulose and MCC, and the degree of crystallinity ranged between 88.8% and 96.3%, dependent on the x-ray methods and experimental iodine absorption method. RS and banana plant wastes were used for the preparation of cellulosic fibres after pretreatment with alkaline–acid pulping (PAC) and PCA followed by bleaching. Moreover, enzyme treatment was applied for the resulting cellulosic fibres in order to prepare MCC. The less content of the Klason lignin after PAC treatment can be due to the fact that alkaline treatment dissolves the lignin which can be separated and results in the black liquor. Also, a low DP was noticed for the resulting cellulosic fibre from the treated banana plant waste. On the other hand, it was noticed by the SEM that both the treatment and the bleaching processes affect the morphological structure of the resulting microfibrillated cellulose (MFC). Moreover, the enzyme treatment has also an effect of the morphological structure of the resulting MCC. In addition, the x-ray diffractograms prove that the MCC samples prepared from bleached banana plant waste are less crystalline compared to those prepared from bleached RS. Furthermore, from the DSC, it was noticed that the amorphous content increases for the resulting cellulosic fibres from bleached banana plant waste compared to that resulting from bleached RS, which can give an indication of the increase of the amorphous content for the MCC resulting from bleached banana plant waste compared to that resulting from bleached RS.

Benítez et al. [28] have studied 'Treatment of banana fiber for use in the reinforcement of polymeric matrices' in 2013. Recent studies on the use of vegetable fibres in polymers have led to the conclusion that the fibres must be treated to improve their adherence to plastic matrices. The present study investigated the effect of physical and chemical treatments on the suitability of fibre from the Canary banana tree for use as reinforcement for polymers in injection moulding processes. This fibre has the advantage of being derived from the vegetable waste that is produced by farms involved in banana cultivation in the Canary Islands. NaOH and MA were used to treat the fibre under different conditions of pressure and temperature, and then the fibre was examined by TGA and FTIR spectroscopy. The best treatment for improving the thermal properties of BF, with no significant decrease in mechanical behaviour, for use in a composite was a combination of 1 N NaOH and saturation pressure. The FTIR results allow us to conclude that BF has a similar spectrum to that of pure cellulose. However, BF also contains substances such as lignin, pectins and hemicellulose that must be removed before it can be used to reinforce polymeric matrices. These substances were removed when the fibre was treated with alkali solutions, and performing the treatment under pressure improved the efficiency of removal of these compounds. It was also observed that although the concentration of alkali influenced the efficiency of elimination of nonfibrous substances, the use of 1 and 4 N NaOH yielded similar results. Previous works have stated that treated fibres with alkaline treatments have lower lignin content, partial removal of wax and oil cover materials and distension of crystalline cellulose order, which seems to coincide with the findings in this study. Finally, MA treatment of fibre that had been treated with different concentrations of NaOH resulted in the chemical modification of the fibre, although it seemed that the fibres were unaffected if they were treated with NaOH under high pressure. In all cases, it was observed that treatment with 4 N NaOH did not result in improved thermal or mechanical behaviour, as compared with lower NaOH concentrations. MA treatment modified the fibre in such a way that the degradation temperature increased but the mechanical properties did not change in a significant manner. Treatment with 1 N NaOH, under high pressure, also increased the degradation temperature but had no significant effect on the mechanical behaviour of the fibre. Treatment with 1 N NaOH and MA did result in a decrease in TS but no significant difference in degradation temperature. In spite of the lack of an improvement in thermal stability, the latter treatment should still be considered, because the use of MA improves adhesion between the fibre and matrix due, perhaps, to the introduction of carbonyl groups on the fibre surface, observed in figures FTIR, although the change in surface energy of fibre appears to be more important. The effects of pressure and NaOH concentration have been determined. At present, it seems that the influence of pressure is more important than NaOH concentration. Finally, it is noted that TS and elastic modulus are higher for Canary BF than for other types of fibre; average values for sisal, jute, flax, cotton or coir are lower than the values obtained for single tensile tests of BF from the Canary Islands, which makes it suitable for use in the reinforcement of polymeric matrices.

El-Meligy et al. [29] have studied 'Study Mechanical, Swelling and Dielectric Properties of Prehydrolysed Banana Fiber – Waste Polyurethane foam Composites', Production of composite from polyurethane foam waste by using prehydrolyzed BFs treated or untreated by MA. Esterification of BFs by using MA reduces swelling and improves both mechanical strength and dielectric properties of composite. Addition of aluminium silicate exhibited enhancements in its dielectric properties at constant lower than 10%, while abrupt increases can be seen in the dielectric constant (ε') and dielectric loss (ε'') for samples contain more than 7.5% aluminium silicate. IR and SEM were studied. For the mechanical measurements, it was found that treatment of the BF by MA, in addition to increase of polyurethane concentrations and application of high pressure, results in production of composite characterized by high-strength properties and lower swelling. For the IR spectroscopy measurements, the treated composite showed a band characterizing C=O of ester group at 1722 cm^{-1}. For SEM measurements, morphology of the treated composite was characterized by a higher compact structure than that obtained for the untreated composite. Adding of aluminium silicate to the treated composite resulted in forming minute cells throughout the composite. For the dielectric property measurements, treatment of the BF by MA in addition to increase of polyurethane concentrations, the dielectric constant (ε') showed increase, while the dielectric loss (ε'') showed decrease, indicating improvements of dielectric properties of the treated composite.

On adding aluminium silicate with concentrations lower than 7.5, the dielectric properties of the treated composite showed decrease. But at higher concentrations, its properties showed abrupt increase.

Twite et al. [30] have studied 'Assessment of Natural Adhesives in Banana Leaf Composite Materials for Architectural Applications'. A preliminary investigation into natural adhesives for use in the manufacture of sandwich panels from banana plant wastes is reported. These panels, named MUSA panels, are designed as a low-cost construction material for rural communities in developing countries. Three natural adhesives—prepared from cassava starch (genus *Manihot*), potato starch (*Solanum tuberosum*) and *Euphorbia tirucalli* latex—have been tested in addition to a PVA synthetic adhesive. Adhesive preparation trials, physicochemical testing of adhesives and mechanical testing of adhesives bonded to dried banana leaves have been completed. The results of these limited tests show *E. tirucalli* latex adhesive to have the best mechanical properties and suitable physicochemical characteristics. Recommendations are made for further work to confirm the suitability of the adhesive for MUSA panel applications. Preliminary work to assess the potential of a limited number of natural adhesives for use in the manufacture of MUSA panels has been completed. From the results of adhesive preparation trials, physicochemical testing of adhesives and mechanical testing of adhesives bonded to dried banana leaves, an adhesive prepared by mixing *E. tirucalli* latex with additions of palm oil while heating was found to provide the most best mechanical properties and suitable physicochemical characteristics. It is recommended that further work is undertaken to investigate fully the suitability of this adhesive in the MUSA panel application. It is recommended that adhesive prepared from *E. tirucalli* latex is provisionally selected for use in the manufacture of MUSA panels and that further work is done to develop the adhesive and fully demonstrate its suitability, including assessment of the costs and feasibility of latex collection and adhesive preparation; assessment of toxicity hazards to humans and domestic animals presented by collection, preparation and service of the *E. tirucalli* latex adhesive including potential for leaching from MUSA panels and assessment of exposure control measures. The conclusions drawn from testing of dried banana leaf face materials also apply to dried banana stem core materials in the composite sandwich panel. If required, selection of low-cost, natural insecticide, FR and moisture sealing treatments and assessment of effects of these on adhesive and panel properties may be done.

17.3　Betel Nut Fibre–Reinforced Composites

Nirmal et al. [31] have studied 'On the effect of different polymer matrix and fibre treatment on single fibre pullout test using betelnut fibres' in 2011. This study is aimed at exploring the possibility of improving the interfacial adhesion strength of betel nut fibres using different chemical treatments, namely, 4% and 6% of HCl and NaOH, respectively. The fibre specimens were partially embedded into different thermosetting polymer matrix (polyester and epoxy) as reinforcement blocks. Single fibre pullout tests were carried out for both the untreated (Ut) and treated betel nut fibres with different resins and tested under dry conditions. SEM was used to examine the material failure morphology. The studies revealed the differences of interfacial adhesion strengths for the various test specimens of betel nut fibres treated with the polyester and epoxy matrix, which followed in the order of N6N4 > H4 > Ut > H6. It was proven that fibres treated with 6% of NaOH exhibit excellent interfacial adhesion properties. The interfacial adhesion shear strength of these fibres using polyester and epoxy has improved by 141% and 115% correspondingly compared to untreated fibre under the same treatment. The effect of interfacial adhesive strength for treated and untreated betel nut fibres embedded in polyester and epoxy matrix has been experimented. Chemical treatment with alkaline and acid has strongly influenced the fibre matrix bond. From the results of the studies, the interfacial adhesion shear strength followed the order of N6 > N4 > H4 > Ut > H6 for both types of thermosetting resins (polyester and epoxy). Treating the fibre with 6% of NaOH revealed excellent interfacial adhesion properties of the fibre with the matrix while 6% of HCl fibre treatment being vice versa. Interfacial adhesion shear strength for 6% NaOH-treated fibres increased by about 141% and 115% when the fibres were embedded in polyester and epoxy resins, respectively, compared to the untreated. Interfacial adhesion shear strength for 6% HCl-treated fibres performed the weakest in the pullout test. The interfacial adhesion shear strength was found to be 4% and 33% lower when the fibres were embedded in polyester and epoxy resins, respectively, compared to the untreated. Finally, the fibre/polymer bond is found to be greater in epoxy resin compared to polyester resin due to lower viscosity of epoxy resin. The interfacial adhesion shear strength of the fibre has enhanced by 66% using epoxy resin compared to polymer resin. The improved results are predominantly due to improved fibre surface wettability, enhanced surface roughness and significant improvement in the presence of trichomes on the fibre's outer surface.

17.4 Cellulose Fibre–Reinforced Composites

Sehaqui et al. [32] have studied 'Wood cellulose bio-composites with fibrous structures at micro- and nanoscale' in 2011. High-strength composites from wood fibre and nanofibrillated cellulose (NFC) were prepared in a semiautomatic sheet former. The composites were characterized by tensile tests, dynamic mechanical thermal analysis, field-emission SEM and porosity measurements. The TS increased from 98 to 160 MPa, and the work to fracture was more than doubled with the addition of 10% NFC to wood fibres. A hierarchical structure was obtained in the composites in the form of a microscale wood fibre network and an additional NFC nanofibre network linking wood fibres and also occupying some of the microscale porosity. Deformation mechanisms are discussed as well as possible applications of this biocomposite concept. Wood cellulose biocomposites with fibre network structures at both micro- and nanoscale were successfully processed by filtering and drying of a fibrous water suspension. The porosity was around 35%, and the structure consisted of collapsed wood cellulose fibres with nanofibres adsorbed to fibres but also filling the microscale pore space. Young's modulus was typically around 10 GPa, and the fibre fibril orientation distribution was random in the plane. The high cellulose content (around 85 wt.%) ensures recyclability and degradability and decreases the risk for odour problems. The presence of nanofibres increased the TS and work to fracture of the composite considerably. The strength increased from 98 MPa (wood fibre reference) to 160 MPa (10% NFC, 90% wood fibres by weight), and the work to fracture was more than doubled (from 1.7 to 4.4 MJ/m^3). The reason for this strong improvement is the presence of fibre network structures at two length scales (micro and nano). The NFC fibril network improves load transfer between wood fibres as damage starts to develop. This delays the development of large-scale damage sites to higher stress and strain. The concept of biocomposites with fibrous networks at two length scales is of interest for the development of new types of binderless fibreboard composites with improved mechanical performance.

Qiu and and Netravali [33] have studied 'Fabrication and characterization of biodegradable composites based on microfibrillated cellulose and polyvinyl alcohol' in 2012. MFC-based thin membrane-like fully biodegradable composites were produced by blending MFC suspension with polyvinyl alcohol (PVA). Desired MFC content in the composites could be easily obtained by varying the PVA solution concentration. Chemical cross-linking of PVA was carried out using glyoxal to increase the mechanical and thermal properties of the composites as well as to make the PVA partially water insoluble. Examination of composite surfaces and fracture topographies indicated that the MFC fibrils were well bonded to PVA and uniformly distributed. IR spectroscopy showed that acetal linkages could be formed in the MFC–PVA composites by a glyoxal cross-linking reaction. Sol–gel and swelling results indicated that cross-linking reaction made PVA partially insoluble and reduced its swelling ability. The MFC–PVA composites had excellent tensile properties which were further enhanced by cross-linking. TGA showed higher thermal stability for MFC–PVA composites compared to PVA. The cross-linked MFC–PVA composites showed even higher thermal stability. DSC indicated that cross-linking increased the glass transition temperature and reduced melting temperature and crystallinity of PVA in MFC–PVA composites. Results also indicated that nano- and microfibrils in MFC inhibit the crystallization of PVA. These composites could be good candidates for replacing today's traditional nonbiodegradable plastics. The biodegradable membrane-like MFC–PVA composites with differing MFC contents were produced using simple immersion method. The SEM images of surface and fracture topographies indicated that MFC was uniformly distributed in PVA and the MFC–PVA bonding was strong. Cross-links can be formed using glyoxal between PVA, MFC and MFC–PVA. Cross-linked PVA was partially water insoluble with decreased swelling ability. PVA and MFC–PVA composites showed higher mechanical properties and thermal stability after cross-linking. MFC reinforcement also was responsible for higher thermal stability of the composites. At the same time, the fracture strain decreased significantly with cross-linking. The DSC also showed higher Tg and lower Tm and lower crystallinity of the PVA in MFC–PVA composites as a result of glyoxal cross-linking.

Pöllänen et al. [34] have studied 'Cellulose reinforced high density polyethylene composites – Morphology, mechanical and thermal expansion properties' in 2013. A comparative study was made to find out how two cellulose-based micron-sized fillers affect the morphological, thermal and mechanical properties of HDPE. The fillers used in this study were commercial MCC and viscose fibres (filler content of 5, 10, 20 and 40 wt.%). The composites were prepared by melt mixing using MAPP (PEgMA) as a polymeric compatibilizer in order to improve the chemical compatibility of the fillers and the matrix. SEM images of the composites revealed that the PEgMA improved adhesion between the cellulose fillers and the HDPE matrix. Cellulose was found to have a clear effect on the thermal dimensional stability of the HDPE. The viscose fibres and the MCC decreased the coefficients of the linear thermal expansion (CLTE) of the HDPE matrix in the flow direction; CLTEs (at 25°C) as low as 37 ppm/°C (40 wt.% of viscose fibre) and 79 ppm/°C (40 wt.% of MCC) were achieved for the composites compared to 209 ppm/°C for the HDPE. Young's modulus and the TS of the HDPE significantly increased, and elongation at break decreased with a filler content. Enhanced adhesion in the presence of the PEgMA increased the mechanical properties of the composites, especially TS. The highest values for Young's modulus and TS and the lowest values for CLTE in the flow direction were achieved with the viscose fibres. The measured Young's moduli for the viscose fibre composites were compared to those calculated by composite theories. The Halpin–Tsai composite theory successfully predicted the elastic moduli for the viscose fibre composites. The influence of the viscose fibres and the MCC on the morphological, mechanical and thermal expansion properties of polyethylene was studied. The interaction between the fillers and the matrix was improved with the polymeric compatibilizer, PEgMA. Additions of viscose fibre or MCC clearly increased Young's modulus and TS of polyethylene; the highest values were achieved with the viscose fibres. The measured Young's moduli of the viscose fibre–reinforced composites agreed well with those values predicted by the Halpin–Tsai theory in the longitudinal direction, indicating the orientation of fibres during the injection moulding process, as suggested by the SEM. The HDPE's thermal dimensional stability in the flow direction was dramatically improved with cellulose fillers in accordance with the threefold to fivefold increase in the HDPE's elastic modulus due to cellulose reinforcement.

Pan et al. [35] have studied 'Biocomposites containing cellulose fibers treated with nanosized elastomeric latex for enhancing impact strength' in 2013. The surface modification of the cellulose fibre by adsorbing the cationic poly(butyl acrylate-co-methyl methacrylate/2-ethylhexylacrylate-co-styrene) [P(BA-co-MMA/EHA-co-St)] latex with core–shell structure on fibre surfaces was conducted in an attempt to enhance the toughness of the PP-based biocomposites. The mechanical, morphological and thermal properties of experimentally manufactured biocomposites based on PP and cellulose fibre modified with the cationic nanosized latex were investigated. The results indicated that the presence of the sulphite fibre treated with 1% (wt on dry fibres) of the matrix-compatible latex increased the unnotched Izod impact strength of the fibre/PP composites, so did the TS and flexural modulus. The latex-treated fibres have strong impact on the nucleation of PP, which might create the synergetic effect with toughening. The cellulose fibres treated with cationic P(BA-co-MMA/EHA-co-St) latex improved the mechanical properties of the fibre-reinforced PP composites significantly. With the addition of P(BA-co-MMA/EHA-co-St) at 1% (wt on dry fibres) to sulphite fibres, the unnotched Izod impact strength of the biocomposites containing 20% of the treated fibres was increased, so did the TS and flexural modulus. However, further increasing latex content tended to deteriorate the mechanical properties of the biocomposites. The corresponding notched Izod impact strength of the biocomposites was only improved slightly, implying the importance of the crack initiation in governing the toughness of biocomposites. The latex layers formed on fibre surface act as interfacial elastic buffers, which not only improve the compatibility between hydrophilic wood fibres (as reinforcements) and hydrophobic PP matrix but also retard the crack propagation. DSC results suggested that the latex-treated fibres improved the nucleation of PP, thus creating the synergetic effect with the compatibility improvement for toughening the biocomposites.

17.5 Date Palm Fibre–Reinforced Composites

Agoudjil et al. [36] have studied 'Renewable materials to reduce building heat loss: Characterization of date palm wood' in 2011. This paper reports the results of an experimental investigation on the thermophysical, chemical and dielectric properties of three varieties of date palm wood (*Phoenix dactylifera* L.) from Biskra oasis in Algeria. The goal is to use this natural material in the manufacture of thermal insulation for buildings. SEM and EDS analyses of the date palm wood were investigated to characterize the microstructure and the chemical composition of the samples. A simultaneous determination of the thermal conductivity and the diffusivity was achieved using a periodic method. The relative permittivity was obtained from capacitance measurements performed at room temperature. The results have shown that the surfaces of the samples are irregular with many filaments, impurities, cells and pores. The effect of the fibre orientation was significant on the relative permittivity when compared to the thermal conductivity of the date palm wood. Furthermore, the thermal conductivity measured in vacuum and at atmospheric pressure showed that the material remains, in both cases, with good properties. This result was confirmed comparing to the thermal conductivity of other natural insulating materials. Hence, the date palm wood is a good candidate for the development of efficient and safe insulating materials. This paper reports the results of an experimental investigation on the thermophysical, chemical and dielectric properties of date palm wood (*P. dactylifera* L.). The goal is to use this natural material in the manufacture of thermal insulation for buildings. Experimental work was conducted to study some physical properties of three varieties of date palm material relating to the poor existence of data in the literature. Petioles and bunch parts of these palms were tested. The effect of pressure and fibre orientation was also studied. As a result, the effect of the fibre orientation, the date palm variety and the date palm part was significant on the relative permittivity when compared to the thermal conductivity of the date palm wood. The thermal conductivity measured in vacuum chamber was half of that measured at atmospheric pressure. The date palm wood is a good candidate for the development of efficient and safe insulating materials when compared to the other natural materials.

Salim et al. [37] have studied 'Biosynthesis of poly(3-hydroxybutyrate-co-3-hydroxyvalerate) and characterisation of its blend with oil palm empty fruit bunch fibers' in 2011. Poly(3-hydroxybutyrate-co-38 mol.%-3-hydroxyvalerate) [P(3HB-co-38 mol.%-3HV)] was produced by *Cupriavidus* sp. USMAA2–4 in the presence of oleic acid and 1-pentanol. Due to enormous production of empty fruit bunch (EFB) in the oil palm plantation and high production cost of P(3HB-co-3HV), oil palm EFB fibres were used for biocomposite preparation. In this study, MA and benzoyl peroxide (DBPO) were used to improve the miscibility between P(3HB-co-3HV) and EFB fibres. Introduction of MA into P(3HB-co-3HV) backbone reduced the molecular weight and improved the thermal stability of P(3HB-co-3HV). Thermal stability of P(3HB-co-3HV)/EFB composites was shown to be comparable to that of commercial packaging product. Composites with 35% EFB fibre content have the highest TS compared to 30% and 40%. P(3HB-co-3HV)/EFB blends showed less chemicals leached compared to commercial packaging. In summary, introduction of MA to the P(3HB-co-38 mol.%-3HV) backbone has demonstrated the improvement of adhesiveness and thermal stability of P(3HB-co-3HV).

Incorporation of EFB fibres to P(3HB-co-3HV) was shown to reduce the TS and affect the thermal behaviour of the composites. Finally, investigation of the possible toxic chemicals leached demonstrated that more phenolic compounds were released from the commercial packaging. On the basis of this study, the possibility of utilizing biodegradable P(3HB-co-3HV)/EFB blends could be reasonably expected for packaging applications.

Shanmugam and Thiruchitrambalam [38] have studied 'Static and dynamic mechanical properties of alkali treated unidirectional continuous Palmyra Palm Leaf Stalk Fiber/jute fiber reinforced hybrid polyester composites' in 2013. Alkali-treated continuous palmyra palm leaf stalk fibre (PPLSF) and jute fibres were used as reinforcement in unsaturated polyester (UP) matrix, and their static and dynamic mechanical properties were evaluated. Continuous PPLSF and jute fibres were aligned unidirectionally in bilayer arrangement, and the hybrid composites were fabricated by compression moulding process. Positive hybrid effect was observed for the composites due to hybridization. Increasing jute fibre loading showed a considerable increase in tensile and flexural properties of the hybrid composites as compared to treated PPLSF composites. SEM of the fractured surfaces showed the nature of fibre–matrix interface. The impact strength of the hybrid composites was observed to be less compared to pure PPLSF composites. Addition of jute fibres to PPLSF and alkali treatment of the fibres has enhanced the storage and loss modulus of the hybrid composites. A positive shift of Tand peaks to higher temperature and reduction in the peak height of the composites was also observed. The composites with higher jute loading showed maximum damping behaviour. Overall, the hybridization was found to be efficient, showing increased static and dynamic mechanical properties. A comparative study of properties of this hybrid composite with other hybrids made out of using natural/glass fibres is elaborated. Hybridization of alkali-treated jute and PPLSF has resulted in enhanced properties, which are comparable with other natural/glass fibre composites, thus increasing the scope of application in manufacturing of lightweight automotive parts. Unidirectionally aligned bilayer hybrid composites were fabricated by reinforcing alkali-treated continuous jute fibres with alkali-treated continuous PPLSF in UP matrix by compression moulding process. Addition of alkali-treated jute fibre with alkali-treated PPLSF has enhanced tensile and flexural properties of the composites. The P25J75 composites had 46% and 65% improvement in TS and tensile modulus, respectively, and the flexural strength and flexural modulus improved by 56% and 19%, respectively, in comparison with P100 composites. A positive hybrid effect on the tensile and flexural strength was observed by addition of jute fibre with PPLSF in polyester matrix. SEM showed the nature of fibre–matrix interface. The impact strength was observed to be high for P100 composites (36 kJ/m^2), and hybridization has reduced the impact strength of the composites. The J100 composites had a least value of impact strength (25 kJ/m^2). The storage modulus (E0) of hybrid composites was found to be less at higher temperature, but it increased on increasing the jute fibre content. Similarly the loss modulus (E00) was observed to increase on increasing the jute content due to effective stress transfer. Both storage modulus (E0) and loss modulus were observed to be high for P25J75 hybrid composites. The value of Tg obtained from the loss modulus curve (E00) is less than Tg from Tand curve. Tg shifting towards the right of the graph indicates the evidence for good interfacial adhesion, which is due to alkali treatment performed on the fibres. Tand is found to be lowest for J100 and P25J75 composites compared to P100, P75J25 and P50J50 composites. Comparing the properties of the unidirectional bilayer PPLSF/jute hybrid polyester composites with other well-known hybrids and single fibre composites, it can be concluded that the new hybrid composites made out of PPLSF/jute can be a potential replacement in place of synthetic fibres such as glass fibres and some popular NFs like sisal, hemp, coir, banana and bamboo, and hence new possibilities of use of this hybrid system can be explored. Also use of these composites will improve opportunity for growth of this particular palmyra palm tree species.

Norul Izani et al. [39] have studied 'Effects of fiber treatment on morphology, tensile and thermogravimetric analysis of oil palm empty fruit bunches fibers' in 2013. The objective of this study was to evaluate the effect of fibre treatment on both morphological and single fibre TS of EFB. EFB fibre was treated with boiling water, 2% NaOH and combination both NaOH and boiling water. Fibre morphology was characterized by SEM. TGA was further used to measure the amount and rate of change in the weight (weight loss) of treated fibre as a function of temperature. Based on the results of this work, it seems that alkali treatment improved most of the fibre properties. NaOH treatment was found to alter the characteristic of the fibre surface topography as seen by the SEM. The thermal stability of NaOH-treated and water boiling–treated EFB fibre was found to be significantly higher than untreated fibre. The best results were obtained for alkali-treated fibre where the TS and Young's modulus increased compared to untreated fibres. The overall results showed that alkali treatment on EFB fibre enhanced the TS and thermal stability of the fibre samples. The influence of fibre treatments by NaOH soaking, water boiling and combination of NaOH soaking and boiling on single fibre TS and thermal analysis were analyzed. The initial and final degradation temperatures for EFB fibres were measured in the temperature range 30°C–500°C. From these values, it can be concluded that the thermal stability of fibres was improved by alkali treatment. The tensile tests on EFB fibres revealed that mechanical properties of alkali-treated fibres are superior to those of untreated fibres. NaOH soaking significantly removed surface impurities on fibres, producing modifications on the surface and enhanced its thermal stability. Results were supported by SEM analysis. Additionally, TGA measurements of untreated and treated fibres revealed that NaOH soaking and water boiling treated had a positive impact on the thermal stability of sample. This study indicated that alkali treatment on EFB fibre enhanced the TS and thermal stability of the fibre.

17.6 Hemp Fibre–Reinforced Composites

Manthey et al. [40] have studied 'Green Building Materials: Hemp Oil Based Biocomposites' in 2012. Novel acrylated epoxidized hemp oil (AEHO)–based bioresins were successfully synthesized, characterized and applied to biocomposites reinforced with woven jute fibre. Characterization of the synthesized AEHO consisted of acid number titrations and FTIR spectroscopy to assess the success of the acrylation reaction. Three different matrices were produced (vinyl ester [VE], 50/50 blend of AEHO/VE and 100% AEHO) and reinforced with jute fibre to form three different types of biocomposite samples. Mechanical properties in the form of flexural and interlaminar shear strength (ILSS) were investigated and compared for the different samples. Results from the mechanical tests showed that AEHO and 50/50-based neat bioresins displayed lower flexural properties compared with the VE samples. However, when applied to biocomposites and compared with VE-based samples, AEHO biocomposites demonstrated comparable flexural performance and improved ILSS. These results are attributed to improved fibre–matrix interfacial adhesion due to surface-chemical compatibility between the NFs and bioresin. In this study, AEHO was synthesized, characterized and applied to jute fibre–reinforced biocomposites. Comparisons with commercial VE-based resin and AEHO/VE 50/50 blended resin systems both in neat resin form and applied to biocomposites were performed. Mechanical properties, flexural and interlaminar shear, were investigated and compared for all three bioresin and biocomposite sample types. Results from the mechanical tests showed that AEHO and 50/50-based neat bioresins exhibited lower flexural properties compared with the VE samples. Conversely when AEHO-based bioresin matrices were reinforced with jute fibre, they displayed comparable flexural performance and improved ILSS. This is thought to be due to surface chemical compatibility between the natural jute fibres and the AEHO-based bioresin. Specifically the greater quantity of hydroxyl groups present in the AEHO bioresin compared with the VE is proposed as the reason for this enhanced fibre–matrix interfacial adhesion. These hydroxyl functional groups present in the AEHO provide functional groups that interact with the hydroxyl groups present in the cellulose of the NFs. The end result is believed to be the formation of strong hydrogen bonds, thereby improving adhesion and ultimately flexural and ILSS performance. Further work in the form of SEM will be undertaken to confirm the enhanced interfacial adhesion of AEHO and NFs. Mechanical testing in the form of tensile and impact and moisture absorption analysis will be undertaken. Furthermore, in conjunction to this, DMA will be performed to determine the thermomechanical properties of the AEHO bioresin and biocomposite systems.

Kabir et al. [41] have studied 'Mechanical properties of chemically-treated hemp fibre reinforced sandwich composites' in 2012. In this study, hemp fabrics were used as reinforcements with polyester resin to form composite skins, while short hemp fibres with polyester as a core for making composite sand-wich structures. To improve the fibre–matrix adhesion properties, alkaliza-tion, silane and acetylation treatments on the fibre surface were carried out. Examinations through FTIR spectroscopy, SEM, DSC, and TGA were conducted to investigate the physical and thermal properties of the fibres. Mechanical properties such as flexural and compressive strengths of the sandwich struc-tures made by treated and untreated hemp fibres were studied. Based on the results obtained from the experiments, it was found that the fibre treated with alkali solution and post-soaked by 8% NaOH exhibited better mechanical strength as compared with other treated and untreated fibre composite sam-ples. Besides, DSC and TGA analysis showed that the thermal stability of all treated fibre was enhanced as compared with untreated samples. There is no doubt that the mechanical strength of NF composites (NFCs) is basically gov-erned by their interfacial bonding between the fibre and matrix. Many studies have addressed that a poor bonding interface appears in the composites. In this study, chopped hemp fibre was used in core and hemp fabric was used in skin in sandwich composite structures. The hemp fibres were treated with different chemical substances to measure the thermal and mechanical proper-ties of the structures. The results obtained from FTIR, DSC and TGA indicated that hydrophilic hydroxyl groups of hemicelluloses, lignin and cellulosic con-stituents on the fibre surface were removed. It therefore enhanced the bonding strength between the fibre and matrix. This phenomenon can be reflected from the results of mechanical property tests and SEM imaging. Alkali, silane and acetylation treatment were used on the hemp fibre surfaces prior to mix with polyester resin. According to the results measured, an alkali-treated hemp fibre pre-soaked with 8% NaOH composite exhibited better mechanical strength as compared with other treated and untreated types. For the 8% NaOH pre-soaked fibre composite, pure shear and tensile failures were observed. This may be due to the improvement of fibre hydrophobicity through the removal of hemicelluloses, lignin and other cellulosic constituents by chemical treatment. For other types of samples, failure due to debond between the upper skin and core material was seen. For acetylation-treated fibre, due to its brittle in nature, poor mechanical properties were found. In this study, it was observed that the mechanical properties of NFCs are highly influenced by their surface condi-tions. An appropriate chemical treatment on the fibre surface can enhance the mechanical strength around 30%.

17.7 Jute Fibre–Reinforced Composites

Das et al. [42] have studied 'Physico-mechanical properties of the jute micro/ nanofibril reinforced starch/polyvinyl alcohol biocomposite films' in 2011. Jute micro/nanofibrils (JNFs) were prepared from jute by acid hydrolysis route. The JNFs were characterized with the help of transmission electron microscope (TEM). Starch/PVA-based biocomposite films reinforced with JNF at different loading of 5, 10 and 15 wt.% were prepared by solution casting method, incorporating glycerol as a plasticizer. These biocomposite films were characterized by mechanical characterization, thermal analysis, moisture uptake behaviour, SEM, and AFM. The 10 wt.% JNF loaded films (SPVA_JNF 10) exhibited best combination of properties. JNFs were prepared from jute felt by acid hydrolysis. JNF-reinforced starch/PVA-based biocomposite films were prepared by solution casting method. JNF content was varied between 0 and 15 wt.%. An optimum combination of TS and % strain was observed in SPVA_JNF 10. The AFM study revealed most uniform dispersion of fillers in SPVA_JNF 10. Moisture uptake decreased significantly in the biocomposites when exposed to 93% relative humidity (RH) condition. After fifth day, a decrease in weight was observed in SPVA_JNF 0, SPVA_JNF 5 and SPVA_JNF 15 due to dissolution in water. But the SPVA_JNF 10 resisted the dissolution in water, indicating the stabilization of the matrix within a cellulose network. This study clearly depicts that jute nanofibres can act as an efficient, environmentally friendly green nanofiller in starch/PVA blend matrix, and such biocomposites can be an effective alternative to the conventional plastic films in the future with immense potential.

Hossain et al. [43] have studied 'Mechanical performances of surface modified jute fiber reinforced biopol nanophased green composites' in 2011. Surface modification of jute fibres was accomplished by performing chemical treatments, including detergent washing, dewaxing, alkali and acetic acid treatment. Morphology of modified surfaces examined using SEM and FTIR spectroscopy revealed improved surfaces for better adhesion with matrix. Enhanced tensile properties of treated fibres were obtained from fibre bundle tensile tests. Using solution intercalation technique and magnetic stirring, 2%, 3% and 4% by weight montmorillonite K10 nanoclay were dispersed into the biodegradable polymer, biopol. Jute fibre–reinforced biopol biocomposites with and without nanoclay manufactured using treated and untreated fibres by compression moulding process showed almost the same volume fraction for all the samples. However, the lower void content was observed in the surface-modified and nanoclay-infused jute biopol composites. Mechanical responses of treated fibre–reinforced biopol composites (TJBC) without nanoclay evaluated using DMA and flexure tests showed 9% and 12% increase in storage modulus and flexure strength, respectively, compared to untreated jute fibre–reinforced biopol composites (UTJBC). The respective values were 100% and 35% for 4% nanoclay-infused TJBC, compared to UTJBC without nanoclay. Lower moisture absorption and better mechanical properties were found in the nanophased composites even after moisture conditioning. Surface treatments resulted in the removal of pectin, hemicelluloses and other noncellulosic substances from the fibres and the higher percentage of celluloses in the final treated fibres. Rougher surface and increased effective surface area of the chemically treated fibres facilitated better interaction between the fibre and matrix. Surface-modified fibres showed better tensile properties compared to untreated fibres for the presence of higher percentage of crystalline celluloses. NFs possess very good specific properties, which are comparable to synthetic fibres. By increasing the fibre volume fraction, modifying fibre and matrix and incorporating nanoparticles, better NF-reinforced composites are being developed such that the properties of these composites are expected to be comparable to the properties of the synthetic fibre–reinforced composites as the technology advances. Thus, these composites might be the alternative to synthetic fibre–reinforced composites in the long run. Instead of compression moulding technique, vacuum-assisted resin transfer moulding (VARTM), extrusion and hand lay-up techniques can also be employed to fabricate NF-reinforced composites depending on the cure requirements. However, in this study, conventional and nanophased composites were fabricated by stacking the biopol film and fibres like a sandwich using the compression moulding process. The fibre volume fraction was about 0.27 in all kinds of composites, and the void content was lower in the nanoclay-infused composites. Treated jute–biopol composites with/without nanoclay showed better storage modulus, loss modulus, flexural strength and modulus compared to untreated jute–biopol composites for the better interaction between fibre and matrix. Properties of the jute-based biocomposites were increased with the increase percentage of nanoclay. Nanoclay acted as a nucleating agent in the biopol as well as void-reducing agent in the composites. The 4% nanoclay-infused composites showed lower moisture absorption, better dynamic mechanical properties and flexural properties compared to the 2%, 3% nanoclay-infused composites as well as the conventional composites.

López et al. [44] have studied 'PP composites based on mechanical pulp, deinked newspaper and jute strands: A comparative study' in 2012. In the present work, PP-based composites reinforced with three types of randomly distributed short lignocellulosic fibres, namely, mechanical pulp (MP), deinked pulp (DIP) and jute strands, were prepared and analyzed. Addition of 6% (wt/wt) of MAPP resulted in a significant enhancement in the TS in line with the improvement of the fibre–matrix interfacial adhesion, making more effective the transfer of stress from the matrix to the rigid reinforcement. The mechanical properties of these composites were analyzed in terms of Bowyer–Bader and Hirsch models to fit the obtained experimental data. From the stress–strain curves and the fibre length distributions, it was possible to access the orientation factor, the interfacial shear strength, the intrinsic TS and the modulus of the fibres. In the present work, three types of natural reinforcements were used for the preparation of PP composites. MP was obtained from softwood, deinked newspaper fibres (DINPs) were obtained from newspaper (consisted of hardwood) after deinking, and jute strands (JSs) were obtained from jute stalks after harvesting. The tensile modulus of the composite attained the highest value with jute fibres followed by DINP and MP, and the mechanical strength in the presence of coupling MAPP was higher with jute followed by MP and DINP. From the Hirsch model, the intrinsic fibre modulus was estimated, adopting a value of the efficiency stress transfer factor b equal to 0.433. Then the orientation factor $v1$, the interfacial shear strength (s) and the intrinsic TS of the fibres were predicted, calling upon Bowyer–Bader model and using the strain–stress curves and the fibre length distributions only as experimental data. From the calculated values, the following remarks were pointed out: (1) the highest intrinsic modulus was attained for jute followed by DINP and MP, (2) the intrinsic TS of jute and MP was quite similar followed by DINP and (3) a significant improvement in the interfacial shear strength (s) by more than 45% was noted following the addition of 6% MAPP, bringing about a reduction in the critical fibre length Lc by more than 30%.

Doan et al. [45] have studied 'Jute fibre/epoxy composites: Surface properties and interfacial adhesion' in 2012. Jute fibres were surface treated in order to enhance the interfacial interaction between jute NFs and an epoxy matrix. The fibres are exposed to alkali treatment in combination with organosilane coupling agents and aqueous epoxy dispersions. The surface topography and surface energy influenced by the treatments were characterized. Single fibre pullout tests combined with SEM and AFM characterization of the fracture surfaces were used to identify the interfacial strengths and to reveal the mechanisms of failure. Jute fibre surface treatments alter the fibre properties (in terms of morphology, topography, wettability and adhesion) and thereby affect the internal interphase properties. AFM investigations on the nanometre scale as well as SEM images and pullout tests on a micrometre scale are utilized to reveal changes in roughness and crack propagation. Untreated fibres show compact structures of a waxy surface, and the fracture occurs in the weak cementing layer of hemicellulose matrix in between the cellulose fibrils. Since the weak cementing layer contains low molecular fats, lignin, pectin and hemicellulose, the fracture in this zone towards the epoxy matrix is characterized by low interfacial shear strength of 43 MPa. Removing this cementing layer is the aim of the alkali fibre treatments. This improves the interfacial shear strengths of jute/epoxy model microcomposites significantly up to 40%. The stronger interaction between NaOH-treated fibre and epoxy matrices is caused by a cleaner and rougher fibre surface due to dewaxing and the formation of a stronger interphase. Moreover, it enhances the mechanical interlocking with the matrix and improves the wetting of the resin on the NaOH-treated fibre surface. The extended surface area could also be a reason for stronger interfacial adhesion. For the epoxy matrix, fibre surface treatment by alkali, organosilane, epoxy dispersions and their combinations enhances the adhesion strength further. Especially the presence of coupling agent leads to improved adhesion in the interface, through chemical bonds between, on the one hand, the fibre and silane coupling agent/film former and, on the other hand, the silane coupling agent/film former and the matrix. Although the interphase material makes up only a relatively small fraction of the total mass, these functional interphases influence the composite properties significantly. Furthermore, the fibre surfaces exhibit a porous structure, and the anchoring coupling agents can penetrate into the pores and form a mechanically interlocked coating. As a consequence, the fracture surface shows a clear longitudinal pattern within its cross-hatched structure. The bonding must therefore be strong enough to shift the fracture from the interphase into the jute fibre. This deformability through relatively softly bonded radial layers is typical for cellulose-based NFs. These flow plastically before the fracture due to a large number of comparatively weak bonds and a tilt of the fibrils, which undergo reorientation when stretched uniaxially. Due to the chemical bonds between fibre and epoxy resin, the interphase has become stronger than the fibre's internal structure, marking a strength limit of NF reinforcement. This limit could be shifted by using higher-strength NFs or fibres modified, for example, by a carbon nanotube coating for higher strength.

Prachayawarakorn et al. [46] have studied 'Effect of jute and kapok fibers on properties of thermoplastic cassava starch composites' in 2013. Since mechanical properties and water uptake of biodegradable thermoplastic cassava starch (TPCS) were still the main disadvantages for many applications, the TPCS matrix was, therefore, reinforced by two types of cellulosic fibres, that is jute or kapok fibres, classified as the low and high oil absorbency characteristics, respectively. The TPCS, plasticized by glycerol, was compounded by internal mixer and shaped by compression moulding machine. It was found that water absorption of the TPCS/jute fibre and TPCS/kapok fibre composites was clearly reduced by the addition of the cellulosic fibres. Moreover, stress at maximum load and Young's modulus of the composites increased significantly by the incorporation of both jute and kapok fibres. Thermal degradation temperature, determined from TGA, of the TPCS matrix increased by the addition of jute fibres; however, thermal degradation temperature decreased by the addition of kapok fibres. Functional group analysis and morphology of the TPCS/jute fibre and TPCS/kapok fibre composites were also examined using FTIR spectroscopy and SEM techniques. The TPCS matrix was reinforced by the jute and kapok fibres in order to prepare the TPCS/jute fibre and TPCS/kapok fibre composites. New hydrogen bond was found by the addition of the jute or kapok fibres as detected by the IR peak shift. Surface wetting of both cellulosic fibres by the TPCS matrix was also observed. Mechanical properties of the TPCS matrix reinforced by jute or kapok fibres were significantly improved. In addition, water absorption of the TPCS matrix was clearly reduced by the incorporation of the jute or kapok fibres, but the TPCS/kapok fibre composite showed higher water absorption. Thermal stability of the TPCS matrix was enhanced by the addition of both types of the fibres. Nevertheless, thermal degradation temperature of the TPCS matrix remained unchanged by the addition of the jute fibres, but it tended to decrease by the addition of the kapok fibres.

17.8 Kenaf Fibre–Reinforced Composites

Bernard et al. [47] have studied 'The effect of processing parameters on the mechanical properties of kenaf fibre plastic composite' in 2011. This paper investigates the effects that processing parameters, including temperature and speed, have on the mechanical properties of kenaf fibre (KF) plastic composite. KF was used to fabricate a composite material along with PP as a binding material. The composite was manufactured using a newly developed compression moulding machine. Tensile and impact tests were performed on the PP/kenaf composite to characterize its mechanical properties. The tensile properties of PP/kenaf composite increased by 10% after the addition of unidirectional kenaf fibre (UKF). However, its impact properties simultaneously deteriorated. DMA was carried out to examine the material properties. Results show that the storage modulus (E0) and loss modulus (E00) increase with the addition of UKF. However, its addition decreases the tan d amplitude. The fracture surface of PP/kenaf composite was investigated by SEM. The newly invented compression moulding machine illustrates a new trend in processing parameters of long KF plastic composite. The processing parameters of PP/kenaf long fibre composite have been studied and optimized. The optimal processing parameters using our compression moulding machine are a temperature of 230°C and a barrel speed of 16 Hz. These processing parameters were found to influence the mechanical properties of the composite. Compression moulding produced a great TS in PP/kenaf composite of 35.1 MPa. In contrast, the impact properties reduced to 130.9 J/M for the unnotched Izod test and 79.5 J/M for the notched Izod test. This result is obviously different from that with pure PP, where the value is 328 J/M. DMA indicates that the Tg of the PP sample has a higher peak compared to the PP/kenaf sample. For storage modulus and loss modulus, PP/kenaf has better properties than PP. SEM analysis reveals that our compression moulding method gives a strong bonding between the fibre and polymer matrix.

Elsaid et al. [48] have studied 'Mechanical properties of kenaf fiber reinforced concrete' in 2011. This paper presents the findings of an experimental research program that was conducted to study the mechanical properties of a NF-reinforced concrete (FRC), which is made using the bast fibres of the kenaf plant. The kenaf plant is quickly developing as a replacement crop for the dwindling tobacco industry in the Southeastern United States. Appropriate mixture proportions and mixing procedures are recommended to produce kenaf FRC (KFRC) with fibre volume contents of 1.2% and 2.4%. The compressive strength, compressive modulus, splitting TS and modulus of rupture (MOR) of KFRC specimens are presented and compared to the properties of plain concrete control specimens. The experimental results indicate that the mechanical properties of KFRC are comparable to those of plain concrete control specimens, particularly when accounting for the effect of the increased water/cement ratio required to produce workable KFRC. Further, the results indicate that KFRC generally exhibits more distributed cracking and higher toughness than plain concrete. SEMs indicate that a good bond between the KFs and the surrounding matrix is achieved. The SEMs also provide interesting information regarding the mechanisms which contribute to the failure and postpeak behaviour of the KFRC, which may be beneficial to future modelling efforts. The research findings indicate that KFRC is a promising *green* construction material which could potentially be used in a number of different structural applications. This paper presents the findings of an experimental research program that was conducted to evaluate the basic characteristics of KFRC. Suitable mixture proportions were established for KFRC with fibre contents of 1.2% and 2.4%. The research indicates that in order to maintain the workability of the KFRC requires the use of a cement-rich mixture and coarse aggregates with a maximum diameter of 9.5 mm. Further, additional water should be added to account for water absorbed by the fibres and to maintain the workability of the fresh KFRC. The findings of the experimental program indicate that at lower fibre contents, the strength of KFRC is similar to that of non-air-entrained plain concrete with a similar water/cement ratio. At higher fibre contents, the compressive strength of KFRC is somewhat lower than that of non-air-entrained plain concrete with a similar water/cement ratio. In general, KFRC specimens exhibited more ductile behaviour with greater energy absorption and more well-distributed cracking patterns, which is typical for FRC. The research further indicates that due to the cement-rich mixture proportions and the use of smaller-sized coarse aggregates, the compressive elastic modulus of KFRC is considerably lower than that of plain concrete with a similar compressive strength and this should be accounted for in design where the serviceability limit state is critical. Results from splitting tension tests and flexural tests indicate that the ACI 318M-08 [27] provisions related to the TS of conventional concrete are also appropriate for KFRC. Further, the results of the flexural tests indicate that KFRC exhibits a ductile failure mode compared to conventional concrete with an average measured toughness approximately three times that of similar plain concrete control specimens. As a result, KFRC could be used in the production of impact resisting members. Moreover, the observed improvement of the cracking behaviour enhances the durability of concrete at relatively low cost compared to other types of fibres. Thus, KFRC could be used for casting slab on grade. SEMs indicate the good bond between the KFs and the surrounding matrix and provide some hints regarding the possible failure mode and damage progression of the KFRC, which could facilitate future modelling efforts. The research findings of this study indicate that kenaf is a promising material for the manufacture of sustainable, *green* concrete by CO_2 sequestration from the atmosphere and represents a good replacement crop for the declining tobacco industry. Unfortunately, the KFRC requires the use of a cement-rich mixture compared to conventional concrete, resulting in more CO_2 emissions during the cement manufacturing process. Future research should focus on the use of cementitious mineral admixtures to off-set the increased cement demand and include a comprehensive life-cycle assessment.

Akil et al. [49] have studied 'Kenaf fiber reinforced composites: A review' in 2011. The development of high-performance engineering products made from natural resources is increasing worldwide, due to renewable and envi-ronmental issues. Among the many different types of natural resources, kenaf plants have been extensively exploited over the past few years. Therefore, this paper presents an overview of the developments made in the area of KF-reinforced composites, in terms of their market, manufacturing methods and overall properties. Several critical issues and suggestions for future work are discussed, which underscore the roles of material scientists and manufacturing engineers, for the bright future of this new *green* mate-rial through value addition to enhance its use. Research on KF-reinforced composite is generating increased attention due to its excellent properties and ecological considerations. A brief discussion of KF-reinforced compos-ites is given along with a review, in the previous study. The aforementioned topics are aimed at bringing scientists to look at the potential of KF as an alternative medium to replace conventional materials or synthetic fibres as reinforcement in composites. Processing techniques for KF-reinforced composite are well documented, and many of their main properties have been studied. In general, the use of KF-reinforced composite can help to generate jobs in both rural and urban areas, in addition to helping to reduce waste and thus contributing to a healthier environment. However, looking at future demands, more crucial studies are required on product commercialization and manufacturing processes, especially for large-scale end products.

Suharty et al. [50] have studied 'Flammability, Biodegradability and Mechanical Properties of Bio-Composites Waste Polypropylene/Kenaf Fiber Containing Nano $CaCO_3$ with Diammonium Phosphate' in 2012. Biocomposites based on waste PP and KF using coupling agent acrylic acid (AA) and cross-linker divinylbenzene (DVB) containing nano-$CaCO_3$ (nCC) with and without DAP as a mixture FR were successfully processed in melt. Flammability of biocomposite was horizontally burning tested, according to ASTM D635. To study the nature of its biodegradability, the biocomposites were technically buried in garbage dump land. The TS prop-erties of biocomposites were measured according to ASTM D638 type V. Effect of 20% total weight flame-retardant [nCC+DAP] ratio 7:13 can effec-tively reduce the BR up to 54% compared to biocomposites without any FR. Biodegradability of biocomposite rPP/DVB/AA/KF/[nCC+DAP] was examined by burying the biocomposite specimens in the garbage soil dur-ing 4 months. The biodegradability of biocomposite was measured by the losing weight (LW) of biocomposite specimens; after burying in the soil for 4 months, it was found up to 11.82%. However, the present of [nCC+DAP] in the biocomposites can marginally decrease the TS, compared to that of without FR. Biocomposites rPP/DVB/AA/KF containing FR mixtures of [nCC+NaPP] or [nCC+DAP] were studied for their structure interactions, flammability, biodegradability and mechanical properties. The biocomposite rPP/DVB/AA/KF containing 20 phr of mixture [nCC+NaPP] with ratio of 7:13 (C8) can effectively increase TTI to 129% and in the same time reduce BR to 49% and heat release to 18% compared to biocomposites without any addition of FR. The biodegradability (LW) of biocomposite C7 was 7%. The biocomposite rPP/DVB/AA/KF containing 20 phr of mixture [nCC+DAP] with ratio of 7:13 (C7) can effectively increase TTI to 149% and in the same time reduce BR to 54% and increase heat release to 15% compared to biocomposites without any addition of FR. The biodegradability (LW) of biocomposite C7 was 11.82%. However, TS of biocomposites in the presence of FR mixture was slightly decreased. At the same time, better properties than rPP were obtained.

Salleh et al. [51] have studied 'Fracture Toughness Investigation on Long Kenaf/Woven Glass Hybrid Composite Due To Water Absorption Effect' in 2012. Water absorption of NFC is of serious concern especially for outdoor application. In this study, long kenaf/woven glass hybrid composite is fabricated in-house using cold press technique. The effect of water absorption on the hybrid composites is investigated at room temperature under three different environmental conditions, that is distilled water, rain water and seawater. The moisture absorption amount is obtained by calculating the different percentage weight before and after the immersion process. The moisture content is found to exhibit non-Fickian behaviour regardless of three different conditions. Liquid exposure of long kenaf/woven glass hybrid composite deteriorates the fracture toughness due to the weakening of interface between fibre and matrix. There are also several recognized modes of humidity ageing found through SEM observation. The effect of water absorption on mechanical properties of long kenaf/woven glass hybrid composite has been successfully investigated. The rates of the moisture uptake by the composites increase with immersion time and exhibit non-Fickian behaviour. Exposure of the NFC material to environmental conditions such as distilled water, seawater and rain water results in decreasing of fracture toughness. The decrement of the fracture toughness may be due to the water absorption characteristic depends on the content of the fibre, fibre orientation, area of exposed surface, permeability of fibre, void content and the hydrophilicity of the individual component.

Meona et al. [52] have studied 'Improving tensile properties of kenaf fibers treated with sodium hydroxide' in 2012. The study on kenaf short fibres compounded with MAPP/MAPE was successfully conducted. Kenaf has potential reinforced fibre in thermosets and thermoplastic composites. Basically, to produce the new type of composite, this project utilized short KF as the main material. The fibre is soaked with 3%, 6% and 9%

of NaOH for a day and then dried at 80°C for 24 h. The composition of short Kenaf used is 100 g. Two sets of combination were produced; combination between KF and MAPP as well as KF together with MAPE. The fabrication processes started when the mixture is poured into the mould and it is compacted until it perfectly fulfilled the mould. The mixture took about 1–2 h to completely dry. The specimens then were cut into standard dimension according to ISO 5275. MAPP and MAPE were used to improve the matrix–filler interaction and tensile properties of the composites. It has been found that the tensile properties of the treated KFs have improved significantly as compared to untreated KFs especially at the optimum level of 6% NaOH. From this research paper, the investigation of the effect of chemical treatment on the mechanical properties of short KFs with MAPP and MAPE had been achieved and presented. It has been found that the alkalization treatment has improved the tensile properties of the short KFs significantly as compared to untreated short KFs. It is also interesting to highlight that 6% NaOH yields the optimum concentration of NaOH for the chemical treatment. Furthermore, by adding MAPP or MAPE as coupling agent together with the short KFs has also amplified the TS in all untreated and treated cases.

Hao et al. [53] have studied 'Kenaf/polypropylene nonwoven composites: The influence of manufacturing conditions on mechanical, thermal and acoustical performance' in 2013. The kenaf/polypropylene nonwoven composites (KPNCs), with 50/50 blend ratio by weight, were produced by carding and needle-punching techniques, followed by a compression moulding with 6 mm thick gauge. The uniaxial tensile, three-point bending, in-plane shearing and Izod impact tests were performed to evaluate the composite mechanical properties. The thermal properties were evaluated using TGA, DSC and DMA. The performance of sound absorption and sound insulation was also investigated. An adhesive-free sandwich structure was found to have excellent sound absorption and insulation performance. Based on the evaluation of end-use performance, the best processing condition combination of 230°C and 120 s was determined, and the correlation between mechanical properties and acoustical behaviour was also verified by the panel resonance theory. The influence of manufacturing conditions was investigated by evaluating the mechanical, thermal and acoustical performance of KPNCs. We found that temperature and time are the most significant processing factors. For the 6 mm KPNC panels, processing at 230°C for 120 s (sample 5/230/120 or 7/230/120) gave the best mechanical properties. In contrast, samples 5/200/60 and 7/200/60, having the lowest moduli, were the best impact energy and sound absorbers and excellent sound barriers due to their panel–felt sandwich structure. In addition, the manufacturing conditions did not significantly affect the composite thermomechanical properties. KPNCs were more thermally stable than virgin PP plastics by adding KF as reinforcement. The sound insulation test verified the relationship between mechanical properties and acoustical behaviour described by the panel resonance theory, indicating that the classic panel resonance theory could be applied to the KPNCs produced in this study.

17.9 Keratin Fibre–Reinforced Composites

Wrześniewska et al. [54] have studied 'Novel Biocomposites with Feather Keratin' in 2007. This work deals with the preparation and characterization of cellulose–keratin biocomposites. A method of manufacturing fibrous composite materials by wet spinning is presented. We used natural polymers, bio-modified cellulose and keratin obtained from chicken feathers. Keratin waste is a potential renewable starting material. Spinning solutions were prepared from these polymers, and after filtration and aeration, they were used for the formation of fibres and fibrids. The investigations included the preparation of bio-modified cellulose–keratin spinning solutions of different keratin content, estimation of the influences of formation speed and drawing on the fibre properties, and estimation of the sorption properties of the composites obtained. The bio-modified cellulose–keratin fibres obtained are characterized by better sorption properties, higher hygroscopicity and smaller wetting angle, than those of cellulose fibres. The introduction of keratin into cellulose fibres lowered their mechanical properties, but they reach a level which enables the application of these fibres for manufacturing composite fibrous materials. Composting tests, carried out at 500C, showed the better biodegradability of composite fibres than that of cellulose fibres. The bio-modified cellulose–keratin fibrids are also characterized by better sorption properties than those of cellulose fibrids. As a result of the lyophilization of composite fibrids, cellulose–keratin sponges were obtained. Cellulose–keratin biocomposites in the shape of fibres and fibrids were obtained as a result of our investigations. It was demonstrated that adding keratin improves the hygroscopic properties. The modification of cellulose fibres by adding keratin, evaluated on the basis of nitrogen content, the changes in hygroscopicity, and the value of the wetting angle, is evident for all cellulose–keratin fibres. The cellulose–keratin fibres obtained are characterized by better sorption properties, higher hygroscopicity and a smaller wetting angle, than those of cellulose fibres. Also, the biodegradability of cellulose–keratin fibres is better than that of cellulose fibres. Introducing keratin into cellulose fibres lowered their mechanical properties, decreased the tenacity by a factor of nearly two, and a significant decrease in elongation at break took place, but both parameters remained at a level which enables these fibres to be used for manufacturing composite fibrous materials. However, a further increase in the amount of keratin in the cellulose spinning solution over the values used in this research work may essentially hinder the fibre formation process. Depending on the keratin content in a ready-to-use spinning solution, the nitrogen content in cellulose–keratin fibrids varies within the range of 0.95%–2.38%. The water retention value changes from 133.5% to 186.47%, as compared to cellulose fibres for which it is 98.87%. Evaluating the view of cellulose–keratin fibrous composite materials by the SEM method, we noted that cellulose–keratin fibres have a more developed surface than that of cellulose fibres, whereas cellulose–keratin fibrids are thinner and shorter than alginate fibrids. Studies in progress indicate that this approach is highly promising and may lead to the production of a new range of fibrous composite materials with innovative properties and based on renewable natural resources. Improving the sorption properties of the biocomposites obtained creates opportunities to use them as hygienic fabrics.

Salhi et al. [55] have studied 'Development of Biocomposites Based of Polymer Metrix And Keratin Fibers: Contribution To Poultry Feather Biomass Recycling. Part I' in 2011. Processing poultry feather biomass into useful products presents interesting opportunities of recycling agricultural waste material. However, few works are relating the poultry feather properties and to their processing. According to a chemical process developed in our laboratory, poultry feathers were converted into wool, fibres or powder so as to meet several applications in many fields. The morphology of a feather is mainly composed of a central shaft called the rachis to which are attached the secondary structures, the barbs. The tertiary structures, the barbules, are attached to the barbs in a manner similar to the barbs being attached to the rachis. The longitudinal and cross-sectional features of the rachis and the barbs are characterized by honeycomb-shaped hollow cells in the cross section. Hence, hollow cells may act as air and heat insulators especially in automotive and building applications. Due to their unique structure and to their physical properties that are not available unlike to most natural or synthetic fibre case, a suitable moulding technique was performed in order to produce highly filled composite samples. The concentration range was varied from 10% to 50% by weight with an increase of 10, because of practical reasons. The morphological, thermophysical and mechanical investigations of polyester matrix filled with several concentration of the keratin fibres show good results. Poultry feathers are converted into useful products according to the chemical technique performed in our laboratory. The feathers are converted into powder, fibres and wool in order to meet several applications. The structure and properties of the converted poultry feather indicate that these products are useful as natural protein fibres. Due to the specific features of the fibres, their extremely low density, the very large availability and low cost of production may make these fibres preferable for composites and textiles. From the morphological point of view, the presence of honeycomb structures makes poultry feather fibres to have low density and also provides air and heat insulating capabilities unlike any other natural or synthetic fibres. The relevant composites showed a decrease in mechanical and thermophysical properties with increasing poultry feather products.

17.10 Other Natural Fibre–Reinforced Composites

Drzal et al. [56] have studied 'Bio-Composite Materials as Alternatives to Petroleum-Based Composites for Automotive Applications' in 2003. Natural/biofibre composites (biocomposites) are emerging as a viable alternative to glass fibre–reinforced composites especially in automotive applications. NFs, which traditionally were used as fillers for thermosets, are now becoming one of the fastest growing performance additives for thermoplastics. Advantages of NFs over man-made glass fibre are low cost, low density, competitive specific mechanical properties, reduced energy consumption, carbon dioxide sequestration and biodegradability. NFs offer a possibility to developing countries to use their own natural resources in their composite processing industries. The combination of biofibres like kenaf, hemp, flax, jute, henequen, pineapple leaf fibre and sisal with polymer matrices from both nonrenewable and renewable resources to produce composite materials that are competitive with synthetic composites requires special attention, that is biofibre–matrix interface and novel processing. NF-reinforced PP composites have attained commercial attraction in automotive industries. Needle punching techniques as well as extrusion followed by injection moulding for NF–PP composites as presently adopted in the industry need a *greener* technology—powder impregnation technology. NF–PP or NF–polyester composites are not sufficiently eco-friendly due to the petro-based source as well as nonbiodegradable nature of the polymer matrix. Sustainability, industrial ecology, eco-efficiency and green chemistry are forcing the automotive industry to seek alternative, more eco-friendly materials for automotive interior applications. Using NFs with polymers (plastics) based on renewable resources will allow many environmental issues to be solved. By embedding biofibres with renewable resource–based biopolymers such as cellulosic plastic, corn-based plastic, starch plastic and soy-based plastic are continuously being developed at Michigan State University. After decades of high-tech developments of artificial fibres like aramid, carbon and glass, it is remarkable that NFs have gained a renewed interest, especially as a glass fibre substitute in automotive industries. New environmental regulations and societal concern have triggered the search for new products and processes that are compatible to the environment. The incorporation of bio-resources into composite materials can reduce further dependency of petroleum reserves. The major limitations of present biopolymers are their high cost. Again renewable resource–based bioplastics are currently being developed and need to be researched more to overcome the performance limitations. Biocomposites can supplement and eventually replace petroleum-based composite materials in several applications, thus offering new agricultural, environmental, manufacturing and consumer benefits. The main advantage of using renewable materials is that the global CO_2 balance is kept at a stable level. Several critical issues related to biofibre surface treatment to make it more reactive, bioplastic modification to make it a suitable matrix for composite application and processing techniques depending on the type of fibre form (chopped, nonwoven/woven fabrics, yarn, sliver, etc.) need to be solved to design biocomposites of commercial interests. Biocomposites are now emerging as a realistic alternative to glass-reinforced composites. Biocomposites being derived from renewable resources; the cost of the materials can be markedly reduced with their large-scale usage. Recent advances in genetic engineering, NF development and composite science offer significant opportunities for improved value-added materials from renewable resources with enhanced support of global sustainability. Thus, the main motivation for developing biocomposites has been and still is to create a new generation of fibre-reinforced plastics with glass fibre reinforced–like or even superior properties that are environmentally compatible in terms of production, usage and removal. NFs are biodegradable, but renewable resource–based bioplastic can be designed to be either biodegradable or not according to the specific demands of a given application. The raw materials being taken from renewable resources; biocomposites are prone to integrate into natural cycle. The general environmental awareness and new rules and regulations will contribute to an increase in the work for more eco-friendly concept in the automotive industry.

Singha and Thakur [57] have studied 'Mechanical properties of natural fibre reinforced polymer composites' in 2008. During the last few years, NFs have received much more attention than ever before from the research community all over the world. These NFs offer a number of advantages over traditional synthetic fibres. In the present communication, a study on the synthesis and mechanical properties of new series of green composites involving *Hibiscus sabdariffa* fibre as a reinforcing material in urea–formaldehyde (UF) resin–based polymer matrix has been reported. Static mechanical properties of randomly oriented intimately mixed *H. sabdariffa* fibre–reinforced polymer composites such as tensile, compressive and wear properties were investigated as a function of fibre loading. Initially UF resin prepared was subjected to evaluation of its optimum mechanical properties. Then reinforcing of the resin with *H. sabdariffa* fibre was accomplished in three different forms: particle size, short fibre, and long fibre by employing optimized resin. The present work reveals that mechanical properties such as TS, compressive strength and wear resistance of the UF resin increase to a considerable extent when reinforced with the fibre. Thermal (TGA/DTA/DTG) and morphological studies (SEM) of the resin and biocomposites have also been carried out. Various test methods were adapted for mechanical characterization of NF-reinforced polymer composites. In case of mechanical behaviour, particle reinforcement of the UF resin has been found to be more effective as compared to short fibre reinforcement. These results suggest that *H. sabdariffa* fibre has immense scope in the fabrication of NF-reinforced polymer composites having vast number of industrial applications.

Dr. Beckwith et al. [58] have studied 'Natural Fibers: Nature Providing Technology for Composites' in 2008. NFs are rapidly emerging in composite applications where glass fibres (predominantly E-glass) have been traditionally used. This is particularly true within the automotive and construction industries. These NFs provide several benefits: low cost, *green* availability, lower densities, recyclable, biodegradable, moderate mechanical properties and abundant. Their uses have found entry into both the thermoset and thermoplastic composite market places. Industries are rapidly learning how to effectively process these natural resources and use them in numerous composite applications. Typically they are used with well-recognized thermoset resin families: polyesters, VEs and epoxies. Thermoplastic resin matrices also are those commonly seen within the commercial markets: PP, LDPE, HDPE, polystyrene, Nylon 6 and Nylon 6,6 systems. Soy-based resin systems also are coming into vogue in some applications as we learn more about its chemistry and processing. NF systems tend to fall into several categories. There is insufficient space and time to discuss all of the processing pros and cons of NFs within this short technical note. However, it is worthwhile looking at least a few of the more common NFs and make some comparisons with traditional commercial fibres (such as E-glass) and aerospace fibres (S-glass and aramid). A number of the critical mechanical and physical properties of common NFs compared to E-glass, S-glass, commercial aramid and standard modulus PAN-based carbon fibre materials were studied. While the strength properties, and often the stiffness properties, are not on comparable levels to S-glass, aramid, and carbon fibres, the NFs do provide a wide range of workable strength and stiffness properties when compared with E-glass fibre. NFs are very typically about 30%–50% lower in density when compared to E-glass fibres and roughly the same as the aramid commercial grade systems. This advantage has made NFs quite attractive to use within the automotive industry across a wide range of applications with both thermoset and thermoplastic resin matrices. In fact, the stiffness of some of these NFs can be higher than or equivalent to that of E-glass (see, e.g., hemp and ramie). Flax, sisal and KFs also tend towards a higher stiffness. Hence, for stiffness-driven designs, these fibres are reasonable options. Their abundance, and a growing understanding of their processibility, makes them attractive to engineers.

Lee et al. [59] have studied 'Fabrication of Long and Discontinuous Natural Fiber Reinforced Polypropylene Biocomposites and Their Mechanical Properties' in 2008. NF-reinforced PP biocomposites were fabricated by blending long and discontinuous (LD) NFs with LD PP fibres. Firstly, random fibre mats were prepared by mixing NFs and PP fibres using a carding process. Then heat and pressure were applied to the mats, such that the PP fibres dispersed in the mats melted and flowed out, resulting in the formation of consolidated sheets upon subsequent cooling. The effect of the fibre volume fraction on the mechanical properties of the biocomposites was scrutinized by carrying out tensile and flexural tests and observing the interface between the fibre and matrix. It was observed that the natural LD fibre content needs to be maintained at less than the nominal fibre fraction of 40% by weight for the composites fabricated using the current method, which is quite low compared to that of continuous or short fibre–reinforced composites. The limited fibre fraction can be explained by the void content in the biocomposites, which may be caused by the nonuniform packing or the deficiency of the matrix PP fibres. LD NF (kenaf and jute)–reinforced PP composites were fabricated using the carding and punching processes followed by hot press compression moulding. It was concluded that the KF-reinforced PP composites have an optimum nominal fibre fraction of 30% by weight at which the tensile and flexural modulus are the highest, while the reduction in strength is minimal. For the jute fibre–reinforced composites, a fibre fraction of 40% (by weight) seems to be the optimum value. The limited fibre fraction was explained by the void content in the biocomposites, which may be caused by the nonuniform packing or the deficiency of the matrix PP fibres. To incorporate more NFs into the biocomposites, the fibre length of both the natural and PP fibres may need to be shortened; however, too short fibres may spoil the processibility of the carding operation; thus, further experimental or theoretical studies are necessary to determine the optimum fibre fraction.

Westman et al. [60] have studied 'Natural Fiber Composites: A Review' in 2010. The need for renewable fibre–reinforced composites has never been as prevalent as it currently is. NFs offer both cost savings and a reduction in density when compared to glass fibres. Though the strength of NFs is not as great as glass, the specific properties are comparable. Currently, NFCs have two issues that need to be addressed: resin compatibility and water absorption. The following preliminary research has investigated the use of kenaf, *Hibiscus cannabinus*, as a possible glass replacement in fibre-reinforced composites. As predicted, the specific properties of kenaf were less than that of the glass composites. This is primarily explained by the interface between the kenaf and the VE resin. The wettability of the chopped glass is significantly higher than the kenaf mats, which leads to stronger samples. Additionally, the kenaf has only bidirectional orientation, while the glass has a multidirectional orientation. This difference in orientation drastically changes how the stress is distributed across the composite. The 24 h water absorption of kenaf and glass composite samples have been done respectively. The percent mass increase was significantly higher for the kenaf samples than the glass. As expected, the tensile samples had the greatest percent of water absorption due to their greater contact area with the water. The kenaf samples were heavily distorted including large bulges on the edges and warping of the surface. This attributed to the lower mechanical properties. Finally, the proceeding data have shown that in their native form, kenaf composites cannot compete with glass composites. While the dry specific properties were only slightly lower for the kenaf composites, the wet samples were drastically lower. Fibre treatments will need to be explored to reduce the water absorption and increase the wettability of the fibres.

Ku et al. [61] have studied 'A review on the tensile properties of natural fiber reinforced polymer composites' in 2011. This paper is a review on the tensile properties of NF-reinforced polymer composites. NFs have recently become attractive to researchers, engineers and scientists as an alternative reinforcement for fibre-reinforced polymer (FRP) composites. Due to their low cost, fairly good mechanical properties, high specific strength, nonabrasive, eco-friendly and bio-degradability characteristics, they are exploited as a replacement for the conventional fibre, such as glass, aramid and carbon. The tensile properties of NF-reinforced polymers (both thermoplastics and thermosets) are mainly influenced by the interfacial adhesion between the matrix and the fibres. Several chemical modifications are employed to improve the interfacial matrix–fibre bonding, resulting in the enhancement of tensile properties of the composites. In general, the TSs of the NF-reinforced polymer composites increase with fibre content, up to a maximum or optimum value; the value will then drop. However, Young's modulus of the NF-reinforced polymer composites increases with increasing fibre loading. Khoathane et al. found that the TS and Young's modulus of composites reinforced with bleached hemp fibres increased incredibly with increasing fibre loading. Mathematical modelling was also mentioned. It was discovered that the rule of mixture (ROM) predicted and experimental TS of different NF-reinforced HDPE composites were very close to each other. Halpin–Tsai equation was found to be the most effective equation in predicting Young's modulus of composites containing different types of NFs. The scientific world is facing a serious problem of developing new and advanced technologies and methods to treat solid wastes, particularly non-naturally-reversible polymers. The processes to decompose those wastes are actually not cost-effective and will subsequently produce harmful chemicals. Owing to the aforementioned ground, reinforcing polymers with NFs is the way to go. In this paper, most of the NFs mentioned were plant based, but it should be noted that animal fibres like silkworm cocoon silk, chicken feather and spider silk have also been used and the trend should go on. Those fibres, both animal and plant based, have provided useful solutions for new materials development, in the field of material science and engineering. NFs are indeed renewable resources that can be grown and made within a short period of time, in which the supply can be unlimited as compared with traditional glass and carbon fibres for making advanced composites. However, for some recyclable polymers, their overall energy consumption during collecting, recycling, refining and remoulding processes has to be considered to ensure the damage of the natural cycle would be kept as minimal. On top of it, NFs are low cost, recyclable, low density and eco-friendly material. Their tensile properties are very good and can be used to replace the conventional fibres such as glass and carbon in reinforcing plastic materials. A major drawback of using NFs as reinforcement in plastics is the incompatibility, resulting in poor adhesion between NFs and matrix resins, which subsequently leads to low tensile properties. In order to improve fibre–matrix interfacial bonding and enhance tensile properties of the composites, novel processing techniques and chemical and physical modification methods are developed. Also, it is obviously clear that the strength and stiffness of the NF-reinforced polymer composites are strongly dependent on fibre loading. The TS and modulus increase with increasing fibre weight ratio up to a certain amount. If the fibre weight ratio increases below the optimum value, load is distributed to more fibres, which are well bonded with resin matrix resulting in better tensile properties. Further increment in fibre weight ratio has resulted in decreased TS as described in the main text. Mathematical models were also found to be an effective tool to predict the tensile properties of NF-reinforced composites. Finally, it can be found that the main weakness to predict the tensile properties of plant-based NFCs by modelling was giving too optimistic values like results. The modelling has to be improved to allow improvements in the prediction of tensile properties of composites reinforced with both plant- and animal-based fibres.

Sahari and Sapuan [62] have studied 'Natural Fibre Reinforced Biodegradable Polymer Composites' in 2011. Currently, numerous research groups have explored the production and properties of biocomposites where the polymer matrices are derived from renewable resources such as PLA, thermoplastic starch (TPS), cellulose and polyhydroxyalkanoates (PHAs). This review is carried out to evaluate the development and properties of NF-reinforced biodegradable polymer composites. They are the materials that have the capability to fully degrade and compatible with the environment. NF-reinforced biodegradable polymer composites appear to have very bright future for wide range of applications. These biocomposite materials with various interesting properties may soon be competitive with the existing fossil plastic materials. However, the present low level of production and high cost restrict them to be applied in industrial application. In addition, its hydrophilic properties make the real challenge to design the product, which can be a good candidate for outdoor applications. Thus, further research and improvement should be conducted so that these fully degraded composites can easily be manipulated and can give benefit to all mankind and environmental issues.

Gironès et al. [63] have studied 'Natural fiber-reinforced thermoplastic starch composites obtained by melt processing' in 2012. TPS from industrial nonmodified corn starch was obtained and reinforced with natural strands. The influence of the reinforcement on physical–chemical properties of the composites obtained by melt processing has been analyzed. For this purpose, composites reinforced with different amounts of either sisal or hemp strands have been prepared and evaluated in terms of crystallinity, water sorption and thermal and mechanical properties. The results showed that the incorporation of sisal or hemp strands caused an increase in the glass transition temperature (Tg) of the TPS as determined by DMTA. The reinforcement also increased the stiffness of the material, as reflected in both the storage modulus and Young's modulus. Intrinsic mechanical properties of the reinforcing fibres showed a lower effect on the final mechanical properties of the materials than their homogeneity and distribution within the matrix. Additionally, the addition of a natural latex plasticizer to the composite decreased the water absorption kinetics without affecting significantly the thermal and mechanical properties of the material. TPS has been obtained from corn starch and glycerol. The TPS obtained has been physicochemical and mechanically characterized. Addition of natural latex has probed to decrease the kinetics of water absorption of the composites, although its effect on water uptake, crystallinity and mechanical properties was negligible. Composite materials obtained through reinforcement of the TPS matrix with sisal and hemp strands have been prepared and characterized. Incorporation of the NF strands, independently of type and content, caused the reduction of VH-type structures. The results demonstrated that both tensile and flexural strengths improved with the percentage of reinforcement fibres. Despite having similar mechanical properties, hemp strands provided composites with better mechanical properties than those obtained with sisal. This effect is assigned to the better fibrillation obtained during the mixing of the TPS with the hemp strands and points out the mechanical anchoring as the main cause of the enhanced mechanical resistance.

Kuranska and Prociak [64] have studied 'Porous polyurethane composites with natural fibres' in 2012. In this work, the methodology and results of the investigations that concern rigid polyurethane foams modified with NFs and oil-based polyol are presented. The goal of the investigations was to obtain the cellular, polyurethane composites with the heat insulating and mechanical properties similar or better as in the case of the reference material. The obtained polyurethane composites had apparent densities of about 40 kg/m^3. The modified composites contained the considerable part of biodegradable components on the base of renewable raw materials. The influence of the rapeseed oil-based polyol, flax and hemp fibres of different length on the cell structure; closed cell content; apparent density; thermal conductivity; and compression strength of the rigid polyurethane composites are analyzed. In the case of application of fibre in the amount of 5% php (per 100 polyols), the foam composites with the highest values of compressive strength and the lowest thermal conductivity were obtained. The replacement of petrochemical polyol with the amount of 30 wt.% of rapeseed oil derivative and the addition of flax fibres with the length of 0.5 mm in the amount of 5 php have beneficial influence on thermal insulation and mechanical properties of rigid polyurethane foams. The increasing of compression strength up to about 18% for the perpendicular direction (Y) and decreasing of thermal conductivity 1.9% can be achieved in comparison to the reference material. The incorporation of flax fibres up to 10 php has no significant effect on the composite cell size; however, considerable influence on cell anisotropy index what reflects the tendency for cell elongating, especially in the foam rise direction. The application of such renewable components as bio-polyol and NFs in the polyurethane formulation allows to increase the content of biocomponents and to improve mechanical and heat insulating properties of such porous composites.

Sobczak et al. [65] have studied 'Polypropylene composites with natural fibers and wood–General mechanical property profiles' in 2012. NFCs and wood polymer composites (WPCs) based on PP have gained increasing interest over the past two decades, both in the scientific community and in industry. Meanwhile, a large number of publications are available, yet the actual market penetration of such materials is rather limited. To close the existing gap between scientific and technical knowledge, on the one hand, and actual market applications, on the other, it is the purpose of this paper to analyze the current state of knowledge on mechanical performance profiles of injection moulded NFCs and WPCs. As the composite properties are a result of the constituent properties and their interactions, special attention is also given to mechanical fibre/filler properties. Moreover, considering that NFCs and WPCs for a variety of potential applications compete with mineral reinforced (MR; represented in this study by talc), short glass fibre (SGF), long glass fibre (LGF) and short carbon fibre (SCF)–reinforced PP, property profiles of the latter materials are included in the analysis. To visualize the performance characteristics of various materials in a comparative manner, the data were compiled and illustrated in so-called Ashby plots. Based on these comparisons, an assessment of the substitution potential of NFCs and WPCs is finally performed, along with a discussion of still open issues, which may help in guiding future material development and market application efforts. As to the fibre properties, conventional fibres such as SGFs, LGFs, and SCFs exhibit significantly higher strength values than even the best NFs (factor 2–4). In terms of modulus, some NFs, like hemp and kenaf, show values similar to SGF/LGF, with SCF modulus values superseding these fibres by a factor of 3. Due to the lower density of NFs compared to glass fibres, the specific properties of NFs, on the one hand, and SGF/LGF, on the other hand, shift closer together. Particularly remarkable is that hemp even supersedes SGF/ LGF in the specific modulus, and flax approaches SGF in specific strength. For the resulting PP composites, in terms of absolute properties, the picture is largely similar, with some remarkable exceptions. First, the relative difference between NFCs and WPCs, on the one hand, and PP–SGF/LGF on the other, is reduced due to ROM-based effects. Second, WPCs supersede PP–talc composites, both in modulus and strength, while NFCs largely overlap with the PP–SGF/LGF range for modulus and approach its lower end for strength. In terms of specific properties, the position of NFCs and WPCs relative to the conventional composites (except for PP–SCF) is again somewhat improved, due to the aforementioned density differences. The perhaps most significant drawback of NFCs and WPCs, compared to the conventional PP composites, is related to their lower IS. While NFCs at least partly overlap with the PP– talc range, all WPCs exhibit inferior impact behaviour. It should be pointed out, however, that one NFC grade, that is PP–Tencel, performs remarkably well, being the only PP composite presented which retains the (unnotched Charpy) IS level of neat PP, thus even exceeding PP–SGF composites. Overall, NFCs may substitute PP–SGF composites when some reduction in strength is accepted. WPCs, on the other hand, may replace PP–talc composites, in applications where impact strength is not critical. Reflecting on the current state of knowledge and technology in the field of NFCs and WPCs, there are several open issues and aspects yet to be addressed. These include the effects of temperature and moisture uptake on mechanical properties and processing behaviour. Moreover, there is a large variability in properties of NFs and wood, depending on growth and harvesting conditions. Also, for a number of advantages usually associated with NFCs and WPCs (i.e. reduced abrasiveness in processing, improved noise damping behaviour, improved overall ecobalance compared to conventional composites), there is a need for quantitative and reliable data in support of these reputed benefits. Finally, based on constituent property considerations, the performance potential of NF and wood composites at this stage may not yet be fully exploited. Hence, further research is warranted on elucidating structure–property relation-ships for these materials to overcome current weaknesses (e.g. impact strength).

Christian et al. [66] have studied 'Moisture diffusion and its impact on uniaxial tensile response of biobased composites' in 2012. Bio-based com-posites made from biopolymers and plant-based fibres are being evaluated for construction applications as replacements for wood or petroleum-based composites and plastics. The bio-based composites studied here have been demonstrated to rapidly biodegrade in anaerobic conditions to methane, thereby reducing construction-related landfill waste and producing a useful end product, namely, fuel for energy or a feedstock to grow more biopoly-mer. To be useful in construction, susceptibility to moisture and eventual moisture resistance are necessary. Diffusion properties and mechanical properties are characterized in various moisture and temperature con-ditions for hemp/cellulose acetate and hemp/poly(b-hydroxybutyrate) (PHB) composites. The composites were observed to follow Fick's second law of diffusion. The tensile moduli of elasticity were found to decrease with full moisture saturation, while the ductility increased and the ulti-mate strength did not change significantly. Measured diffusion coefficients are compared to petroleum-based and other bio-based composites. The rate of moisture diffusion into two bio-based composite materials and how that moisture affects composite mechanical behaviour have been investigated. Specifically, cellulose acetate and PHB biopolymer matrices with hemp fab-ric reinforcement were studied. The diffusion and tensile properties were measured for these bio-based composites conditioned in a moisture cham-ber at 100% RH at three different temperatures. The diffusion coefficients, calculated using Fick's law, for both bio-based composites were comparable to wood and other bio-based composites and higher than the coefficients for synthetic composites. The cellulose in the hemp fibre is hydrophilic and absorbs a significant amount of water similar to the cellulose in wood but in contrast to synthetic fibres such as glass. The hemp/cellulose acetate composite absorbed 40%–60% more moisture than the hemp/PHB composite. This observation is attributed to the cellulose acetate itself being hydrophilic, whereas PHB is both hydrophobic and more crystalline than cellulose acetate. The rate of diffusion for both materials was found to increase with increasing temperature. The behaviour was found to be Fickian until near saturation at which point there was some material loss, which is attributed primarily to damage from the swelling of the NFs. To determine the effects of moisture on mechanical behaviour, specimens conditioned by saturation and by saturation followed by drying were tested in tension and compared to unconditioned specimens tested in tension. The hemp/cellulose acetate composites were more affected by moisture absorption than the hemp/PHB composites. The initial moduli of elasticity decreased upon saturation by 75% and 29% for the hemp/cellulose acetate and hemp/PHB composites, respectively. Strain at failure increased by 70% for the hemp/PHB and 108% for the hemp/cellulose acetate. The strength properties degraded by 16% for the hemp/cellulose acetate and increased by 8% for the hemp/PHB composite. The absorption of moisture likely plasticized the cellulose acetate (possibly leading to hydrolytic decomposition) and also weakened its interfacial bond to the fibres, resulting in degraded mechanical properties. In the PHB composite, there was less damage to the PHB itself when saturated, but still a loss of composite stiffness attributed to changes in interfacial bond and fibre swelling. The strength increase for the hemp/PHB composite may have been the result of increased fibre–matrix interaction due to differential moisture swelling: the fibre swelled while the matrix did not, thus increasing the mechanical bond and frictional resistance when saturated. To determine if saturation may cause permanent damage to the bio-based composites, specimens were saturated and then allowed to dry in the unconditioned environment prior to tensile testing. Upon drying, some of the lost properties were recovered, but permanent damage was apparent, as the stress–strain behaviour did not return to the unconditioned behaviour. The strength of the hemp/PHB composites decreased by 14% upon drying from the saturated state. Permanent damage was attributed to damage to the interface from the moist to dry cycle and cracking or crazing of the matrices due to stresses from fibre swelling. The mechanical properties of bio-based composites conditioned to full saturation at different temperatures were compared, and the temperature within the conditioning chamber was found to have no effect on tensile behaviour. The degradation of mechanical properties upon introduction to humid environments limits the potential applications of these bio-based composites. For these bio-based composites to be used widely within the construction industry, the original motivation for the study, they must be protected from moisture, for example through sealants and/or fibre treatments. Due to its hydrophobic behaviour, the PHB matrix will likely be more suitable for use in humid environments than the cellulose acetate matrix. The results of this research are currently being used to guide the optimal design of new composites for both moisture-resistant in-service performance as well as rapid anaerobic biodegradation out of service.

Kabir et al. [67] have studied 'Chemical treatments on plant-based natural fibre reinforced polymer composites: An overview' in 2012. This paper provides a comprehensive overview on different surface treatments applied to NFs for advanced composite applications. In practice, the major drawbacks of using NFs are their high degree of moisture absorption and poor dimensional stability. The primary objective of surface treatments on NFs is to maximize the bonding strength so as the stress transferability in the composites. The overall mechanical properties of NF-reinforced polymer composites are highly dependent on the morphology, aspect ratio, hydrophilic tendency and dimensional stability of the fibres used. The effects of different chemical treatments on cellulosic fibres that are used as reinforcements for thermoset and thermoplastics are studied. The chemical sources for the treatments include alkali, silane, acetylation, benzoylation, acrylation, acrylonitrile grafting, maleated coupling agents, permanganate, peroxide, isocyanate, stearic acid, sodium chlorite, triazine, fatty acid derivate (oleoyl chloride) and fungal. The significance of chemically treated NFs is seen through the improvement of mechanical strength and dimensional stability of resultant composites as compared with a pristine sample. NFs are gaining interest to be used as reinforcement in polymer composites due to its potential mechanical properties, processing advantages and environmental benefits. However, hydrophilic nature of the fibres lowers the compatibility with the matrix. This incompatibility results poor mechanical properties of the composites. Chemical treatment is an essential processing parameter to reduce hydrophilic nature of the fibres and thus improves adhesion with the matrix. Pretreatments of fibre change its structure and surface morphology. Hydrophilic hydroxyl groups are removed from the fibre by the action of different chemicals. More reactive hydroxyl groups make an effective coupling with the matrix. Significant improvements in the mechanical properties of the composites are reported by using different chemical treatment processes on the reinforcing fibre.

Ho et al. [68] have studied 'Critical factors on manufacturing processes of natural fibre composites' in 2012. Elevated environmental awareness of the general public in reducing carbon footprints and the use nonnaturally decomposed solid wastes has resulted in an increasing use of natural materials, biodegradable and recyclable polymers and their composites for a wide range of engineering applications. The properties of NF-reinforced polymer composites are generally governed by the pretreated process of fibre and the manufacturing process of the composites. These properties can be tailored for various types of applications by properly selecting suitable fibres, matrices, additives and production methods. Besides, due to the complexity of fibre structures, different mechanical performances of the composites are obtained even with the use of the same fibre types with different matrices. Some critical issues like poor wettability, poor bonding and degradation at the fibre–matrix interface (a hydrophilic and hydrophobic effect) and damage of the fibre during the manufacturing process are the main causes of the reduction of the composite's strength. In this paper, different manufacturing processes and their suitability for NFCs, based on the materials, mechanical and thermal properties of the fibres and matrices, are discussed in detail. The development of NF-reinforced polymer composites has been a hot topic recently due to the increasing environmental awareness on reducing the use of fossil fuel and its related products. NF can be classified for plant based and animal based. The selection criteria are highly dependent on their type, application and cost. However, there is still uncertain on which type of manufacturing processes that are suitable for producing these composites as their materials and mechanical characteristics are different as compared with traditional carbon and glass fibre composites in general. Some processes, their original design, were not targeted for NFs, while their technologies have been well developed for fast and reliable composite production. This paper addresses a comprehensive review on different types of composites, manufacturing process and their effects to the NF and its composites.

Azwa et al. [69] have studied 'A review on the degradability of polymeric composites based on natural fibres' in 2013. The applications of NF/polymer composites in civil engineering are mostly concentrated on non-load-bearing indoor components due to its vulnerability to environmental attack. This paper evaluates the characteristics of several NFCs exposed to moisture, thermal, fire and UV degradation through an extensive literature review. The effects of chemical additives such as fibre treatments, FRs and UV stabilizers are also addressed. Based on the evaluation conducted, optimum fibre content provides strength in a polymer composite, but it also becomes an entry point for moisture attack. Several fibre treatments are also being used to improve fibre–matrix interface, thereby increasing moisture durability. However, the treated fibres were found to behave poorly when exposed to weather. The addition of UV stabilizers and FRs is suggested to enhance outdoor and fire performance of NF/polymer composite but compromises its strength. Therefore, from the collected data and various experimental results, it was concluded that an optimum blend ratio of chemical additives must be employed to achieve a balance between strength and durability requirements for NFCs. The degradability of polymer composites based on NFs was reviewed in this article. This includes the degradation due to moisture, thermal effects, fire and UV rays. NFs are susceptible to biodegradation; thus composites based on them face higher risk of degradation when subjected to outdoor applications as compared to composites with synthetic fibres. Different cell wall polymers of lignocellulosic fibres have different influence on their properties and degradability. For instance, cellulose is responsible for strength of fibres, hemicelluloses for thermal, biological and moisture degradation, while lignin for UV degradation and char formation.

Fibre content is the major factor affecting water absorption of composites as it enhances matrix porosity by creating more moisture path into the matrices. Poor adhesion between fibre particles and polymer matrix generates void spaces around the fibre particles. Higher fibre volume composites immersed in water generally have greater decrement in tensile and flexural properties compared to dry samples. Moisture absorption can be reduced through fibre modifications such as alkalization and addition of coupling agents. Polymer matrix decomposes at 300°C–500°C. Flame retardant is one of the methods used to improve its thermal durability. The best potential for flame retardant of natural FRP is by combining char-forming cellulosic material with an intumescent system. FRs from inorganic compounds such as metallic hydroxide additives are preferred due to environmental and health safety reasons, with magnesium hydroxide showing very good results. Weathering causes degradation of polymer composite through photoradiation, thermal degradation, photooxidation and hydrolysis. These processes result in changes in their chemical, physical and mechanical properties. Water enhances rate of degradation through swelling of fibre, and this leads to further light penetration. Weathered chemically treated composites show a relatively greater extent of decrease in TS, which proves that treated composites undergo more degradation compared to untreated composites.

Arrakhiz et al. [70] have studied 'Mechanical and thermal properties of natural fibers reinforced polymer composites: Doum/low density polyethylene' in 2013. Polymer composite materials with vegetable fibres were an attractive field for many industries and researchers; however, these materials required the issues of compatibility between the fibres and the polymeric matrix. This work evaluates the thermal and mechanical properties of doum fibres reinforcing an LDPE composite to follow the effect of adding fibres into polymer matrix. Doum fibres were alkali treated to clean the fibre surface and improve the polymer/fibre adhesion. The doum fibres were compounded in LDPE matrix at various contents and extruded as continuous strands. An enhance on mechanical properties of composites was found, a gain of 145% compared to neat polymer at 30 wt.% fibre loading in Young's modulus, a gain of 135% in flexural modulus at 20 wt.% fibre loading, and a gain of 97% in torsional modulus at 0.1 Hz. Thermal properties were evaluated, and the results show a slight decrease with increase of added doum. The objective of this study was to define the benefits of reinforcing LDPE with doum fibres. To improve the surface properties of the fibres, they have received a traditional surface treatment to bleach them and clean their surface from hydrogen bonds and other wastes. All samples were prepared in a twin screw extruder and injected in a moulding injection machine. After addition of the doum fibres, a significant increase in Young's modulus, the flexural modulus and the torsion modulus was observed. Thermal analysis showed a slight decrease in properties from neat matrix polymer; the crystallinity and the melt of composites are mainly influenced by the NF loading.

Prasanna et al. [71] have studied 'Modification, Flexural, Impact, Compressive Properties & Chemical Resistance Of Natural Fibers Reinforced Blend Composites' in 2013. In the present work, intercross-linked networks of UP-toughened epoxy blends were developed, and this blend was used as matrix. The palmyra–BFs reinforced into this matrix blend and the matrix blend hybrid biocomposites were fabricated by hand lay-up technique. The objective of work is to optimize the process for the production of high-performance, low-cost and less weight NF-reinforced blend hybrid biocomposite for the automotive and transportation industry applications. The variation of impact and flexural and compressive properties of 10%, 20% and 30% volume content of untreated and treated palmyra–BF-reinforced blend hybrid biocomposites were studied. The effect of different percentage of fibre content with and without alkali treatment on mechanical properties was investigated. The mechanical test results clearly show that these properties were found to be higher when alkali-treated banana–palmyra fibres were used in the matrix blend composites. The mechanical properties were optimally increased at 20% volume content of alkali-treated banana–palmyra fibres when compared with 10% and 30% volume content of treated fibres and 10%, 20% and 30% volume content of untreated fibre–reinforced composites. Chemical resistance was also significantly improved for all the chemicals except toluene and carbon tetrachloride. In the case of morphological features, results clearly show that when polymer resin matrix was reinforced with fibres of different % volume content, morphological changes take place depending on the % volume content of fibres. The morphology of fractured surface indicates good bonding between matrix and fibre reinforcement. The variation of impact and flexural and compressive properties for the fibre-reinforced UP blended with epoxy composites for 0%, 10%, 20% and 30% fibres content were studied as the function of with and without alkali treatment. It was observed that composites having 20% treated fibre content posses higher values for aforementioned properties than untreated composites, 10% and 30% treated fibre composites. The interfacial area plays a major role in determining the strength of polymer composite material because each fibre forms an individual interface with the matrix. After the alkali treatment, it was found that treated composites possess higher values of the aforementioned mechanical properties than the untreated composites, because the alkali treatment improves the adhesive characteristics of the surface of the banana and palmyra fibres by removing hemicellulose, waxes, impurities and lignin from the fibres, leading to higher crystallinity of palmyra and BFs. In the present work, it was found that optimum values and significant improvements were at 20% treated fibre–reinforced composites than untreated composites. It was observed that all the composites have resistance to almost all chemicals except toluene and carbon tetrachloride. Morphological changes take place depending upon the interfacial interaction between matrix blend and varying percentage of untreated and treated fibres.

17.11 Silk Fibre–Reinforced Composites

Zhao et al. [72] have studied 'Silkworm silk/poly(lactic acid) biocomposites: Dynamic mechanical, thermal and biodegradable properties' in 2010. Silkworm silk/PLA biocomposites with potential for environmental engineering applications were prepared by using melting compound methods. By means of DMA, DSC, TGA, coefficient of thermal expansion (CTE) test, enzymatic degradation test and SEM, the effect of silk fibre on the structural, thermal and dynamic mechanical properties and enzymatic degradation behaviour of the PLA matrix was investigated. As silk fibre was incorporated into PLA matrix, the stiffness of the PLA matrix at higher temperature (70°C–160°C) was remarkably enhanced, and the dimension stability also was improved, but its thermal stability became poorer. Moreover, the presence of silk fibres also significantly enhanced the enzymatic degradation ability of the PLA matrix. The higher the silk fibre content, the more the weight loss. Before glass transition, there is no obvious difference for the storage modulus E0 between the neat PLA and silk/PLA biocomposites. However, at higher temperature (70°C–160°C), the E0 increases with increasing silk fibre content due to the cooperation of the hydrodynamic effects and the mechanical restraint introduced by the silk fibres, which reduce the mobility and deformability of the PLA matrix. Following the increase of E0, the loss modulus E00 also increases with increasing silk fibre content at higher temperature. The damping (tand) decreases with increase of silk fibre content. By investigating the effect of frequency on the dynamic mechanical properties of the neat PLA and silk/PLA biocomposites, it is found that the E0 (after the glassy region) and the Tg of all the sample increase with increase of frequency, namely, the relaxation movement of PLA chains can be delayed when high frequency is used. And silk/PLA biocomposites can behave stiffer than it can be, if the frequency is chosen to be high enough. From DSC analysis, it is found that the DHcc and DHm of all the silk/PLA biocomposites is higher than that of the neat PLA and the Tcc of the PLA decreases as the incorporation of silk fibre over 3 wt.%. This suggests that the crystallization ability and crystallinity of the PLA matrix can be enhanced due to the presence of silk fibres, which may behave as nucleating agent. The Tg of all the silk/PLA biocomposites is somewhat lower than that of the neat PLA. And the addition of silk fibre over 3 wt.% into PLA matrix can slightly decrease melting temperature Tm of the PLA matrix. As can be seen from the TGA results, T_D^i, $T_D^{1/2}$ and Tp of silk/PLA biocomposites will decrease with increasing silk fibre content in the major degradation region from 265°C to 390°C. This indicates the thermal stability of silk/PLA biocomposites is worse than that of the neat PLA in this region, which is attributed to the poor thermal stability of silk fibre and the fact that the water molecules remained on silk fibre surface may promote the ester bond scission of the PLA matrix. On the other hand, the thermal stability of the PLA matrix can be slightly improved in the region from 390°C to 600°C. Moreover, the CTEs of all the silk/PLA biocomposites are slightly lower than that of the neat PLA. Therefore, the addition of silk fibre can provide the composites with relatively good dimension stability. In the enzymatic degradation test, the water absorption ratio of silk/PLA biocomposites increases with increasing silk fibre content and degradation time owing to the hydrophilicity of silk fibres. The absorbed water will increase the free volume within the PLA matrix and thus facilitate the enzymatic attack. Furthermore, from SEM images, it is found that the enzymatic hydrolysis not only occurs on the surface but also at the interface between the silk fibres and PLA matrix, which allow the PLA chain in the composites to become more susceptible to enzymatic degradation. Moreover, the sericin remained on silk fibre may be degraded into sericin peptides or partially hydrolyzed. Therefore, the weight loss of silk/PLA biocomposites increases with increasing silk fibre content and degradation time, that is to say, the presence of silk fibres can enhance the enzymatic degradation ability of the PLA matrix. This also suggests that the hydrolysis ability of the PLA matrix can be controlled by altering the amount of the added silk fibres. From the results of DMA, DSC, TGA, CTE and biodegradation measurements, there is a tendency that the interaction between silk fibre and PLA matrix will decrease with increasing of silk fibre content in PLA matrix. Moreover, based on our previous work and the results of this work, 5 wt.% is considered as the optimum condition of silk fibre in PLA matrix.

Ho et al. [73] have studied 'Characteristics of a silk fibre reinforced biodegradable plastic' in 2011. Silk fibre is one kind of well-recognized animal fibres for biomedical engineering and surgical operation applications because of its biocompatible and bioresorbable properties. Recently, the use of silk fibre as reinforcement for some biopolymers to enhance the stiffnesses of scaffolds and bone fixators has been a hot research topic. However, their mechanical and biodegradable properties have not yet been fully understood by many researchers, scientists and biomedical engineers although these properties would govern the usefulness of resultant products. In this paper, a study on the mechanical properties and biodegradability of silk fibre–reinforced PLA composites is conducted. It has been found that Young's modulus and flexural modulus of the composites increased with the use of silk fibre reinforcement, while their tensile and flexural strengths decreased. This phenomenon is attributed to the disruption of inter- and intramolecular bonding on the silk fibre with PLA during the mixing process and consequent reduction of the silk fibre strength. Moreover, biodegradability tests showed that the hydrophilic properties of the silk may alter the biodegradation properties of the composites compared to that of a pristine PLA sample. As the use of silk fibre–reinforced PLA composites is aimed at developing a suitable material for bone fixation, their moduli and biodegradation rate are the leading parameters for the design of implant plates. Before the degradation test, Young's modulus and flexural modulus of PLA increased over 27%

and 2%, respectively, with the use of 5 wt.% of silk fibre as reinforcement. The reduction of strength is mainly because of the poor interfacial bonding. However, the limited fibre content in injection moulding process and the poor fibre–matrix interface are the significant factors for strength decrement. After immersing samples into PBS solution to simulate their exposure in a liquidized environment, like human body, their mechanical properties were altered with immersion time. The declining rate of mechanical properties of the samples was demonstrated faster than that of pure PLA. It further proves that the hydrophilic effect of the silk fibre does affect the water absorbability in its related composites. In the current study, it is found that the use of silk fibre as reinforcement can enhance the modulus of PLA as well as alter its rate of biodegradability, in which these properties are the primary parameters for the design of implants for bone fixation. The biodegradation rate is crucial in which it have to be compromised with the cell growth rate of bone inside the human body. Besides for the successful use of silk fibre/PLA composites in bone implant applications, the mechanical properties of the composite should withstand the loads which human bone suffered daily. Nevertheless, too stiff bone implant causes stress shielding of the bone, and consequently the bone may become osteoporosis. Therefore, it should require further investigation of the bone plate with desirable biodegradation rate while at the same time possess proper mechanical properties (high modulus 20 GPa, close to natural bone properties, with moderate strength) for supporting a load that an undamaged bone should withstand.

Ho et al. [74] have studied 'Interfacial bonding and degumming effects on silk fibre/polymer biocomposites' in 2012. Silk fibre has been popularly used for biomedical engineering and surgically operational applications for centuries because of its biocompatible and bioresorbable properties. Using silk fibre as reinforcement for biopolymers could enhance the stiffness of scaffoldings and bone implants. However, raw silk fibre consists of silk fibroin that is bound together by a hydrophilic glue-like protein layer called *sericin*. Degumming is a surface modification process for sericin removal which allows a wide control of the silk fibre's properties, making the silk fibre possible to be properly used for the development and production of novel biocomposites with specific mechanical and biodegradable properties. Some critical issues such as wettability, bonding efficiency and biodegradability at the fibre–matrix interface are of interesting topics in the study of the degumming process. Therefore, it is a need to detailedly study the effect on different degumming processes to the properties of the silk fibre for real-life applications. Silkworm silk fibre is a renewable protein biopolymer, which is not only valuable in the textile industry but also for the medical application because of its superior mechanical properties and biocompatibility. Preprocessing of silk commonly known as degumming is an essential process to obtain an ideal fibre because of its fibre structure. Silk degumming process scours the sericin and some impurities from silk fibres. This paper addresses a comprehensive review on different types of degumming process and their effects to the silkworm silk fibre. Theoretical analysis also retrieves that the stress transfer properties are also affected by the effectiveness of the degumming process. Degradation of the surrounding matrix (biodegradable polymer) would also influence the result properties of silk fibre–reinforced polymer composites.

17.12 Sisal Fibre–Reinforced Composites

Milanese et al. [75] have studied 'Thermal and mechanical behaviour of sisal/phenolic composites' in 2012. This research proposes the development of polymeric composites reinforced with NFs to become stronger the damaged timber structures and proposes thermal and mechanical characterization of these composites. Fibres with larger structural applications are glass and carbon fibres, but the use of NFs is an economical alternative and possesses many advantages such as biodegradability, low cost and a renewable source. Woven sisal fabric was submitted to heat treatment before moulding, and the influence of moisture content of fibres on the composite behaviour was observed. The paper presents mechanical characterization by tensile and flexural strength of woven sisal fabric composites, with and without thermal treatment (at 60°C for 72 h) on the fabric, thermal characterization by TGA and the manufacturing process by compression moulding. Experimental results show a TS and a flexural strength value of 25.0 and 11.0 MPa, to sisal/phenolic composites respectively, independent to the use of sisal fibres (SFs) with or without thermal treatment. Based on tensile and flexural tests, the phenolic resin behaves as a fragile material. This material presents low tensile and flexural strengths besides low elongation and low deflection. After polymerization, it was observed a great volumetric retraction on the specimens, and it was possible to verify the existence of heterogeneity into the sample by SEM. TS and flexural strength of sisal/phenolic composites are 25.0 and 11.0 MPa, respectively, independent to the use of SFs with or without thermal treatment. But if it were considered tensile and flexural behaviour, the thermal treatment of sisal reinforcement is indicated to production of structural phenolic composites because the heat treatment causes a decrease on the standard deviation and the coefficient of variation besides increases the elongation and the deflection. Experimental results, shown in this paper, indicate that the use of phenolic composites to reinforce timber structures is viable once these composites present considerable strength to application according to the practical use.

Kaewkuk et al. [76] have studied 'Effects of interfacial modification and fiber content on physical properties of sisal fiber/polypropylene composites' in 2013. SF/PP composites were prepared at fibre content of 10, 20 and 30 wt.%, and their mechanical, thermal, morphological and water absorption properties were characterized. The effects of fibre treatment (alkalization and heat treatment) and adding a compatibilizer (MAPP) on the properties of the PP composites were comparatively studied. The fibre treatment and adding MAPP led to an improvement of mechanical properties, cellulose decomposition temperature and water resistance of the PP composites. SEM micrographs revealed that the interfacial modification enhanced the interfacial adhesion between the fibre and the PP matrix. Alkalization and heat treatment provided comparable enhancement in the mechanical properties of the SF/PP composites. Adding MAPP showed the most effective improvement of the mechanical properties of the PP composites. With increasing fibre content, TS and modulus of the PP composites increased, but impact strength and elongation at break decreased. Water absorption of PP composites also increased with fibre content. The interfacial modifications led to improved mechanical properties of the PP composites. Mechanical properties of the PP composites treated with alkalization and heat treatment were similar. This suggested that these techniques for fibre treatment were comparable. Adding MAPP provided the most effective enhancement in mechanical properties of the PP composites. With increasing fibre content, TS and Young's modulus of the PP composites increased, while elongation at break and impact strength decreased. SEM micrographs revealed that the interfacial modifications enhanced the interfacial adhesion between the fibre and the PP matrix. The interfacial modifications resulted in increased cellulose decomposition temperature of the PP composites. As fibre content increased, cellulose decomposition temperature of the PP composites decreased. An increase in water absorption of the PP composites was observed with increasing fibre content. The interfacial modifications resulted in a reduction in water absorption of the PP composites. MAPP-modified composite displayed the lowest water absorption.

17.13 Other Fibre-Reinforced Composites

Yu et al. [77] have studied 'Polymer blends and composites from renewable resources' in 2006. This article reviews recent advances in polymer blends and composites from renewable resources and introduces a number of potential applications for this material class. In order to overcome disadvantages such as poor mechanical properties of polymers from renewable resources or to offset the high price of synthetic biodegradable polymers, various blends and composites have been developed over the last decade. The progress of blends from three kinds of polymers from renewable resources – (1) natural polymers, such as starch, protein and cellulose; (2) synthetic polymers from natural monomers, such as PLA; and (3) polymers from microbial fermentation, such as PHB – is described with an emphasis on potential applications.

The hydrophilic character of natural polymers has contributed to the successful development of environmentally friendly composites, as most natural fibres and nanoclays are also hydrophilic in nature. Compatibilizers and the technology of reactive extrusion are used to improve the interfacial adhesion between natural and synthetic polymers. The study and utilization of natural polymers is an ancient science. The use of these materials has rapidly evolved over the last decade primarily due to the issue of the environment and the shortage of oil. Modern technologies provide powerful tools to elucidate microstructures at different levels and to understand the relationships between the structure and properties. However, there is still a long way to go in research to obtain ideal polymeric blends and composites from renewable resources. This review outlines the significance of the research and development that has been undertaken in the field of polymer blends from renewable resources. The three general classifications of this materials group include (1) natural polymers, such as starch, protein and cellulose; (2) synthetic polymers from natural monomers, such as PLA; and (3) polymers from microbial fermentation, such as PHB. A wide range of naturally occurring polymers that are derived from renewable resources are available for various materials applications. One of the main disadvantages of biodegradable polymers obtained from renewable sources is their dominant hydrophilic character, fast degradation rate, and, in some cases, unsatisfactory mechanical properties particularly under wet environments. Since most natural polymers are water soluble, the methods of both melting and aqueous blending have been used in many natural polymer blends. Aliphatic polyesters have been recognized for their biodegradability and their susceptibility to hydrolytic degradation. Among the family of biodegradable polyesters, PLAs have been the focus of much attention because they are produced from renewable resources such as starch, they are biodegradable and compostable, and they have very low or no toxicity and high mechanical performance comparable to those of commercial polymers. The higher price of aliphatic polyesters limits their general application. Blending hydrophilic natural polymers and aliphatic polyesters is of significant interest, since it could lead to the development of a new range of biodegradable polymeric materials. However, aliphatic polyesters and hydrophilic natural polymers are thermodynamically immiscible, leading to poor adhesion between the two components. Various compatibilizers and additives have been developed to improve their interface. One alternative way is the technology of multilayer extrusion. NFs are of basic interest since they have the ability to be functionalized and also have advantages from the point of view of weight and fibre–matrix adhesion, specifically with polar matrix materials. They have good potential for use in waste management due to their biodegradability and their much lower production of ash during incineration. Starch reinforced by cellulose is a typical example of natural polymer composites. Nanocomposites consisting of nanoclays and biodegradable polymers have also been investigated more recently and have shown to exhibit superior mechanical and thermal properties. The dispersions of nanoclays in polymers from renewable resources have been enhanced via chemical surface modification and the use of novel ultrasonic methods. The future growth and sustainability of polymers and composites from renewable resources are reliant on continued research, in particular in the fields of compatibilizing mechanisms, surface modification and advanced processing techniques, and it is through an understanding of these that they are expected to replace more and more petroleum-based plastics.

Christian et al. [78] have studied 'Sustainable Biocomposites for Construction' in 2009. Typical construction materials and practices have a large ecological footprint. Many materials are energy intensive to produce, and construction and demolition debris constitute a large percentage of US landfill volume. Biocomposites are structural materials made from renewable resources that biodegrade in an anaerobic environment after their useful service life to produce a fuel or feedstock to produce a biopolymer for a new generation of composites. These materials are being researched and developed to replace less eco-friendly structural and nonstructural materials used in the construction industry. In this study, the mechanical behaviour of biocomposites made from cellulose acetate and PHB matrices and hemp fibre (fabric) has been characterized experimentally. The data show that these biocomposites have mechanical properties similar to structural wood. The biocomposites studied have the potential to be used for scaffolding, formwork, flooring, walls and for many other applications within buildings, as well as temporary construction. Mechanical testing of hemp/cellulose acetate and hemp/PHB composites was performed and demonstrated that these biocomposites have strength properties comparable to structural lumber and higher than plywood. The moduli of elasticity of the biocomposites are lower than that for lumber and plywood parallel to grain. Due to the low modulus of elasticity (MOE), deflection limits are expected to control the design of hemp biocomposite components. Biocomposite components can be shaped to have high geometric stiffness to meet deflection limits while minimizing material use. Finally, from preliminary analysis, it appears that a 15.625 mm (5/8″) plywood sheathing for formwork could be replaced by a similar thickness of biocomposite to meet deflection limits for formwork for slabs.

Avella et al. [79] have studied 'Eco-Challenges of Bio-Based Polymer Composites' in 2009. In recent years, bio-based polymer composites have been the subject of many scientific and research projects, as well as many commercial programs. Growing global environmental and social concern, the high rate of depletion of petroleum resources and new environmental regulations have forced the search for new composites and green materials, compatible with the environment. The aim of this article is to present a brief review of the most suitable and commonly used biodegradable polymer matrices and NF reinforcements in ecocomposites and nanocomposites, with special focus on PLA-based materials. PLA-based materials are a new class of materials that in recent years has aroused an ever-growing interest due to the continuously increasing environmental awareness throughout the world. They can be considered as the *green* evolution of the more traditional ecocomposites, essentially consisting of synthetic polymer-based composites reinforced with NFs or other micro- or nanofiller. From a technological point of view, some doubts about the performance of these new materials, as well as the higher costs of biodegradable polymer matrices with respect to other polymers, still prevent their wider commercial diffusion. With regard to the still higher commercial prices of biodegradable polymers with respect to commodities, other factors, such as the lower costs for disposal, should also be taken into account. Moreover, the price of biodegradable polymers, and in particular those derived from natural sources, is expected to further decrease in the coming years due to innovative manufacturing practices, so biodegradable composites can be considered a valid alternative to traditional composites and ecocomposites: in fact, because of their compostability, they can represent an effective solution to the waste disposal problem of polymer-based materials. Their unique properties should be a solid base to develop new applications and opportunities for biocomposites in the twenty-first-century *green* materials world.

Tan et al. [80] have studied 'Characterization of polyester composites from recycled polyethylene terephthalate reinforced with empty fruit bunch fibers' in 2011. Unsaturated polyester resin (UPR) was synthesized from recycled polyethylene terephthalate (PET), which acted as a matrix for the preparation of UPR/EFB fibre composite. Chemical recycling on fine pieces of PET bottles was conducted through glycolysis process using ethylene glycol. The UPR was then prepared by reacting the glycolyzed product with MA. FTIR analysis of glycolyzed product and prepared UPR showed that cross-links between UP chain and styrene monomer occurred at the unsaturated sites, which resulted in the formation of cross-linking network. The preparation of UPR/EFB composite was carried out by adding EFB into prepared UPR matrix. The effects of surface treatment on EFB with NaOH, silane coupling agent and MA were then studied. The experimental results showed that treated EFB has higher values of tensile and impact strengths compared with untreated EFB. The best results were obtained for silane treatment followed by MA and NaOH treatments where the TS was increased by about 21%, 18% and 13%, respectively. SEM micrographs of the tensile fracture surfaces of UPR/EFB composite also proved that treatment on EFB has increased the interfacial adhesion between the fibre and UPR matrix compared to the untreated UPR/EFB composite. Chemical recycling of waste PET bottles was successfully performed through the glycolysis process of PET wastes to produce UPR, which was found suitable to be a matrix. The FTIR spectrum of the prepared resin showed that the absorption at 1645 cm^{-1} but did not appear for glycolyzed product. This absorption peak is associated with the stretching of C=C group in polyester, which is absent in the chemical formula of the glycolyzed product. Studies on the effect of EFB fibre surface treatment on the strength of polyester composites have been conducted, and treated composites showed better mechanical properties compared to the untreated fibre composite. Among the three types of chemical treatments that were invoked, silane treatment has given the highest value of mechanical strength compared to other composite samples with the increment in TS is 21% followed by MA and NaOH treatments with 18% and 13% increment, respectively. SEM morphology study also proves that silane treatment improved the interfacial compatibility and adhesion of the EFB fibre and the UPR matrix. TGA showed that treated fibre composites gave higher decomposition temperature and lower residue compared to untreated fibre composite. The residue of the treated composites and untreated composite was higher than that of UPR matrix.

Lu et al. [81] have studied 'Friction and wear behaviour of hydroxyapatite based composite ceramics reinforced with fibers' in 2012. The hydroxyapatite (HA)-based composites reinforced with multicomponent fibres were prepared by hot pressing. The friction properties of the composite at different temperatures were investigated by a block-on-disc tester. The results show that the addition of Cu has a significant positive effect on the friction and wear behaviours of HA-based composites. The improvements in the friction and wear properties of HA-based composites depend on the formation of an interfacial layer. The plastic deformation and the mending effect of Cu benefit the formation of the interfacial layer. The wear mechanism of the composites changes from the delamination and abrasive wear to the adhesive wear with increasing Cu. The friction and wear behaviours of HA-based composites were studied. The HA-based composites produced in this study show good thermal stability. With increasing Cu, the density of the composites increases, while the hardness decreases. With increasing Cu, both low friction coefficient and wear rate are obtained, and a stable friction and wear is reached. The improvements in the friction and wear properties of HA-based composites are due to the formation of an interfacial layer. Plastic deformation contributes to the spread and unification of the interfacial layer on the worn surface. With increasing Cu, the wear mechanisms of the composites convert from the delamination and abrasive wear to the adhesive wear.

Kaewtatip and Tanrattanakul [82] have studied 'Structure and properties of pregelatinized cassava starch/kaolin composites' in 2012. Pregelatinized cassava starch/kaolin composites were prepared using compression moulding. The morphology of the fractured surfaces, retrogradation behaviour, thermal decomposition temperatures and mechanical properties of the composites were investigated using SEM, XRD, TGA and tensile testing, respectively. The TSs and thermal degradation temperatures of the composites were higher than for TPS. The retrogradation behaviour of the composites was hindered by kaolin. The water absorption was measured after ageing for 12 and 45 days at an RH of 15% and 55%. It indicated that all the composites displayed lower water absorption values than TPS. Compression-moulded TPS was prepared from pregelatinized cassava starch using kaolin as an additive. The results from XRD confirmed that the retrogradation behaviour effect was dependent on the water absorption properties of the samples. Kaolin prevented the retrogradation and water absorption of TPS as it hindered the intermolecular and intramolecular hydrogen bonds between starch molecules and at the same time expelled water molecules from crystalline zones during storage. The maximum TS of TPS/kaolin composites (1.19 MPa) was derived when 10 wt.% of kaolin was used. Moreover, when the amount of kaolin increased further, the composites became very brittle. From the SEM results, it seems that kaolin is prone to form aggregate structures in the composites when the amount of kaolin was increased. The thermal stability of TPS was improved by the addition of the kaolin because it acts as a heat barrier.

Kaewtatip and Thongmee [83] have studied 'Studies on the structure and properties of thermoplastic starch/luffa fiber composites' in 2012. TPS/luffa fibre composites were prepared using compression moulding. The luffa fibre contents ranged from 0 to 20 wt.%. The TS of the TPS/luffa fibre composite with 10 wt.% of luffa fibre had a twofold increase compared to TPS. The temperature values of maximum weight loss of the TPS/luffa fibre composites were higher than for TPS. The water absorption of the TPS/luffa fibre composites decreased significantly when the luffa fibre contents increased. The strength of adhesion between the luffa fibre and the TPS matrix was clearly demonstrated by their compatibility presumably due to their similar chemical structures as shown by SEM micrographs and FTIR spectra. The luffa fibre reduced the water absorption and increased the thermal stability and TSs of the TPS. Maximum TS (1.24 MPa) was obtained with a 10 wt.% of luffa fibre. The results obtained from TGA confirmed that luffa fibre can improve the thermal stability of TPS. Moreover, the water absorption of the TPS/luffa fibre composites was lower than the TPS and reduced when the luffa fibre content increased. SEM micrographs of the TPS/luffa fibre composite did not detect any gap between the luffa fibre and TPS matrix, so the two materials were completely compatible as their chemical structures were similar as shown by the FTIR technique. Luffa fibre can improve the TS and thermal properties and reduce water absorption of TPS due to the luffa fibre forming a strong adhesion with the TPS matrix.

Hirose et al. [84] have studied 'Novel Epoxy Resins Derived from Biomass Components' in 2012. Biomass has received considerable attention because it is renewable and offers the prospect of circulation of carbon in the ecological system. The concept *biorefinery* has been developed rapidly in order to establish sustainable industries. Recently, new types of epoxy resins with polyester chains, which can be derived from saccharides, lignin and glycerol, have been investigated. In the aforementioned studies, the relationship between chemical structure and physical properties was investigated. In the present review, the features of the preparation system and the action of biomass components in epoxy resin polymer networks are described. The glass transition temperatures of the epoxy resins increased with increasing content of biomass components in epoxy resin polymer networks. Thermal decomposition temperatures were almost constant regardless of the content of biomass component contents in epoxy resins. Mass residue at 500°C increased with increasing contents of biomass components in epoxy resins. It was found that the thermal properties can be controlled by changing the contents of biomass components. The relationship between the chemical structure of biomass components and thermal properties such as Tg, Td and MR500 of the obtained epoxy resins was discussed. Tg values can be controlled by changing the amount of biomass components. The aforementioned epoxy resins show excellent thermal stability, that is Td values are ca. 340°C. Accordingly, the preparation system studied in the present investigation provides new epoxy resins, which can be derived from various biomass components such as saccharides, lignin and glycerol. Furthermore, the obtained epoxy resins have greater potential for various practical uses. One example is epoxy resin composites filled with nanoclay, which show excellent thermal stability and barrier properties.

Liu et al. [85] have studied 'Flexural properties of Rice Straw and Starch Composites' in 2012. The main goal of this work was to use RS in the production of environmentally sound composites using corn-based adhesives (CA). Treatments of RS with NaOH and hot water were undertaken to evaluate the effect of such treatments on the performance of produced composites. The influence of composite density, starch content and varieties of starch (cornstarch, cassava starch, potato starch) on flexural properties of composites was investigated. The microstructure of fractured surfaces was further observed. Results showed that cornstarch-based composites had higher flexural properties. Composites made from hot water–treated straw and cornstarch had better interface and higher flexural properties and flexural strength and flexural elastic modulus reached peak values at starch content of 10% and composite density of 0.7 g/cm^3. The composites developed from this work may have potential application for ceiling panels and bulletin boards. Cornstarch-based composites had higher flexural properties. Composites from hot water–treated straw and cornstarch had better interface and higher flexural properties and flexural strength and flexural elastic modulus reached peak values at starch mass fraction of 10% and density of 0.7 g/cm^3.

Zhang et al. [86] have studied 'Synthesis and Characterization of PVA-HA-Silk Composite Hydrogel by Orthogonal Experiment' in 2012. PVA–HA–silk composite hydrogel was synthesized with PVA, nano-HA, and natural silk by using the method of repeated freezing and thawing. A series of tests were performed to study water content, stress relaxation behaviour, elastic modulus and creep characteristics of PVA–HA–silk composite hydrogel. Orthogonal experimental design method was used to analyze the influence degree of PVA, HA and silk (three kinds of raw materials) on mechanical properties and water content of the PVA–HA–silk composite hydrogel to select the best material ratio according to their overall performance. The results demonstrate that the mass percentage of PVA has the greatest impact on the water content, followed by HA and silk. Compression stress–strain variation of PVA–HA–silk composite hydrogel presents a nonlinear relationship, which proves that it is a typical viscoelastic material. Comparing the mechanical properties of 16 formulas, the formula of PVA–HA–silk composite hydrogel with mass percentage of PVA mass of 15%, HA mass of 2.0% and silk mass of 1.0% is the best. The morphologies of PVA–HA–silk composite hydrogel reveal that the silk exists in the hydrogel substrate from the surface to the interior in the forms of an independent individual and crossing with each other. The silk has no agglomeration in the composite hydrogel, and the distribution is uniform. The mass percentage of PVA has the strongest impact on the water content, followed by the HA and silk. PVA among three raw materials has also the strongest impact on the stress relaxation rate and the creep properties of the hydrogel, the impact of silk is second, and HA is the weakest. Silk has the strongest impact on the elastic modulus of PVA–HA–silk composite hydrogel, followed by PVA and HA. The compression stress–strain variation presents a nonlinear relationship, which proves that the hydrogel is a typical viscoelastic material. Comparison of mechanical properties of 16 formulas demonstrates that PVA–HA–silk composite hydrogel has optimum mechanical properties by repeated freezing–thawing for 9 cycles, with PVA mass of 15%, HA mass of 2.0% and silk mass of 1.0%.

Srubar III et al. [87] have studied 'Mechanisms and impact of fiber–matrix compatibilization techniques on the material characterization of PHBV/oak wood flour engineered biobased composites' in 2012. Fully bio-based composite materials were fabricated using a natural, lignocellulosic filler, namely, oak wood flour (OWF), as particle reinforcement in a biosynthesized microbial polyester matrix derived from poly(b-hydroxybutyrate)-co-poly(b-hydroxyvalerate) (PHBV) via an extrusion injection moulding process. The mechanisms and effects of processing, filler volume percent (vol.%), a silane coupling agent and an MA grafting technique on polymer and composite morphologies and tensile mechanical properties were investigated and substantiated through calorimetry testing, SEM and micromechanical modelling of initial composite stiffness. The addition of 46 vol.% silane-treated OWF improved the tensile modulus of neat PHBV by 165%. Similarly, the tensile modulus of MA-grafted PHBV increased 170% over that of neat PHBV with a 28 vol.% addition of untreated OWF. Incorporation of OWF reduced the overall degree of crystallinity of the matrix phase and induced embrittlement in the composites, which led to reductions in ultimate tensile stress and strain for both treated and untreated specimens. Deviations from the Halpin–Tsai/Tsai–Pagano micromechanical model for composite stiffness in the silane and MA-compatibilized specimens are attributed to the inability of the model both to incorporate improved dispersion and wettability due to fibre–matrix modifications and to account for changes in neat PHBV and MA-grafted PHBV polymer morphology induced by the OWF. The effects of OWF vol.%, a silane coupling agent, and an MA grafting technique on the mechanical, morphological and thermodynamic properties of PHBV-based composites were investigated. SEM micrographs confirm a phenomenological increase in overall composite density due to the collapse or matrix infill of OWF during the injection moulding process. The introduction of OWF led to an increase in composite stiffness and to decreases in strength and elongation to break. From the experimental results, it can be concluded that the compatibilization techniques employed caused an increase in fibre wettability and distribution, as well as enhanced mechanical interlock between fibres and the matrix phase, which resulted in improved mechanical properties. The crystallinity of neat PHBV decreased after MA grafting via reactive extrusion, as evidenced by the marked softening of polymer stiffness, as well as the reduction in the calorimetric heat of melting, and the overall crystallinity of neat PHBV and MA-grafted PHBV decreased with the incorporation of OWF. Attributed to the migration of MA chains to the fibre–matrix interface, the crystallinity of MA-treated PHBV composite matrices decreased less than silane-treated and untreated composite matrices. Furthermore, the initial modulus softening of MA-PHBV was overcome with as little as 17 vol.% OWF. The Halpin–Tsai/Tsai–Pagano micromechanical model for initial composite stiffness showed good correlation with neat PHBV/untreated OWF composites. Disparities in the model were evident for composites with higher fibre loadings and for composites that were compatibilized with silane and MA. The differences in predicted and experimental values are attributed to the model's inability to account for increased wettability and therefore dispersion of the particulate filler and changes in the crystalline morphology of the polymer. Micromechanical models that incorporate the phenomenological alterations of the microstructure need be developed to accurately predict the global compatibilized constituent composite behaviour.

Conzatti et al. [88] have studied 'Polyester-based biocomposites containing wool fibres' in 2012. Biocomposites based on a biodegradable polyester containing different amounts of wool fibres (up to 40 wt.%) were prepared by melt blending in an internal batch mixer. Wool fibres were used as received or pretreated in order to preserve a high aspect ratio and increase adhesion with polymer matrix. Morphological, thermal, mechanical and dynamic mechanical properties of the ensuing composites were investigated focusing the attention on fibre length and their distribution as well as on fibre–matrix interaction in order to correlate these aspects with polymer reinforcement. Data from mechanical and dynamic mechanical analysis were also compared with theoretical models. Homogeneous biocomposites based on a biodegradable polyester containing high amounts of wool fibres (up to 40 wt.%) were successfully prepared by melt blending. A certain degree of fibre–matrix adhesion was obtained in particular for PVA-treated fibres. This resulted in an increment of Young's modulus up to 500% with respect to the neat polyester matrix. On the other hand, since the length of the wool fibres in the composites was mostly shorter than the critical length, the effective reinforcing action of fibres on matrix was limited, as indicated by the low values of strength at break. A certain degree of consistency was found between the data obtained from mechanical and dynamic mechanical investigations and those calculated with appropriate theoretical models. In particular, a certain degree of fibre orientation in the plane upon the preparation of the film was observed, and the improved fibre–matrix interaction was confirmed for PVA-treated fibres.

Porras and Maranon [89] have studied 'Development and characterization of a laminate composite material from polylactic acid (PLA) and woven bamboo fabric' in 2012. This article presents the development and mechanical characterization of a composite material fabricated from both renewable resources and biodegradable materials: bamboo woven fabric as reinforcement and PLA as resin matrix. The laminate composites were produced using a film stacking method. The physical, thermal and mechanical properties of bamboo fabric, PLA matrix and laminate composites were investigated. It is shown that the breaking force of the plain woven bamboo fabric in the weft direction was greater than in the warp direction. Further, the tensile, flexure and impact properties of PLA increased when weft direction bamboo fabric reinforcement is used. In addition, SEM examination of laminate composite showed good bonding between bamboo fibre and PLA resin. In summary, laminated composites based on PLA and bamboo fabric display excellent energy absorption capability, which can be exploited for the development of engineering structural applications. In this work, bamboo fabric and PLA resin were used to fabricate fully biodegradable *green* laminate, an attractive combination of mechanical and physical properties together with their environmental friendly character. In terms of mechanical properties, the laminated direction-2 (2 is transverse direction that is 90° weft side) presented excellent energy absorption capabilities and advantage for engineering applications. In the design ballistic armour systems, the potential of this kind of laminate could be an attractive future study. The fabric characterization showed that the plain woven bamboo fabric presented higher breaking force in the transverse direction, as compared to the axial direction. In reference to the characterization of the laminate was observed that tensile, flexure and impact properties of PLA increased when direction 2 (2 is transverse direction that is 90° weft side) bamboo fabric reinforcement is used. Also, the composite showed good bonding between bamboo fibre and PLA resin.

Malmstein et al. [90] have studied 'Hygrothermal ageing of plant oil based marine composites' in 2013. In this paper, the effect of hygrothermal ageing on the flexural properties of glass/epoxy, glass/linseed oil and glass/castor oil composites is reported. Plant oil–based resins offer renewable and potentially less toxic alternatives to conventional largely petroleum-based marine composites. The long-term performance of these novel composites needs to be investigated and understood before using them in structural applications. In this research, it was found that in the unaged condition, the flexural properties of glass/epoxy were significantly higher than both glass/castor oil and glass/linseed oil composites. After ageing in water at 40°C for 46 weeks, the properties of glass/castor oil were comparable to glass/epoxy, while the properties of glass/linseed oil were remarkably lower. The decrease in glass/linseed oil performance was explained in terms of the changes in the failure modes caused by moisture uptake. The moisture uptake of glass/castor oil and glass/linseed oil differs from that of glass/epoxy. Glass/castor oil and glass/linseed oil absorbed more water and had not reached a moisture equilibrium condition after 46 weeks. Blisters increase the moisture content in glass/linseed oil by approximately 2%. After 20+ weeks of ageing, the flexural strength of glass/castor oil was comparable to glass/epoxy. The glass/linseed oil specimens lost 72% of their dry strength; this may be due to the UV curing and chemical reactions between water and linseed oil resin and/or fibre–matrix interface. The properties of glass/epoxy kept degrading, while the moisture content stayed in a level, indicating chemical degradation of the composite. Glass/linseed degraded rapidly during the first 2 weeks of immersion and after 4 weeks stayed on the same level until the end of the testing period. The moisture uptake of glass/linseed oil kept increasing, suggesting that the effects of water and temperature are limited. Blisters had no effect on the strength of glass/linseed oil specimens after 26 weeks of ageing. The failure modes of composites are severely affected by water. CT and SEM images showed that the glass/epoxy failure became less brittle after ageing, resulting in less cracking inside the material. However, the failure remained largely fibre/interface dominated. The failure of glass/linseed oil specimens changed from compression to tension combined to compressive failure, suggesting that after only 3 days, the interface was unable to transfer loads from matrix to fibres.

Yan-Hong et al. [91] have studied 'Effect of fiber morphology on rheological properties of plant fiber reinforced poly(butylene succinate) composites' in 2013. SFs, steam exploded sisal fibres (SESFs) and steam exploded bagasse fibres (SEBFs), which have different fibre morphologies, were mixed with poly(butylene succinate) (PBS) using a torque rheometer. The rheological properties of these plant fibre–reinforced PBS composites were evaluated. Results show that the fibre morphology has a large effect on rheological behaviour. At the same fibre content (e.g. 10 and 30 wt.%), the non-Newtonian index n of composites reinforced by flexible fibres with a higher aspect ratio and larger contact area with the matrix is smaller. In general, n decreases with increasing fibre content, but when the fibre content is too high (e.g. 50 wt.%), the aggregation of fibres is too extensive so that the actual contact area between fibres and matrix becomes much lower, n increase instead. At the same fibre content (e.g. 10 and 30 wt.%), the consistency indices of fibrous filler-reinforced composites are larger than those of powder-filled composites; the larger the actual contact area between the matrix and the fibres, the greater the consistency index of the composite. The rheological properties of three different types of plant fibre–reinforced PBS composites were evaluated using a torque rheometer. The results show that even with the same fibre content, the rheological properties of the composites reinforced by fibres of different morphologies differed. The non-Newtonian index n is related to the degree of change of the orientation, disentanglement and fracture of fibres caused by a change in rotation speed. When the fibre content is relatively low, the fibres can be uniformly dispersed in the matrix, and the fibres become more oriented with the increase in rotation speed. Therefore, compared with the n value of PBS, the decrements in the n values of the SF PBS and SESF/PBS composites (in which the reinforcements are fibrous) are larger than that of the SEBF/PBS composite, which includes many small fragments of cells other than BF fibre cells. At medium fibre content, the degree of fibre entanglement is more extensive, increasing the interaction among these fibres. With an increase in rotation speed, an increased degree of fibre disentanglement, orientation and fracture occurs because of the increase in stress applied to the fibres, therefore causing a further reduction in the values of n. When the fibre content is relatively high, there is a high tendency for the fibrous SFs and SESFs to tangle into aggregates within the matrix. In addition, the viscosity of the matrix is too low to generate sufficient stress to evenly disperse the fibres in the matrix, so these still exist in the matrix even at increased shear rate. As a result, the actual contact area between matrix and fibres decreases, causing the shear viscosities in these two types of composites to become less sensitive to changes in shear rate. Therefore, the n values of these two composites are higher than that of the SEBF/PBS composite, in which there are not as many aggregates because of the shorter average aspect ratio of SEBFs and the presence of cell fragments other than BF fibre cells. The consistency index of the composite is also closely related to fibre morphology. Generally speaking, at the same fibre content, the consistency indices of fibrous filler-reinforced composites are larger than those of powder-filled composites. The larger the fibrous filler surface area, the higher the consistency index of the composite. The larger the actual contact area between the matrix and the fibres, the greater the consistency index of the composite.

Imre and Pukánszky [92] have studied 'Compatibilization in bio-based and biodegradable polymer blends' in 2013. The production and use of biopolymers increase continuously with a very high rate; thus, all information on these materials is very important. This feature article first defines the terms used in the area and then discusses the distinction between degradation and biodegradation as well as their importance for practice. Biopolymers often have inferior properties compared to commodity polymers. Modification is a way to improve properties and achieve property combinations required for specific applications. One technique is blending which allows considerable improvement in the impact resistance of brittle polymers. However, further study is needed on the miscibility–structure–property relationships of these materials to utilize all potentials of the approach. The chemical structure of biopolymers opens up possibilities to their reactive modification. Copolymerization, grafting, transesterification and the use of reactive coupling agents have all been utilized with success to achieve polymers and blends with improved properties. Several examples are shown for the various approaches and their outcome. Biopolymers and their blends are applied successfully in several areas from agriculture to consumer goods, packaging and automotive. Biopolymers are in the centre of attention; their production and use increase continuously at a very high rate. However, they are surrounded with much controversy, and even terms used in the area need further clarification. Biopolymers themselves, both natural polymers and plastics produced from natural feedstock by synthetic routes, often have inferior properties compared to commodity polymers. Modification is a way to improve properties and achieve property combinations required for specific applications. Blending is one of the approaches to modify the properties of biopolymers; the impact strength of inherently brittle polymers, mainly aliphatic polyesters, can be improved considerably by the approach. Further study is needed on the miscibility–structure–property relationships of these materials to utilize all potentials of blending. Their chemical structure opens up possibilities to the reactive modification of these polymers. Copolymerization, grafting, transesterification and the use of reactive coupling agents have all been utilized with success to achieve polymers and blends with advantageous properties. The possibilities are unlimited, and further progress is expected in the field. Biopolymers and their blends are applied successfully in several areas already.

Reddy et al. [93] have studied 'Tensile and structural characterization of alkali treated Borassus fruit fine fibers' in 2013. *Borassus* fine fibres possess superior tensile properties and are an important renewable, natural reinforcement material for composites. This paper reports the improved tensile properties of NFs extracted from *Borassus* fruit. Changes occurring in *Borassus* fibres when treated with a 5% concentration NaOH solution for different periods (1, 4, 8 and 12 h) were characterized using tensile testing, chemical analysis, FTIR spectroscopy, XRD, SEM, and TGA. The tensile properties (strength, modulus and % elongation) of the fibres improved by 41%, 69% and 40%, respectively, after 8 h of alkali treatment. Based on the properties determined for the *Borassus* fibres, we expect that these fibres will be suitable for use as a reinforcement in green composites. In this research, *Borassus* fine fibres were treated with 5% NaOH solution for 1, 4, 8 and 12 h. The tensile properties of the fibres were found to increase after alkali treatment due to improved fibre structure and found to be best for 8 h alkali treatment. It is attributed to the large amount of hemicellulose dissolution. The elimination of amorphous hemicellulose of the fibres on alkali treatment was proved by chemical analysis and FTIR studies. The removal of hemicellulose from fibre cells releases the internal constraint, and the fibrils become more capable of rearranging themselves in a compact manner, leading to a closer packing of cellulose chains. This closer packing of cellulose also improves the fibre crystallinity after alkali treatment as evident by XRD analysis. The surface morphology of the fibre becomes rough after alkali treatment, and roughness increases with an increase in the treatment period due to the removal of hemicellulose. The thermal stability of the fibres improved slightly by alkali treatment. Based on improved tensile properties, renewability and environment friendly nature, *Borassus* fine fibres can be favourably considered as a reinforcement in green composites.

El-Kassas and Mourad [94] have studied 'Novel fibers preparation technique for manufacturing of rice straw based fiberboards and their characterization' in 2013. The present work introduces a novel fibre preparation technique for manufacturing of dry-formed RS-based medium density fibreboard (RSMDF). The properties and the quality of the produced RSMDF were evaluated. The technique is based only on a mechanical method without pretreatment (wetting and chemical additives) before defibreation to loosen up the straw structure as usually followed in the conventional processes. The components of the manufacturing system were especially designed and fabricated. The commercial UF resin was used as binder in manufacturing of the RSMDF. Different content levels of the conventional UF resin (16%, 18% and 19%) were targeted. The RSMDF was produced at different thicknesses (7, 9, and 16 mm) and average densities in the range of 728–941 kg/m^3. The RSMDF mechanical properties (internal bond [IB], MOR and MOE) and thickness swilling were measured and analyzed as a function of density and resin content. The results revealed that the properties of the produced fibreboards are dependent on the average density and the resin contents. The RSMDF panels were produced successfully using the new RS fibre preparation technique, and their properties met the requirements of MDF standard. New technique, based on mechanical method only, without any straw pretreatment (with heat, hot water, steam, chemical additives, etc.) was used to produce RS-based fibres and RSMDF panels. The fibreboards were produced with different UF resin contents (16%, 18% and 19%) and at different densities (728–941 kg/m^3) and thicknesses (7, 9 and 16 mm). RS (which is inexpensive agricultural residue and annual plant) was used as non-wood raw material with UF resin, which is more frequently used and less expensive commercial resin as a binder. The average density and the resin content are probably the most pronounced parameters influencing the RSMDF panel's mechanical properties (IB, MOR and MOE). The TS is also strongly dependent on the density and declines with increasing density. The used method for preparation of the RS fibres improved the diffuse and penetration of the resin into straw fibres and resulted in higher bond strength and consequently lower TS/higher water absorption resistance. In general, increasing the density and resin content improves most of typical fibreboard properties due to the increase in the numbers of fibre–fibre contact points and the higher level of cross-links between resinated fibres. The newly introduced fibre preparation technique, in the current work, has many advantages from industrial point of view, such as the reduction of the pretreatment process (no heat, hot water, steam chemicals, black liquor and drainage), smaller allocated area for the manufacturing system, less numbers of equipment, and consequently an expected reduction in the general cost. The new technique also maintains the natural mechanical properties of the rice fibres. Good mat stability and pressing process without adhesion (sticking) of the fibreboards to the press plate problems were observed. The produced RSMDF panels met the MDF standard requirements. The produced panels with an average density more than 745 kg/m3 have IB level above 0.60 MPa, MOR above 26 MPa, MOE value above 3.1 GPa and TS less than 12%. Therefore, the RS is a promising candidate as a raw material for the manufacture of RSMDF panels.

Xie et al. [95] have studied 'Starch-based nano-biocomposites' in 2013. The last decade has seen the development of green materials, which intends to reduce the human impact on the environment. Green polymers are obviously tendency subset of this stream, and numerous bio-sourced plastics (bioplastics) have been developed. Starch as an agro-sourced polymer has received much attention recently due to its strong advantages such as low cost, wide availability, and total compostability without toxic residues. However, despite considerable commercial products being available, the fundamental properties (mechanical properties, moisture sensitivity, etc.) of plasticized starch-based materials have to be enhanced to enable such materials to be truly competitive with traditional petroleum-based plastics over a wider range of applications. Regarding this, one of the most promising technical advances has been the development of nano-biocomposites, namely, dispersion of nano-sized filler into a starch biopolymer matrix. This paper reviews the state of the art in the field of starch-based nano-biocomposites. Various types of nanofillers that have been used with plasticized starch are discussed such as phyllosilicates (montmorillonite, hectorite, sepiolite, etc.), polysaccharide nanofillers (nanowhiskers/nanoparticles from cellulose, starch, chitin and chitosan) and carbonaceous nanofillers (carbon nanotubes, graphite oxide and carbon black). The main preparation strategies for starch-based nano-biocomposites with these types of nanofillers and the corresponding dispersion state and related properties are also discussed. The critical issues in this area are also addressed. A wide variety of nanofillers have been examined with starch. Phyllosilicates (especially MMT of the smectite group) have been mostly utilized due to their advantages such as wide availability, low cost and high aspect ratio and thus vast exposed surface area (and also the swelling nature). In addition, polysaccharide nanofillers represent the second most popular group due to their abundance in nature, the biological sources and the chemical similarity to starch. Nevertheless, the preparation of these bio-nanoparticles is time consuming and involves acid hydrolysis in multiple steps, which is not eco-friendly. Furthermore, many other nanofillers such as carbonaceous nanofillers, metalloid oxides, metal oxides and metal chalcogenides have been used. One of the advantages in utilizing such nanofillers is that they can provide new functionalities to starch-based materials in addition to the general reinforcement. With the incorporation of the nanofiller, starch-based materials generally show improvement in some of their properties such as mechanical properties (typically, modulus of elasticity E), glass-transition temperature (Tg), thermal stability, moisture resistance, oxygen barrier property and biodegradation rate. The improvement can be fundamentally ascribed to the homogeneous dispersion of the nanofiller in the matrix and the strong interface adhesion, which can contribute to the formation of a rigid nanofiller network and influence the molecular and crystalline structures in the matrix. To realize these, the nanofiller–matrix compatibility is the key point to address, which mainly depends on the surface chemistry of the nanofiller and is usually achieved by hydrogen bonding, although more factors such as the plasticizer(s)/additive(s), the starch type and chemical modification, the presence of other polymer(s) and the processing and annealing conditions also have strong influences. But above all, a major role is played by the nanofiller itself, of which the aspect ratio/surface area, chemistry and mechanical properties could be influenced by its preparation and modification. Nevertheless, how the nanofiller affected the crystalline structure and crystallinity of the starch matrix has not been unambiguously elucidated across the literature. These can be highly affected by the formulation (e.g. the amylose content of the starch and the type and content of the plasticizer), the processing conditions (e.g. temperature, pressure, shearing and orientation) and the storing conditions (e.g. time, temperature and RH). Besides, phase separation of the plasticizer, the starch and/or the nanofiller may exist in the system, with the different domains showing different recrystallization/anticrystallization behaviours. These reasons may account for the discrepancies in some of the results such as Tg and moisture resistance. With improved properties that are comparable to those of traditional petroleum-based polymers such as polyethylene and PP, the current applications of starch-based materials can be greatly enhanced and widened. The renewable resource and inherent environmental friendliness of such materials can justify its wide use for a sustainable future. Particularly, the use of starch-based nano-biocomposites as new packaging materials would be based on their biodegradation and improved barrier and mechanical properties. In the future research, it is still very important to test new nanofillers to be incorporated into starch for developing promising nano-biocomposites with excellent performance and new functionalities to be competitive in the materials world. In addition, the heterogeneous dispersion of the nanofiller and the phase separation issue existed in some of the past studies should be addressed in the future. While the manipulation of chemistry might help to some extent, the future research should also emphasize the importance in using processing techniques like extrusion which are more aligned to the efficient industrial production. Thus, research is also needed regarding how thermomechanical treatment in this kind of processing can assist in achieving a well-dispersive structure without adding a detrimental effect to the final properties due to starch molecular degradation.

Koronis et al. [96] have studied 'Green composites: A review of adequate materials for automotive applications' in 2013. This study provides a bibliographic review in the broad field of green composites seeking out for materials with a potential to be applied in the near future on automotive body panels. Hereupon, materials deriving from renewable resources will be preferred as opposed to the exhaustible fossil products. With the technical information of biopolymers and natural reinforcements, a database was created with the mechanical performance of several possible components for the prospect green composite. Following the review, an assessment is performed where aspects of suitability for the candidate elements in terms of mechanical properties are analyzed. In that section, renewable materials for matrix and reinforcement are screened accordingly in order to identify which hold both adequate strength and stiffness performance along with affordable cost so as to be a promising proposal for a green composite. The application of green composites in automobile body panels seems to be feasible as far as green composites have comparable mechanical performance with the synthetic ones. Conversely, green composites seem to be rather problematic due to their decomposable nature. The biodegradability issue is one problem that needs to be addressed when aiming to 100% bio-based composite application, especially when dealing with structural parts of exterior panels for future vehicles. More aspects have to be considered such as reproducibility of these composites' properties and their long life cycle as parts of the exterior body parts. Unfortunately, to the present the bio-thermoplastics' cost is a major barrier for their generalized use in the automotive industry, but it is expected that soon manufacturers of these materials will turn up affordable solutions as their demand in industrial-scale applications will no doubt tend to decrease their prices to more affordable levels. The trend can also be reversed in the sense that the necessity for environmentally conscious solutions can overturn the value chain and put a premium price on environmental impact of current solutions. An essential point is whether these materials can be combined in the best way to reach the level of performance of their predecessors while having the lowest possible cost. This methodology could be the first step in the vast area of multifactor decision making. Aspects that have to do with manufacturability and/or supply chain were not taken into account while still very critical and will be included in future studies.

References

1. K.J. Wong, S. Zahi, K.O. Low, and C.C. Lim, Fracture characterisation of short bamboo fibre reinforced polyester composites, *Materials and Design*, 31, 2010, 4147–4154.
2. D. Liu, T. Zhong, P.R. Chang, K. Li, and Q. Wuc, Starch composites reinforced by bamboo cellulosic crystals, *Bioresource Technology*, 101, 2010, 2529–2536.
3. H.P.S. Abdul Khalil, I.U.H. Bhat, M. Jawaid, A. Zaidon, D. Hermawan, and Y.S. Hadi, Bamboo fibre reinforced biocomposites: A review, *Materials and Design*, 42, 2012, 353–368.

4. B. Deepa, E. Abraham, B.M. Cherian, A. Bismarck, J.J. Blaker, L.A. Pothan, A.L. Leao, S.F. de Souza, and M. Kottaisamy, Structure, morphology and thermal characteristics of banana nano fibers obtained by steam explosion. Elsevier Ltd. 2010.
5. K.S. Wickramarachchi and S.L. Ranamukhaarachchi, Preservation of fiber-rich banana blossom as a dehydrated vegetable, *Science Asia*, 31, 2005, 265–271.
6. M.A. Belewu and K.Y. Belewu, Cultivation of mushroom (*Volvariella volvacea*) on banana leaves, *African Journal of Biotechnology*, 4(12), December 2005, 1401–1403.
7. R. Kumar, V. Choudhary, S. Mishra, and I.K Varma, Banana fiber-reinforced biodegradable soy protein composites, *Frontiers of Chemistry in China*, 3(3), 2008, 243–250, 2008.
8. J.R. Memon, S.Q. Memon, M.I. Bhanger, and M.Y. Khuhawar, Banana peel: A green and economical sorbent for Cr(III) removal, *Pakistan Journal of Analytical and Environmental Chemistry*, 9(1), 2008, 20–25.
9. A.V. Kiruthika and K. Veluraja, Experimental studies on the physico-chemical properties of banana fibre from various varieties, *Fibers and Polymers*, 10(2), 2009, 193–199.
10. P. Wachirasiri, S. Julakarangka, and S. Wanlapa, The effects of banana peel preparations on the properties of banana peel dietary fibre concentrate, *Songklanakarin Journal of Science and Technology*, 31(6), November–December 2009, 605–611.
11. S.K. Nayak, Degradation and flammability behaviour of PP/banana and glass fiber-based hybrid composites, *International Journal of Plastics Technology*, 13(1), 2009, 47–67.
12. A.C.H. Barreto, M.M. Costa, A.S.B. Sombra, D.S. Rosa, R.F. Nascimento, S.E. Mazzetto, and P.B.A. Fechine, Chemically modified banana fiber: Structure, dielectrical properties and biodegradability, *Journal of Polymers and the Environment*, 18, 2010, 523–531.
13. S.K. Majhi, S.K. Nayak, S. Mohanty, and L. Unnikrishnan, Mechanical and fracture behavior of banana fiber reinforced Polylactic acid biocomposites, *International Journal of Plastics Technology*, 14(Suppl 1), 2010, S57–S75.
14. K. Sathasivam and M.R.H.M. Haris, Adsorption kinetics and capacity of fatty acid-modified banana trunk fibers for oil in water, *Water Air and Soil Pollution*, 213, 2010, 413–423.
15. Y.-F. Shih and C.-C. Huang, Polylactic acid (PLA)/banana fiber (BF) biodegradable green composites, *Journal of Polymer Research*, 18, 2011, 2335–2340.
16. Akmal Hadi Ma' Radzi, Noor Akmal Mohamad Saleh, Banana Fiber Reinforced Polymer Composites, UMTAS 2011.
17. N. Venkateshwaran, A. Elaya Perumal, and R.H. Arwin Raj, Mechanical and dynamic mechanical analysis of woven banana/epoxy composite, *Journal of Polymer Environment*, 20, 2012, 565–572.
18. F.R. Oliveira, L. Erkens, R. Fangueiro, and A.P. Souto, Surface modification of banana fibers by DBD plasma treatment, *Plasma Chemistry and Plasma Processing*, 32, 2012, 259–273.
19. P.J. Jandas, S. Mohanty, and S.K. Nayak, Renewable resource-based biocomposites of various surface treated banana fiber and poly lactic acid: Characterization and biodegradability, *Journal of Polymer Environment*, 20, 2012, 583–595.

20. A.V. Kiruthika, T.R.K. Priyadarzini, and K. Veluraja, Preparation, properties and application of tamarind seed gum reinforced banana fibre composite materials, *Fibers and Polymers*, 13(1), 2012, 51–56.

21. K. Sathasivam and M.R.H.M. Haris, Thermal properties of modified banana trunk fibers, *Journal of Thermal Analysis and Calorimetry*, 108, 2012, 9–17.

22. N. Venkateshwaran and A. Elaya Perumal, Mechanical and water absorption properties of woven jute/banana hybrid composites, *Fibers and Polymers*, 13(7), 2012, 907–914.

23. K.N. Indira, P. Jyotishkumar, and S. Thomas, Thermal stability and degradation of banana fibre/PF composites fabricated by RTM, *Fibers and Polymers*, 13(10), 2012, 1319–1325.

24. N. Khurana, N. Kanoongo, and R.V. Adivarekar, Statistical modelling on grafting and hydrolysis for prediction of absorbency of banana fibres, *European International Journal of Science and Technology*, 2(1), 153–160, February 2013.

25. H.U. Zaman, M.A. Khan, and R.A. Khan, Banana fiber-reinforced polypropylene composites: A study of the physico-mechanical properties, *Fibers and Polymers*, 14(1), 2013, 121–126.

26. P.J. Jandas, S. Mohanty, and S.K. Nayak, Thermal properties and cold crystallization kinetics of surface-treated banana fiber (BF)-reinforced poly(lactic acid) (PLA) nanocomposites, *Journal of Thermal Analysis and Calorimetry*, 114, 2013, 1265–1278.

27. M.M. Ibrahim, W.K. El-Zawawy, Y. Jiittke, A. Koschella, and T. Heinze, Cellulose and microcrystalline cellulose from rice straw and banana plant waste: Preparation and characterization, Springer Science+Business Media Dordrecht 2013.

28. A.N. Benítez, M.D. Monzón, I. Angulo, Z. Ortega, P.M. Hernández, and M.D. Marrero, Treatment of banana fiber for use in the reinforcement of polymeric matrices, *Measurement*, 46, 2013, 1065–1073.

29. M.G. El-Meligy, S.H. Mohamed, and R.M. Mahani, Study mechanical, swelling and dielectric properties of prehydrolysed banana fiber—Waste polyurethane foam composites, *Carbohydrate Polymers*, 80(2), 12 April 2010, 366–372.

30. M. Twite, E. Kovaleva, J. Munyaneza, and V. Habimana, Assessment of natural adhesives in banana leaf composite materials for architectural applications, *Second International Conference on Advances in Engineering and Technology*, India.

31. U. Nirmal, N. Singh, J. Hashim, S.T.W. Lau, and N. Jamil, On the effect of different polymer matrix and fibre treatment on single fibre pullout test using betelnut fibres, *Materials and Design*, 32, 2011, 2717–2726.

32. H. Sehaqui, M. Allais, Q. Zhou, and L.A. Berglund, Wood cellulose biocomposites with fibrous structures at micro- and nanoscale, *Composites Science and Technology*, 71, 2011, 382–387.

33. K. Qiu, and A.N. Netravali, Fabrication and characterization of biodegradable composites based on microfibrillated cellulose and polyvinyl alcohol, *Composites Science and Technology*, 72, 2012, 1588–1594.

34. M. Pöllänen, M. Suvanto, and T.T. Pakkanen, Cellulose reinforced high density polyethylene composites—Morphology, mechanical and thermal expansion properties, *Composites Science and Technology*, 76, 2013, 21–28.

35. Y. Pan, M.Z. Wang, and H. Xiao, Biocomposites containing cellulose fibers treated with nanosized elastomeric latex for enhancing impact strength, *Composites Science and Technology*, 77, 2013, 81–86.

36. B. Agoudjil, A. Benchabane, A. Boudenne, L. Ibosc, and M. Foisc, Renewable materials to reduce building heat loss: Characterization of date palm wood, *Energy and Buildings*, 43, 2011, 491–497.

37. Y.S. Salim, A.A. Abdullah, C.S. Sipaut, M. Nasri, and M.N.M Ibrahim, Biosynthesis of poly(3-hydroxybutyrate-co-3-hydroxyvalerate) and characterisation of its blend with oil palm empty fruit bunch fibers, *Bioresource Technology*, 102, 2011, 3626–3628.

38. D. Shanmugam and M. Thiruchitrambalam, Static and dynamic mechanical properties of alkali treated unidirectional continuous palmyra palm leaf stalk fiber/jute fiber reinforced hybrid polyester composites, *Materials and Design*, 50, 2013, 533–542.

39. M.A. Norul Izani, M.T. Paridah, U.M.K. Anwar, M.Y. Mohd Nor, and P.S. H'ng, Effects of fiber treatment on morphology, tensile and thermogravimetric analysis of oil palm empty fruit bunches fibers, *Composites: Part B*, 45, 2013, 1251–1257.

40. N.W. Manthey, F. Cardona, G.M. Francucci, and T. Aravinthan, Green building materials: Hemp oil based biocomposites, *World Academy of Science, Engineering and Technology*, 6, 2012, 10–21.

41. M.M. Kabir, H. Wang, K.T. Lau, F. Cardona, and T. Aravinthan, Mechanical properties of chemically-treated hemp fibre reinforced sandwich composites, *Composites: Part B*, 43, 2012, 159–169.

42. K. Das, D. Ray, N.R. Bandyopadhyay, S. Sahoo, A.K. Mohanty, and M. Misra, Physico-mechanical properties of the jute micro/nanofibril reinforced starch/ polyvinyl alcohol biocomposite films, *Composites: Part B*, 42, 2011, 376–381.

43. M.K. Hossain, M.W. Dewan, M. Hosur, and S. Jeelani, Mechanical performances of surface modified jute fiber reinforced biopol nanophased green composites, *Composites: Part B*, 42, 2011, 1701–1707.

44. J.P. López, S. Boufi, N.E. El Mansouri, P. Mutjé, and F. Vilaseca, PP composites based on mechanical pulp, deinked newspaper and jute strands: A comparative study, *Composites: Part B*, 43, 2012, 3453–3461.

45. T.-T.-L. Doan, H. Brodowsky, and E. Mäder, Jute fibre/epoxy composites: Surface properties and interfacial adhesion, *Composites Science and Technology*, 72, 2012, 1160–1166.

46. J. Prachayawarakorn, S. Chaiwatyothin, S. Mueangta, and A. Hanchana, Effect of jute and kapok fibers on properties of thermoplastic cassava starch composites, *Materials and Design*, 47 2013, 309–315.

47. M. Bernard, A. Khalina, A. Ali, R. Janius, M. Faizal, K.S. Hasnah, and A.B. Sanuddin, The effect of processing parameters on the mechanical properties of kenaf fibre plastic composite, *Materials and Design*, 32, 2011, 1039–1043.

48. A. Elsaid, M. Dawood, R. Seracino, and C. Bobko, Mechanical properties of kenaf fiber reinforced concrete, *Construction and Building Materials*, 25, 2011, 1991–2001.

49. H.M. Akil, M.F. Omar, A.A.M. Mazuki, S. Safiee, Z.A.M. Ishak, and A. Abu Bakar, Kenaf fiber reinforced composites: A review, *Materials and Design*, 32, 2011, 4107–4121.

50. N.S. Suhartya, I.P. Almanarb, Sudirmanc, K. Dihardjod, and N. Astasaria, Flammability, biodegradability and mechanical properties of bio-composites waste polypropylene/kenaf fiber containing nano $CaCO_3$ with diammonium phosphate, *Procedia Chemistry*, 4, 2012, 282–287.

51. Z. Salleh, Y.M. Taib, K.M. Hyie, M. Mihat, M.N. Berhan, and M.A.A. Ghani, Fracture toughness investigation on long kenaf/woven glass hybrid composite due to water absorption effect. *Procedia Engineering*, 41, 2012, 1667–1673.

52. M.S. Meona, M.F. Othmana, H. Husaina, M.F. Remelia, and M.S.M. Syawala, Improving tensile properties of kenaf fibers treated with sodium hydroxide, *Procedia Engineering*, 41, 2012, 1587–1592.

53. A. Hao, H. Zhao, and J.Y. Chen, Kenaf/polypropylene nonwoven composites: The influence of manufacturing conditions on mechanical, thermal, and acoustical performance, *Composites: Part B*, 54, 2013, 44–51.

54. K. Wrześniewska-Tosik, D. Wawro, M. Ratajska, and W. Stęplewski, Novel biocomposites with feather keratin, *Fibres & Textiles in Eastern Europe*, 15(5–6), January/December 2007, 64–65.

55. A. Salhi, S. Kaci, A. Boudene, and Y. Candau, Development of biocomposites based of polymer matrix and keratin fibers: Contribution to poultry feather biomass recycling. Part I, *Conference and Training School Multiphase Polymer and Composite Systems: From nanoscale to Macro Composites*, Paris – Est, France, June 7–10, 2011.

56. L.T. Drzal, A.K. Mohanty, and M. Misra, Bio-composite materials as alternatives to petroleum-based composites for automotive applications, Composite Materials and Structures Center, Michigan State University, East Lansing, MI.

57. A.S. Singha and V.K. Thakur, Mechanical properties of natural fibre reinforced polymer composites, *Material Science*, 31(5), October 2008, 791–799.

58. S.W. Beckwith, Natural fibers: Nature providing technology for composites, *SAMPE Journal*, 44(3), 64–65, May/June 2008.

59. B.-H. Lee, H.-J. Kim, and W.-R. Yu, Fabrication of long and discontinuous natural fiber reinforced polypropylene biocomposites and their mechanical properties, *Fibers and Polymers*, 10(1), 2009, 83–90.

60. M.P. Westman, S.G. Laddha, L.S. Fifield, T.A. Kafentzis, and K.L. Simmons, *Natural Fiber Composites: A Review*, March 2010.

61. H. Ku, H. Wang, N. Pattarachaiyakoop, and M. Trada, A review on the tensile properties of natural fiber reinforced polymer composites, *Composites: Part B*, 42, 2011, 856–873.

62. J. Sahari and S.M. Sapuan, Natural fibre reinforced biodegradable polymer composites, *Reviews on Advanced Materials Science*, 30, 2011, 166–174.

63. J. Gironès, J.P. López, P. Mutjé, A.J.F. Carvalho, A.A.S. Curvelo, and F. Vilaseca, Natural fiber-reinforced thermoplastic starch composites obtained by melt processing, *Composites Science and Technology*, 72, 2012, 858–863.

64. M. Kuranska and A. Prociak, Porous polyurethane composites with natural fibres, *Composites Science and Technology*, 72, 2012, 299–304.

65. L. Sobczak, R.W. Lang, and A. Haider, Polypropylene composites with natural fibers and wood—General mechanical property profiles, *Composites Science and Technology* 72, 2012, 550–557.

66. S.J. Christian and S.L. Billington, Moisture diffusion and its impact on uniaxial tensile response of biobased composites, *Composites: Part B*, 43, 2012, 2303–2312.

67. M.M. Kabir, H. Wang, K.T. Lau, and F. Cardona, Chemical treatments on plant-based natural fibre reinforced polymer composites: An overview, *Composites: Part B*, 43, 2012, 2883–2892.

68. M.-P. Ho, H. Wanga, J.-H. Lee, C.-K. Ho, K.-T. Lau, J. Leng, and D. Hui, Critical factors on manufacturing processes of natural fibre composites, *Composites: Part B*, 43, 2012, 3549–3562.

69. Z.N. Azwa, B.F. Yousif, A.C. Manalo, and W. Karunasena, A review on the degradability of polymeric composites based on natural fibres, *Materials and Design*, 47, 2013, 424–442.

70. F.Z. Arrakhiz, M. El Achaby, M. Malha, M.O. Bensalah, O. Fassi-Fehri, R. Bouhfid, K. Benmoussa, and A. Qaiss, Mechanical and thermal properties of natural fibers reinforced polymer composites: Doum/low density polyethylene, *Materials and Design*, 43, 2013, 200–205.

71. G.V. Prasanna and K.Venkata Subbaiah, Modification, flexural, impact, compressive properties & chemical resistance of natural fibers reinforced blend composites, *Malaysian Polymer Journal*, 8(1), 2013, 38–44.

72. Y.-Q. Zhao, H.-Y. Cheung, K.-T. Laub, C.-L. Xu, D.-D. Zhao, and H.-L. Li, Silkworm silk/poly(lactic acid) biocomposites: Dynamic mechanical, thermal and biodegradable properties, *Polymer Degradation and Stability* 95, 2010, 1978e1987.

73. M.-P. Ho, K.-T. Lau, H. Wang, and D. Bhattacharyya, Characteristics of a silk fibre reinforced biodegradable plastic, *Composites: Part B*, 42, 2011, 117–122.

74. M.-P. Ho, H. Wanga, K.-T. Lau, J.-H. Lee, and D. Hui, Interfacial bonding and degumming effects on silk fibre/polymer biocomposites, *Composites: Part B*, 43, 2012, 2801–2812.

75. A.C. Milanese, M.O.H. Cioffi, and H.J.C. Voorwald, Thermal and mechanical behaviour of sisal/phenolic composites, *Composites: Part B*, 43, 2012, 2843–2850.

76. S. Kaewkuk, W. Sutapun, and K. Jarukumjorn, Effects of interfacial modification and fiber content on physical properties of sisal fiber/polypropylene composites, *Composites: Part B*, 45, 2013, 544–549.

77. L. Yua, K. Deana, and L. Li, Polymer blends and composites from renewable resources, *Progress in Polymer Science*, 31, 2006, 576–602.

78. S. Christian and S. Billington, Sustainable biocomposites for construction, Composites & Polycon 2009, American Composites Manufacturers Association, January 15–17, 2009.

79. M. Avella, A. Buzarovska, M. Emanuela Errico, G. Gentile and A. Grozdanov, Eco-challenges of bio-based polymer composites, *Materials*, 2, 2009, 911–925; doi:10.3390/ma2030911.

80. C. Tan, I. Ahmad, and M. Heng, Characterization of polyester composites from recycled polyethylene terephthalate reinforced with empty fruit bunch fibers, *Materials and Design*, 32, 2011, 4493–4501.

81. Z. Lu, Y. Liu, B. Liu, and M. Liu, Friction and wear behavior of hydroxyapatite based composite ceramics reinforced with fibers, *Materials and Design*, 39, 2012, 444–449.

82. K. Kaewtatip and V. Tanrattanakul, Structure and properties of pregelatinized cassava starch/kaolin composites, *Materials and Design*, 37, 2012, 423–428.

83. K. Kaewtatip and J. Thongmee, Studies on the structure and properties of thermoplastic starch/luffa fiber composites, *Materials and Design*, 40, 2012, 314–318.

84. S. Hirosea, T. Hatakeyamab, and H. Hatakeyamab, Novel epoxy resins derived from biomass components, *Procedia Chemistry*, 4, 2012, 26–33.

85. L.I.U. Junjun, J.I.A. Chanjuan, and H.E. Chunxia, Flexural properties of rice straw and starch composites, *AASRI Procedia* 3, 2012, 89–94.

86. D. Zhang, K. Chen, L. Wu, D. Wang, and S. Ge, Synthesis and characterization of PVA-HA-Silk composite hydrogel by orthogonal experiment, *Journal of Bionic Engineering*, 9, 2012, 234–242.
87. W.V. Srubar III, S. Pilla, Z.C. Wright, C.A. Ryan, J.P. Greene, C.W. Frank, and S.L. Billington, Mechanisms and impact of fiber–matrix compatibilization techniques on the material characterization of PHBV/oak wood flour engineered biobased composites, *Composites Science and Technology*, 72, 2012, 708–715.
88. L. Conzatti, F. Giunco, P. Stagnaro, M. Capobianco, M. Castellano, and E. Marsano, Polyester-based biocomposites containing wool fibres, *Composites: Part A*, 43, 2012, 1113–1119.
89. A. Porras and A. Maranon, Development and characterization of a laminate composite material from polylactic acid (PLA) and woven bamboo fabric, *Composites: Part B*, 43, 2012, 2782–2788.
90. M. Malmstein, A.R. Chambers, and J.I.R. Blake, Hygrothermal ageing of plant oil based marine composites, *Composite Structures*, 101, 2013, 138–143.
91. F. Yan-Hong, L. Yi-Jie, X. Bai-Ping, Z. Da-Wei, Q. Jin-Ping, and H. He-Zhi, Effect of fiber morphology on rheological properties of plant fiber reinforced poly(butylene succinate) composites, *Composites: Part B*, 2013, 44, 193–199.
92. B. Imre and B. Pukánszky, Compatibilization in bio-based and biodegradable polymer blends, *European Polymer Journal*, 49, 2013, 1215–1233.
93. K.O. Reddy, C. Uma Maheswarib, M. Shukla, J.I. Song, and A. Varada Rajulu, Tensile and structural characterization of alkali treated Borassus fruit fine fibers, *Composites: Part B*, 44, 2013, 433–438.
94. A.M. El-Kassas and A-H.I. Mourad, Novel fibers preparation technique for manufacturing of rice straw based fiberboards and their characterization, *Materials and Design*, 50, 2013, 757–765.
95. F. Xiea, E. Pollet, P.J. Halleya, and L. Avérous, Starch-based nano-biocomposites, *Progress in Polymer Science*, 38 (10–11), October–November 2013, 1590–1628.
96. G. Koronis, A. Silva, and M. Fontul, Green composites: A review of adequate materials for automotive applications, *Composites: Part B*, 44, 2013, 120–127.

18

Recent Researches and Developments
of Magical Materials

18.1 Porous and Foam Materials

18.1.1 Recent Trends in the Development of Porous Metals and Metallic Foams [1]

Cellular metals and metallic foams are metals with pores that are deliberately integrated in their structure. The terms cellular metals or porous metals refer to metals having large volume of porosities, while the terms foamed metal or metallic foams apply to porous metals produced with processes where foaming takes place. Porous metals and metallic foams have combinations of properties that cannot be obtained with dense polymers, metals and ceramics or polymer and ceramic foams. For example the mechanical strength, stiffness and energy absorption of metallic foams are much higher than those of polymer foams. They are thermally and electrically conductive and maintain their mechanical properties at much higher temperatures than polymers. They are more stable also in harsh environments than polymer foams. As opposed to ceramics, they have the ability to deform plastically and absorb energy. If they have open porosity, they are permeable and can have very high specific surface areas, characteristics required for flow-through applications or when surface exchange is involved.

Sintered powder and meshes were the first porous metals used in engineering applications. They have been used with success for the fabrication of filters, batteries and self-lubricated bearing still used today in high-volume applications.

18.1.1.1 Different Types of Metallic Foams

The selection of the material structure is generally based on the end applications. Metallic foams may broadly be classified into the following two categories:

1. Closed-cell materials
2. Open-cell materials

Closed-cell materials provide good mechanical properties but do not allow access to their internal surface. Therefore, they are mostly used in structural, load-bearing applications. In contrast, open porosity is generally required for functions associated with the interior of the material. Accordingly, *open-cell foams* are mostly used in functional applications where load-bearing capability is not the primary goal. The present section is divided into these two groups of applications while keeping in mind that dual or multiple uses are possible. Efforts to develop foams with other metals are nevertheless ongoing, including steel, superplastic zinc, magnesium and even heavy elements such as gold.

18.1.2 Closed-Cell Metallic Foams [1]

Foams are complex mixtures of solid, liquid and gaseous phases. Closed-cell metallic foams can be manufactured in many ways. In most cases, a metallic melt is stabilized with nonmetallic particles, foamed by a gas and solidified. Gas bubbles are created by gas injection or by decomposing a chemical blowing agent in the melt. These bubbles do not merge or rupture due to the presence of stabilizing particles.

Foam stabilization can be obtained by adding ceramic particles into the metallic melt, which adhere to the gas/metal interfaces during foaming and prevent pore coalescence. One foam production process (also referred to as the *Alcan process*) uses liquid metal matrix composites (MMCs) containing 10–20 vol.% particles (typically 10 μm silicon carbide or alumina particles) into which a blowing gas is injected. Very regular and highly porous metal foams can be produced with this method. However, the high particle content makes the solid foams very brittle and hard to machine. Another area of investigation involves the development and improvement of the blowing agents. TiH_2 is presently considered the most powerful blowing agent available for the production of aluminium and magnesium alloy foams. But the hydride is costly. Since 0.5–1.5 wt.% TiH_2 is needed to produce an aluminium foam, the blowing agent contributes significantly to the final cost of the material. Replacing TiH_2 by a less expensive blowing agent, for example $CaCO_3$, is under investigation.

18.1.2.1 Metal Foam–Based Composites

Combining metal foams with other materials offers unique opportunity to tailor material properties. The simplest approach is to fill the hollow sections with foam or to produce sandwich structures with a metal foam core and a variety of face sheets. The potential for improvements is still large as demonstrated by the combination of aluminium foam and glass fibre–reinforced polymer composites and aluminium alloy sheets or by composites containing aluminium foam spheres bonded together with a polymeric adhesive. A related approach is to deposit a high-strength coating on the struts of a foam, as demonstrated with nanocrystalline Ni–W-coated aluminium foams showing improvements in strength and energy absorption [1].

18.1.2.2 Iron-Based Materials

Iron-based foams are considered as an alternative to aluminium foams [1]. It is because the steel has higher strength and higher capability to absorb energy and is generally cheaper than aluminium. However, steel is denser than aluminium, and its much higher melting temperature makes the low-cost production of foams a challenge. Various processes to overcome these problems have been proposed such as to foam the cold-pressed precursors made of pure (electrolytic) iron, graphite and hematite powders between the solidus and liquidus temperatures. Iron foams with porosity of approximately 55% are obtained using the process. An alternative method uses less expensive powder (i.e. water atomized powders) and a slip reaction sintering process to produce steel foams with compressive strength significantly higher than that of aluminium foams. The material density (2.3–3.0 g/cm^3) is, however, also higher than that of aluminium foams.

High-strength stainless steel foams are also produced with powder metallurgy approaches when the high mechanical strength and corrosion resistance are required. Besides, steel hollow spheres with high strength have also been prepared. The densities of most iron-based foams are still high, and additional work is required to develop steel foams with lower density.

18.1.2.3 Metallic Hollow Spheres

The metallic hollow spheres are used in various researches to produce high-strength steel hollow spheres [1]. Composites and syntactic structures have also been developed with hollow spheres for structural applications. A process to produce hollow Cu_2O and NiO nanoparticles via the oxidation of Cu and Ni nanoparticles has also been presented. More details are discussed later in this section.

18.1.2.4 Amorphous Metallic Foams [1]

Porosity can be created by precipitation of dissolved hydrogen during cooling, gas entrapment in the melt followed by expansion in the supercooled liquid state, infiltration of a bed of hollow spheres to create syntactic foams (SFs) and infiltration of salt space holder particles, which are removed by dissolution in acidic solutions. Various liquid-state methods have been demonstrated to create Pd- or Zr-based bulk metallic glass (BMG) foams. Solid-state methods have also been used: dissolution of crystalline phase from an extruded amorphous-crystalline composite and selective dissolution of one of the two amorphous phases of an alloy. Besides reducing the density, the pores in amorphous metals improve their compressive ductility, from near zero in the bulk to values as high as 80% for cellular architectures.

18.1.2.5 Wire Mesh Structures

In this technique, the fabrication is based on the assembly of helical wires in six directions. The structures are periodic and very uniform, have good specific properties and are highly permeable. Different materials such as iron, steel and aluminium have been produced with this process. The properties of these structures were compared to other periodic cellular structures. The results show that the effective modulus of the specimen is insensitive to the specimen size, while the peak stress rapidly increases with a reduction of the specimen size.

18.1.3 Applications of Porous Materials [1]

Porous metals and metallic foams have been commercially used for many years for different applications. Important commercial developments took place on closed-cell aluminium foams and sandwich panels for structural applications. The most important commercial applications of porous metals and metallic foams remain open-cell materials for the production of filters, gas flow controlling devices, batteries, biomedical implants and bearings. Due to their good deformability and high energy absorption capability, the aluminium foams have been considered for the production of crash energy absorbers for many years. A crash bumper is used in light city railways for some years. Other applications are such as an Alporas metal foam crash element in the front tip of the chassis of a racing car.

18.1.3.1 Materials with Elongated Pores [1]

Various materials such as Cu, Ni, Mg, Al, Si, TiAl and alumina are now produced on laboratory scale. Recent work showed promising results on the use of chemical foaming agents to avoid using large volume of pressurized hydrogen, that represent potential hazard risks for the large-scale manufacturing with the traditional high-pressure gas methods.

Koh et al. presented an alternative process to produce anisotropic porous tubular structures by extruding copper and aluminium wires, followed by leaching of the aluminium to produce tubes with long elongated pores. The resulting structures have low tortuosity and high permeability. A similar approach, using polymer or organic binders instead of aluminium followed by extrusion, has been demonstrated for various metals including copper, with applications for heat sinks and heat exchangers. A related approach consists of sintering titanium powders around steel wires, which are then removed electrochemically. Titanium with aligned, elongated pore has also been produced by directional freeze casting of aqueous slurries. Ice dendrites grow directionally in a temperature gradient, pushing Ti particles into inter-dendritic spaces. After removal of the ice by freeze-drying, the powders are sintered, forming walls separating elongated pores which replicate the dendrites.

18.1.3.2 Nanoporous Materials [1]

Nanoporous metallic structural materials, characterized by very high specific surface area and fine pore size, have been considered in electrodes, catalyst, sensors, actuators and filtration applications. Most research is presently done using leaching or dealloying techniques. The dissolution of silver and diffusion of silver and gold during Ag–Au dealloying allowed for reproducing the mechanisms of dealloying (morphology and kinetics). In addition to dealloying, other processes are being developed to produce materials with submicron porosities, for example a selective dissolution technique to leach Ni-based superalloys that contain γ'-precipitates in a γ-matrix. The resulting material has fine and regular submicron porosity, is permeable and is mechanically strong. The materials could be suited in membrane applications where fine porosity, thermal stability, mechanical strength and ductility are required.

18.1.3.3 High Temperature–Resistant Cellular Materials [1]

The development of cellular metals for high-temperature filtration and solid-oxide fuel cells operating at elevated temperatures is a recent development. Filters trap soot particles produced by the diesel engines, and catalysts convert various toxic gases and lower the soot combustion temperature allowing the filter to regenerate. As the efficiency of the engine is affected by the pressure drop at the exhaust, materials with good permeability are required. Metallic foams with good corrosion resistance at temperatures up to 600°C–800°C, especially NiCrAl-based alloys, are being considered for these applications.

Different processes are being developed to produce new high temperature–resistant foams. These are a process to produce high temperature–resistant Fe–Ni–Cr–Al foams using the deposition and transient liquid phase sintering of fine prealloyed particles onto thin commercially available nickel foam strip. Filters are produced either by coiling the foam strip or by sintering the sheets in stacks. Compared to the state-of-the-art wall-flow filters, the regeneration of the foam filters occurs faster and at lower temperatures because of the improved soot-filter wall contact provided by the turbulent gas flow:

- To produce heat-resistant foams using vapour deposition of Al and Cr (pack aluminization and chromizing) on nickel foam strips followed by a homogenization treatment to produce Ni–Cr and Ni–Cr–Al foams.

- A process to produce Inconel 625 foams using slurries of particles. They showed that the material maintains its compressive properties up to 600°C and can withstand thermal shock up to 1000°C. A strategy proposed to enhance the chemical stability of the material by deposing TiN on the surface using chemical vapour infiltration (CVI) processes.
- Casting replication of an oxide space holder has also been demonstrated to create sponges from an oxidation-resistant Ni–Mo–Cr alloy developed for fuel cell interconnects.

18.1.3.4 Use of Porous Coating in Biomaterials [1]

Porous coatings have been used with bone cement to produce a rough surface that increases the friction between the implant and the surrounding bone. After implantation, the bone grows into the porous surface and helps to secure the long-term stability of the implant. These coatings are now extensively used in various orthopaedic applications. The development nowadays is from thin porous bead coatings, sintered mesh and thermal-sprayed rough coatings to metallic foams. These materials are used in the development of various new treatments (i.e. bone augmentation and graft-free vertebra fusion for the treatment of degenerative disc diseases). Presently, the materials being developed are titanium alloys and nitinol. Directional freeze casting has also been applied to produce titanium foams with 60 vol.% aligned pores.

18.1.4 Various Properties of Porous Materials [1]

18.1.4.1 Permeability

Permeability is an important property for flow-through foam applications such as thermal applications, filtration, electrochemistry, acoustic absorption and porous implants. Recent investigations have shown that the classical models such as Darcy's or Forchheimer's models do not describe appropriately the pressure drop in porous solid, especially at high fluid velocity. It was shown that the pressure drop cannot be normalized by the material thickness since entrance and exit effects can significantly affect the pressure drop.

18.1.4.2 Mechanical Properties of Porous Materials

Most foam applications are primarily load bearing, for example for sandwich structures. The foams whose main properties are functional, for example acoustic, thermal or surface area, require minimal mechanical properties to prevent damage or failure. Age-hardening of aluminium foams improves strength of foams with large cell sizes (400 mm) but not with fine cell size (75 mm), damage accumulation. Density-graded aluminium foams exhibit a smoothly rising plateau stress, unlike uniform foam with a near constant plateau stress.

18.1.4.3 Thermal Properties

Metallic foams are conductive and permeable and have high surface area. This combination of properties makes them attractive for various thermal applications like heat exchangers, heat sink and heat pipes. Heat exchanges and conduction in metallic foams are complex phenomena. Heat exchange efficiency is affected by the conductivity of the foam, the heat exchange between the foam and the surrounding fluid and the pressure drop in the foam. These characteristics are all affected by various structural parameters such as density, pore size distribution, cell connectivity, tortuosity, strut size, density and geometry and surface roughness.

18.1.4.4 Acoustic Properties

Metallic foams are being used in various acoustic applications. The unique structure of these materials provides good sound absorption characteristics. Their acoustic properties can be combined thermal and chemical stability and mechanical properties of the metallic foams to make them more attractive than common sound absorbers such as mineral wool or polymer foams. Standard sound absorbers are generally much cheaper than most metallic foams.

18.1.5 Open-Cell Materials [1]

Electrodes for NiMH and NiCd batteries are probably the largest industrial applications of metallic foams nowadays. With the development of lithium ion batteries, demand will eventually be reduced. But as these Ni foams are currently produced in large volume with uniform and controllable structure and properties at acceptable cost, they will surely find other large-scale applications in the coming years. Another important commercial application of metallic foams and porous metals is the production of biomedical implants. Porous coatings (sintered beads and meshes) have been used with success for the last 20 years in orthopaedic applications. In the automotive industry, the open-cell metallic foams are used for the production of catalytic converters for diesel engines. Porous stainless steel laminates have found applications for the thermal management, flow distribution, membrane support and current collection components of fuel cell stacks. Fuel cell is likely to become a large-scale application of metallic foams.

18.1.6 Porous Metals and Metal Foams Made from Powders [2]

Porous metals are metals having a large volume of porosity, typically 75%–95%, and metal foams are metals with pores deliberately integrated into their structure through a foaming process depending on their structure. Porous materials are found in natural structures such as wood, bone, coral, cork and sponge and are synonymous with strong and lightweight structures. The man-made porous materials, made from polymers and ceramics, have been widely exploited. The unique combination of physical and mechanical properties offered by porous metals, combinations that cannot be obtained with dense metals, or either dense or porous polymers and ceramics, makes them attractive materials for exploitation. Interest mainly focuses on exploiting their ability to be incorporated into strong, stiff lightweight structures, particularly those involving Al foams as the *filling* in sandwich panels; their ability to absorb energy, vibration and sound; and their resilience at high temperature coupled with good thermal conductivity.

Closed-cell foams have gas-filled pores separated from each other by metal cell walls. They have good strength and are mainly used for structural applications. Open-cell foams, which contain a continuous network of metallic struts and the enclosed pores in each strut frame are connected (in most cases, these materials are actually porous or cellular metals), are weaker and are mainly used in functional applications where the continuous nature of the porosity is exploited.

18.1.6.1 Fields of Applications of Porous Metals and Foams

These are displayed in Table 18.1.

TABLE 18.1

Potential Application Areas for Porous Metals and Metal Foams

Field of Application	Behaviour
• For lightweight structures	Excellent stiffness-to-weight ratio when loaded in bending.
• For vibration control	Foamed panels have higher natural flexural vibration. Frequencies than solid sheet of the same mass per unit area.
• For energy absorbers/ packaging	Exceptional ability to absorb energy at almost constant pressure.
• For mechanical damping	Damping capacity is larger than solid metals by up to 10×.
• For acoustic absorption	Open-cell metal foams have sound-absorbing capacity.
• For filters	Open-cell foams for high-temperature gas and fluid filtration.
• For biocompatible inserts	Cellular texture stimulates cell growth.
• For heat exchangers	Open-cell foams have large accessible surface area and high cell wall conduction giving exceptional heat transfer ability.

Source: Modified from Kennedy, A., *Powder Metallurgy*, Dr. Katsuyoshi Kondoh (Ed.), InTech.

18.1.7 Processing Methods for Metal Foams [2]

There are many different ways to produce porous metals and metallic foams. These methods are usually classified into four different types of production, using

- Liquid metals
- Powdered metals
- Metal vapour
- Metal ions

The use of powdered metals as the starting material for foam production offers the same types of advantages and limitations as conventional powder metal-lurgical processes. If a particular metal or alloy can be pressed and sintered, there is a likelihood that it can be made into a porous metal or metal foam.

18.1.7.1 Porous Metals Produced by Pressureless Powder Sintering [2]

Loose pack, pressureless or gravity sintered metal powders are the first form of porous metals and are still widely used as filters and as self-lubricating bearings. The porosity in these components is simply derived from the incomplete space filling of powders poured into and sintered in a die. With packing densities broadly in the range of 40%–60%, the porosities in these structures are well below those for most porous metals. The simplicity of the process means that porosity can be included in a wide range of metals. The process is most commonly used to sinter bronze powders to make bearing.

18.1.7.2 Gas Entrapment [2]

Internal porosity can be developed in metal structures by a gas expansion (or foaming) based on hot isostatic pressing. Initially, a gas-tight metal can is filled with powder and evacuated. The can is then filled with argon gas at pressures between 3 and 5 bar before being sealed, isostatically pressed at high temperature and then worked to form a shaped product, normally a sheet. Porosity is generated by annealing the part. When holding at elevated temperature, the pressurized argon gas present within small pores in the structure causes the material to expand (foam) by creep. It is an effective method for sintering. This process can be used for different metal powders. These types of structures are ideal for lightweight construction, but the dis-advantages are low porosity and irregular-shaped pores.

18.1.7.3 Reactive Processing [2]

In highly reactive multicomponent powder systems such as those which undergo self-propagating high-temperature synthesis (SHS), the porosity is evolved much more rapidly when foaming occurs. The highly exothermic reactions, initiated either by local or global heating of compacted powder mixtures to the reaction ignition temperature, lead to vapourization of hydrated oxides on the powder surfaces and the release of gases dissolved in the powder. The reacting powder mixture heats up rapidly to form a liquid containing mostly hydrogen gas bubbles and when the reaction is complete, cools rapidly, entrapping the gas to form a foam. This process is simple, but the production of foams is limited to combinations of materials that react exothermically, namely, some metal–metal systems, metals and carbon or carbides or metals and oxides.

18.1.7.4 Addition of Space-Holding Fillers [2]

Porous metals are produced by mixing and compacting metal powders with a space holder, which is later removed either during or after sintering, by dissolution or thermal degradation, to leave porosity. This simple method has the advantage that the morphologies of the pores and their size are determined by the characteristics of the space holder particles and the foam porosity can be easily controlled by varying the metal/space holder volume ratio. Addition levels of the space holder are typically between 50% and 85%. Above 85%, the structure of the struts is unlikely to be continuous, and below 50%, residual space filler will be enclosed within the structure, making removal very difficult.

18.1.7.5 Metal Powder Slurry Processing

In this processing method, the following slurries are used:

1. Metal powder slurries
2. Slurry coating of polyurethane foams
3. Slip reaction foam sintering

18.1.8 Aluminium Foam: A Potential Functional Material [3]

Cellular material has emerged as a new class of engineering material with porous structure (range of porosity, 40%–98%). The pores may be either interconnected or closed depending upon the synthesis process. These materials have exceptionally light weight with unique combination of properties such as high energy absorption capacity, high specific strength and stiffness, non-linear deformation characteristics, high surface area and electrical insulating properties. Because of these characteristics, cellular material finds applications in making shock and impact absorber for automobiles, defence, heat exchanger and flame arrestor, for domestic application and sandwich structure and for packaging industries. Cellular material exhibits low-density, high bending stiffness and specific strength compared to solid metallic material because of its porous nature, which enables them suitable for sandwich construction and energy absorption applications.

The properties of cellular material considerably depend on its density and morphology of pores. For example Al cellular material with density ranges 0.5–0.6 g/cm^3 is suitable for energy absorption, in automobile, and also as a cushion material for engine. However, Al cellular material with the density range 0.25–0.4 g/cm^3 has best acoustical properties and heat conductivity, which is suitable for door filling and ceiling and partitions of car engine. The thermal performance of cellular material mainly depends on the cell structure. For example open cell is suitable for heat exchanger, whereas closed cell used is for heat insulator. The cellular material (closed pore) can convert much of the impact energy into plastic energy and absorb more energy than the bulk material. The energy absorption capacity of Al cellular material up to a compressive deformation of 50% ranges from 6,850 to 39,200 kJ/m^3.

18.1.8.1 Fabrication, Testing and Results

Al foam is synthesized by melt route using SiC particles as the thickening agent and metal hydride as foaming agent. The structural parameters such as cell size (1–4 mm), cell wall thickness (200–400 µm) and pore size distributions are found out using Materials-Pro image analysis software. Compression deformation behaviour is studied at room temperature to 200°C, at different strain rates (0.01–10/s). As the deformation temperature increases, plateau stress decreases due to softening of the metallic cells and increases with strain rate due to work hardening. Crash boxes are made and are tested using drop test and the energy absorption is found out to be 7 MJ/m^3. The mechanical damping of Al foam, Al foam–filled sections and empty tube is evaluated. The results show that damping capacity of tube could be enhanced by 85%–90% by incorporation of Al foam. Noise attenuation property of Al foam is also improved significantly. Prototype components such as Al foam–filled inverter bracket and sandwich panels are made.

Al foams having density of 0.6 g/cm^3 and Young's modulus of 4.5 GPa could be made through stir casting technique using metal hydride as foaming and SiC as thickening agent. The cell shape in the foam is found to be elliptical and its size is in the range of 1–4 mm. The compressive deformation behaviour of these foams is found to be strain rate insensitive at room temperature. But at higher temperature, the plateau stress and energy absorption are very sensitive to strain rate and are potential materials for energy absorption. The damping behaviour of these foams, as compared to dense steel bars, hollow steel tubes and foam-filled tubes, is ~20–30 times greater. Foam-filled tubes exhibit damping capacity ~9–10 times higher than the dense steel bars or hollow steel tube. The sound absorption coefficient is also found improved by 34% as compared to reference material. Crash box, foam-filled structure and brackets are made successfully and the performance evaluation showed encouraging results.

18.1.9 Trends in the Research of Syntactic Foams

SFs are a novel class of materials having unique characteristics. They are the particulate composites made by dispersing the hollow inclusions. Due to their versatility of behaviour, they are now used as deep sea and offshore construction materials, energy absorption materials, in marines, aerospace and automobiles. This article aims at elaborating the development of such materials by compiling the ongoing researches in this field. In this regard, the research developments of some newer syntactic materials by different investigators have been presented. These include brief details of the development of ceramic sphere (or cenosphere)–filled aluminium alloy SF, shape memory polymer (SMP)–based self-healing SF, polymer-matrix SFs for marine applications, titanium–cenosphere SF for biomedical uses and carbon nanofibre–reinforced multiscale SFs. In these elaborations, it is shown by the respective investigators that these SFs can be effectively used as instrument panels; self-healing, cushion materials for packaging; fire resistance materials; bone implants and various other applications.

18.1.9.1 Introduction

SFs are composite materials synthesized by filling a metal, polymer or ceramic matrix with the hollow particles known as microballoons. These foams exhibit excellent combination of physical and mechanical properties. They have high damage tolerance through high energy absorbing capabilities, high specific stiffness, improved strength and high acoustic and mechanical damping capacities. These properties make SFs an attractive choice for various applications such as core in sandwich structures and for packaging/fire proof, crash safety, damping panels and underwater buoyant structures.

The ceramic particle–filled SFs exhibit excellent thermal insulation and low thermal expansion coefficient due to significant volume fraction of ceramic phase in the matrix. Although SFs containing hollow ceramic spheres, that is cenosphere, have higher densities than conventional aluminium foams, they have the advantages of higher strengths, isotropic mechanical properties and excellent energy absorbing capacities due to extensive strain accumulation at relatively high stresses. Various kinds of SFs are in the list of emerging materials. Main among them are discussed in the forthcoming articles. In the next sections, the trends in research of these SF have been presented.

18.1.9.2 Ceramic Sphere (or Cenosphere)–Filled Aluminium Alloy Syntactic Foam [4]

In order to use SFs in the advanced applications such as crash or impact, blast resistance, aeronautical and space structures, it is crucial to understand their behaviour under high rate of loading. Quasistatic compression behaviour of SFs has been studied in the past by a few researchers for assessing their energy absorption capacity. High strain rate deformation behaviour has been examined on commercially pure aluminium cenosphere SF [4]. However, aluminium alloys are preferred in engineering and other applications because they provide improved strength and toughness. In aluminium alloy foams, the strain rate sensitivity is dependent on matrix materials, whereas in case of MMCs, the particles govern the strain rate sensitivity instead of matrix. SF exhibits a structure combination of both metal foam and MMCs and hence their deformation behaviour under varying strain rate is highly complex and is not consistent in nature. In quasistatic condition, the strain rate sensitivity is reported to be very low [4].

18.1.9.3 Shape Memory Polymer–Based Self-Healing Syntactic Foam [5]

SFs, a class of composites made of a polymeric matrix filled with hollow particulate inclusions, renowned for its low density and high specific mechanical properties, are currently enjoying continuous growth in various civilian and military sectors. Their typical applications range from buoyancy materials for submarines to cushion materials for packing. The versatility and applicability of such composites attract interest from numerous researchers and thus propel many characteristic: improvement studies such as functionally grading SFs and toughened SFs. Because of the improved mechanical/functional properties, a tendency is found that SFs are being increasingly used in high-performance load-bearing structures such as foam cored sandwich and grid stiffened foam cored sandwich. However, like other composite structures, foam cored or hybrid foam cored sandwich structures are still vulnerable to impact damage. Hence, the self-healing or self-mending of structural damage is desired.

One of the grand challenges facing self-healing community is how to heal macroscopic or structural scale damage such as impact damage autonomously, repeatedly, efficiently and at molecular length scale. For the existing self-healing schemes such as microcapsules, hollow fibres, microvascular networks, ionomers, thermally reversible covalent bonds and thermal plastic particle additives, they are very effective in self-healing microlength scale damages. However, they face tremendous challenges when they are used to heal large, millimetre-scale, structural damage. The reason for this is that the structural damage needs a sufficient amount of healing agent to fill in the crack. But the incorporation of a large amount of healing agent such as thermoplastic particles will significantly alter the physical/mechanical properties of the host structure. Also, large capsules or thick hollow fibres themselves may become potential defects when the encased healing agent is released.

Although ionomer has been proved to self-heal ballistic impact damage, it inherently utilized the rebound of the broken ionomers. Without the elastic rebound, the broken pieces cannot be brought into contact and ionomer cannot self-heal itself, either. Because of this, we propose that a sequential two-step scheme be used by mimicking the biological healing process such as human skin: seal then heal (STH). In STH, the structural scale crack will be first sealed or closed by a certain mechanism before the existing self-healing mechanisms such as microcapsule or thermoplastic particles take effect.

SMP, due to its autonomous, conformational entropy-driven shape recovery properties, can be utilized to achieve the self-sealing purposes. In a previous study, it has been proved that the confined shape recovery of an SMP-based SF is able to seal impact damage repeatedly, efficiently and almost autonomously. It is believed that if the existing microlength scale repairing scheme such as microcapsules or thermoplastic particles is incorporated into the SMP matrix, a two-step (STH) self-healing scheme could be achieved and the healing would be achieved repeatedly, efficiently, almost autonomously and at molecular length scale.

18.1.9.4 Fire-Resistant Syntactic Foam [6]

The wide use of polymer composites in structural applications raised the potential for fire hazards significantly. Although fire cannot be completely eliminated, it can be mitigated to reduce the loss of life and property. Extensive research is being conducted to improve the fire safety of composite materials for various applications. A fire-resistant core material called *Eco-Core* has been produced for sandwich structures [6]. The sandwich panel acts as a fire constrainer and stops the growth and spread of fire. Eco-Core has the compressibility of about 40%–50% due to crushing of hollow ceramic micro-bubbles. It has a potential to be used in marine, aerospace and transport industries as a fire and toxicity safe structural core. All transportation structures are subjected to vibration or cyclic loading. Therefore, understanding of the fatigue strength, fatigue life and the associated failure modes is of great importance under various stresses. Understanding the fatigue performance of Eco-Core under compression, tension, flexure and shear stress states is very important before the material can be used in structural applications.

Fatigue properties of polymeric and aluminium alloy foams were studied by many researchers. Fatigue failure of the open-cell and closed-cell aluminium alloy foams was also studied. Compression–compression fatigue of open-cell aluminium foams demonstrated that the foam structure when subjected to repeated loading rapidly loses its strength and the fatigue degradation. In these materials, fatigue damage is associated with the formation of compressive failure bands (in case of compression) and macroscopic cracks (in case of tension). The progressive shortening of specimen in compression–compression fatigue and progressive lengthening in tension–tension fatigue were reported. Relatively high-density foams under compression failed by formation of 45° shear planes to the loading direction also highlighted that the aspect ratio affects the shear failure.

The density of SF affects failure mechanism. For lower-density foam, the failure is by longitudinal splitting and for high-density foam, it is by layered crushing. All those studies were on SFs that are made by manufactured micro-bubbles. The influences of filler content, reinforcement type and their weight fraction on flexural fatigue behaviour of SFs concluded that in case of unre-inforced foams, endurance limit of the foam based on one million of cycles changes with changing filler content. Eco-Core is a special class of SF that has very low binder content (about 5% by volume) and microbubbles extracted from fly ash that is produced by coal burn electric thermal power plants.

18.1.9.5 Titanium–Cenosphere Syntactic Foam for Biomedical Uses [7]

Titanium foams are finding a wide range of applications in high-temperature energy absorptions, catalysts, crashworthiness, lightweight sandwich struc-tures, high-temperature heat exchangers, bone scaffolds and bone implanta-tions. However, most significant efforts have been made to prepare open-cell Ti-foam compatible to the characteristics of bones for its use as bone implants. Ti-foam with 57%–67% porosity has been prepared through freeze casting of aqueous slurry of Ti-powder. It is also reported by these investigators that the size of Ti-powder plays an important role in the synthesis of Ti-foam. The strength of these foam ranges between 40 and 60 MPa, which is equivalent to that of human bone. The sintering and densification kinetics also depend on the addition of foreign particles like TiC in the Ti-powder mixture. Attempt was made to prepare Ti-foam with functionally graded characteristics to make it more closely compatible with bone tissues. The ingrowth behav-iour of Ti-foam implants into hard tissue could be improved with bioactive polymer or hydroxyapatite coating. In a study, it was found that the coating with autologous osteoblasts accelerates and enhances the osseointegration of Ti-foam implants.

18.1.9.6 Ti-Foam as Futuristic Material for Bone Implant

Attempts have been made by several investigators to synthesize Ti-foam using different techniques, and these foams are futuristic material for bone implants. Different methods include controlled compaction and sintering of irregular Ti-powders with and without ZrO_2 addition through plasma rotating electrode process, spark plasma sintering of loose Ti-powder com-pact, sintering of hot compacted Ti–SiC fibre monotapes at temperature between 800°C and 900°C, entrapment of argon gas in a cold compacted Ti-powder followed by liquid stage sintering, sintering of leachable salt and Ti-powder compacts followed by dissolution of salts and sintering of Mg and Ti-powder compacts, followed by removal of Mg by heating at con-trolled atmosphere at temperature higher than the melting point of Mg. Ti6 Al4 V foam was also prepared with 60% porosity and pore size using space holder technique through powder metallurgy route.

18.1.9.7 Polymer-Matrix Syntactic Foams for Marine Applications [8]

Polymer-matrix SFs are widely used in marine and aerospace applications as core material in sandwich structures. These lightweight composites can be tailored to have better compressive properties than the matrix resin. The closed-cell microstructure defined by porosity enclosed within thin-walled hollow inclusions, called microballoons, provides low moisture absorption and high buoyancy that are beneficial for marine applications. Enhancing the breadth of applications of SFs in marine structures requires a thorough understanding of the water uptake mechanisms. Additionally, the effect of moisture-induced degradation on mechanical properties such as Young's modulus and flexural strength that strongly characterize their performances as core materials in sandwich structures needs to be further explored.

These studies describe the long- and short-term diffusion process in composites through observations of weight gain or changes in buoyancy and thermal and electrical properties. It is therein shown that the combined effect of water uptake and hot ageing conditions increases the thermal and electrical conductivity of SFs. Many experimental investigations focus on the effect of water environment on the fracture toughness as well as compressive, tensile and thermal properties of epoxy matrix SFs. For example it is shown that a decrease or increase in compressive strength of the composite may occur and that testing temperature and water type play a role in determining the extent of property variation.

18.1.9.8 Carbon Nanofibre–Reinforced Multiscale Syntactic Foams [9]

Hollow particle–filled composites, called SFs, are widely studied in recent times due to increasing interest in lightweight materials. These particulate composites contain porosity enclosed inside thin shells, called microballoons, dispersed in the matrix material. Filling high volume content of microballoons in the matrix material allows for significant weight saving that is crucial in marine structures, where buoyancy can be increased, and aerospace structures, where payload capacity can be improved. However, higher volume fraction of hollow particles leads to lower strength of SFs. Thus, it is of great interest to develop methods of reinforcing SFs without decreasing the particle packing limit in the material structure. Among different compositions of SFs, epoxy resins filled with glass microballoons are widely studied due to the extensive use of epoxy resins in aerospace applications. The present study [9] explores the possibility of reinforcing epoxy matrix with nanofibres for developing high-performance SFs.

Reinforcement of SFs through embedded fibres is investigated as a viable means for enhancing the strength and stiffness. Nevertheless, microfibres may limit the microballoon volume fraction levels and thus impact SF weight. In addition, in high microballoon volume fraction composites, glass fibres can increase the particle failure during composite processing. On the other hand, the use of nanoscale fillers is beneficial in reinforcing SFs because nanoparticles can modify the properties of the matrix and exist in the spaces between microballoons, without decreasing the microballoon packing limit. The use of nanoclay and carbon nanofibres for matrix modification in SFs is explored.

Experimental results on epoxy systems show increase in the modulus that may be attributed to the large surface area of nanoclay platelets available to bond with the resin. Typically, the increase in the modulus is followed by a simultaneous decrease in the SF strength. Such effects also influence the quality of nanoclay dispersion. On the other hand, the use of nanofibres can be beneficial since large aspect ratio nanofibres can bridge cracks and reduce their propagation rate to simultaneously improve tensile strength and modulus. Carbon nanofibres are presently used in reinforcing epoxy, polyethylene, poly(methyl methacrylate) and several other polymers. A recent study demonstrates that tensile strength of SFs is improved by incorporation of a small volume fraction of CNFs. This effort is focused on finding the effect of microballoon volume fraction and wall thickness on the properties of CNF-reinforced composites and shows promise in such reinforcement scheme.

18.1.9.9 Conclusions

From the aforementioned studies, the following conclusions can be drawn:

1. The present SF has about 10%–30% higher dynamic compressive strength than that of quasistatic condition. The compressive strength of these foams reaches to the maximum at a critical strain rate of 750/s. When the strain rate exceeds this critical value, it reduces further with the increase in strain rate. The strength and energy absorption are also a strong function of cenosphere size. These parameters decrease with increase in cenosphere size. The energy absorption efficiency of aluminium cenosphere SF is noted to be about 80% or more.

2. It is found that while volume reduction during programming is the key for the foam to have self-closing functionality, the volume reduction must be within a certain limit (54.34%); otherwise, the foam will lose its shape memory functionality and the capability for self-closing cracks.

3. The study showed that the Eco-Core has well-defined failure modes and associated fatigue lives. Eco-Core's failure can be classified into three types, namely, damage onset is characterized by formation of a single crush band at the middle or edge of the specimen, damage progression is characterized by crush band propagation and final failure characterized by 7% compliance change.

4. Cenospheres can be used as space holders to make Ti–cenosphere SF. But cold compaction pressure has to be applied cautiously.

An applied pressure of 75 MPa will be sufficient during cold compaction to minimize crushing of cenosphere shell, getting a porosity of 62%. The crushing of cenosphere, porosity fraction and density of Ti–cenosphere SF are strong function of applied pressure. Ti–cenosphere could be excellent and new champion materials for replacement of Ti and its alloys for structural, shock and energy absorbing applications. The porosity fraction as low as 52% can be achieved in Ti–cenosphere SF. The pores are observed primarily due to the presence of unbroken cenosphere.

5. The effects of moisture absorption on the flexural properties of polymer-matrix SFs are conducted on vinyl ester–glass hollow particle–filled composites that are increasingly used in marine applications. The influence of particle wall thickness and volume fraction on the flexural properties of water-exposed SFs is evaluated. Moisture-exposed SFs exhibit a Young's modulus reduction that can be as large as 35% and 30% for DIW and SW environments, respectively. Generally, the entity of the reduction increases as particle volume fraction increases. On the other hand, moisture absorption generally increases the flexural strength of exposed SFs, and this effect is more evident for SW environments and thick hollow particles. Results show that particle volume fraction plays an important role in the diffusion properties of SFs, while they indicate that the particle wall thickness is a secondary parameter.

6. Microballoons of two different wall thicknesses are used in two volume fractions to obtain four types of composites. Results show that the tensile properties are enhanced by the presence of nanofibres. The specific tensile modulus of SFs is found to be considerably higher than that of CNF/epoxy composites. The specific tensile strength of several foam compositions is higher than the CNF/epoxy composites. Compared to the plain SFs of similar particles, the CNF SFs having 30 vol.% particles showed 8%–15% lower tensile modulus and strength. However, CNF SFs containing 50 vol.% particles showed 10%–47% higher tensile strength and modulus than the plain SFs. The contribution of particles in the mechanical properties of both classes of SFs is similar.

18.2 Max Phase Materials

18.2.1 Introduction [10]

MAX phase materials are the ternary carbides and nitrides having the general formula of $M_{n+1}AX_n$ (MAX) in which $n = 1, 2$ or 3, M is an early transition

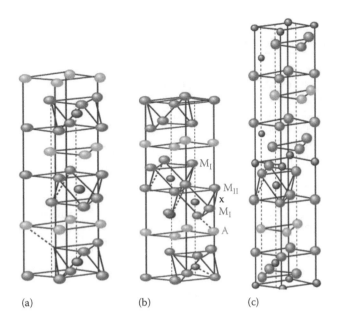

FIGURE 18.1
MAX phase unit cells having (a) M_2AX or 211 phases, (b) M_3AX_2 or 312 phases, and (c) M_4AX_3 or 413 phases. (From Barsoum, M.W. and Radovic, M., *Annu. Rev. Mater. Res.*, 41, 195, March 30, 2011.)

metal (TM), A is an A-group element (a subset of group 13–16 elements), and X is C and/or N. MAX phase materials are a new class of solids. More than 60 MAX phases of different chemical compositions share a common structure, that is $M_{n+1}X_n$ layers interleaved with pure A-group element layers, as illustrated in Figure 18.1 for structures with $n = 1$–3. MAX phase materials exist in both powder form and bulk form. The growing interest in $M_{n+1}AX_n$ phases lies in their unusual and unique properties. Because of their chemical and structural similarities, the MAX phases and their corresponding MX phases share many physical and chemical properties.

Elastically, the MAX phases are in general quite stiff and elastically isotropic. The MAX phases are relatively soft having a hardness of 2–8 GPa, are most readily machinable and are damage tolerant. Some of them are also lightweight and resistant to thermal shock, oxidation, fatigue and creep. In addition, they behave as nonlinear elastic solids dissipating 25% of the mechanical energy during compressive cycling loading of up to 1 GPa at room temperature. At higher temperatures, they undergo a brittle-to-plastic transition, and their mechanical behaviour is a strong function of deformation rate [10].

18.2.2 Crystal Structure [10]

The hexagonal MAX unit cells have two formula units per unit cell (Figure 18.1). Table 18.2 lists most of the known MAX phases. The M_2AX

TABLE 18.2

Some of the Known $M_{n+1}AX_n$ Phases

Column IIIA Materials	
Al, Ti_2AlC, V_2AlC, Cr_2AlC, Nb_2AlC, Ta_2AlC, Ti_2AlN, Ti_3AlC_2, Ti_4AlN_3, α-Ta_4AlC_3, β-Ta_4AlC_3, Nb_4AlC_3, $V_4AlC_{3-1/3}$	
Ga, Ti_2GaC, V_2GaC, Cr_2GaC, Nb_2GaC, Mo_2GaC, Ta_2GaC, Ti_2GaN, Cr_2GaN, V_2Ga	
In, Sc_2InC (?), Ti_2InC, Zr_2InC, Nb_2InC, Hf_2InC, Ti_2InN, Zr_2InN	
Tl, Ti_2TlC, Zr_2TlC, Hf_2TlC, Zr_2TlN	
Column IVA Materials	
Si, Ti_3, SiC_2	Sn, Ti_2SnC, Zr_2SnC, Nb_2SnC, Hf_2SnC, Hf_2SnN
Ge, Ti_2GeC, V_2GeC, Cr_2GeC, Ti_3 GeC_2	Pb, Ti_2PbC, Zr_2PbC, Hf_2PbC
Column VA Materials	
P, V_2PC, Nb_2PC	As, V_2AsC, Nb_2AsC
Column VIA Materials	
S, Ti_2SC, Zr_2SC, $Nb_2SC0.4$, Hf_2SC	Se

Source: Modified from Barsoum, M.W. and Radovic, M., *Ann. Rev. Mater. Res.*, 41, 195, 2011.

phases are also known as 211 phases (Figure 18.1c), M_3AX_2 phases are known as 312 (Figure 18.1b) phases and M_4AX_3 as 413 phases (Figure 18.1c).

The A-group elements are mostly IIIA and IVA (i.e. groups 13–16). The most versatile of these elements is Al because it forms nine compounds, including two nitrides, one 312 phase, and four 413 phases. Ga also forms nine 211 phases, six of which are carbides and three are nitrides. The unit cells of the MAX phases are characterized by near close-packed M-layers interleaved with layers of a pure A-group element, with the X atoms filling the octahedral sites between the former. The M_6X octahedra are identical to those found in the rock salt structure of the corresponding binary MX carbides. The A-group elements are located at the centres of trigonal prisms, which are slightly larger and thus able to accommodate the larger A atoms than the octahedral sites.

When $n=1$, the A-layers are separated by two M-layers (Figure 18.1a). When $n=2$, they are separated by three layers (M_3AX_2 in Figure 18.1b). When $n=3$, they are separated by four layers (M_3AX_2 in Figure 18.1c). MAX phases with more complex stacking sequences, such as M_5AX_4, M_6AX_5, and M_7AX_6, also have been reported.

When $n=2$, as in Ti_3SiC_2, two M-layers separate the A-layers (Figure 18.1b). For $n=3$, as in Ti_4AlN_3, there are four M-layers (Figure 18.1c). The 413 phases exist in two polymorphs, α and β. The layering in the α polymorph is ABABACBCBC, and that of the β polymorph is ABABABABA, where the underlined and italicized letters denote the stacking order of the A-layer. With the exception of Ta_4AlC_3 which exists in bulk form in both polymorphs,

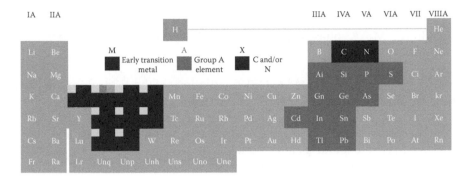

FIGURE 18.2
Unit cells of the $M_{n+1}AX_n$ phases for $n=1$ or M_2AX, $n=2$ or M_3AX_2, and $n=3$ or M_4AX_3 phases and M, A, and X elements that form the MAX phases. (From Barsoum, M.W. and El-Raghy, T., *Am. Sci.*, 89, 334.)

the remaining 413 phases crystallize in bulk form in the α polymorph. There are many possible solid solution permutations and combinations. Solid solutions can be formed on the M sites, the A sites, the X sites and combinations thereof.

In addition to the *pure* MAX phases that contain one of each of the M, A and X elements highlighted in Figure 18.2, the number of possible solid solutions is quite large. Solid solutions have been processed and characterized with substitution on

- M sites, for example $(Nb,Zr)_2AlC$, $(Ti,V)_2AlC$, $(Ti,Nb)_2AlC$, $(Ti,Cr)_2AlC$, $(Ti,Hf)_2InC$ and $(Ti,V)_2SC$
- A sites, for example $Ti_3(Si,Ge)C_2$ and $Ti_3(Sn,Al)C_2$
- X sites, for example $Ti_2Al(C,N)$ and $Ti_3Al(C,N)_2$

The number of MAX phases and their solid solutions continues to expand. For example in the family of the MAX phases, none expands to compounds with magnetic properties that contain later transition metal substitutions on the M sites, such as $(Cr, Mn)_2AlC$.

18.2.2.1 Atomic Bonding in the MAX Phases

Bonding in the MAX phases is a combination of ionic, covalent and metallic bonds. The following conclusions of the theoretical studies are noteworthy:

1. There is a strong overlap between the p levels of the X elements and the d levels of the M atoms, leading to strong covalent bonds.
2. The density of states at the Fermi level is substantial.
3. The p orbitals of A atoms overlap the d orbitals of the M atoms.

4. The electronic states at the Fermi level are mostly d–d M orbitals.

5. In the M_2AlC phases, there is a net transfer of charge from A to X atoms.

A large body of work devoted to DFT calculations of the electronic structures and chemical bonding in the MAX phases shows that

- Similar to the MX phases, MAX phase bonding is a combination of ionic, covalent and metallic bonds
- The M and X atoms form strong directional covalent bonds in the M–X layers that are comparable to those in the MX binaries
- M–d–M–d metallic bonding dominates the electronic density of states at the Fermi level
- In most MAX phases, the M–A bonds are relatively weaker than the M–X bonds

18.2.2.2 Imperfections in the MAX Phases [10]

The fundamental difference between the MAX phases and their corresponding MX phases is that in the former, the basal plane dislocations are numerous, multiply and are mobile at temperatures as low as 77 K and higher:

1. In typical ceramics at room temperature, the number of independent slip systems is essentially zero, that is the critical resolved shear stresses needed for dislocation motion are higher than the stresses at which they fail in a brittle manner.

2. The MAX phases are pseudoductile in constrained deformation modes, in highly oriented microstructures and/or at higher temperatures.

3. In unconstrained deformation and especially in tension at lower temperatures, they behave in a more brittle fashion.

4. The dislocations arrange themselves either in walls, for example as low-angle or high-angle grain boundaries, normal to the basal planes or in arrays parallel to the basal planes. The walls have both tilt and twist components. To account for both components, the boundary is interpreted to be composed of parallel, alternating, mixed, perfect dislocations with two different Burgers vectors lying in the basal plane at an angle of 120° relative to one another.

18.2.2.3 Elastic Properties [10]

The MAX phases are elastically quite stiff for the most part. The densities of some MAX phases are relatively as low as 4–5 g/cm^3 (Table 18.2); their specific stiffness values can be high. For example the specific stiffness of Ti_3SiC_2 is comparable to that of Si_3N_4 and is roughly three times that of Ti. Poisson's ratio for all the MAX phases is approximately 0.2.

In general, the In-, Pb- and Sn-containing MAX phases are less stiff than those composed of lighter A elements. For example at 178, 216 and 237 GPa, the Young's moduli E of Zr_2SnC, Nb_2SnC and Hf_2SnC, respectively, are all lower than those of any of the Al-containing ternaries or that of Ti_3SiC_2. At 127 GPa, the bulk modulus B of Zr_2InC is the lowest reported to date. At the other extreme is Ta_4AlC_3 for which at 260 GPa, its B is the highest reported.

In contrast to other layered solids that are elastically quite anisotropic, such as graphite and mica, the MAX phases are mildly so.

In generalized Hooke's law with a C_{11} of 308 GPa and a C_{33} value of 270 GPa, Ti_2AlC is slightly more anisotropic. For Ti_2SC, two of the stiffest 211 phases known to date, C_{11} and C_{33}, are predicted to be ≈338 and ≈348 GPa, respectively.

18.2.2.4 Physical Properties [11]

Most of the MAX phases are excellent electrical conductors having the electrical resistivities in the narrow range of 0.2–0.7 $\mu\Omega \cdot m$ at room temperature. Like other metallic conductors, their resistivities increase with increasing temperatures. Ti_3SiC_2 and Ti_3AlC_2 conduct better than titanium metal. Even more interesting and intriguing, many of the MAX phases appear to be compensated conductors, wherein the concentrations of electrons and holes are roughly equal, but their mobilities are about equal, too:

- Several MAX phases, most notably Ti_3SiC_2, have very low thermoelectric or Seebeck coefficients. Solids with essentially zero thermopower can, in principle, serve as reference materials in thermoelectric measurements, for example as leads to measure the absolute thermopower of other solids.

- The optical properties of the MAX phases are dominated by delocalized electrons. Magnetically, most of them are Pauli paramagnets, wherein the susceptibility is, again, determined by the delocalized electrons and, thus, is neither very high nor temperature dependent.

- Thermally, the MAX phases share much in common with their MX counterparts, that is they are good thermal conductors because they are good electrical conductors. At room temperatures, their thermal conductivities fall in the 12–60 W/(m·K) range. The coefficients of thermal expansion (CTE) of the MAX phases fall in the 5–10 μK^{-1} range and are relatively low as expected for refractory solids. The exceptions are some chromium-containing phases with CTEs in the 12–14 μK^{-1} range.

- At high temperatures, the MAX phases do not melt congruently but decompose peritectically to A-rich liquids and $M_{n+1}X_n$ carbides or nitrides. Thermal decomposition occurs by the loss of the A element and the formation of higher n-containing MAX phases and/or MX. Some MAX phases, such as Ti_3SiC_2, are quite refractory with decomposition temperatures above 2300°C.

- Because of their excellent electrical, thermal and high-temperature mechanical properties, some MAX phases currently are being considered for structural and nonstructural high-temperature applications.

- Their oxidation resistance, however, determines their usefulness in air. In most cases, MAX phases oxidize according to Equation 18.1:

$$M_{n+1}AX_n + bO_2 = (n+1)MO_{x/n+1} + AO_y + X_nO_{2b-x-y}.$$ (18.1)

- The most oxidation-resistant MAX phase is Ti_2AlC, because it forms a stable and protective Al_2O_3 layer that can withstand thermal cycling up to 1,350°C for 10,000 cycles without spallation or cracking (Figure 18.3). The oxidation resistance of Cr_2AlC also is superb because it also forms a protective Al_2O_3 layer; however, the oxide spalls off during thermal cycling.

- Elastically, the MAX phases are quite stiff, with near-isotropic room temperature Young's and shear moduli in the 178–362 GPa and 80–142 GPa ranges, respectively. Because the densities of some of the MAX phases are as low as 4–5 g/cm^3, their specific stiffness values can be quite high. For example the specific stiffness of Ti_3SiC_2 is comparable to Si_3N_4 and roughly three times that of titanium metal.

18.2.2.5 *Processing of MAX Phases [12]*

There are many methods for processing the MAX phases as bulk materials, powders, porous foams, coatings and thin films. Some of the methods are widely used. For example MAX phases in any form usually are fabricated

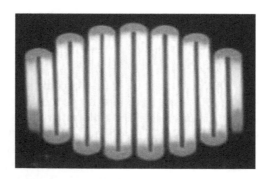

FIGURE 18.3
Ti_2AlC-based heating element resistively heated to 1, 450°C in air. (From Barsoum, M.W. and El-Raghy, T., *Am. Sci.*, 89, 334. With permission.)

from elemental powders and/or binary carbides. Thus, their price is determined by the price of those powders. Sandvik Materials Technology (Hallstahammar, Sweden) has manufactured Ti_3SiC_2 and Ti_2AlC powders and parts since the late 1990s under its MAXthal brand. This price is significantly higher than the price of Al_2O_3, SiC and Si_3N_4 powders used to make other structural and high-temperature ceramics. However, pressureless sintering in inert atmospheres can yield fully dense MAX phase parts without using sintering aids.

18.3 Superplastic Materials

18.3.1 Different Types of Deformations

Deformation of a material means a change in its shape under applied load or stress. Change in the shape may be linear, angular or a combination of both. This deformation may be temporary (recoverable) or permanent (irrecoverable) in nature. Both deformations may be time dependent or time independent. Based on the behaviour of material in response to an externally applied stress, the deformations are classified as follows:

1. Elastic deformation
2. Elastomeric deformation
3. Plastic deformation
4. Anelastic deformation
5. Viscoelastic deformation

Elastic deformation is small, time independent and fully recoverable and obeys Hooke's law. It occurs in metals within their elastic limits.

Elastomeric deformation is too large, time independent and fully recoverable and does not obey Hooke's law. It occurs in elastomers.

Plastic deformation is large, permanent and time independent and does not obey Hooke's law. It occurs in metals beyond their elastic limits.

Anelastic deformation is small and fully recoverable but time dependent. It may or may not obey Hooke's law. It occurs in rubber, plastics and metals due to thermoelastic phenomenon.

Viscoelastic deformation is time dependent, partially elastic and partially permanent. It obeys Hooke's law along with Newton's law for viscous flow. It occurs in polymers.

18.3.1.1 Superplastic Materials

Some materials deform by several hundred percent strain without necking. They exhibit *superplastic behaviour* and are called *superplastic materials*.

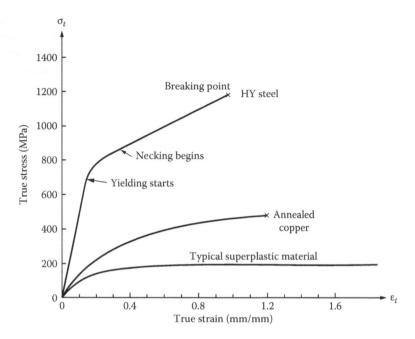

FIGURE 18.4
Curves showing true stress–true strain and superplastic behaviours.

The stress–strain behaviour of a typical superplastic material is shown in Figure 18.4:

- Ultrafine grained alloys of aluminium and stainless steels are such examples.
- Long glass rods are made due to this behaviour of glass above certain temperature.

The superplastic behaviour of materials can be explained with the help of true stress–true strain curve. The true stress σ_t and true strain ε_t^n curves for high yield steel (HYS) and annealed copper are shown in Figure 18.4. The portion of the elastic range is very small since true stress and strain are primarily applicable to plastic range. Curve beyond yield point is governed by power law equation:

$$\sigma_t = K\varepsilon_t^n \tag{18.2}$$

where
K is the material's constant or strength coefficient
n is the *strain hardening coefficient*

The plot of log σt versus log εt (not shown) will result in a straight line whose slope gives the value of n. The value of K may be noticed in this curve at log $\varepsilon_t = 1$. For heat-treated steel, the value of n is about 0.15 and it is ≈ 0.5 for copper and brass and 0.2 for aluminium. Materials having larger values of n can be substantially deformed in plastic action.

18.4 High-Temperature Metals and Alloys

18.4.1 High-Temperature Metals and Alloys

High-temperature metals and alloys (HTMA) are those which can effectively function above 500°C. They are required for use in the following fields and applications:

- Energy generation
- Engines (cars, trains, airplanes, ships)
- Chemical industry
- Metallurgy
- Mechanical engineering
- Jet engines
- Stationary gas turbines
- Waste-to-energy plant

Before selecting a suitable HTMA, it is essential to know the maximum sustainable temperatures of different materials. These are given in the following.

S. No.	Category of Material	Maximum Service Temperature (°C)	Mode of Failure (Deformation/Damage)
1.	Polymer	Up to about 300	Melting, decomposing
2.	Glass	Up to about 800	Viscous flow
3.	Metals	Fe basis (coated) up to 1100 Fe-ODS up to 1300 Ni- and Co-base up to 1200 Pt-base up to 1600 refractory metals in inert atmosphere above 1600 $MoSi_2$ up to 1800	Creep, dislocation climb, grain boundary sliding
4.	Ceramics	SiC up to 1600	Viscous flow, glass transition temperature, grain boundary sliding
5.	Composites	(SiC/C) up to 1600	Complex phenomenon

Comparative usable strength as a function of temperature for different materials is shown in Figure 18.5. Thus, we see that the usable strength at

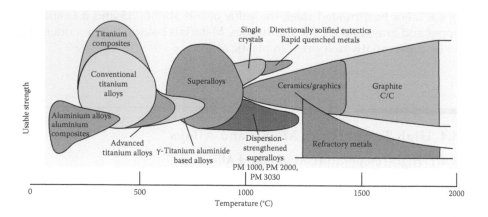

FIGURE 18.5
Comparative usable strength as a function of temperature. (From Dr.-Ing. and Glatzel, U., *Metals and Alloys*, University Bayreuth SS, 2012.)

higher temperature is better in refractory materials, ceramic–graphite and graphite composites.

18.4.1.1 High-Temperature Creep-Resistant Materials

The permanent deformation (strain) of a material under steady load as a function of time is called *creep*. Although the creep occurs at room temperature in many materials such as lead, zinc, solder wire (an alloy of Pb and Sn), white metals, rubber, plastics and leather, it is a thermally actuated process and hence is influenced by temperature. It is, however, appreciable at temperatures above $0.4T_m$ where T_m is the melting point of material in degree kelvin. Consideration of creep is very important in applications such as

1. Industrial belts
2. Blades of gas turbines
3. Blades of steam turbines
4. Pistons of internal combustion (IC) engines
5. Rockets and missiles
6. Nuclear reactors
7. Tubes of heat exchangers
8. Polymeric and elastomeric constructions

Creep is sometimes desirable also as in metal forming operations such as hot rolling, extrusion and forging. These processes are carried out at high temperatures where deformation follows power creep law. The forces required in operations are reduced due to raised temperatures.

Machine and structural part functioning at higher temperatures must be creep resistant. Pressure vessels and heat exchangers in oil refinery and chemical industries operate at elevated temperatures. Heat engines need to operate at higher operating temperatures to achieve enhanced thermal efficiency. This necessitates the creep-resistant materials to have high melting points. Some of the probable materials may be as follows:

1. Refractories
2. Tungsten-based alloys
3. Nickel-based alloys and nickel superalloys
4. Cobalt-based alloys
5. Steel-based alloys
6. Monocrystal titanium
7. Thoria (ThO_2) dispersed nickel

Of these, the refractories are brittle and cannot take purposeful tensile load. Tungsten and titanium are costly metals. Tungsten is also heavy. Nickel-based alloys, cobalt-based alloys and steel-based alloys are suitable for use from different viewpoints.

18.4.1.2 Advanced Creep-Resisting Materials

Nickel using thoria by dispersion hardening method is a very good creep-resistant material. It can maintain its strength up to a temperature of about 0.9 Tm. Some of the latest materials as given in the following are also useful:

1. Silicon nitride (Si_3N_4) for piston rings and cylinder heads
2. Sialons (alloys of Si_3N_4 and Al_2O_3) for gas turbine blades up to 1300°C

Finer-grained materials having small crystals are undesirable for use as creep-resistant materials.

18.4.2 High-Temperature Oxidation-Resistant Materials

Oxidation is a high-temperature environmental effect on materials. The problem of oxidation becomes more severe at the increasing functional temperatures. An adherent oxide layer offers good oxidation resistance. The addition of nickel, chromium and aluminium in steel improves its oxidation resistance. Examples of components and materials that resist oxidation are given as follows:

- The superheater tubes in boilers and turbine blades should resist attack by gases at high temperatures.

- Turbine blade in service reacts with SO_2, H_2S and other combustion products and in due course gets oxidized.
- Ceramics and alkali halides (NaCl, NaBr, KCl, etc.) are stable but metals except gold are not stable against oxidation.
- Polymers like PTFE survive long durations at high temperatures and are stable.
- Other polymers and composites based on polymers are unstable against oxidation.

Rate of oxidation: While designing with oxidation-prone materials, it is desirable to know how fast or slow the oxidation process is going to be. The rate of oxidation may be very fast as in tungsten and molybdenum or very slow as in tin, silver and gold. This rate varies between a few minutes to many million hours. Table 18.3 shows oxidation time for some materials.

Mo and W lose weight linearly on oxidation. It is due to the volatile nature of their oxides MoO_3 and WO_3. At high temperatures, these oxides evaporate

TABLE 18.3

Density, Young's Modulus E, and Poisson's Ratio ν of Some MAX Phases

MAX Phases Solid	Density (g/cm³)	Young Modulus E (GPa)	Poisson's Ratio ν
Ti_2AlC	4.1	277	0.19
$Ti_2AlC_{0.5}N_{0.5}$	4.2	290	0.18
V_2AlC	4.81	235	0.20
Cr_2AlC	5.24	245	0.20
	5.1	288	
Nb_2AlC	6.34	286	0.21
Ta_2AlC	11.46	292	
Ti_3SiC_2	4.52	343–339	0.20
Ti_3GeC_2	5.02	340–347	0.19
$Ti_3(Si,Ge)C_2$	4.35	322	0.18
Ti_3AlC_2	4.2	297	0.20
Ti_3AlCN	4.5	330	0.21
Cr_2GeC	6.88	245	0.29
Ti_2SC		290	0.16
Nb_2SnC		216	
Zr_2SnC		178	
Hf_2SnC		237	
Nb_2AsC	8.05		
Nb_4AlC_3	6.98	306	
β-Ta_4AlC_3	13.2	324	
Ti_4AlN_3	4.7	310	0.22
$TiC_{0.96}$	4.93	≈500	0.19
Zr_2SC	6.20		

just after their formation and therefore do not offer any barrier to oxidation. Due to loss of oxide, the material loses its weight.

Methods to reduce oxidation: Oxidation of materials follows linear and parabolic laws. Their rates can be expressed by an Arrhenius-type equation. Oxidation rates can be decreased by the addition of some oxidation-resistant elements. Rate of oxidation in steel decreases by about

1. 30 times on adding 5% Al
2. 20 times on adding 5% Si
3. 100 times on adding 18% Cr at 900°C

18.4.2.1 Use of Corrosion- and Oxidation-Resistant Materials

Many alloys offer good resistance to corrosion in different environments. For example

1. Copper alloys, brass (Cu + Zn) and bronze (Cu + Sn) have good resistance against the environments of water and salty air.
2. Titanium and zirconium resist chlorine environment.
3. 18–8 stainless steel (18% Cr and 8% Ni) is excellent against all types of environments.
4. The addition of niobium in steel offers very good corrosion resistance.
5. Chromium, nickel and aluminium as alloying elements in steel provide excellent resistance at elevated temperatures. Various combinations of steel alloys showing their important applications are depicted in Table 18.4.

TABLE 18.4

Oxidation Time for Some Materials, Oxidized to 0.1 mm Depth at $0.7T_m$ in Air

Material	Oxidation Time (h)	Melting Point (K)	Material	Oxidation Time (h)	Melting Point (K)
W	Very short	3680	Zn	>10^4	695
Mo	Very short	2880	Mg	>10^5	925
Zr	0.2	2125	Pt	1.8×10^5	2040
Fe	24	1810	Sn	Very long	505
Cu	25	1355	Al	Very long	935
Ni	600	1725	Ag	Very long	1235
Cr	1600	2150	Au	Infinite	1335

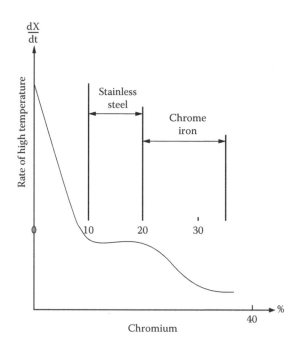

FIGURE 18.6
Effect of chromium content on rate of high-temperature corrosion (oxidation).

Figure 18.6 shows the effect of chromium content on high-temperature corrosion rate of iron. It shows that rate of corrosion drops down on increasing the chromium content.

18.4.3 Thermal Shock-Resisting Materials

A situation in the material when there is a severe and sudden temperature change is known as thermal shock. The capability of a material to withstand ill effects of such drastic change is called thermal shock resistance. In these regards, the following points are to be noted:

- Ductile materials are better than brittle materials in withstanding thermal shocks.
- Ductile materials are excessively deformed due to high thermal stresses, whereas the brittle materials fracture.
- Thermal fatigue reduces plastic deformability, thereby causes formation of thermal cracks.
- These cracks are responsible for failure.
- Failure of coating in coated refractory metals (Ni_3, Nb, etc.) used for gas turbine blades is due to thermal fatigue.

High thermal shock-resisting materials: The following are the high thermal shock-resisting materials:

- Graphite
- Cermets
- Glass–ceramics
- Lithium–ceramics
- Fused silica
- Pyrex glass

18.4.3.1 Thermal Protection

Special techniques and specific materials are used for thermal protection of structures. Thermal insulating materials used to control the heat transfer rate have low thermal conductivity:

1. Such materials are cellular, granular and fibrous with air entrant arrangements.
2. Numerous fine pores enclosing air are desired to decrease thermal conductivity.
3. The pores may be closed as in foam glass and foamed plastics or may be left open.
4. The presence of water (moisture) is undesired in these pores as its conductivity is about 20 times more than that of the air.

Technique of reflective insulation: The technique of reflective insulation is employed in tankers transporting liquid oxygen, liquid hydrogen and liquid nitrogen:

- In this technique, highly reflective surfaces are separated by large width of airspace.
- Conduction and convection are minimum at about 20 mm width of air.
- Aluminium foils on paper, reflective aluminium surfaces separated by glass fibre lamina and Mylar (aluminized plastic films) under high vacuum are used for this purpose.
- Air is not suitable for thermal insulation at cryogenic temperatures as it solidifies at 81.3 K. Hence, the insulating layers are kept free from air and are sealed.

18.4.3.2 High-Temperature Effects

Modern advancements in engineering and technology demand the development of high temperature-resisting materials. There is a need of about 1000°C in gas turbines, above 1000°C in automobile engines, and many 1000°C in fast-breeder reactors in their functioning. The problem of high temperature-resisting material is more serious in rockets, missiles and supersonic aircrafts. The effects of high temperatures on materials are as follows:

1. Loss of strength
2. Gain in ductility
3. Reduced stiffness
4. Lower yield strength
5. Polymorphic transformations
6. Decrease in hardness

Variation in ultimate strength of materials as a function of time: Young's modulus and yield strength of several materials as a function of temperature are shown in Figures 18.7a through c.

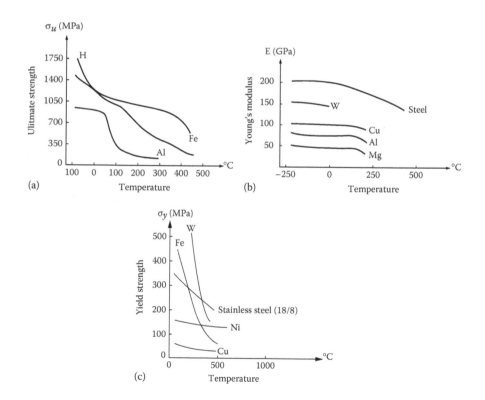

FIGURE 18.7
Effect of temperature on (a) ultimate strength, (b) Young's modulus, and (c) yield strength of various materials.

TABLE 18.5

Oxidation-Resistant Alloy Steels

Name of Alloy	Composition of Chromium and Others in Steel	Application
Chromium steel	Cr up to 10%	Oil refinery
Chromel	Cr 10% and Ni Cr 12%–17%	Thermocouples
Stainless steel	Cr 16%, Ni 76%	Gas and steam turbine blades, ovens and furnaces, exhaust valves of IC engines
Inconel	Cr 18%, Ni 8%	—
18-8 Stainless steel	Cr 20%, Ni 80%	Utensils, pressure vessels, heat exchangers
Nichrome	Cr 24%, A 16%, Co 2%	Strain gauges
Kanthal	Al 2%, Mn 2%,	Windings of ovens and furnaces
Alumel	Si 1%, Ni	Heat-resistant wires

18.4.3.3 Effect of Temperature and Time on Structural Changes

Several structural changes in materials occur slowly at normal temperatures but many takes place at higher temperatures. Rate of reaction in them is therefore influenced by slow or fast reactions. The time t for occurrence of reaction decreases rapidly when the temperature T of reaction is raised. This time–temperature relation may be expressed by Arrhenius-type equation given as follows:

$$\frac{1}{t} = Ae^{-B/T} \quad \text{or} \quad t = \frac{e^{B/T}}{A} \cdots \tag{18.3}$$

Here, A and B are material-related constants. Generally, the service life of a material is reduced to half if it is serviced under increase in temperature by about 25°C–50°C. Effect of sufficient time available at higher temperatures on changes in structures and materials is summarized in Tables 18.5 and 18.6).

18.5 Nanostructured Materials

18.5.1 Nanostructured Steels

Steel is one of the most used engineering materials by alloying and process control. It is possible to obtain its strengths ranging from 100 MPa to over 2 GPa by alloying. These advances have taken place by control of microstructure during processing. Many of the microstructure events are controlled at

TABLE 18.6

High Temperature-Sufficient Time Effect on Structure of
Different Materials

Material/Structure	Altered Form of Structure at High Temperature When Exposed for Sufficient Time
Metals	Oxidized form/products
Organic materials	Oxidized form/products
Fine-grained materials	Coarse-grained materials
White cast iron	Malleable iron
Cold-worked alloys	Recrystallized alloys
Age-hardened alloys	Overaged alloys
Pearlite	Spheroidite
Martensite	Spheroidite and tempered martensite
Tempered martensite	Spheroidite

the micron level, though others such as precipitation hardening are clearly at the nanoscale level of control. Steel is one of the most widely used engineering materials in the world. A wide variety of engineered microstructures with superior properties can be produced in it. Nanoengineered steels for structural applications [14] are an effort in this direction. Currently, there is a growing awareness about the potential benefits of nanotechnology and research in the area of nanostructured steels. The theoretical strength of steel is 27.30 GPa. There are two ways of achieving ultrahigh strength in steels. The first one is to reduce the size of a crystal to such an extent that it is devoid of any defects, like in the case of a whisker. The second alternative is to introduce a very large density of defects in a metal sample that act as an obstacle to the motion of dislocations. The strengthening arises due to the presence of nanoscale cementite/ferrite lamellar structure. The ferrite phase in this structure contains very high dislocation density and supersaturated carbon atoms, and the cementite phase contains amorphous and nanocrystalline regions. To meet this challenge, a number of innovative approaches are being developed to produce nanostructured steels, as shown in Figure 18.8.

Different processing strategies and alloy development aspects being currently explored for the manufacture of nanostructured steels are briefly outlined in this section. *Severe plastic deformation (SPD) processing* is one of the promising routes for grain size refinement to nanoscale levels. Nontraditional processes, such as equal channel angular processing (ECAP), accumulative roll bonding, torsion under very high pressures and multiple compressions, have been developed for this purpose. The ultrafine (<1 μm) grain sizes lead to exceptionally high strengths in conventional steels. However, there is a drastic reduction in tensile ductility, especially uniform elongation in tension. Therefore, a number of processing routes are being developed for the improvement of ductility.

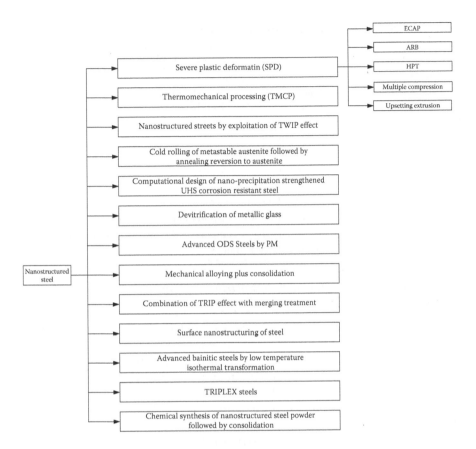

FIGURE 18.8
Routes to produce nanostructured steels. (From Mahajan, Y.R., *Issue of Nanotech Insights*, December 2013.)

Advanced bainitic steels: New-generation bainitic steels (e.g. Fe, 0.98; C, 1.46; Si, 1.89; Mn, 0.26; Mo, 1.26; Cr, 0.09V) are designed using detailed phase transformation theory for the bainitic reaction [2]. The bainitic transformation occurs at low temperatures (200°C–300°C), which avoids the diffusion of iron or any substitutional solutes. As a consequence, the plates of bainite are extremely slender, 20–40 nm thick, making the steel very strong.

TRIPLEX steels: TRIPLEX steels are designed on the basis of Fe–Mn–C–Al with Al > 8%. Mn is usually >19% [14]. The alloy consists of an austenitic face centered cubic (FCC) matrix and about 8% ferrite and nanosize k-carbides regularly distributed in the FCC matrix in an orderly fashion. The TRIPLEX alloys exhibit low-density, high-strength level, excellent formability and high energy absorption capacity.

18.5.1.1 Applications of Nanostructured Steels

Area of Application	Example of Application
Aerospace	Aircraft landing gears
Automotive	Automobile body, structural and safety parts, chassis and suspension parts and TWBs
Consumer	Razors for shaving machine
Defence	Ballistic armour
Industrial	Cutting tools and bearings
Infrastructure	Concrete reinforcing rebars
Medical	Surgical needles and clips
Nuclear	Fuel cladding tubes for nuclear reactors
Oil and gas	Pipeline steels for transportation of natural gas
Power	Advanced exhaust components for heavy-duty diesel engines
Sports	Mountain bicycle frames

18.5.1.2 Low-Carbon Steel with Nanostructured Surface [15]

Nanostructured surface layer can be obtained in a low-carbon steel plate by means of the high-energy shot peening (HESP) technique. The average grain size in the top surface layer can be as small as a few nanometres and gradually increases with the distance from the surface. The thickness of the nanostructured surface layer is not less than 20 mm. The yield strength and the strengthening of material after the HESP treatment can be attributed to work hardening and formation of surface nanostructure. The present work shows the possibility to significantly improve mechanical property of engineering materials by generation of a nanostructured surface layer induced by the HESP technique.

Mechanical properties of engineering materials can be improved by refining their microstructures. In the past, various kinds of techniques were developed to synthesize the ultrafine-grained materials such as equal channel angular pressing, cold-rolling and torsion straining. However, most of these techniques are still difficult for practical application to conventional engineering materials. It is known that most failures of engineering materials are very sensitive to the structure and properties of material surface, and in most cases, material failures occur on the surface. Therefore, optimization of surface structure and properties may effectively improve the behaviour of material. With the increasing evidences of unique properties for nanostructured materials, it is expected to achieve surface modification by generation of a nanostructured surface layer, that is surface nanocrystallization (SNC). Using the ultrasonic shot peening (USP) method, the nanostructured surface layers have been successfully obtained on 316L stainless steel and the pure Fe section samples. In this, the synthesis of a nanostructured surface layer on a low-carbon steel by using a HESP is presented.

18.5.1.3 Nanostructure Improvement in Corrosion Resistance of Steels [16]

Steels are the most widely used metallic materials in industrial life. This is due to their superior properties such as high strength, good ductility, high hardness and low costs. However, steels often suffer from staining, rusting and corroding in different environments, which may cause serious problems or even failure of the components made of steels. Usually, general corrosion occurs in carbon steels or low-alloyed steels since there are no dense protective films formed on these steels. Therefore, the steels are continuously corroded. The so-called stainless steel usually contains Cr of more than 11 wt.%. As a result of the high Cr content, a dense protective Cr oxide layer can be formed on the surface in corrosive environments, which will prevent the further corrosion of the bulk material. Nevertheless, stainless steel can be affected by pitting in the presence of halide ions, particularly the chloride ion. Therefore, corrosion resistance is of first concern for different kinds of steels to be used in industries. The addition of alloying elements such as Cr into steels is a way to improve the corrosion resistance of steels. However, alloying of the bulk material will also change the mechanical properties of the steels. Furthermore, alloying is also very costly.

To compromise the corrosion, mechanical properties and the cost, the ideal solution would be to use surface modification techniques to generate a protective layer at the near-surface region or on the surface of the steels. In the past decades, different surface modification techniques have been applied to improve the corrosion resistance of steels. It is found that carburizing, boriding and nitriding can be used to create a hardened layer on the steel surface containing mainly carbides, borides or nitrides, respectively. These hardened layers may also have better corrosion resistance than their substrates. By using laser surface remelting and cladding, the corrosion resistance of laser-treated steels can be also greatly improved. This was mainly attributed to a homogenization of the elements in the remelted surface layer or the addition of protective alloying elements/compounds. Other methods, mainly film deposition techniques such as physical vapour deposition (PVD), chemical vapour deposition (CVD), solgel methods and electroless plating, have also been introduced to modify the steel surfaces. For instances TiN, CrN, NbN and TiCN films or multilayer films can be deposited onto steel surfaces by using various film deposition techniques. Although these films always have better corrosion resistance, the surface properties strongly depend on the quality of the films such as pinholes and impurities, which are not easy to control. Furthermore, the films often suffer from poor adhesion with the substrates. Ion implantation is also testified to be able to improve the corrosion resistance of steels through introducing alloying elements into the very top surface layers. However, the ion-implanted layers, usually less than 1 μmin thickness, are too thin to sustain long-time corrosion.

Low-energy high-current pulsed electron beam (LEHCPEB) technique is a recently developed technique for surface modification of metallic materials.

The pulsed electron irradiation induces (1) very rapid heating, melting, solidification and cooling of the surface together with (2) the formation of thermal stress waves. As a result, improved surface properties of the material, often unattainable with conventional surface treatment techniques, can be obtained fairly easily. This is particularly true for tribological properties. Previous investigations have suggested that the LEHCPEB technique could be also used to improve the corrosion resistance of precipitates containing alloys. The pulsed electron irradiation generated by this surface treatment technique induces rapid melting of the surface followed by extremely fast solidification. This process leads to the dissolution of second-phase particles and, after sufficient number of pulses, to the formation of a 2–3 μm thick homogeneous melted surface. In the case of the steels, the subsequent ultra-fast solidification leads to a structure containing nanostructured domains. It will be shown in the following sections that those effects induced by the LEHCPEB treatments can significantly improve the corrosion resistance of steels.

18.6 Nanostructured Materials for Renewable Energy Applications [17]

Hydrogen can be generated by a variety of means including the electrolysis of water using electricity derived from wind power and photovoltaic or by thermochemical processing of biomass. Hydrogen can then be reacted with oxygen in fuel cells to generate electricity, combusted in an engine to generate mechanical energy or simply burned to generate heat. In each of these cases, water is produced in a virtually pollution-free process. Unfortunately, before hydrogen can be employed in the transportation sector, numerous technical hurdles must still be overcome. Recent theoretical studies have shown that by complexing fullerenes with a TM, H_2 dihydrogen ligands may be bound with binding energies appropriate for onboard vehicular storage. In an optimal structure, scandium has been predicted to complex with all of the twelve five-membered rings in a fullerene. The $C_{60}[ScH_2(H_2)_4]_{12}$ complex has a reversible hydrogen capacity of 7.0 wt.%.

Recently, hot-wire chemical vapour deposition (HWCVD) has been employed as an economically scalable method for deposition of crystalline tungsten oxide nanorods and nanoparticles [17]. Under optimal synthesis conditions, only crystalline nanostructures with a smallest dimension of approximately 10–50 nm are observed with extensive transmission electron microscopy (TEM) analyses. The incorporation of these particles into porous films led to profound advancement in state-of-the-art electrochromic (EC) technologies. The development of durable inexpensive EC materials could make them suitable for large area window coatings and lead to decreased energy consumption in air conditioning. HWCVD has also been employed to produce crystalline molybdenum oxide nanorods, particles, and tubes at high density. It is also possible to fabricate large area porous films containing these MoO_3 nanostructures. Furthermore, these films have been tested as the negative electrode in lithium-ion batteries and a surprisingly high reversible capacity as well as dramatically improved charging and discharging insertion.

18.7 Emerging Scope of Hybrid Solar Cells in Organic Photovoltaic Applications by Incorporating Nanomaterials [18]

In recent years, semiconductor nanomaterials have been extensively studied, and reports are available for their preparation methods, physical and chemical properties of nanoparticles, and their characterization techniques. Because of their potential applications, ZnS nanoparticles are recently the major area of research. It is an important inorganic material for a variety of applications including photoconductors, solar cells, field effect transistors, sensors, transducers, optical coatings, and light-emitting materials. Inorganic nanoparticles have found potential application in various electronic devices. Synthesis, shape, and size control are important issues for nanoparticle research. Various *nanostructured materials* have found potential applications in optical and electrical devices such as photoconductors, LEDs, solar cells, field effect transistors, and optical coatings. ZnS has a wide bandgap ranging from 3.5 to 3.8 eV at room temperature, and the bandgap can be tuned in the UV region by controlling the size of the nanoparticles. In the present section, the synthesis of ZnS nanoparticles and their characterization to investigate various properties such as size, structure, bandgap, and luminescence via different characterization tools are discussed. The particles were then used as acceptors for fabrication of organic hybrid solar cells.

The synthesis and application of nanoparticles are some of the most interesting fields of research from a basic and applied points of view. The particle size of the nanoparticle ranges between 1 and 50 nm but is strongly dependent of the surfactant employed. Different compositions and sizes can be obtained using different synthesis processes. Owing to the interesting structure and chemical and physical properties, which are different from those of the bulk materials, the semiconductor nanoparticles have generated great R&D interest in recent years.

The synthesis of materials at nanometrical scale has become an important research line in the last years due to the strong dependency of the size of the material particles over their properties: optical, mechanical, and electrical. This knowledge has been employed in the development of new technologies in different areas such as the medicine, energy, environmental engineering, and biology. Semiconductors deserve special attention in a new world of functional materials and numerous research works are devoted to them. Nanoparticles of CdS, CdSe, CdTe, ZnS, and ZnSe are getting attention due to their potential application in photocatalytic processes and solar cells. ZnS is one of the most applied semiconductors in optical devices due to its high refraction index and high transmission within the visible range [8].

References

1. L.P. Lefebvre, J. Banhart, and D.C. Dunand, Porous metals and metallic foams: Current status and recent developments, *Advanced Engineering Materials*, 10(9), 2008, 775–787.
2. A. Kennedy, Porous metals and metal foams made from powders. In Dr. Katsuyoshi Kondoh (ed.), *Powder Metallurgy*, InTech.
3. S. Das, Aluminium Foam: A Potential Functional Material, Advanced Functional Materials and Structures (AFMS-2012) Organised by: Applied Mechanics Department, MNNIT Allahabad, INDIA In Collaboration with: University of Missouri (MU), Columbia.
4. M.D. Goel, M. Peroni, G. Solomos, D.P. Mondal, V.A. Matsagar, A.K. Gupta, M. Larcher, and S. Marburg, Dynamic compression behavior of cenosphere aluminium alloy syntactic foam, *Materials and Design*, 42, 2012, 418–423, doi: http://dx.doi.org/10.1016/j.matdes.2012.06.013.
5. W. Xu and G. Li, Constitutive modeling of shape memory polymer based self-healing syntactic foam, *International Journal of Solids and Structures*, 47, 2010, 1306–1316.
6. M.M. Hossain and K. Shivakumar, Compression fatigue performance of a fire resistant syntactic foam, *Composite Structures*, 94, 2011, 290–298.
7. D.P. Mondal, J.D. Majumder, N. Jha, A. Badkul, S. Das, A. Patel, and G. Gupta, Titanium-cenosphere syntactic foam made through powder metallurgy route, *Materials and Design*, 34, 2012, 82–89.
8. G. Tagliavia, M. Porfiri, and N. Gupta, Influence of moisture absorption on flexural properties of syntactic foams, *Composites: Part B*, 43, 2012, 115–123.
9. M. Colloca, N. Gupta, and M. Porfiri, Tensile properties of carbon nanofibre reinforced multiscale syntactic foams, *Composites: Part B*, 44, 2013, 584–591.
10. M.W. Barsoum and M. Radovic, Elastic and mechanical properties of the MAX phases, *Annual Review of Material Research* 41, March 30, 2011, 195–227, doi: 10.1146/annurev-matsci-062910-100448.
11. M. Radovic. and M.W. Barsoum, MAX phases: Bridging the gap between metals and ceramics, *American Ceramic Society Bulletin*, 92(3), April 2013, 20–26.
12. M.W. Barsoum. and T. El-Raghy, The MAX phases: Unique new carbide and nitride materials, *American Scientist*, 89, 2001, 334–343.
13. Dr.-Ing and U. Glatzel, *Metals and Alloys*, University Bayreuth SS, Bayreuth, Germany, 2012.

14. Y.R. Mahajan, *CKMNT*, Nano-engineered steels for structural applications, *Issue of Nanotech Insights*, December 2013.
15. G. Liu, S.C. Wang, X.F. Lou, J. Lu, and K. Lu, Low carbon steel with nanostructured surface layer induced by high-energy shot peening, *Scripta Materialia*, 44, 2001, 1791–1795.
16. K.M. Zhang, J.X. Zou, and T. Grosdidier, Nanostructure formations and improvement in corrosion resistance of steels by means of pulsed electron beam surface treatment, Hindawi Publishing Corporation, *Journal of Nano Materials*, Article ID: 978568, 2013, 8p.
17. C. Dillon, E.S. Whitney, C. Curtis, C. Engtrakul, M. Davis, Y. Zhao, Y.-H. Kim et al., Novel nanostructured materials for a variety of renewable energy applications National Renewable Energy Laboratory, 1617 Cole Blvd., Golden, CO 80401.
18. N. Gupta and K.M. Gupta, Emerging scope of hybrid solar cells in organic photovoltaic applications by incorporating nanomaterials, *Advanced Materials Research*, 548, 2012, 143–146, © Trans Tech Publications, Switzerland.

Index

Micro-/nanoelectromechanical systems
(MEMs/NEMs), 223
Micro–robot fish, 225–226
Micro technology and devices, 4
Minesweeper, 194–195
Mini-VIPeR model, 234
Mixed oxides, 90–91
Mode I fracture toughness test, 152–154,
158–159
Modulus of rigidity, 38
Molybdenum, 344
Monomers, 55–56; *see also* Expanding
monomers (EMs)
Mosquito bite, 242–243
Moth eye–inspired biomimetic
materials, 238–239
Mylar, 117

N

Nano-electromechanical system
(NEMS), 27
Nanofibrillated cellulose (NFC),
474–475
Nanoporous metallic structural
materials, 539
Nanostructured materials
renewable energy applications,
574–575
steels
applications, 572
corrosion resistance, 573–574
HESP technique, 572
processing routes, 571
strengthening, 570
Nanotechnology
applications, 24–26
atomic-size devices, 27
history, 23
nanodevices, 5, 27
nanomaterials, 26–27
NEMS, 27
preparation process, 24
working philosophy, 24
National Bureau of Standards (NBS)
smoke test, 198
National Emission Standards for
Hazardous Air Pollutant
(NESHAP), 202

Natural (palmyra) fibre–reinforced
composite, *see* Palmyra fibre–
reinforced composite (PFRC)
Natural fibres (NFs), 110, 113–115
bamboo fibres, 452–453
banana, 439
flax, 416, 424
palmyra, 424
New-generation bainitic steels, 571
Newton's law for viscous flow, 559
NFC, *see* Nanofibrillated
cellulose (NFC)
NFs, *see* Natural fibres (NFs)
Nickel, 338
Nickel alloys, 320–321
Niobium, 346
Nonconventional machining
materials, 265
Non-ferrous metals and alloys, 3
aluminium, 334–337
bearing metals, 341
beryllium, 345
contact materials, 341–343
copper (*see* Copper)
fine silver, 343
hafnium, 345
lead, 339–340
magnesium, 336–337
molybdenum, 344
nickel, 338
niobium, 346
palladium, 344
platinum, 343–344
rhodium, 345
tantalum, 345
thermocouple materials, 346
tin, 340
titanium, 339
tungsten, 344
zinc, 338–339
Nuclear industry
cooling materials, 370
fuel materials, 370
moderator–reflector
materials, 370
reaction control materials, 370
shielding materials, 371
structures, 371
Nylon fibres, 63